Encyclopaedia of Mathematical Sciences

Volume 16

Editor-in-Chief: R.V. Gamkrelidze

V. I. Arnol'd   S. P. Novikov (Eds.)

Dynamical Systems VII

Integrable Systems
Nonholonomic Dynamical Systems

With 9 Figures

Springer-Verlag Berlin Heidelberg GmbH

Title of the Russian edition:
Itogi nauki i tekhniki, Sovremennye problemy matematiki,
Fundamental'nye napravleniya, Vol. 16, Dinamicheskie sistemy 7
Publisher VINITI, Moscow 1987

Mathematics Subject Classification (1991):
58F05, 58F06, 58F07, 70Exx, 70Fxx, 70Hxx

ISBN 978-3-642-05738-0 ISBN 978-3-662-06796-3 (eBook)
DOI 10.1007/978-3-662-06796-3

Library of Congress Cataloging-in-Publication Data
Integrable systems, nonholonomic dynamical systems /
V. I. Arnol'd, S. P. Novikov, eds. p. cm. – (Dynamical systems ; 7)
Translated from the Russian.
Includes bibliographical references and index.

1. Differentiable dynamical systems. I. Novikov, Sergeĭ Petrovich.
II. Series: Dinamicheskie sistemy. English ; 7.
QA805D5613 1994 no.7 [QA614.8] 531 s–dc20 [515'.352] 92-24315

Originally published by Springer-Verlag Berlin Heidelberg New York in 1994.
Softcover reprint of the hardcover 1st edition 1994

Typesetting: Asco Trade Typesetting Ltd., Hong Kong
41/3140 - 5 4 3 2 1 0 - Printed on acid-free paper

List of Editors, Authors and Translators

Editor-in-Chief

R.V. Gamkrelidze, Russian Academy of Sciences, Steklov Mathematical Institute, ul. Vavilova 42, 117966 Moscow, Institute for Scientific Information (VINITI), ul. Usievicha 20a, 125219 Moscow, Russia

Consulting Editors

V. I. Arnol'd, Steklov Mathematical Institute, ul. Vavilova 42, 117966 Moscow, Russia

S. P. Novikov, L.D. Landau Institute for Theoretical Physics, Russian Academy of Sciences, Vorobevskoe Shosse 2, 117334 Moscow, Russia

Authors

A.T. Fomenko, Department of Mathematics and Mechanics, Moscow University, 119899 Moscow, Russia

V. Ya. Gershkovich, Department of Mathematics and Mechanics, St. Petersburg University, Staryj Petergof, 198904 St. Petersburg, Russia

M. A. Olshanetsky, Institute for Theoretical and Experimental Physics, Moscow, Russia

A. M. Perelomov, Institute for Theoretical and Experimental Physics, Moscow, Russia

A. G. Reyman, Steklov Mathematical Institute, Fontanka 27, 191011 St. Petersburg, Russia

M. A. Semenov-Tian-Shansky, Steklov Mathematical Institute, Fontanka 27, 191011 St. Petersburg, Russia

V. V. Trofimov, Department of Mathematics and Mechanics, Moscow University, 119899 Moscow, Russia

A. M. Vershik, Department of Mathematics and Mechanics, St. Petersburg University, Staryj Petergof, 198904 St. Petersburg, Russia

Translators

A. G. Reyman, Steklov Mathematical Institute, Fontanka 27, 191011 St. Petersburg, Russia

M. A. Semenov-Tian-Shansky, Steklov Mathematical Institute, Fontanka 27, 191011 St. Petersburg, Russia

Contents

I. Nonholonomic Dynamical Systems, Geometry of Distributions and Variational Problems

A.M. Vershik, V.Ya. Gershkovich

Translated from the Russian
by M.A. Semenov-Tian-Shansky

Contents

Introduction

0. A nonholonomic manifold is a smooth manifold equipped with a smooth distribution. This distribution is in general nonintegrable. The term 'holonomic' is due to Hertz and means 'universal', 'integral', 'integrable' (literally, *ὁλος* – entire, *νομος* – law). 'Nonholonomic' is therefore a synonym of 'nonintegrable'.

A nonholonomic manifold is the geometric (or, more precisely, kinematic) counterpart of a nonholonomic dynamical system with linear constraints. As we shall see later, there are two main ways to construct dynamics on nonholonomic manifolds. They will be referred to (somewhat conventionally) as mechanical and variational, respectively.

The aim of the present survey is to give a possibly self-consistent account of the geometry of distributions and to lay down a foundation for a systematic study of nonholonomic dynamical systems. Our review is somewhat different in its character from other reviews in the present series. The reasons for this are rooted in the peculiar history of the subject as will be explained below.

Due to lack of space, the present volume includes only the first part of our survey which comprises geometry of distributions and variational dynamics. Geometry and dynamics of nonholonomic mechanical systems, nonholonomic connections, etc., will be described separately. However, in this introduction we will discuss all these topics, at least in historical aspect. Several problems of nonholonomic mechanics have already been considered in Volume 3 of this Encyclopaedia (Arnol'd, Kozlov and Nejshtadt [1985]). In this survey we give a modern exposition of some earlier results as well as new results which were not published previously.

1. Nonholonomic geometry and the theory of nonholonomic systems are the subject of numerous papers. A few of them are due to prominent geometers of the beginning of our century, others and still more numerous date back to the early past-war period. Nevertheless, this theory has not gained popularity in a broader mathematical audience. For instance, in most textbooks on Riemannian geometry and the calculus of variations there are hardly any facts on these subjects, save perhaps for the classical Frobenius theorem. Even the term 'nonholonomic' is scarcely mentioned. The exceptions are rather rare (e.g. the recent book of Griffiths [1983] where such results as the Chow-Rashewsky theorem are exposed).

There are many reasons to feel dissatisfied with this state of affairs. First of all, nonholonomic systems always held a sufficiently important place in mechanics. Mathematicians have always been thoroughly studying classical dynamics and the mathematical structures that are inherent to it. In view of this long-standing tradition the neglect of nonholonomic problems is almost striking. It is even not commonly known that nonholonomic mechanical problems cannot be stated as variational problems (cf. below).

Secondly, nonholonomic variational problems have much in common with optimal control problems which have been the subject of so many papers in recent years. This similarity has not been noticed until recently. Although the statements of nonholonomic variational problems are classical, some features of their solutions (e.g. the structure of the accessibility set, cf. Chapter 2) makes them quite close to non-classical problems. However, in textbooks on variational calculus even existence theorems for the simplest nonholonomic problems are lacking. The corresponding theorem in Chapter 2 of our survey is based on a recent observation made in connection with non-classical problems.

Thirdly, problems of thermodynamics (Gibbs, Carathéodory) and of quantum theory (Dirac) also lead to nonholonomic variational problems. A modern mathematical treatment of this kind of problems is still to be given.

Fourthly, nonholonomic problems are closely connected with the general theory of partial differential equatons. The best known results that display this connection are the theorems of L. Hörmander and A.D. Aleksandrov on hypoellipticity and hypoharmonicity. The study of these questions from a nonholonomic point of view has been actively pursued in recent years (we included some additional references to make this translation more up-to-date).

Finally, the general mathematical theory of dynamical systems (now in its maturity, as confirmed by the present edition) may well find in nonholonomic dynamics a vast source of new problems, examples, and paradoxes. We hope to support this view by the present survey.

2. We shall now give a brief review of the history of our subject. We believe that it will also explain the isolation of nonholonomic theory from the rest of mathematics which has continued up to the present time. A systematic development of the nonholonomic theory was started in the twenties and thirties, following a pattern which dates back to the turn of the century. This pattern was shaped in a series of classical papers on nonholonomic mechanics, due to many mathematicians and physicists: Hertz, Voss, Hölder, Chaplygin, Appell, Routh, Woronets, Korteweg, Carathéodory, Horac, Volterra, to mention a few. See Aleksandrov [1947], and Synge [1936] for a review. As mentioned by Grigoryan and Fradlin [1982], nonholonomic mechanical problems were already treated by Euler. However, it was not until the turn of the century that a clear understanding of their special features was gained. Hertz's name (Prinzipien der Mechanik, 1894) should be ranked first in this respect. A less known source of the theory comes from physics, namely, from the works of Gibbs and Carathéodory on the foundations of thermodynamics. They deal with contact structure which is the simplest example of nonholonomic structure. As for pure mathematics itself, in particular, the theory of distributions, which is an indispensable part of nonholonomic theories, we should begin with the theory of Pfaffian systems and subsequent works on the general theory of differential equations. The contribution of E. Cartan to this domain was of particular importance. He was the first to introduce differential forms and codistributions. Unfortunately, these tools were not widely used in nonholonomic problems.

Finally, we mention one trend of research in geometry going back to Issaly (1880), or, perhaps, even earlier. We mean the studies of nonholonomic surfaces, i.e. of nonholonomic 2-dimensional distributions in 3-dimensional space generalizing the ordinary theory of surfaces. This trend has been developed further by D.M. Sintsov and his school and by some other geometers in the Soviet Union. However, these works were not sufficiently known even to the experts and did not play a major role.

3. In the twenties when Levi-Civita and H. Weyl defined the notions of Riemannian and affine connections and discovered deep relations between mechanics and geometry, it became clear that nonholonomic mechanics should also serve as a source of new geometrical structures which, in turn, provide mechanics and physics with a convenient and concise language. This mutual interaction was started by Vranceanu and Synge. In the Soviet Union nonholonomic problems were activey advertized for by V. Kagan. In 1937 he proposed the following theme for the Lobachevsky prize competition (Vagner, [1940]): "To lay down the foundation of a general theory of nonholonomic manifolds. ⟨...⟩ Applications to mechanics, physics, or integration of Pfaffian systems are desirable".

The most important results on nonholonomic geometry and its connections with mechanics were obtained in the pre-war years and are due to Vranceanu and Synge, and also to Schouten and V. Vagner. In two short notes and an article a Romanian mathematician Vranceanu [1931] gave the first precise definition of a nonholonomic structure on a Riemannian manifold and outlined its relation to the dynamics of nonholonomic systems. Synge [1927, 1928] has studied the stability of the free motion of nonholonomic systems. In his work he anticipated the notion of the curvature of a nonholonomic manifold. It was formally introduced somewhat later and in two stages. First, Schouten defined what was later to be called truncated connection, i.e. parallel transport of certain vectors along certain vector fields. The geodesics of this connection are precisely the trajectories studied by Synge. Finally, a Soviet geometer V. Vagner made the next important step in a series of papers which won him the Lobachevsky prize of Kazan University for young Soviet mathematicians in 1937. He defined (in a very complicated way) the general curvature tensor which extends the Schouten tensor and satisfies all the natural conditions (e.g. it is zero if and only if the Schouten-Vranceanu connection is flat). In his subsequent papers Vagner extended and generalized his results. (Notably, Schouten was in the jury when Vagner defended his thesis.)[1]

4. The geometry of the straightest lines (i.e. classical mechanics of nonholonomic systems) is the subject of quite a few papers written mainly between the wars. By contrast, much less was done on the geometry of the shortest lines, i.e. on the variational theory of nonholonomic systems. The main

[1] Recently, a more modern exposition of Vagner's main work has appeared (Gorbatenko [1985].)

contributions bearing on this subject may be listed quite easily. The starting point for the theory was a paper of Carathéodory [1909] in which he proves that any two points on a contact manifold may be connected by an admissible curve. (This statement was already mentioned without proof in earlier papers, e.g. by Hertz.) It is interesting to notice that Carathéodory needed this theorem in connection with his work on foundations of thermodynamics, namely in order to justify the definition of thermodynamical entropy. Although this theorem has a kinematic nature, it may be used to define a variational, or nonholonomic, metric sometimes referred to as the Carnot-Carathéodory metric (see Chapter 3). An extension of this theorem to arbitrary totally nonholonomic manifolds was proved independently by Chow [1939] and by Rashevsky [1938]. Several results of the classical calculus of variations were extended to the nonholonomic case by Schoenberg who was specially studying variational problems. A comparison of mechanical and variational problems for nonholonomic manifolds was given by Franklin and Moore [1931]. An interpretation of nonholonomic variational problems in mechanical or optical terms has been proposed quite recently (Arnol'd, Kozlov and Nejshtadt [1985], Karapetyan [1981], Kozlov [1982a, b, 1983]).

5. Before we come to describe the contents of this paper it is worth commenting on the reasons which possibly account for the contrast between the importance of these subjects and their modest position in "main-stream" mathematics. The point is that most papers which bear on the subject were written extremely vaguely, even if one allows for the usual difficulties of "coordinate language". To put it in a better way, at that time it was practically impossible to give a clear exposition of the theory (which is far from being simple by any standards). The key concept that was badly needed was that of a connection on a principal bundle in its almost full generality. However, nothing of the kind was used at that time. The absence of an adequate language was really painful and resulted in enormous and completely unmanageable texts. It was difficult to extract from them even the main notions, to say nothing of theorems. As a consequence, these papers were not duly understood.[2] One needed such tools as jets, germs, groups of germs of diffeomorphisms, etc. This was already clear to E. Cartan. Although he never studied this subject specially, he frequently stresses that connections on principal fibre bundles should be used in nonholonomic problems (ironically, he quotes this idea in the same volume dedicated to the Lobachevsky prize competition that we already mentioned (Vagner [1940])).

All this may be the reason why the fundamental reshaping of differential geometry in a coordinate-free manner has left aside the nonholonomic theory. In the fifties and sixties this theory was already out of fashion and was to remain

[2] V. Vagner wrote in 1948: "The lack of rigour which is typical for differential geometry is reflected also in the absence of precise definitions of such notions as spaces, multi-dimensional surfaces, etc. Differential geometry is certainly dropping behind and this became even more dangerous when it lost its direct contacts with theoretical physics".

in obscurity for many years to come. Although many of the more recent papers on nonholonomic theory already used modern language, they were isolated from the fertilizing applications which had served as the starting point for geometers of the prewar time.

After a complete renewal of its language in the fifties and sixties modern differential geometry became one of the central parts of contemporary mathematics. Along with topology, the theory of Lie groups, the theory of singularities, etc., it has created a genuine mathematical foundation of mechanics and theoretical physics in general. An invariant formulation of dynamics permitted to apply various powerful tools in this domain. Gradually, this process has brought to bear on nonholonomic theory. Quite recently, the Schouten-Vranceanu connection was rediscovered by Vershik and Faddeev [1975]. (See also Godbillon [1969], Vershik [1984].) In this paper nonholonomic mechanics was exposed systematically in terms of differential geometry. In particular, it was shown that the local d'Alembert principle regarded as a precise geometric axiom implies the above mentioned theorem on geodesics in nonholonomic theory.

A good deal of the authors' efforts was aimed to extract geometrical ideas and constructions from the papers of the past years and to present them in a modern form. This goal has not been fully achieved, but it seems indispensable in order to develop systematically the qualitative and geometric theory of nonholonomic dynamical systems, in analogy with other theories of dynamical systems (e.g. Hamiltonian, smooth, ergodic, etc.). We tried to give the basic definitions and to describe the simplest (3-dimensional) examples. Many mathematicians and physicists have helped us by pointing out various scattered papers on the subject. We are particularly indebted to A.D. Aleksandrov, V.I. Arnol'd, A.M. Vassil'ev, A.M. Vinogradov, A.V. Nakhmann, V.V. Kozlov, N.V. Ivanov, Yu.G. Lumiste, Yu.I. Lyubich, N.N. Petrov, A.G. Chernyakov, V.N. Shcherbakov, and Ya.M. Eliashberg.

6. Let us now turn to a brief description of the general structure of the survey. Nonholonomic dynamics is based on the geometry of distributions which is the subject of Chapter 1. The simplest and best known example of a nonholonomic manifold is the contact structure, i.e. a maximally nonholonomic distribution of codimension 1. Since in the existing literature very little is available on distributions of larger codimension, we present the main definitions and the most important examples of distributions in Section 1 of Chapter 1. In Section 2 we study generic distributions and the classification problem. In particular, we present results on the existence of functional moduli of distributions for almost all growth vectors. We also briefly mention the notion of nilpotentization which is of particular importance, especially in the recent advances of the theory. (For more information consult the list of references which was enlarged to make this translation more up-to-date.) As already mentioned, there are two completely different dynamics associated with a nonholonomic Riemannian manifold: the dynamics of the 'straightests', or mechanical, and the dynamics of the 'shortests', or variational. The terms 'straightest' and 'shortest' were first intro-

duced in connection with mechanics by Hertz. The difference between them is briefly as follows. Using the distribution we may introduce the so called truncated connection (Schouten [1930]). The study of its geodesics and of the corresponding flow is connected with mechanics of systems with linear constraints, e.g., with the problem of rolling, etc. (The general theory comprises also nonlinear restrictions, cf. for instance Vershik and Faddeev [1975], Vershik and Gershkovich [1984].) These questions will be considered separately. If we restrict the metric to the distribution, we get a new metric on the manifold. Its geodesics (the shortests) are the subject of variational theory discussed in detail in Chapter 2. The phase space of a nonholonomic variational problem is the so called mixed bundle, i.e. the direct sum of the distribution regarded as a subbundle of the tangent bundle and of its annihilator regarded as a subbundle of the cotangent bundle. As already mentioned, variational problems also admit a proper mechanical interpretation.

In Section 1 of Chapter 2 we present the main notions and constructions related to nonholonomic variational problems, such as nonholonomic geodesic flow, nonholonomic metrics, the nonholonomic exponential mapping, wave fronts, etc. In Section 2 we compute the accessibility sets for nonholonomic problems (or control sets, in the language of control theory).

In Chapter 3 we consider nonholonomic variational problems on Lie groups and homogeneous spaces. As usual, problems on Lie groups offer the most important class of examples, as well as a training field to develop constructions and methods which may be then extended to the general setting. In Section 1 we discuss local questions: the wave front and the ε-sphere of a nonholonomic Riemannian metric. In Section 2 we present a complete description of the dynamics of systems associated with the nonholonomic geodesic flow on homogeneous spaces of 3-dimensional Lie groups. Our approach is based on the wide use of geometry (more precisely, nonholonomic Riemannian geometry) and of the theory of nilpotent Lie groups. These two sources provide a better understanding of various domains connected with the study of distributions, such as nonholonomic mechanics, the theory of hypoelliptic operators, etc. At the same time this approach leads to new problems in geometry and in the theory of Lie groups.

Chapter 1
Geometry of Distributions

§ 1. Distributions and Related Objects

1.1. Distributions and Differential Systems. In the sequel without further notice all objects, such as manifolds, functions, mappings, distributions, vector fields, forms, etc., are supposed to be infinitely differentiable.

Definition 1.1. Let X be a real smooth manifold without boundary, and let TX be its tangent bundle. A subbundle $V \subset TX$, i.e. a family $\{V_x\}_{x \in X}$ of linear subspaces $V_x \subset T_x$ of the tangent spaces which depend smoothly on the point $x \in X$, is called a *distribution* on X. If X is connected, the number $\dim V_x \equiv \dim V$ is called the *dimension* of the distribution.

In the simplest case a distribution has the following structure: there is a decomposition of X into submanifolds (leaves), and V_x is the tangent space to the leaf passing through x. In this case the distribution is said to be *integrable* and determines a foliation. Its leaves are called maximal integral submanifolds of the distribution; their dimension is equal to that of V. If $\dim V = 1$, V is always integrable and its integral submanifolds are (locally) integral curves of a vector field that generates V.

In this survey we shall be mainly interested in the opposite case of nonintegrable, or nonholonomic distributions. The simplest example of a nonholonomic distribution is provided by two-dimensional distributions in $\mathbb{R}^3$, for instance, given by $V_x = \mathrm{Lin}\{\partial/\partial x_1, -\partial/\partial x_2 + x_1 \partial/\partial x_3\}$, $x = (x_1, x_2, x_3)$. As is frequently done, the distribution is defined here as the linear span of vector fields. Another way to define the same distribution is as follows: V is the null-space of a 1-form $x_1\, dx_2 + dx_3$ which defines a contact structure on $\mathbb{R}^3$. The description of a distribution as the set of null-spaces of a system of differential forms will also be frequently used.

A k-dimensional distribution on X may be regarded as a section of the *Grassmann bundle* associated with TX, i.e. of the bundle of k-dimensional subspaces of the tangent spaces. This construction equips the space $\mathscr{V}_k(X)$ of such distributions with the natural topology of C^∞-sections. If X is an open ball, then, for $k \geqslant 2$, nonintegrable distributions (and even maximally nonintegrable distributions, see Section 2) form an open dense subset in $\mathscr{V}_k(X)$. On the contrary, the integrability of a distribution, i.e. the existence of foliation, is an extremely rare (nowhere dense) event.

Definition 1.2. We shall say that a vector field ξ on X is *subordinate* (or *belongs*) *to* $V = \{V_x\}$ if $\xi_x \in V_x$ for all $x \in X$. If $V_x = \mathrm{Lin}\{\xi_x^i, i = 1, \ldots, n\}$, the vector fields ξ^i are said to *generate* V. An integral curve γ of a vector field belonging to V is called *admissible* (with respect to V): $\dot{\gamma}_x \in V_x$, $x \in X$.

Recall that the linear space (over $\mathbb{R}$) of smooth vector fields Vect X is a Lie algebra with respect to the Lie bracket $[\xi, \eta] = \xi\eta - \eta\xi$. (Vector fields may be regarded as derivations, and their product means composition of derivations.) Moreover, Vect X is a $C^\infty(X)$-module, since each $\xi \in$ Vect X may be multiplied by $f \in C^\infty(X)$.

If V is a distribution, the set of all vector fields belonging to V (i.e. the set of its sections) is a C^∞-submodule in Vect X. We introduce the following

Definition 1.3. A *differential system* on X is a linear space of vector fields on X which is a $C^\infty(X)$-module[1].

As we have explained above each distribution V gives rise to a differential system $N(V)$. However, there exist other differential systems that correspond to distributions with singularities, i.e. to fields of linear subspaces in TX of non-constant dimension. Such distributions appear quite naturally. For instance, the Lie bracket of two distributions may already have singularities, which motivates the necessity of the above definition.

Let F be a differential system. The set of all vectors $v \in T_xX$ for which there is a vector field $\xi \in F$ such that $\xi_x = v$ is a linear subspace $V_x \subset T_x$. If dim $V_x =$ const, F is generated by a distribution $\{V_x\} = V$, $F = N(V)$; otherwise such a distribution does not exist.

Proposition. *A differential system on a smooth manifold X is the space of sections of a distribution if and only if it is a projective $C^\infty(X)$-module.*

Recall that if X is an open ball, any projective module is free, hence, in local problems, differential systems that are free $C^\infty(X)$-modules are the same as distributions. (A free module is the direct sum of several copies of $C^\infty(X)$, a projective module is a direct summand of a free module).

Definition 1.4. A distribution V (a differential system N) is *involutive* if $N(V)$ (respectively, N) is a Lie algebra; in other words, the Lie bracket of two vector fields that are subordinate to V (respectively, belong to N) also belongs to V (respectively, to N).

In the sequel we shall maninly deal with local problems, and so it is useful to introduce local versions of the main definitions using the language of germs and jets. (Concerning the notions of germs, jets, etc, see Bröcker and Lander [1975], Golubitsky and Guillemin [1973].) Let W_n^r, $1 < r \leqslant \infty$, $n = 1, \ldots,$ be the space of *r-jets of vector fields* at $0 \in \mathbb{R}^n$, let ω_n be the space of germs of vector fields in a neighborhood of $0 \in \mathbb{R}^n$. The spaces $W_n^\infty \equiv W_n$ and ω_n are Lie algebras with respect to the Lie bracket of vector fields. Moreover, these spaces are modules over the ring of jets J_n^∞ and the ring of germs of functions E_n, respectively.

---

[1] Some authors use the term 'differential systems' for distributions. Since the latter term is well established in Russian literature we shall use the term 'differential system' for a different notion. Recall that the notion of $C^\infty(X)$-module means that vector fields from F may be multiplied by an arbitrary element of $C^\infty(X)$: $\forall \xi \in F\, \forall f \in C^\infty(X)\, f\xi \in F$.

The ring E_n of germs of C^∞-functions is a local algebra, i.e. it contains a unique nontrivial maximal ideal $\mathfrak{M}_n$ generated by the germs of functions that vanish at $0 \in \mathbb{R}^n$. The ideal of flat relations $\mathfrak{M}_n^\infty = \bigcap_k \mathfrak{M}_n^k$ has a very complicated structure. The quotient ring $E_n/\mathfrak{M}_n^\infty$ is called the ring of jets and is described by the following theorem (cf. Bröcker and Lander [1975]).

Borel Theorem.

$$J_n^\infty = E_n/\mathfrak{M}_n^\infty \cong \mathbb{R}[[y_1, \ldots, y_n]].$$

Hence this ring, as a ring of power series, is Noetherian, i.e. it contains only finite sequences of strictly increasing ideals.

We shall keep the name "differential system" for submodules both in W_n and in ω_n. Thus in these terms a *jet* (*a germ*) *of a distribution* means a free submodule in W_n (or in ω_n).

In complete analogy with the global statement above, a differential system F is generated by the germ of a distribution (and is a free module over E_n) if and only if dim V_x is the germ of a constant function (here $V_x = \{\xi(x); \xi \in F\}$).

1.2. Frobenius Theorem and the Flag of a Distribution. The classical Frobenius theorem (cf. Sternberg [1964]) asserts that each involutive distribution is integrable. Integrability of a distribution means that there exists a foliation such that the distribution consists of tangent spaces to its leaves. We shall give a proof of this theorem that focuses on the algebraic side of the question (cf. also Treves [1980]). Since the statement of the Frobenius theorem is local, we shall consider only the local case.

Frobenius Theorem. *The germ* (*or the jet*) *of an involutive distribution is generated by commuting vector fields. More precisely, let N be an involutive differential system* (*either in germs, or in jets*) *which is a subalgebra and a free E_n-module of dimension m. Then N is generated, as a module, by m commuting vector fields.*

Corollary. *All jets* (*germs*) *of involutive distributions of a given dimension constitute a single orbit of the group of jets* (*germs*) *of diffeomorphisms of $\mathbb{R}^n$. In other words, the jets* (*germs*) *of two involutive distributions of the same dimension may be transformed one into another by a jet* (*germ*) *of a diffeomorphism.*

Hence the orbit is defined by the 1-jet of a distribution, since involutivity is a 1-jet condition.

The final part of Frobenius theorem in its usual form (construction of leaves) is now reduced to the standard existence theorems for ordinary differential equatons. Indeed, the integral curves for commuting vector fields may be determined successively, which yields the desired leaf.

Proof. Let V be the germ of an involutive distribution of dimension m in $\mathbb{R}^n$ generated by $\xi_1, \ldots, \xi_m \in W_n$. We shall prove by induction over m that V is generated by commuting vector fields $\Psi_1, \ldots, \Psi_m$. Choose a coordinate system

$\{x_i\}_{i=1}^n$ in $\mathbb{R}^n$ in such a way that $\xi_1 = \frac{\partial}{\partial x_1}$. Let us decompose ξ_i with respect to $\frac{\partial}{\partial x_i}$: $\xi_i = \sum_{j=1}^n c_{ij}\frac{\partial}{\partial x_j}$, $c_{ij} \in E_n$, where E_n is the ring of germs of smooth functions in $\mathbb{R}^n$.

The vector fields $\tilde{\xi}_i = \xi_i - c_{i,1}\xi_1$, $i = 2, \ldots, m$, form an involutive system. By the induction hypothesis, there exist commuting vector fields $\Psi_2, \ldots, \Psi_m$ such that $V(\tilde{\xi}_2, \ldots, \tilde{\xi}_m) = V(\Psi_2, \ldots, \Psi_m)$. Choose a coordinate system $\{y_i\}$ in $\mathbb{R}^n$ in such a way that $\Psi_i = \frac{\partial}{\partial y_i}$, $i = 2, \ldots, m$, and decompose ξ_1, with respect to $\frac{\partial}{\partial y_i}$: $\xi_1 = \sum_{i=1}^n b_i \frac{\partial}{\partial y_i}$. Put $\tilde{\xi}_1 = \xi_1 - \sum_{i=2}^m b_i\xi_i$ $(b_i \in E_n)$. Then $\tilde{\xi}_1 = b_1\frac{\partial}{\partial y_1} + \sum_{j=m+1}^n b_j \frac{\partial}{\partial y_j}$. There exists $j \in \{1, m+1, m+2, \ldots, n\}$ such that $b_j(0) \neq 0$. After a suitable reordering of coordinates we may assume that $j = 1$. Put $\Psi_1 = b_1^{-1}\tilde{\xi}_1$. Then $V(\Psi_1, \ldots, \Psi_m) = V(\xi_1, \ldots, \xi_m)$ and the fields Ψ_i commute with each other. The only claim that requires some comment is that $[\Psi_i, \Psi_1] = 0$ for $i \geqslant 2$. Indeed,

$$[\Psi_i, \Psi_1] = \sum_{j=m+1}^n \frac{\partial}{\partial y_i}\left(\frac{b_j}{b_1}\right)\frac{\partial}{\partial y_j}. \qquad (*)$$

On the other hand, there exist functions c_i^k, $k = 1, \ldots, m$, such that

$$[\Psi_i, \Psi_1] = \sum_{k=1}^m c_i^k \Psi_k = \sum_{k=1}^m c_i^k \frac{\partial}{\partial y_k} + \sum_{k=m+1}^n c_i^1 b_1^{-1} b_k \frac{\partial}{\partial y_k} \qquad (**)$$

Comparing the coefficients in $(*)$ and $(**)$ we get $c_i^k = 0$ for all k, which concludes the proof. □

A submanifold $Y \subset X$ is called an *integral submanifold* of the distribution V if $T_x Y \subset V_x$, $x \in Y$. If the distribution is not involutive, and hence nonintegrable, there are no integral submanifolds of dimension equal to the dimension of the distribution. However, integral submanifolds exist (e.g., the integral curves of a vector field which is subordinate to V). One can show that in the case of generic distributions of codimension $\geqslant 3$ there are no integral submanifolds of dimension greater than one at any point (we shall discuss this question later in more detail).

Let N be a differential system in W_n (respectively, in ω_n).

Definition 1.5. The *flag of a differential system* N (or simply the flag of N) is the sequence of differential systems $N_1 = N$, $N_2 = [N, N]$, $\ldots$, $N_l = [N_{l-1}, N]$. Here $[A, B]$ is the C^∞-submodule generated by $[\alpha, \beta]$, $\alpha \in A$, $\beta \in B$.

Proposition. *The sequence $N_1, N_2, \dots \subset W_n$ stabilizes, i.e. there exists an integer r such that $N_{r-1} \neq N_r = N_{r+1} = \cdots$, moreover, N_r is a Lie subalgebra of* Vect X.

Proof. The set W_n of jets of vector fields is a finitely generated J_n-module. According to the Borel theorem (see the end of Section 1.1.), J_n is a Noetherian ring, hence W_n is a Noetherian module, so that an increasing sequence of its submodules must stabilize.

If $N_r = W_n$, the differential system N is called *totally nonholonomic*, and the smallest r such that $N_r = W_n$ is called the *nonholonomicity degree* of N (this degree depends on the point of the manifold).

Let V be a distribution, let $N = N(V)$ be the differential system associated with V, and let $N(V) = N_1 \subset N_2 \subset \cdots$ be its flag. In general, the Pfaffian systems N_i are not generated by distributions (see Section 1.1). If this is the case for all i, we may define the flag of the distribution $V = V_1 \subset V_2 \subset \cdots$.

Definition 1.6. A distribution V is called *regular* if its flag is well defined.

Let $n_i = \dim V_i$. The vector $n_1 < n_2 < \cdots$ is called the *growth vector* of the distribution V. Clearly, the sequence V_i stabilizes. If $V_r = TX$ for some r, the distribution V is *totally nonholonomic* and r is called its *nonholonomicity degree*. If the sequence V_i stabilizes for $i = 1$, so that $V_1 = V_2$, then V is a subalgebra and we are dealing with the integrable case.

1.3. Codistributions and Pfaffian Systems. All the notions introduced above admit a dualization: thus, distributions correspond to codistributions and differential systems to Pfaffian systems. This correspondence may be established both globally and locally, for jets and germs. Classics (E. Cartan, for instance) used the language of forms and codistributions much more frequently than the language of vector fields. For instance, in mechanics and geometry a distribution is usually defined as the annihilator of a codistribution. However, some authors, e.g. Vessiot, wrote in the same years (1926) that many concepts of Cartan's theory might be expressed more naturally (at least from the algebraic point of view) by using the language of vector fields. The advantage of differential forms is the existence of exterior derivative and wedge product while the advantage of vector fields is the existence of the Lie algebra structure. In the sequel, $\Omega^1(X)$ denotes the space of 1-forms on X, i.e. of smooth sections of T^*X.

Definition 1.7. A *codistribution* on X is a subbundle of the cotangent bundle T^*X or, in other words, a field of subspaces of the cotangent spaces which depend smoothly on the point of X. A *Pfaffian system* is a submodule of the $C^\infty(X)$-module $\Omega^1(X)$.

The annihilator of a distribution V, i.e. $V^\perp = \{\omega \colon \langle \omega, \xi \rangle = 0 \ \forall \xi \in V\}$, is a codistribution. In a similar fashion, the annihilator of a codistribution is a distribution. In the sequel we shall use both the language of vector fields and the language of differential forms. Below we give a table of parallel notions. The transition from a notion to its dual is based on the duality between the space of vector fields (or the jets (germs) of vector fields) and the space of 1-forms (respectively, of their jets, etc.) regarded as modules over the ring of functions (respectively, the ring of jets (germs) of functions). Let W_n^* denote the space of ∞-jets of 1-forms at $0 \in \mathbb{R}^n$, and ω_n^* the space of their germs. Let $\langle \xi, \omega \rangle$ be the pairing which maps vector fields and forms into functions (e.g. $W_n \times W_n^* \to J_n$ and so on) The following assertion is dual to the proposition in Section 1.1.

Proposition 1. *A Pfaffian system is generated by a codistribution if and only if it is a projective $C^\infty(X)$-module.*

Recall that the space $\Omega^*(X)$ of all differential forms on X is equipped with a graded commutative operation of wedge product which makes $\Omega^*(X)$ an algebra (and a superalgebra), and with the exterior derivative d (see Manin [1984]). We shall use them to introduce the notion of an involutive codistribution.

A Pfaffian system, and in particular a codistribution, defines an ideal $J(V^*)$ in the exterior algebra $\Omega^*(X)$. An ideal $J \subset \Omega^*$ is called a *differential ideal* if $dJ = \{d\omega | \omega \in J\} \subset J$. Consider a distribution V and assume that $V_2 = [V, V]$ is again a distribution. Let us compute its annihilator $V_2^\perp$. Since $V \subset V_2$, we have $V_2^\perp \subset V^\perp$. Let us check that $V_2^\perp = U$, where

$$U = \{\omega \in V^\perp: d\omega \in J(V^\perp)\}.$$

Indeed, let $\omega \in V^\perp$, $d\omega \in J(V^\perp)$, then the Maurer – Cartan formula implies that for $\xi, \eta \in V$

$$\omega[\xi, \eta] = d\omega(\xi, \eta) - \xi\omega(\eta) + \eta\omega(\xi) = d\omega(\xi, \eta) = 0,$$

since $d\omega = \sum_i \omega_i \wedge \alpha_i$, where $\omega_i \in V^\perp$. Thus $V_2^\perp \supset U$. The opposite inclusion is checked in an equally simple way. We arrive at the following assertion:

Proposition 2. *Let V be a distribution and V^* its annihilator. Then V is involutive or, in other words, $N(V)$ is a Lie subalgebra if and only if $J(V^*)$ is a differential ideal.*

Indeed, if the distribution V is integrable, then

$$V_2 = V \Leftrightarrow (\forall \omega \in J(V^*): d\omega \in J(V^*)) \Leftrightarrow dJ(V^*) \subset J(V^*),$$

i.e. $J(V^*)$ is a differential ideal. □

This proposition is sometimes called *Cartan's Theorem.* If $V \subset V_2 \subset \cdots$ is the flag of a distribution (see Section 1.2), we can successively compute the elements of the coflag of the codistribution $V^\perp \supset V_2^\perp \supset \cdots$, where $V_l^\perp$ is the annihilator of V_l. This may be done by similar formulae using the differentials of the forms from $V^\perp$. This construction is covered by a more general scheme which gives an independent (and, in fact an earlier) definition of the *coflag of a Pfaffian system* (see Griffiths [1983]). Let N^* be a Pfaffian system (in the previous case $N^* = N^*J(V^\perp)$), i.e. the $C^\infty(X)$ – module of 1-forms that annihilate V. Put $N_1^* = N^*$. Assume that $N_1^*, \ldots, N_i^*$ are already constructed. Consider the mapping d: $N_i^* \to \Omega^*$ and put $N_{i+1}^* = d^{-1}N_i^*$. Thus we obtain a decreasing chain of ideals in Ω^*: $N_i^* \supset N_2^* \supset \cdots$. An ideal $J \subset \Omega^*$ is called *special* if it is generated (as an ideal) by the set of 1-forms which it contains. Observe that a coflag always consists of special ideals. Hence we may regard a coflag as a sequence of Pfaffian systems.

Let us describe the duality properties of a flag and a coflag. Let N be a differential system, N^* the annihilator of N. Let N^{**} be the biannihilator of N. In general $N^{**} \supset N$ and N does not coincide with N^{**}. However, the following assertion is valid.

Proposition 3. *Let N be a differential system, $\{N_i\}$ its flag, $N^* = M$ the annihilator of N in Ω, $\{M_i\}$ the coflag of the Pfaffian system N^*. Then $M_i^* = N_i^{**}$ and $N_i^* = M_i^{**}$.*

Proposition 3 and the proposition of Section 1.1 imply:

Proposition 4. *The coflag $\{M_i\}$ of any Pfaffian system M stabilizes.*

A distribution V is totally non-holonomic (i.e. $V_k = TX$ for some k) if and only if its coflag shrinks to zero (i.e. $N_k^* = 0$).

The table below sums up the system of dual notions:

1. *Distribution*	*Codistribution* (*a subbundle in T^*X*)
2. *Differential system*	*Pfaffian system* (*a submodule in the space of* 1-*forms on the manifold*)
3. *Involutive differential system*	*Pfaffian system which is closed with respect to exteriour derivative* (*a differential ideal in* $\Omega(X)$)
4. *Flag of a distribution*	*Coflag of a distribution*

Let us now introduce a new notion which will play a key role in the sequel.

Definition 1.8. Let X be a smooth manifold, V a distribution on X, $V^\perp$ the codistribution annihilating V. The fiber bundle $\text{Ken}\, V = V \oplus V^\perp$ over X is called the *mixed bundle* (the '*centaur*') associated with V.

The fiber over $x \in X$ is $(V_x, V_x^\perp)$, i.e. the set of pairs consisting of a vector from V_x and a covector from $V_x^\perp$. The mixed bundle is not associated with any principal bundle over X.

Mixed bundles are phase spaces for nonholonomic dynamical systems. Physically the component lying in $V^\perp$ is interpreted as the constraint reaction, or the Lagrange multiplier (see Sections 2, 3).

1.4. Regular Distributions. Distributions are much more manageable geometrical objects than differential systems. However, as already mentioned, the class of distributions is not closed with respect to the Lie bracket and hence one has to deal with a broader notion of differential systems. But it is possible, on the contrary, to restrict the class of distributions and to consider *regular distributions* (see Section 1.2). Annihilator and biannihilator of a regular distribution are regular. Hence for regular distributions Proposition 3 of Section 1.3 takes a simpler form: *the flag and the coflag of a regular distribution are set into duality.*

Propositions of Section 1.1 allow to reformulate the definitions of regularity for each class of objects.

1. *A distribution N on X is regular if, for all i, N_i is a projective module over $C^\infty(X)$.*

2. *The germ of a distribution N is regular if all N_i are free modules over E_n, or, equivalently, if* $\dim N_i(y)$ *is the germ of a constant function for any i.*

3. *The jet of a distribution V is regular if V_{i+1}/V_i are free modules over J_n^∞.*

Besides its convenience, the class of regular distributions is particularly important, since it represents the generic case. Moreover, it includes important special cases: connections on Riemannian manifolds and left invariant distributions on Lie groups.

As already mentioned, the dimension of a regular distribution is a well defined notion. Let us define the notion of its *basis.* A set of germs (jets) of vector fields $\xi_1, \ldots, \xi_{n_1}$ ($n_1 = \dim V$) is called a *basis* of a regular distribution V if

(a) *$\{\xi_1, \ldots, \xi_{n_1}\}$ is a basis of V regarded as a C^∞-module.*

(b) *The Lie brackets of length l of basic vector fields span V_l/V_{l-1}.*

Proposition. *The germ of each regular distribution has a basis.*

We have shown in Section 1.2 that by a change of variables a basis of an integrable distribution may be reduced to the standard form $\xi_i = \partial_{x_i}$. A basis of a regular distribution may be reduced to a special form. This reduction plays roughly the same role in the theory of distributions as the reduction of a matrix to upper triangular form in linear algebra.

Let k be the nonholonomicity degree of a distribution V. Let $\xi_1, \ldots, \xi_{n_1}$ be its basis. Put $n_i = \dim V_i$, $n_0 = 0$. Choose a set of $n_2 - n_1$ elements $\xi_{n_1+j} = [\xi_{i_1^j}, \xi_{i_2^j}]$, $j = 1, \ldots, n_2 - n_1$, among the Lie brackets $[\xi_{i_1}, \xi_{i_2}]$ of basic vector fields which generate V_2 modulo V_1. Next choose $n_3 - n_2$ elements (i_1^j, i_2^j, i_3^j) among the Lie brackets of length three of basic vector fields such that $\xi_{n_2+j} = [\xi_{i_1^j}, [\xi_{i_2^j}, \xi_{i_3^j}]]$ generate V_3 modulo V_2 and so on. Define a function φ on the positive integers by setting $\varphi(i) = j$ for $n_{j-1} < i \leqslant n_j$.

Lemma (on quasitriangular form). *Let $\xi_1, \ldots, \xi_m$ be a basis of a regular distribution V. One can choose a frame in $\mathbb{R}^n$, $x^i\colon \mathbb{R}^n \to \mathbb{R}$, $i = 1, \ldots, n$, such that the jets of basic vector fields have the form*

$$J^{r-1}\xi_{i_0} = \frac{\partial}{\partial x_{i_0}} + \sum_T \mu_{i_{l+1},\ldots,i_1,i_0} \prod_{j=1}^{l} x_{i_j} \frac{\partial}{\partial x_{i_{l+1}}}$$

with $\mu_{i_{l+1},\ldots,i_1,i_0} \in \mathbb{R}$. (Here r is the nonholonomicity degree of V.) The set T consists of multiindices $(i_{l+1}, \ldots, i_1, i_0)$ such that $l \leqslant r - 1$ and the following three conditions hold:

(1) $i_1 < i_0$,

(2) $i_{j+1} \geqslant i_j$ *for* $j > 0$,

(3) $\varphi(i_{l+1}) \leqslant \sum_{j=0}^{l} \varphi(i_j)$.

The frame described in the lemma will be called *compatible with the basis*; a regular distribution with a fixed compatible basis will be called *a rigged distribution* (cf. Varchenko [1981]).

1.5. Distributions Invariant with Respect to Group Actions and Some Canonical Examples. Let G be a Lie group which acts freely on a smooth manifold X. Then G is acting naturally on the tangent bundle TX, and also on the algebras of jets and germs of vector fields on X, on the cotangent bundle, and on the exteriour algebra of differential forms $\Omega^*(X)$ (cf. Sternberg [1964]). Let us denote all these actions in a uniform way, e.g. $g\colon TX \to TX$, etc. We say that a distribution V (a codistribution, a differential system) is invariant with respect to the action of G if $gV = V$ for all $g \in G$. Cleary, the distribution has the same structure at all points lying on the same G-orbit (i.e. it has the same growth, the same nonholonomicity degree, etc.). The action of g maps the Lie flag of a distribution V at the point x onto the corresponding flag at the point gx. If V is regular at x, the same is true for all points gx, $g \in G$. A basis of V at x is mapped by g onto a basis of V at gx. Left-invariant distributions on groups and homogeneous spaces are one of the principal objects of study in nonholonomic mechanics. For instance, the position of a mechanical system is frequently described by a point of a Lie group G, while the (nonholonomic) constraint imposed on the system restricts its velocity to a fixed subspace (which is not a Lie subalgebra) of the corresponding Lie algebra (see Chapter 2). Moreover, invariant distributions provide good models for more general problems (see below).

Another important example of invariant distributions are connections in principal and associated bundles (see Section 1.6).

We shall consider two classes of examples:

1. Invariant distributions on Lie groups and homogeneous spaces.
2. Canonical distributions on certain manifolds.

Distributions on Lie Groups and Homogeneous Spaces. Left invariant distributions on Lie groups (homogeneous spaces) are distributions that are invariant with respect to the action of the Lie group on itself (on a homogeneous space) by left translations. More precisely, let $\mathfrak{g}$ be the Lie algebra of G and $A \subset \mathfrak{g}$ its subspace, let L_g be a left translation on G:

$$L_g h = gh, \quad dL_g\colon TG \to TG, \quad V(A) = \{dL_g(A)\}_{g \in G}.$$

Since in this case there is only one G-orbit, the distribution $V(A)$ is regular and has the same growth and the same nonholonomicity degree at all points. Conversely, each left-invariant distribution may be obtained in this way, and hence the theory of left-invariant distributions on Lie groups is reduced to the study of subspaces of Lie algebras. The set of m-dimensional left-invariant distributions on a Lie group G may be identified with the Grassmann manifold $\mathrm{Gr}(m, \mathfrak{g})$, i.e. the set of all m-dimensional planes in $\mathfrak{g}$. The distribution $V(A)$ is integrable if A is a Lie subalgebra in $\mathfrak{g}$, it is totally nonholonomic if A generates $\mathfrak{g}$.

The Lie flag of a left-invariant distribution is obtained by left translations from the increasing sequence $A_1 \subset A_2 = A_1 + [A_1, A_1] \subset \cdots$ of linear subspaces in $\mathfrak{g}$. A basis of a left invariant distribution may be chosen to consist of left invariant vector fields obtained by left translations of the vectors of a linear basis in $V \subset \mathfrak{g}$.

Let us consider invariant distributions on certain Lie groups:

(a) Let $G = SO(3)$ be the group of orientation preserving rotations in $\mathbb{R}^3$. Let $\mathfrak{g} = \mathfrak{so}(3)$ be its Lie algebra and $A \subset \mathfrak{g}$ an arbitrary two-dimensional subspace in $\mathfrak{g}$. The left-invariant distribution $V(A)$ generated by A is non-holonomic (its growth vector is (2, 3)). The pair $(G, V(A))$ is the configuration space for the motion of a rigid body with a fixed point and with one component of the velocity vector in the moving frame being identically zero.

(b) Put $G = GL(n, \mathbb{R})$. Clearly, a generic subspace $A \subset \mathfrak{gl}(n, \mathbb{R})$, $2 \leqslant \dim A = t \leqslant n$, generates the whole Lie algebra and hence a left-invariant distribution $V = V(A)$ is totally nonholonomic. Notice that up to certain $k = k(n)$ the growth vector $\{n_l\}$, i.e. the dimensions of V_l, for a generic A are the same as if A were the n_1-dimensional space of generators of the free Lie algebra (with n_1 generators); only the dimensions of the last components are different due to the finite dimensionality. The same is true for other simple groups (cf. § 2).

(c) Let $G = SO(n + 1, \mathbb{R})$. A generic n-dimensional subspace $A \subset \mathfrak{g} = \mathfrak{so}(n + 1, \mathbb{R})$ generates $\mathfrak{g}$ in two steps, i.e. $A + [A, A] = \mathfrak{g}$. The distribution $V(A)$ on G is totally nonholonomic and has the nonholonomicity degree 2. In this example n may be set equal to ∞.

(d) The following example is of particular importance to us. Let $G = N$ be the three-dimensional Heisenberg group of upper triangular matrices with unit diagonal. In this case there exists a unique distribution (up to a Lie algebra automorphism) which generates the whole Lie algebra. For instance,

$$V = \left\{ \begin{pmatrix} 0 & \alpha & 0 \\ 0 & 0 & \beta \\ 0 & 0 & 0 \end{pmatrix}, \alpha, \beta \in \mathbb{R} \right\}.$$

(See Section 1.7 for the classification of nonholonomic planes in three-dimensional Lie algebras.)

(e) Let us generalize the previous example. Let N be an arbitrary nilpotent Lie algebra. Consider its lower central series $N = N_1 \supset N_2 \supset \cdots \supset N_k \supset 0$, where $N_2 = [N, N]$, $N_i = [N_{i-1}, N]$. Choose an arbitrary subspace V_1 which is complementary to N_2. Let $\{V_i\}$ be its flag. Then V_i is complementary to N_{i+1} and the nonholonomicity degree is equal to the length of the lower central series. This determines a left invariant distribution on the corresponding Lie group.

Canonical Distributions on Certain Manifolds. Certain manifolds are equipped with distributions due to their inner structure. The simplest example is the contact structure on the manifold of 1-jets (cf. Arnol'd [1988], Vinogradov et al. [1981], Griffiths [1983]). The following example is particularly useful.

(f) *Flag manifolds.* Fix some integers $0 < n_1 < \cdots < n_k < n$ and consider all possible chains of linear subspaces

$$V_1 \subset V_2 \subset \cdots \subset V_k \subset V, \quad \dim V_i = n_i, \quad i = 1, \ldots, k.$$

These chains are called $(n_1, \ldots, n_k)$-flags, or simply flags. They form the *flag manifold* $\mathcal{F}\ell(\{n_i\})$ which is the quotient space of $\mathrm{GL}(n, \mathbb{R})$ over a parabolic subgroup. In the flag manifold there is a canonical distribution. For concreteness, we shall describe it in the most interesting case $n_i = i$, $i = 1, \ldots, n-1$, i.e. for the space of complete flags $\mathcal{F}\ell$. The general case may be treated similarly. We have $\mathcal{F}\ell = \mathrm{GL}(n, \mathbb{R})/B$ where B is the Borel subgroup (the maximal solvable subgroup, i.e. the subgroup of upper triangular matrices). The distribution W on $\mathcal{F}\ell$ is defined as follows. Let $x = e$ and B the standard Borel subgroup. Then $W_x \subset \mathfrak{g}/\mathfrak{b}$ where $\mathfrak{b}$ is the Lie algebra of B (the Borel subalgebra); it coincides with the image of the $(n-1)$-dimensional space $\gamma \subset \mathfrak{gl}(n)$ consisting of matrices (a_{ij}) with $a_{ij} = 0$, $i \neq j+1$, and $a_{j+1,j}$ arbitrary, $i = 1, \ldots, n-1$. Since $[\mathfrak{b}, \gamma] \subset \gamma + \mathfrak{b}$, this defines an invariant distribution W on G/B. It is not difficult to describe this distribution in an invariant way using simple roots of $\mathfrak{g}$. More important, however, is the following geometric description of W which makes sence for arbitrary flag manifolds: admissible infinitezimal transformations of a flag $f = (V_1, \ldots, V_{n-1})$ leave each V_i inside V_{i+1}, $i = 1, \ldots, n-1$. In other words, the basis in W consists of matrices $a_1, \ldots, a_{n-1}$, where a_i rotates V_i inside V_{i+1}.

It is clear from the algebraic description of W that it is totally nonholonomic and its growth vector for complete flags is

$$(n-1, (n-1)+(n-2), (n-1)+(n-2)+(n-3), \ldots).$$

This example may be generalized without any change to the case when flags of subspaces are replaced by flags of submanifolds, jets of submanifolds, etc.

(g) *Cartan distributions.* The following example is of particular importance. Although its special cases (see below) were defined long ago (Franklin and Moore [1931], Vinogradov et al. [1981]), its systematic study in full generality just begins. There are several versions of the definition; we consider only that one which is a direct generalization of the well known special case, the contact structure in the space of 1-jets. Let B, Y be smooth manifolds, $J^k(B, Y) \to B$ the k-jet bundle of smooth maps from B into Y. Let $f\colon B \to Y$ be such a map and $j^k(f)$ its k-jet extension. It may be regarded as a submanifold in $J^k(B, Y)$, or, more precisely, as a section of the bundle $J^k(B, Y) \to B$. Let $j_0 = (b_0, j^k(f)b_0)$ be a point of $j^k(f)$. Let $C_j(f)$ be the tangent space to $j^k(f)$ at j_0, and C_j the linear hull of $C_j(f)$ with f ranging over all maps such that $j \in j^k(f)$. The distribution $\{C_j; j \in J^k(B, Y)\}$ is called *the Cartan distribution* in $J^k(B, Y)$. Put $B = \mathbb{R}$, $k = 1$, and let Y be n-dimensional manifold. It is easy to check that in this case the Cartan distribution is the common null set of 1-forms which are locally given by $v_i\,dt - dy$ where $(t, y, v) \in J^1(\mathbb{R}, Y)$. Its growth vector is $(n+1, 2n+1)$. If we project the Cartan distribution onto TY (i.e. eliminate the t variable), it gives the well known affine dsitribution $\dot{x}_i = v_i$ which determines the so-called special curves in TY.

Another important example is $B = \mathbb{R}^n$, $Y = \mathbb{R}^1$, $k = 1$. Then $J^1(\mathbb{R}^n, \mathbb{R}^1) \simeq \mathbb{R}^{2n+1}$, and the Cartan distribution is the contact structure defined by the 1-form $du - \sum p_i \, dx_i$ (see Arnol'd and Givental' [1985], Vinogradov et al. [1981].). It is possible to modify this example, for instance, to consider k-jets of submanifolds of fixed dimension in a given manifold (this is done by Vinogradov et al. [1981]). Restriction of the Cartan distribution to the submanifolds of k-jets of certain special mappings (e.g. diffeomorphisms, immersions, etc.) gives another important class of distributions.

The role of the Cartan distribution consists in that it may be used to reduce variational and other problems for k-jets to functional problems in J^k with nonholonomic restrictions. (This point of view is reflected by Griffiths [1983].) It is for this reason that this example is so universal. For example, the simplest problem of variational calculus

$$\int_0^1 f(t, x(t), x'(t)) \, dt$$

may be regarded as a problem on the space of 1-jets

$$J^1(\mathbb{R}; \mathbb{R}^n) = \{(t, x, v)\}$$

with restrictions $v_i \, dt - \dot{x}_i = 0$, $i = 1, \ldots, n$, i.e. as a nonholonomic problem with restrictions determined by the Cartan distribution. In Vinogradov et al. [1981] this distribution is considered from the point of view of the general theory of nonlinear differential equations (cf. also Vosilyus [1983].)

1.6. Connections as Distributions. The most geometrical way to define a connection (of the vast total number) is the following.

Definition 1.9. Let $E \to B$ be a principal bundle with structure group G. A distribution H on E which is invariant under the action of G and is complementary to the vertical distribution (i.e. $H_x + V_x = T_x E$, $H_x \cap V_x = \{0\}$ where V_x is the vertical subspace in $T_x E$ (the tangent space to the fiber)) is called the *horizontal distribution of a connection.*

If we define now a G-invariant $\mathfrak{g}$-valued 1-form on E as a linear mapping $\omega_x \colon T_x E \to V_x \simeq \mathfrak{g}$ such that $\operatorname{Ker} \omega_x = H_x$ ($\mathfrak{g}$ is the Lie algebra of G), it is easy to check that ω satisfies all the conditions imposed on a connection form and hence the distribution H uniquely determines a connection on E in the standard sense, and vice versa. A connection is called flat if H is involutive. The flag of H is called the *connection flag* (in general, it is a flag in the sense of differential systems, cf. Section 1.2), and its nonholonomicity degree is called the *connection degree.*

A connection determines parallel transport of a tangent vector along any curve on the manifold B. By definition, the infinitezimal holonomy group of a connection at the point $x \in B$ is the group of linear transformations of the tangent space $T_x B$ generated by parallel transports along sufficiently small loops passing through x; see Kobayashi [1963, 1969] for a more comprehensive treatment. It is easy to prove the following statement.

Proposition. *The Lie algebra of the infinitezimal holonomy group coincides with the image of the involutive envelope of H under its mapping into* $\mathfrak{g}$ *defined by the connection form. Hence the structure group of a connection is not reducible if and only if its horizontal distribution is totally nonholonomic.*

For Riemannian connections, i.e. connections on a fiber bundle with structure group $O(n)$, there is a geometrical description of the meaning of maximal non-integrability of the horizontal distribution.

Theorem (A.G. Chernyakov). *Let M be a connected Riemannian manifold,* $\mathcal{O}(M)$ *the fiber bundle of orthonormal frames,* $\dim M = n$, $\dim \mathcal{O}(M) = \frac{n(n+1)}{2}$. *The horizontal distribution of the Riemannian connection in* $\mathcal{O}(M)$ *has the nonholonomicity degree 2 (and hence is of maximal growth) if and only if the Gaussian curvature on M has a constant sign (is nowhere zero).*

In a slightly different way this condition may be stated as follows: the curvature form (which in the present case coincides up to a factor with the Frobenius form (cf. above)) is non-degenerate, i.e. defines a surjection onto the Lie aglebra $\mathfrak{o}(n)$. In this form the statement holds for connections in any principal bundle with the same dimension of fibers $\frac{n(n-1)}{2}$ as above. The notions of the flag of a connection, the nonholonomicity degree, and other invariants of a distribution are well suited for a geometric theory. The flag of a connection may be described in terms of covariant derivatives (see Kobayashi [1963, 1969]).

It is somewhat more difficult to characterize connections in associated fibre bundles in terms of their horizontal distributions. This description was given by G.F. Laptev and his school (see Vosilyus [1983], Loomiste [1966] for a review).

Another notion which is important for applications in mechanics is that of a *connection over a distribution* introduced in a series of papers, in particular, in Manin [1984]. Let X be a manifold, $\pi: E \to X$ a fibre bundle, V a distribution on X. A connection over the distribution V is a G-invariant distribution H in E which is transversal to the vertical distribution and such that $\pi H = V$.

It is a fundamental problem to extend a connection defined over a totally non-holonomic distribution to a connection on the whole manifold. This problem goes back to classical papers of Cartan and others.

1.7. A Classification of Left Invariant Contact Structures on Three-Dimensional Lie Groups. A left invariant non-holonomic structure on a three-dimensional Lie group G is specified by a plane V in its Lie algebra $\mathfrak{g}$, or, more precisely, by an arbitrary 2-dimensional subspace V which is not a subalgebra. The set of all left-invariant nonholonomic structures is an open subset of the Grassmann manifold $\mathrm{Gr}_2(\mathfrak{g})$ which we shall denote by $\mathrm{Gr}_2^0(\mathfrak{g})$. A pair $(\mathfrak{g}, V)$ is called a non-holonomic Lie algebra. We shall say that $(\mathfrak{g}_1, V_1)$ is isomorphic to $(\mathfrak{g}_2, V_2)$ if there is an isomorphism $\varphi: \mathfrak{g}_1 \to \mathfrak{g}_2$ such that $\varphi(V_1) = V_2$.

The set $\mathscr{C}\ell\mathfrak{g}$ of classes of isomorphic nonholonomic 3-dimensional Lie algebras coincides with the space of orbits $\mathrm{Gr}_2^0(\mathfrak{g})/\mathrm{Aut}\,\mathfrak{g}$, where $\mathrm{Aut}\,\mathfrak{g}$ is the group of automorphisms of $\mathfrak{g}$. Let us determine $\mathscr{C}\ell\mathfrak{g}$ for all 3-dimensional Lie algebras. A classification of 3-dimensional real Lie algebras up to an isomorphism is well known (see e.g. Dubrovin et al. [1979], Auslander et al. [1963]). Their list is given below.

(1) The abelian algebra t_3.

(2) The nilpotent algebra, which is the Heisenberg algebra N defined by the following generators and relations:

$$N = \mathrm{Lin}\{\xi_1, \xi_2, \xi_3\}, \quad [\xi_1, \xi_2] = \xi_3, \quad [\xi_1, \xi_3] = [\xi_2, \xi_3] = 0.$$

(3) Solvable Lie algebras r_α:

$$[\xi_1, \xi_2] = a_{11}\xi_1 + a_{12}\xi_3, \quad [\xi_1, \xi_2] = 0,$$
$$[\xi_2, \xi_3] = a_{21}\xi_1 + a_{22}\xi_3,$$
$$\alpha = \begin{pmatrix} a_{11} & a_{12} \\ a_{21} & a_{22} \end{pmatrix} \in \mathrm{SL}_2\,\mathbb{R}.$$

The class of solvable algebras is divided into the following subclasses depending on the eigenvalues of α:

(3a) α has different real eigenvalues,

$$a_{11} = \lambda_1, \quad a_{22} = \lambda_2, \quad a_{12} = a_{21} = 0, \quad \lambda_1 \neq \lambda_2;$$

(3b) α is conjugate to a rotation,

$$a_{11} = \cos\varphi, \quad a_{12} = \sin\varphi, \quad a_{21} = -\sin\varphi, \quad a_{22} = \cos\varphi;$$

(3c) α is diagonal,

$$a_{11} = a_{22} = 1, \quad a_{12} = a_{21} = 0;$$

(3d) α is conjugate to the Jordan matrix,

$$a_{11} = a_{22} = a_{12} = 1, \quad a_{21} = 0.$$

(4) Semisimple Lie algebras. Up to an isomorphism, there are two different real semisimple Lie algebras:

(4a) The algebra $\mathfrak{so}(3)$ of 3×3 real skew-symmetric matrices with basis ξ_1, ξ_2, ξ_3 and relations

$$[\xi_1, \xi_2] = \xi_3, \quad [\xi_3, \xi_1] = -\xi_2, \quad [\xi_2, \xi_3] = \xi_1.$$

(4b) The algebra $\mathfrak{sl}_2(\mathbb{R})$ of real traceless 2×2 matrices with basis ξ_1, ξ_2, ξ_3 and relations

$$[\xi_1, \xi_2] = \xi_3, \quad [\xi_1, \xi_3] = 2\xi_1, \quad [\xi_2, \xi_3] = -2\xi_2.$$

The following assertion describes the set $\mathscr{C}\ell\mathfrak{g}$ for each of the 3-dimensional Lie algebras.

Proposition. (1) *There are no non-holonomic 2-dimensional left-invariant distributions either on the abelian group T_3, or on the solvable group R_α with* $\alpha = \begin{pmatrix} 1 & 0 \\ 0 & 1 \end{pmatrix}$.

(2) *For the Lie groups listed below all 2-dimensional nonholonomic left invariant distributions lie on the same orbit of the group of automorphisms of the Lie algebra:*

- *The Heisenberg group N.*
- *The solvable groups R_α of the three remaining types.*
- *The special orthogonal group.*

(3) *The set of non-holonomic left-invariant distributions on $SL_2\,\mathbb{R}$ splits into two orbits.*

Remark. The representatives of the orbits for N and SO(3) are given by $V = \mathrm{Lin}(\xi_1, \xi_2)$ in the bases described above. For $SL_2\,\mathbb{R}$ the representatives of orbits are given by $V_1 = \mathrm{Lin}(\xi_1, \xi_2)$ and $V_2 = \mathrm{Lin}(\xi_3, \xi_1 + \xi_2)$.

§ 2. Generic Distributions and Sets of Vector Fields, and Degeneracies of Small Codimension. Nilpotentization and Classification Problem

In Section 2.1 we list some important properties of generic distributions (the maximality of growth, etc). These results may be regarded as a first step towards a classification of distributions.

Concerning the general classification of distributions, it should be noted that it is meaningful only in local setting, i.e. for germs or jets. As one may anticipate, even here the situation is quite complicated. First of all, classification of germs of distributions under the most general assumptions contains functional moduli even for relatively small dimensions. Varchenko [1981] has shown that such moduli are present already for growth vectors (8, 11), i.e. the set of orbits of generic 8-dimensional distributions in an 11-dimensional space is parametrized by real functions of 11 variables. As a matter of fact, functional parameters do occur even for lower dimensions. This means that a classification of germs of ditributions (even of generic ones) up to an isomorphism does not make sense. Nevertheless, for certain dimensions the situation is different. For instance, the classical Darboux theorem asserts that up to a diffeomorphism there is a unique germ of generic codimension one distribution (contact structure). It seems likely that germs of codimension two distributions are also manageable (although much less so than in the previous case). There are numerous papers on germs of distributions with special growth vector $(n - 2, n - 1, n)$ (see Goursat [1922] and other papers which aim mainly at problems in the theory of differential equations).

However, for most local questions of nonholonomic dynamics and geometry it is sufficient to have information only on finite order jets of distributions. In

Section 2.2 it is shown that $(k-1)$-jets of generic distributions lie on a single orbit of the group of jets of diffeomorphisms of $\mathbb{R}^n$.

The study of dynamics on compact manifolds requires the study of germs of distributions with some degeneracies, since in general on such manifolds there are no distributions which have maximal growth everywhere. On smooth n-dimensional manifolds degeneracies of codimension up to n may occur (Sections 2.3, 2.5). In this section we study degeneracies of codimensions up to $n - \sqrt{n}$. For such small codimensions there exists a simple connection between the codimension of the degeneracy and the growth defect of a distribution which is based on transparent geometrical constructions (see Section 2.5). Notice that there is another and no less natural class of distributions, namely regular distributions of nearly maximal growth. We shall not consider it in the present review. Such distributions form a class of infinite codimension and hence are highly non-generic. However, in the presence of symmetries (e.g. for left-invariant distributions) this case is of special interest.

2.1. Generic Distributions. We shall define a partial ordering on the set of growth vectors of distributions at some point x. We say that a distribution V grows more rapidly at x than W if

(a) $\forall i, n_i^V \geqslant n_i^W$,

(b) $\exists i_0, n_{i_0}^V > n_{i_0}^W$.

We shall speak of *maximal growth distributions* in the sense of this partial ordering. The set of germs of n_1-dimensional distributions in $\mathbb{R}^n$ may be naturally identified with the set $S_n^{n_1}$ of germs of sections $s\colon \mathbb{R}^n \to \mathbb{R}^n \times \mathrm{Gr}_n^{n_1}$, where $\mathrm{Gr}_n^{n_1}$ is the Grassmann manifold of n_1-dimensional subspaces in $\mathbb{R}^n$ (cf. Section 1). The space $S_n^{n_1}$ is equipped with the C^∞-Whitney topology.

The following theorem (see Gershkovich [1984]) describes generic distributions.

Theorem (on generic distributions).

(1) *Maximal growth distributions form an open dense subset in $S_n^{n_1}$ in the C^∞-Whitney topology.*

(2) *Each maximal growth distribution is regular and totally nonholonomic.*

(3) *All maximal growth distributions in $S_n^{n_1}$ have the same growth (and hence also the some nonholonomicity degree which we shall denote by $k(n, n_1)$).*

(4) *For each j $(1 \leqslant j \leqslant k-1)$ the component n_j^V of the growth vector is equal to* $\dim F_j(n_1)$ *where $F_j(n_1)$ is the linear space generated by all words of length $\leqslant j$ in the free Lie algebra $F(n_1)$ with n_1 generators.*

Let us give explicit formulae for $\tilde{n}_j = \dim F_j(n_1)$. The number of words of length j in the free Lie algebra with n_1 generators (i.e. $\tilde{n}_j - \tilde{n}_{j-1}$) is given by the following expression (see Bourbaki [1970]):

$$(\tilde{n}_j - \tilde{n}_{j-1}) = \frac{1}{j} \sum_{d \mid j} \mu(d) n_1^{(j/d)},$$

where μ is the Moebius function (Bourbaki [1972]).

The leading term is given by

$$(\tilde{n}_j - \tilde{n}_{j-1}) = \frac{1}{j} n_1^j + \frac{1}{j} \sum_{\substack{d|j \\ d>1}} \mu(d) n_1^{(j/d)} = \frac{1}{j} n_1^j + O(n_1^{j/2} \log_2 n_1);$$

the estimate for the remainder is based on the fact that $n_1^{j/d} \leqslant n_1^{j/2}$ for $d > 1$ and that the number of divisors of j is $O(\log_2 j)$.

Corollary. *The nonholonomicity degree $k(n, n_1)$ of an n_1-dimensional maximal growth distribution on an n-dimensional manifold behaves asymptotically as $\log_{n_1} n = \dfrac{\ln n}{\ln n_1}$ for $n \to \infty$ and n_1 fixed.*

Remark. Choose n_1 generic matrices in the Lie algebra $\mathrm{gl}(n)$ with $n \gg n_1$. Then the dimensions of the linear subspaces spanned by their Lie brackets of length $\leqslant k$ coincide with the dimensions of the corresponding subspaces in the free Lie algebra as long as it is possible, i.e. as long as these latter do not exceed the dimension of $\mathrm{gl}(n)$. Hence every element of $\mathrm{gl}(n)$ may be represented as a linear combination of Lie brackets of these matrices of length not exceeding $k(n, n_1) \sim \ln n/\ln n_1$.

2.2. Normal Forms of Jets of Basic Vector Fields of a Generic Distribution. The last assertion of the theorem of Section 2.1. shows that until the stabilization moment the maximal growth, i.e. the growth of a generic distribution, coincides with the growth of the number of words in the free Lie algebra. This coincidence has the following reason. A maximal growth distribution is regular and hence (cf. Section 1.4) has a basis consisting of vector fields $\xi_1, \ldots, \xi_{n_1}$. Under the homomorphism φ of the free Lie algebra F into W_n which maps the basis $\{g_i\} \subset F$ into $\{\xi_i\}$, the homogeneous linear basis of F_j/F_{j-1} is mapped onto the basis $\xi_{n_{j-1}+1}, \ldots, \xi_{n_j}$ of the quotient distribution V_j/V_{j-1} (for $j \leqslant k-1$). This allows to construct the normal form of jets of vector fields ξ_j as the quasinormal form of jets of a basis of a regular distribution (cf. Section 1.4) satisfying the additional condition

$$J^0 \xi_j = \frac{\partial}{\partial x_j}, \quad j = 1, \ldots, n.$$

To fulfil this plan we need a description of a homogeneous linear basis in the free Lie algebra. Let us describe the construction of a linear basis in the free Lie algebra with n_1 generators $f_1, \ldots, f_{n_1}$.

Let Z be the set of n_1 elements $x_1, \ldots, x_{n_1}$, let $M(Z)$ be the free monoid over Z, $l(h)$ the length of a word $h \in M(Z)$, $M^i(Z)$ the set of all length i words.

Definition (cf. Bourbaki [1972]). A *Hall family* over Z is an arbitrary linearly ordered subset $H \subset M(Z)$ such that

(1) if $u, v \in H$ and $l(u) < l(v)$, then $u < v$;

(2) $Z \subset H$;

(3) $H \cap M^2(Z) = \{xy | x, y \in Z, x < y\}$;
(4) $H \backslash (M^2(Z) \cup Z) = \{\omega = a(bc) | a, b, c \in H, b \leqslant a < bc, b < c\}$.

Define a mapping θ: $M(Z) \to F$ as follows: take a word from $M(Z)$ and replace in it x_i with f_i and all round brackets with Lie brackets. The following assertion holds.

Theorem (Hall-Witt, cf. Bourbaki [1972]). *θ maps H into a homogeneous linear basis in F.*

The construction of a linear basis of the free Lie algebra is completed. We may now sharpen the last assertion of the theorem of Section 2.1.

Proposition 1. *Let V be the germ of a maximal growth distribution in $\mathbb{R}^n$,* dim $V = n_1$. *Let $\xi_1, \ldots, \xi_{n_1}$ be its basis. Then*
(1) *For $j \leqslant k - 1$ the mapping φ: $F_j \to V_j$ such that $\varphi(f_i) = \xi_i$ is an isomorphism of linear spaces; $\{\varphi(\theta(M^j(Z) \cap H))\}$ is a linear basis in $V_j(x)$.*
(2) *The mapping φ: $F \to V =$* Vect $\mathbb{R}^n$ *is surjective.*

It is natural to label the linear basis of V_j by elements of the Hall family of length $\leqslant j$ in such a way that $\{\xi_{u_i}; u_i \in H \cap M_i(Z)\}$ is a relative linear basis in V_j/V_{j-1}. Let us look for a coordinate system $\{x_i\}$ in the neighborhood of x in which the jets of vector fields ξ_u have the simplest possible form. In particular, it is natural to demand that the vectors $\xi_{u_1}(0), \ldots, \xi_{u_n}(0)$ which form a basis in $T_x\mathbb{R}^n$ be tangent to the coordinate lines in $\mathbb{R}^n$; the coordinates themselves are also naturally labelled by elements of the Hall family. We use this remark in the following theorem which describes the normal forms of $(k-1)$-jets of vector fields ξ_u.

Theorem (on the everywhere dense orbit for distributions). (1) *The jets of the set of basic vector field of all n_1-dimensional maximal growth distributions in $\mathbb{R}^n$ of order $k = k(n, n_1)$ lie on a single orbit of the group of $(k-1)$-jets of diffeomorphisms of $\mathbb{R}^n$.*
(2) *The orbit of maximal growth distributions has a representative, the $(k-1)$-jets of basic vector fields for which are described below.*
Denote by $\tilde{H}$ the first n elements of H (with respect to the linear order in H). Then

$$J^{(k-1)}\xi_{u_0} = \frac{\partial}{\partial x_{u_0}} + \sum_{S_1} x_{u_m} \ldots x_{u_1} \frac{\partial}{\partial x_{u_{m+1}}} + \sum_{S_2} \mu_{u_{m+1}, \ldots, u_1, u_0} x_{u_m} \ldots x_{u_1} \frac{\partial}{\partial x_{u_{m+1}}}, \quad (1)$$

where

$$u_i \in \tilde{H}, \quad S_1 = \left\{(u_{m+1}, \ldots, u_1, u_0) \middle| \sum_{i=1}^{m} l(u_i) \leqslant k - 1, u_{m+1} = u_m(\ldots(u_1, u_0)\ldots)\right\},$$

$$S_2 = \left\{(u_{m+1}, \ldots, u_1, u_0) \middle| \sum_{i=1}^{m} l(u_i) \geqslant k\right\}.$$

2.3. Degeneracies of Small Codimension. The theorem of Section 2.2 completely describes the normal forms of jets of generic distributions. As we shall see, the same technique allows to obtain normal forms of jets of distributions with degeneracies not exceeding certain order and provides a transparent relation between the codimension of the degeneracy and the *growth defect of the distribution* (i.e. the difference between the growth of the number of words in the free Lie algebra and the growth of the distribution). This relation is very simple in the case when the codimensions of degeneracies do not exceed $n - \sqrt{n}$ (such degeneracies are called small). Below we give the formulae relating the order of degeneracy and the growth defect of the distribution and describe their geometric interpretation.

Let $u_1 < u_2 < \cdots < u_n$ be the smallest elements of a Hall family with respect to the linear order on H. For generic distributions the vector fields $\xi_{u_i} = \varphi \cdot \theta(u_i)$ are linearly independent at x. In degenerate cases they become linearly dependent. For small codimension degeneracies there is a simple relation between the step on which a linear dependence appears (i.e. the growth drops down) and the codimension of the degeneracy. This relation is given by the following assertion. Put $\tilde{n}_j = \dim F_j^{n_i}$. Let $S_n^{n_1}$ be the set of n_1-dimensional distributions in $\mathbb{R}^n$. Put $k = k(n, n_1)$, $\rho = \left[\dfrac{k-1}{2}\right]$. Denote by Σ_j^i the subset of $S_n^{n_1}$ consisting of distributions V such that $V_j = \dim F_j^{n_1} - i$. Let $\mathrm{cd}(j, i)$ be the codimension of Σ_j^i in $S_n^{n_1}$.

Proposition 1. *The codimension of Σ_j^i is described by the following formulae*:

(1) $\mathrm{cd}(j, i) \geqslant [n - \sqrt{n}]$ *for* $j < \rho$, $i > 0$.

(2) *Let* $\rho < j \leqslant k - 1$ *and* $i < \tilde{n}_{j+1} - \tilde{n}_j$. *Then*

$$\mathrm{cd}(j, i) = i(n - \tilde{n}_j + i). \tag{2}$$

(3) *Let* $l_1 = n - \tilde{n}_{k-1}$, $i_1 \leqslant l_1$. *Then*

$$\mathrm{cd}(k, i) = i(\tilde{n}_k - n + i). \tag{3}$$

The geometric meaning of these formulae is explained by Proposition 2.

Proposition 2. *Let $n_1 \leqslant n_2 \leqslant \cdots \leqslant n_k$ be the growth vector of a distribution V, and $\tilde{n}_1 \leqslant \tilde{n}_2 < \cdots < \tilde{n}_k$ that of a maximal growth distribution. Then*:

(1) *For all $V \in S_n^{n_1}$, except for a subset of codimension $n - \sqrt{n}$ in $S_n^{n_1}$, the first half of the components of the growth vectors coincide, i.e. $n_i = \tilde{n}_i$ for $i < \rho$.*

(2) *The codimension of the set of those distributions V for which $n_j \leqslant \tilde{n}_j - i$ coincides with the codimension of the set of such n_j-tuples of points in $\mathbb{R}^n$ that all points of each of them lie in a fixed plane $\mathbb{R}^{n-i} \subset \mathbb{R}^n$ in the set of all n_j-tuples of points in $\mathbb{R}^n$.*

Remark. An explanation of the simple form of the relation between the codimension of a degeneracy and the growth defect which is valid for small codimensions and of the distinguished value $n - \sqrt{n}$ which limits "small" codimensions is based on the study of normal forms of sets of vector fields and will be given in Section 2.5.

Proposition 2 and the results of the previous section imply the following assertion concerning the jets of maximal growth distributions.

The dimension n is called regular if $n = \tilde{n}_k$. Denote by Σ_k^0 the set of all maximal growth distributions in $S_n^{n_1}$.

Proposition 3. (1) *The set* Σ_k^0 *is connected (in the* C^∞ *Whitney topology).* (2) *If* n *is not regular, then*

$$\operatorname{codim}(S_n^{n_1} \backslash \Sigma_k^0) \geqslant 2.$$

For degeneracies of codimension greater than $n - \sqrt{n}$ the situation is more complicated and formulae (2), (3) break down.

Let us now turn to the study of sets of vector fields.

2.4. Generic Sets of Vector Fields. The aim of this section is to describe generic sets of vector fields $\xi_1, \ldots, \xi_n$. The natural assumption imposed on generic vector fields is as follows (cf. Section 2.1): we must take vector fields coinciding with the basic fields $\xi_1, \ldots, \xi_{n_1}$ or with their Lie brackets and demand that their values at x be as linearly independent as possible. As already noticed (cf. Section 2.3), it is convenient to choose as these vector fields the images of the elements of a linear basis of the free Lie algebra F^{n_1} under the homomorphism $\varphi: F^{n_1} \to W_n$ which sends the generators f_i into ξ_i. The vector fields so chosen correspond to a subset Q of a Hall family H. It is natural to assume that Q satisfies the following condition: if ξ_q, $q \in Q$, is the Lie bracket of two vector fields, $\xi_q = [\xi_{q_1}, \xi_{q_2}]$, $q_1, q_2 \in H$, then $q_1, q_2 \in Q$, i.e. Q is an ideal of the partially ordered set $(H, \prec)$. The partial order on H is defined as follows.

Definition. Fix a Hall family H. A subset $Q \subset H$ is called an *ideal* of H if the following condition is satisfied:

$$(\alpha\beta \in H) \Rightarrow (\alpha \in H \;\&\; \beta \in H)$$

(in other words, Q is an ideal of H with respect to the following partial ordering on H:

$$\alpha \prec \gamma \Leftrightarrow ((\exists \beta: \alpha\beta = \gamma) \vee (\exists \beta: \beta\alpha = \gamma))).$$

Let $Q \subset H$ be an ideal. Then the linear ordering on H determines a chain of sets $\varnothing \subset Q_1 \subset \cdots \subset Q_l$, $l = \# Q$, such that $Q_j = Q_{j-1} \cup \{q_j\}$ and $q < q_j$ for all $q \in Q_{j-1}$. The chain of sets Q_j gives rise to the chain of linear subspaces

$$L_j = \operatorname{Lin}\{(\varphi\theta(q))x \mid q \in Q_j\}, \quad 0 \subset L_1 \subset \cdots \subset L_l.$$

The manifold consisting of sets of germs of vector fields $\xi_1, \ldots, \xi_{n_1}$ in $\mathbb{R}^n$ may be identified with the manifold of sections of the trivial bundle $\mathbb{R}_n^{n_1} = \{r: U \to [TU]^{n_1}\}$, where U is a neighborhood of the origin in $\mathbb{R}^n$, TU is its tangent bundle and $[TU]^{n_1}$ is the Whitney sum of n_1 copies of TU (see Sternberg [1964]).

Proposition 1. *Let* Q *be an ideal of the Hall family* H, $\# Q = n$. *Then for an open dense subset* $\mathbb{R}_0 \subset \mathbb{R}_n^{n_1}$ *(in the Whitney topology)* $\dim L_n = n$, *i.e.* $\{L_i\}$ *is a complete flag of the linear space.*

If a set $\{\xi\}$ of vector fields has the property described in Proposition 1 we shall say that this set has Q-maximal growth.

The theorem of Section 2.1 may be reduced to Proposition 1 if we choose for Q the ideal A constructed as follows.

Ideal A. Include into A all words in H of length $\leqslant k-1$. Moreover, choose an arbitrary set of $n - n_{k-1}$ words in $M^k(Z)\backslash M^{k-1}(Z) \cap H$ and also include them into A.

Let us now extend the theorem of Section 2.2 to the case of arbitrary ideals of a Hall family.

Theorem 1. *Let $Q \subset H$ be an ideal, $\# Q = n$. Assume that a set $\{\xi_i\}$ of vector fields has the Q-maximal growth. Then there is a coordinate system x_n: $\mathbb{R}^n \to \mathbb{R}$ in $\mathbb{R}^n$ such that*

$$J^\infty \xi_{u_0} = \sum_{S_1} \left(\prod_{i=1}^{l} x_{u_i} \right) \frac{\partial}{\partial x_{u_{l+1}}} + \sum_{S_2} \mu_{u_{l+1}, \dots, u_1, u_0} \prod_{i=1}^{l} x_{u_i} \frac{\partial}{\partial x_{u_{l+1}}},$$

where $u_i \in Q$, and

$$S_1 = \{(u_{l+1}, \dots, u_1, u_0) | u_{l+1} = u_l(u_{l-1}(\dots(u_1 u_0)\dots) \in Q\},$$

$$S_2 = \{(u_{l+1}, \dots, u_1, u_0) | u_l(\dots(u_1 u_0)\dots) \in H \backslash Q\}.$$

As we see, Theorem 1 allows to determine the "jet coefficients" of vector fields that correspond to the ideal Q. Choosing $Q = A$ yields the theorem of Section 2.2 on the normal form of $(k-1)$-jets of basic vector fields of a distribution V. Another natural question is to determine the maximal r such that there is an open dense orbit in the space of r-jets of sets of vector fields. This question is resolved by Theorem 1 if we choose $Q = B$, where B is the ideal described as follows.

Ideal B. We shall construct B inductively. Put $B_1 = Z$; $B_{j+1} = \{h_2 \cdot h_1 \in H | h_1 \in B_j\}$. Put $m_i = \# B_i$ and let $r = r(n, n_1)$ be an integer for which $m_{r-1} < n < m_r$. Put $B = B_{r-1} \cup \tilde{B}$, where $\tilde{B}$ consists of the $n - m_{r-1}$ smallest elements of $B_r \backslash B_{r-1}$.

Putting $Q = B$ we deduce from Theorem 1 the following assertion.

Theorem 2 (on the everywhere dense orbit for vector fields. (1) *The $(r-1)$-jets of sets of n_1 vector fields lie on a single orbit of the group of jets of diffeomorphisms of $\mathbb{R}^n$.*

(2) *The normal forms of $(r-1)$-jets of vector fields are given by*

$$J^{r-1} \xi_{u_0} = \sum_{l=0}^{r} \left(\prod_{i=1}^{l} x_{u_i} \right) \frac{\partial}{\partial x_{u_{l+1}}}, \quad \text{where } u_{l+1} = u_l(\dots(u_1 u_0)\dots).$$

Remark. Theorem 2 shows that jets of order $r-1$ lie on a single orbit of the group of jets of diffeomorphisms. The order of jets cannot be increased. Jets of higher order contain moduli whose number increases exponentially with the growth of the order.

The next assertion gives the asymptotics of $r = r(n, n_1)$ for large n.

Proposition 2. *For fixed n_1 we have*

$$r(n, n_1) = \log_2 \log_{n_1} n(1 + o(1)) \quad \text{for} \quad n \to \infty.$$

Indeed, since $k(n, n_1) = \log_{n_1} n(1 + o(1))$ (see Section 2.1, Remark), our proposition is reduced to the asymptotic formula $r(n, n_1) \sim \log_2 k(n, n_1)$ which follows from the obvious assertion showing that asymptotically the length of almost all elements in $B_l \backslash B_{l-1}$ is equal to 2^{l-1} (i.e. is the maximal possible one for words in B_l).

Assertion. *Put $\tilde{B}_l = \{h \in B_l \backslash B_{l-1}: l(h) = 2^{l-1}\}$. Then*

$$\lim_{l \to \infty} \frac{\# \tilde{B}_l}{\#(B_l \backslash B_{l-1})} = 1.$$

This concludes the description of the generic case. Let us now consider the degeneracies of small codimension.

2.5. Small Codimension Degeneracies of Sets of Vector Fields. We shall formulate a relation between the growth defect and the codimension of a degeneracy for sets of vector fields which is similar to that described in Propositions 1, 2 (Section 2.3) for distributions and is valid for degeneracies of small codimension.

Let $\xi_1, \ldots, \xi_{n_1}$ be a set of vector fields in $\mathbb{R}^n$, Q an ideal in a Hall family H, $0 \subset L_1 \subset L_2 \subset \cdots \subset L_n$ a chain of linear subspaces which corresponds to Q (see Section 2.2). Let $v_j^i \subset \mathbb{R}_n^{n_1}$ be the subspace consisting of those sets of vector fields for which $\dim v_j \leqslant j - i$. Let $cd(j, i)$ be the codimension of v_j^i in $\mathbb{R}_n^{n_1}$.

Proposition 1. *The codimensions of v_j^i are given by the following formulae:*
(1) $\mathrm{cd}(j, 1) = n - j + 1$ *for* $1 \leqslant j \leqslant n$.
(2) *Let*

$$[\sqrt{n}] + 1 \leqslant j, \quad i \leqslant j - [\sqrt{n}] - 1, \quad i < j/2.$$

Then

$$\mathrm{cd}(j, i) = i(n - j + 1).$$

Remark. If we choose $Q = A$ and consider the formulae above only for $j = \tilde{n}_l$, our assertion reduces to Proposition 1 of Section 2.2.

Let us describe the normal forms of jets of sets of vector fields for small codimension degeneracies. Denote by Diff_n^r the group of r-jets of diffeomorphisms of $\mathbb{R}^n$ (here r is the number from Theorem 2, Section 2.4.)

Let $\mathfrak{N}_x$ be the stabilizer in Diff_n^r of the set of r-jets of vector fields. Let M_{ij}, $i < j/2$, be the set of $i \times (j - i)$-matrices (i.e. matrices with i rows and $j - i$ columns) of rank i.

Proposition 2. *Let $[\sqrt{n}] + 1 < j$, $i < j - \sqrt{n}$, $i < j/2$. Then there exists an open dense subset $\mathring{v}_j^i \subset v_j^i$ such that:*
(1) *$\mathring{v}_j^i$ is invariant under the action of* Diff_n^r.
(2) *For any two points $x, y \in \mathring{v}_j^i$ the stabilizers $\mathfrak{N}_x$, $\mathfrak{N}_y$ are conjugate in* Diff_n^r.
(3) *$\mathring{v}_j^i$ is the semi-direct product $M_{j,i} \times (\mathrm{Diff}_n^r/\mathfrak{N})$.*

Remarks 1. The results of Section 2.4 allow to describe the stratification of the space of r-jets of sets of vector fields in R^n up to codimension $n - \sqrt{n}$.

2. We have bounded the domain of small codimensions with the value $n - \sqrt{n}$. Let us explain its origin. Recall Theorem 2 of Section 2.2 which gives the normal form of sets of generic vector fields. The monomials which enter into the decomposition of jets of vector fields $J^{r-1}u_0$ with respect to the basic fields $-\partial/\partial x_{u_{i+1}}$ have the following form

$$\left(\prod_{i=1}^{p} x_{u_i}\right)\cdot\frac{\partial}{\partial x_{u_{p+1}}}, \quad u_{p+1} = u_p(u_{p-1}\dots(u_1 u_0)\dots), \tag{4}$$

where u_i are elements of a Hall family H. To proceed further we need the following assertion.

Proposition 3. *For $u \in H$ let $N(u)$ be its number with respect to the linear order in H. Then for each $i \leqslant p$ the elements $u_i \in H$ in* (4) *satisfy $N(u_i) < \sqrt{n}$.*

Indeed, put $v_i = u_i(u_{i-1}\dots(u_1 u_0)\dots)$. Then

$$v_{p+1} = u_{p+1}, \quad l(v_i) \geqslant l(u_i) + l(v_{i-1}) \geqslant 2l(u_i)$$

and $l(v_i) < l(v_{p+1})$. Thus we get

$$l(u_i) \leqslant \tfrac{1}{2} l(u_{p+1}). \tag{5}$$

From the formulae for the growth of $\tilde{n}_i$ (see Section 2.1) we get

$$N(u_i) < n_1^{l(u_i)} < n_1^{(1/2)l(u_{p+1})} < n^{1/2}.$$

Hence $N(u_i) \leqslant \sqrt{n}$. Now we are able to explain the origin of the upper bound for small codimensions. The coordinates x_{u_i} in $\mathbb{R}^n$ were taken as the directions of the vector fields ξ_{u_i} at x. In the generic case the vectors $\xi_{u_i}(x)$ are linearly independent. In degenerate cases linear dependences may occur. Propositions 1 and 3 imply that for degeneracies of codimension smaller than $n - \sqrt{n}$ the vector fields ξ_{u_i} for such u_i that $N(u_i) < \sqrt{n}$ remain linearly independent. The normal forms reduce to

$$J^{r-1}\xi_{u_0} = \sum_{l=1}^{r-1}\left(\prod_{i=1}^{l} x_{u_i}\right)\frac{\partial}{\partial y_{u_{l+1}}},$$

where the directions $\partial/\partial y_{u_{l+1}}$ are already not necessarily linearly independent. Let us stress again that both the simple relation between the codimension of a degeneracy and the growth defect and the simple description of moduli are due to the fact that the coordinates x_{u_i} which enter into the product in (4) satisfy $N(u_i) < \sqrt{n}$ and hence correspond to the vector fields ξ_{u_i} which remain linearly independent for "small" degeneracies. The general case requires already the study of the combinatorial type of the configuration of generating vector fields and their Lie brackets.

2.6. Projection Map Associated with a Distribution. Let V be the germ of a maximal growth distribution of dimension n_1 in $\mathbb{R}^n$. Let k_V be the nonholonomicity degree of V. As we have seen, the spaces V_i, $i = 1, \ldots, k-1$, that form the flag of V behave in the same way for all maximal growth distributions (of fixed dimension): their dimensions are growing at a maximal rate, i.e. like the dimensions of the corresponding subspaces in the free Lie algebra (see Section 2.1), and the Lie brackets of length j of vector fields from V remain independent for $j \leqslant k-1$ as long as their number does not exceed the dimension n. For larger j the Lie brackets of vector fields from V become linearly dependent. The set of linear dependences is an invariant of the distribution. Let us give its description.

As before, we denote by F the free Lie algebra with n_1 generators $f_1, \ldots, f_{n_1}$. Let F_i be the i-th homogeneous component of F, $F^S = \bigoplus_{i=1}^{S} F_i$. The nondegenerate linear mapping $\pi\colon F_1 \to V$ defines an epimorphism $\pi_i(x)\colon F^i \to V_i(x)$. In particular, we get the mapping $\pi_{k+1}(x)\colon F^{k+1} \to T_x V$ which we shall denote by $P_V(x)$ and refer to it as the projection map of V at the point x.

Let D be the germ of $\mathbb{R}^n$ at the origin. Consider the trivial bundle $D(F)$ over D with fiber F. Let $\mathrm{GL}(E_{n_1})$ be the group of invertible matrices of order n_1. Its natural action on F_1 extends to the action on $D(F)$ which preserves the fibers.

Two mappings $\pi^{(1)}$, $\pi^{(2)}$ are said to be equivalent if $\pi^1(\cdot) = \pi^{(2)}(g(\cdot))$, $g \in \mathrm{GL}(E_{n_1})$. Let $\mathfrak{P}_V$ be the $\mathrm{GL}(E_{n_1})$-orbit of $P_V = \pi_{k+1}|_{F^{k+1}}$. Obviously, we have

Proposition 1. *The correspondence $V \mapsto \mathfrak{P}_V$ establishes a bijection between the orbits of projection maps and the germs of maximal growth distributions of dimension n_1.*

Given a basis of vector fields belonging to V which is extended to a basis of TX, the projection map P_V is determined by the set of the structure coefficients

$$c_{ij}^l = \langle [\xi_i, \xi_j], \xi_l \rangle, \quad 1 \leqslant i \leqslant n_1, \quad n_{k-2} \leqslant j \leqslant n, \quad 1 \leqslant l \leqslant n.$$

Let us say that the germs of two nonholonomic distributions V_1, V_2 in $\mathbb{R}^n$ are isomorphic if there is a germ of diffeomorphism $\varphi \in \mathrm{Diff}_n$ such that $\varphi(V_1) = V_2$.

The case when the structure coefficients do not depend on x is of particular interest, since this means that the distribution is invariant with respect to a certain group.

Proposition 2. (1) *If the structure coefficients with respect to a basis $\{\xi_i\}$ of V do not depend on x, there exists a local Lie group G and a left-invariant distribution $\tilde{V}$ on G such that $(\mathbb{R}^n, V)$ is isomorphic to $(G, \tilde{V})$.*

(2) *If the structure constants of the projection map are identically zero, the group G is nilpotent.*

The projection map will be used in the sequel to describe certain invariants of the distribution.

2.7. Classification of Regular Distributions. Recently Vershik and Gershkovich [1988] obtained the following result. Let W_n^k be the space of germs of regular k-dimensional distributions in $\mathbb{R}^n$, let $W_n^k(l)$ be the space of l-jets of these distributions. There is a natural action of the group of l-jets of diffeomorphisms of $\mathbb{R}^n$ on $W_n^k(l)$. Put $j_n^k(l) = \dim W_n^k(l)$ and let $c_n^k(l)$ be the codimension of a generic orbit in $W_n^k(l)$. Then

$$[k(n-k)-n]\frac{l^n}{n!} \preccurlyeq c_n^k(l), \tag{1}$$

where $a_n \preccurlyeq b_n$ means that $\overline{\lim}\,\frac{a_n}{b_n} < \infty$.

Corollary. *With the exception of the three cases listed below, the space of regular distributions for all pairs (n, k) contains functional moduli, i.e. the space of orbits is infinite dimensional. The three exceptional cases are:*

(1) $k = 1$ (*one-dimensional distributions, i.e. direction fields*);

(2) $k = n - 1$ (*the Darboux case; for odd n this corresponds to the contact structure*);

(3) $k = 2$, $n = 4$ (*the Engel case*): *generic two-dimensional distributions in the four-dimensional space which may be realized as* $V = \operatorname{Lin}(\xi_1, \xi_2)$, $\xi_1 = \partial/\partial x_1$, $\xi_2 = \partial/\partial x_2 + x_1\frac{\partial}{\partial x_3} + \frac{x_1^2}{2}\frac{\partial}{\partial x_4}$.

M.Ya. Zhitomirsky has proved that estimate (1) is exact.

A finer classification makes sense if we give a more detailed information on the distribution than its dimension (n, k). For instance, for certain fixed growth vectors one can find other cases when there exists a single generic orbit (this is true e.g. when the growth vector is $(n-2, n-1, n)$. Still, as a rule, the space of orbits of regular distributions with given growth vector is infinite-dimensional. Only the specification of configuration invariants of the distribution allows to make the spaces of orbits finite-dimensional.

A detailed exposition will be given elsewhere.

2.8. Nilpotentization and Nilpotent Calculus. The study of distributions has lead the authors to introduce a special calculus which we call nilpotent calculus. It deals with functions of graded variables, quasihomogeneous functions and polynomials, their quasi-jets, etc. Elements of this calculus are exposed by Vershik and Gershkovich [1989]. One may also call it the calculus of functions of non-commuting variables which generate a nilpotent Lie algebra. Let us consider here only one important concept of nilpotentization. Namely, we associate with each distribution a sheaf of homogeneous nilpotent Lie algebras. Let V be a distribution on M, $\{V_i\}$ its flag, k its non-holonomicity degree, $x \in M$. It is easy to prove that

$$\mathfrak{g}_x = \bigoplus_{i=1}^{k} V_i/V_{i-1}$$

is a well defined Lie algebra. Note that $\mathfrak{g}_x$ is a graded nilpotent Lie algebra of dimension $n = \dim M$ and degree k.

To prove that the definition is unambiguous we need an elementary lemma.

Lemma. *Let ξ_1, ξ_2 be two vector fields, $x \in M$.*
Let $\tilde{\xi}_i = \xi_i + \eta_i$, $i = 1, 2$, where $\xi_i \in V_{l_i}$, $\eta_i \in V_{l_i+1}$. Then

$$[\tilde{\xi}_1, \tilde{\xi}_2] - [\xi_1, \xi_2] \in V_{l_1+l_2-1}.$$

As a linear space $\mathfrak{g}_x$ may be identified with T_xM. Only the restriction of this isomorphism to V_x is canonical.

In general the sheaf $\mathfrak{g} \overset{\mathfrak{g}_x}{\to} M$ is not locally trivial. However, in certain cases this is true. For instance, if (M, V) is a contact structure, then $\mathfrak{g}_x = \mathfrak{N}$ is the Heisenberg algebra; if V is a generic distribution and the dimension n is regular (see Section 2.3), then $\mathfrak{g}_x$ is isomorphic to the free nilpotent Lie algebra of degree k, etc. The growth vector is one of the invariants of $\mathfrak{g}_x$. If the distribution is regular, the growth vector of $\mathfrak{g}_x$ is the same for all x. Since there may exist many non-isomorphic nilpotent Lie algebras with the same growth vector, it is clear that nilpotentization provides more fine invariants of distributions.

The sheaf $\mathfrak{g} \overset{\mathfrak{g}_x}{\to} M$ may be called the 0-quasijet of the distribution. For more details and for applications to the study of hypoelliptic Laplace operators see Vershik and Gershkovich [1989], Gershkovich and Vershik [1988].

Chapter 2
Basic Theory of Nonholonomic Riemannian Manifolds

§ 1. General Nonholonomic Variational Problem and the Geodesic Flow on Nonholonomic Riemannian Manifolds

In this section we consider the nonholonomic variational problem on the minimum of length on a Riemannian manifold with constraints given by a nonholonomic distribution. The solutions to this problem, the nonholonomic geodesics, satisfy the Euler-Lagrange equations of a conditional problem. They generate a nonholonomic geodesic flow defined on the mixed bundle which is the direct sum of the distribution and its annihilator in the cotangent bundle (see Section 1.3). This flow allows to extend to the nonholonomic case the notion of exponential mapping.

1.1. Rashevsky-Chow Theorem and Nonholonomic Riemannian Metrics (Carnot-Carathéodory Metrics). Let X be a smooth manifold, V a differential system or a distribution on X. A curve $\gamma: \mathbb{R} \to X$ is called admissible if $\dot{\gamma} \in V$. The following theorem was proved independently and almost simultaneously by Rashevsky [1938] and Chow [1939].

Theorem. *Assume that a differential system V is totally nonholonomic at each point of a smooth manifold X. Then each two points $x, y \in X$ may be connected by an admissible piecewise smooth curve of finite length.*

The paper of Chow [1939] extends a similar result obtained by Carathéodory [1909] for distributions of codimension one in connection with his studies on the foundations of thermodynamics. Rashevsky was probably inspired by the vigorous research which was centered at that time in the seminars of V.F. Kagan and S.P. Finikov (V.V. Vagner, etc.). The existence of an admissible curve connecting any two points of a manifold was later proved by Sussmann [1973] under weaker conditions on the distributions (as compared to the total nonholonomicity at all points).

Thus for totally nonholonomic distributions one can consider the two-point variational problem with an arbitrary smooth Lagrangian:

$$\inf \int_0^1 L(x, \dot{x})\, dt: \quad \dot{x}(t) \in V_{x(t)}, \quad x(0) = a, \quad x(1) = b. \tag{1}$$

Let L be the length functional on a Riemannian manifold:

$$L(x, \dot{x}) = \langle \dot{x}, \dot{x} \rangle^{1/2}. \tag{2}$$

Then formula (1) defines a *nonholonomic Riemannian metric* on X:

$$\rho_V(a, b) = \inf \int_0^1 \langle \dot{x}, \dot{x} \rangle^{1/2}\, dt, \quad \dot{x}(t) \in V_{x(t)}, \quad x(0) = a, \quad x(1) = b. \tag{3}$$

For contact structures this metric was considered by Carathéodory and is sometimes called the *Carnot-Carathéodory metric*. Admissible curves which realize the infimum for any two sufficiently close points lying on them are called *nonholonomic geodesics*. Denote by $D_r^V(x)$ the *nonholonomic ball* of radius r with center x and by S_r^V the *nonholonomic sphere* of radius r.

This way to construct a metric on a nonholonomic manifold from the metric on the initial manifold admits a far reaching generalization. Consider an arbitrary Lagrangian which is positively homogeneous of degree 1 in $\dot{x}$. Then

$$\rho^L(x, y) = \inf \int_0^1 L(x, \dot{x})\, dt, \quad x(0) = x, \quad x(1) = y, \tag{4}$$

defines a metric on X, and the corresponding function $\rho_V^L(x, y)$ given by (3) defines a nonholonomic metric ρ_V^L. In a similar way one can consider a more complicated metric, for instance, depending on higher jets of the curve (the Kawaguti spaces) and obtain from them a nonholonomic metric.

1.2. Two-Point Problem and the Hopf-Rinow Theorem. Consider the conditional variational problem (1) with Lagrangian (2). It is easy to construct an example of a differential system V for which the extremum is not attained. The simplest example of this kind is the problem on the minimum of length in $\mathbb{R}^2$ for curves connecting the points $(-1, 0)$ and $(1, 0)$ with constraints imposed by the differential system

$$V = \mathrm{Lin}(\xi_1, \xi_2), \quad \text{where} \quad \xi_1 = x_2 \frac{\partial}{\partial x_1}, \quad \xi_2 = \frac{\partial}{\partial x_2}.$$

The following lemma, due to Filippov [1959], provides powerful sufficient conditions for the existence of solutions of variational problems with constraints. It may also be used for nonholonomic problems. We shall formulate its special case.

Lemma. *Let X be a geodesically complete Riemannian manifold, N a totally nonholonomic differential system on X. Let D_1X be the subbundle of unit balls in TX. Then if $N \cap D_1X$ is semi-continuous from above, i.e.*

$$\lim_{x \to x_0} (D_1X \cap N)_x \subset (D_1X \cap N)_{x_0},$$

each conditional two-point variational problem on X with constraint N has a solution (in general, non-smooth).

In particular, the conditions of the Lemma are satisfied if N is a distribution, and we get an analogue of the Hopf-Rinow theorem for the nonholonomic case.

Theorem (Hopf-Rinow). *Let X be a complete Riemannian manifold, V a totally nonholonomic distribution on X. Then each two points of X may be connected with a shortest nonholonomic geodesic (in general, non-smooth).*

There are interesting recent examples showing that in the non-holonomic case the Euler equations may be not necessary conditions of the extremum, due to the lack of admissible curves which are close to the given one. See the recent papers by R. Montgomery.

1.3. The Cauchy Problem and the Nonholonomic Geodesic Flow. In the study of a nonholonomic variational problem, we have to face the following effect. At first glance, the space of initial data for such a problem consists of pairs (x, v), where $x \in M$ and v_x is an admissible vector at the point x, $\dim V_x = m < n = \dim M$. However, the results of Section 1.2 imply that any two-point problem is solvable. Thus the number of parameters in the space of geodesics (with fixed initial point) is equal to n. One might think, therefore, that the solution of the Cauchy problem is not unique. The point is, however, that the initial data of the Cauchy problem include not only the initial velocity, i.e. an admissible vector $v \in V_x$ but also a covector $\omega \in V^\perp$, since the equations for nonholonomic geodesics (Arnol'd [1974]) are differential with respect to the Lagrange multiplier λ. In coordinate language fixing ω means fixing the initial value of λ.

Recall (cf. Chapter 1) that the *mixed bundle* is the bundle $\operatorname{Ken} V = V \oplus V^\perp$ over M, where V is a distribution on M regarded as a subbundle of the tangent bundle and $V^\perp$ is its annihilator regarded as a subbundle of the cotangent bundle.

The fiber V_x is the set of admissible velocities and $V_x^\perp$ is the set of Lagrange multipliers.

Theorem. *The Euler-Lagrange equations for a nonholonomic problem determine a vector field in the mixed bundle.*

The flow which corresponds to this field gives the solution of the Cauchy problem. If the flow for the unconditional problem exists and the conditions of the Filippov lemma are satisfied, the flow for the conditional problem also exists. In particular, for any totally nonholonomic distribution on a complete Riemannian manifold there exists a global flow on $V \oplus V^{\perp}$. This flow is called *the nonholonomic geodesic flow.*

Remarks. 1. There is a different way to construct the theory. The idea is to exclude the Lagrange multipliers from the Euler-Lagrange equations by appropriate differentiations and to replace them with a system of higher order. In this way the uniqueness in the Cauchy problem is again restored due to the presence of higher derivatives. In an invariant language this approach is similar to the generalization of Riemannian geometry in which the length of a curve depends on its jets of higher order. Instead of the mixed bundle one has to consider certain submanifolds in higher tangent bundles. Apparently, this approach is connected with the one developed by Vosilyus [1983].

2. We defined the nonholonomic geodesic flow and the nonholonomic metric on a Riemannian manifold M with a given totally nonholonomic distribution V. Actually, the nonholonomic geodesic flow and the nonholonomic metric depend only on the restriction of the metric tensor g to the distribution V. If the restrictions to V of two metric tensors coincide, the associated nonholonomic flows and nonholonomic metrics also coincide, although the Euler-Lagrange equations are formally different.

3. The geodesic flow for a Finslerian metric is defined in a similar way.

1.4. The Euler-Lagrange Equations in Invariant Form and in the Orthogonal Moving Frame and Nonholonomic Geodesics. Let V be an n_1-dimensional distribution on a smooth n-dimensional manifold X. Fix a function L on $TX \oplus \mathbb{R}$ (the Lagrangian). Choose a basis $\omega_{n_1+1}, \ldots, \omega_n$ of the codistribution $V^{\perp}$.

In a local coordinate system $\{x_i\}$ the *Euler-Lagrange equations* for problem (3) have the following form

$$\frac{d}{dt}L_{\dot{x}} - L_x = \dot{\lambda}_j\omega_j - \lambda_j(\dot{x} \lrcorner d\omega_j).^1 \tag{5}$$

The admissibility condition for the curve $x(t)$ is given by

$$(\dot{x}, \omega_j) = 0, \quad j = n_1 + 1, \ldots, n.$$

It is an important property of equation (5) that it is differential also with respect to λ. (See, however, the remark following the Hopf-Rinow theorem in Section 1.2.)

Equations (5) may be rewritten in invariant form, following the pattern of Vershik and Gershkovich [1986] and Griffiths [1983]. To this end replace the

---

[1] Here $\dot{x} \lrcorner d\omega$ is the 1-form $(\dot{x} \lrcorner d\omega)\xi = \omega(\dot{x}, \xi)$, cf. Sternberg [1964].

Lagrangian $L(x, \dot{x}, t)$ of the unconditional problem with the Lagrange function

$$L_V(x, \dot{x}, t) = L(x, \dot{x}, t) + \sum_j \lambda_j \langle \omega_j, \dot{x} \rangle.$$

Starting from this point we assume that $L = \langle \dot{x}, \dot{x} \rangle^{1/2}$ is a Riemannian metric; in other words, we restrict ourselves to the study of *nonholonomic Riemannian manifolds*. Using the covariant derivative of the Riemannian connection (see Gromoll et al. [1968], Kobayashi and Nomizu [1963, 1969]) the equations of geodesics may be written as follows:

$$\begin{cases} \langle \dot{\gamma}, \omega_i \rangle = 0, \quad i = n_1 + 1, \ldots, n, \\ \nabla_{\dot{\gamma}} \dot{\gamma} = \sum_j (\dot{\lambda}_j \omega_j - \lambda_j (\dot{\gamma} \lrcorner d\omega_j)). \end{cases} \tag{6}$$

(We identify 1-forms with vector fields with the help of the metric.) It is convenient to write down the equations of geodesics by expanding $\dot{\gamma}$ with respect to an orthonormal basis of vector fields of the distribution V (i.e. with respect to the moving orthogonal frame, to use the terminology of Cartan [1926, 1935], Finikov and others). Let $\xi_1, \ldots, \xi_{n_1}$ be an orthonormal frame of V, $\xi_{n_1+1}, \ldots, \xi_n$ an orthonormal frame of its orthogonal complement $V^\perp$ (we may identify the orthogonal complement of V with its annihilator $V^\perp$ in the space of 1-forms using the Riemannian metric). Since γ is admissible, we have $\dot{\gamma} = \sum_{i=1}^{n_1} v_i \xi_i$ and the system of equations (6) takes the form

$$\begin{cases} \dot{\gamma} = \sum_{i=1}^{n_1} v_i \xi_i, \\ \dot{v}_l + \sum_{i,j=1}^{n_1} \Gamma_{ij}^l v_i v_j - \sum_{\substack{i=1,\ldots,n_1 \\ j=n_1+1,\ldots,n}} c_{il}^j v_i \lambda_j = 0, \quad l = 1, \ldots, n_1, \\ \dot{\lambda}_l - \sum_{i,j=1}^{n_1} \Gamma_{ij}^l v_i v_j - \sum_{\substack{i=1,\ldots,n_1 \\ j=n_1+1,\ldots,n}} c_{il}^j v_i \lambda_j = 0, \quad l = n_1 + 1, \ldots, n, \end{cases} \tag{7}$$

where $\Gamma_{ij}^l = \langle \nabla_{\xi_j} \xi_i, \xi_l \rangle$ are the Christoffel symbols, and $c_{ij}^l = \langle [\xi_i, \xi_j], \xi_l \rangle$ are the structure coefficients of the set of vector fields ξ_i. Equations (7) are obtained from the vector equation (the second equation in (6)) by orthogonal projection on the directions $\xi_1, \ldots, \xi_n$. Equations (7) give the coordinate form of the Euler-Lagrange equations for an arbitrary distribution. For regular distributions it is natural to divide equations (7) into k groups associated with the flag of the distribution (see Section 1.2) Denote by $\{n_i\}$ the growth vector of V. Choose a basis $\xi_{n_1+1}, \ldots, \xi_n$ of $V^\perp$ in such a way that $\xi_{n_i+1}, \ldots, \xi_{n_{i+1}}$ is a relative basis of V_{i+1} modulo V_i. Then equations (6) break up into k groups (where k is the nonholonomicity degree of V). The m-th group includes equations with indices l such that $n_{m-1} + 1 \leqslant l \leqslant n_m$. Equations in this group have the form

$$\dot{\lambda}_l + \sum_{i,j=1}^{n_1} \Gamma_{ij}^l v_i v_j + \sum_{j=1}^{n_m} c_{ij}^l v_i \lambda_j = 0. \tag{8}$$

1.5. The Standard Form of Equations of Nonholonomic Geodesics for Generic Distributions. Let V be a generic distribution, in particular, V has maximal growth. In this case by a suitable choice of vector fields $\xi_1, \ldots, \xi_n$ the equations for nonholonomic geodesics may be reduced to a much simpler form.

Choose the vector fields in the following way:

(1) $\xi_1, \ldots, \xi_{n_1}$ is an arbitrary orthonormal basis of V.
(2) Choose an arbitrary relative basis $\xi_{n_1+1}, \ldots, \xi_{n_2}$ of the space V_2 modulo V_1 (recall that V_2 is spanned by the Lie brackets $[\xi_i, \xi_j]$, $\xi_i \in V$).
(3) Choose an arbitrary relative basis $\xi_{n_2+1}, \ldots, \xi_{n_3}$ of V_3 modulo V_2, etc.

A distribution V with the set of vector fields $\{\xi_i\}_{i=1}^n$ chosen in such a way is called a *rigged distribution.*

Since nonholonomic geodesics are determined by the restriction of the metric tensor to the distribution, we may assume that the basis $\{\xi_1, \ldots, \xi_n\}$ is orthonormal. The metric on X defined in this way is said to be induced by the basis $\{\xi_i\}$.

Proposition 1. *In the basis of vector fields described above the equations of nonholonomic geodesics have the following form*:

$$\begin{cases} \dot{v}_l = \sum\limits_{\substack{i \leqslant n_1 \\ n_1 \leqslant j \leqslant n_2}} c_{il}^j v_i \lambda_j, \quad 1 \leqslant l \leqslant n_i, \\ \dot{\lambda}_l = - \sum\limits_{\substack{i \leqslant n_1 \\ n_m \leqslant j \leqslant n_{m+1}}} c_{il}^j v_i \lambda_j, \quad n_{m-1} < l \leqslant n_m, \quad 2 \leqslant m \leqslant k-1, \\ \dot{\lambda}_l = \sum\limits_{1 \leqslant i,j \leqslant n_1} c_{il}^j v_i v_j + \sum\limits_{\substack{i \leqslant n_1 \\ j \geqslant n_{k-1}}} c_{il}^j v_i \lambda_j, \quad n_{k-1} \leqslant l \leqslant n_k. \end{cases} \tag{9}$$

Remarks. 1. For a generic set of vector fields the structure coefficients $c_{ij}^l(x)$ are constant for $i \leqslant n_1$ & $j \leqslant n_{k-2}$. (They do not depend on the choice of $\{\xi_i\}_{i=1}^{n_1}$ but depend on the choice of the basis in the subspaces spanned by their Lie brackets.) Thus all equations, except for those from the two last groups, depend neither on the choice of an orthonormal basis in V, nor on its extension, in the way described above, to a full basis. Moreover, c_{ij}^l are integers for $l < n_{k-2}$ and satisfy certain special conditions.

2. The set of the structure coefficients c_{ij}^k for $i \leqslant n_1, j \geqslant n_{k-2}$ is determined by the projection map of the distribution (See Chapter 1, Section 2.6).

Proposition 2. (1) *If the structure coefficients mentioned in Remark* 2 *are constants, i.e. the projection map of the distribution is constant in a neighborhood of* x, *then the system* (9) *has constant coefficients, and in that case the germ of the nonholonomic Riemannian manifold* (X, V) *is isometric to the germ of a nonholonomic Lie group* (G, V) *with a left-invariant metric.*

(2) *If the projection map of the distribution is zero in a neighborhood of* x, *the nonholonomic Lie group is nilpotent and the distribution is the orthogonal complement of* $[\mathfrak{g}, \mathfrak{g}]$.

Hence the germs of all maximal growth distributions with zero projection map lie on the same orbit of the group of germs of diffeomorphisms of $\mathbb{R}^n$.

1.6. Nonholonomic Exponential Mapping and the Wave Front. The nonholonomic geodesic flow allows to define the exponential mapping in the same way as for ordinary Riemannian manifolds (cf. Gromoll et al. [1968], Kobayashi and Nomizu [1963, 1969]).

Let X be a complete Riemannian manifold, V a regular distribution of dimension n_1 on X. Define *the nonholonomic exponential mapping*

$$\exp_{V,\delta}: V_x \oplus V_x^{\perp} \to X$$

in the following way: fix $v_0 \in V_x$, $\omega_0 \in V_x^{\perp}$ and let γ_{v_0,ω_0} be the nonholonomic geodesic with the origin x, the initial velocity v_0 and the initial value of the Lagrange multiplier ω_0. Put

$$\exp_{V,\delta}(v_0, \omega_0) = \gamma_{v_0,\omega_0}(\delta).$$

The mapping $\exp_{V,\delta}$ plays in nonholonomic geometry the role which is similar to that of the ordinary exponential mapping; however, its structure is much more complicated. The ordinary exponential mapping is a diffeomorphism of a neighborhood of zero in the tangent space onto some neighborhood of $x \in X$. The exponential mapping defined above is not a diffeomorphism on any neighborhood of zero in $V_x \oplus V_x^{\perp}$. For instance, $\exp_{V,\delta}(0, \omega) = x$ for any $\omega \in V_x^{\perp}$. Another feature is that in an arbitrarily small neighborhood of $0 \in V_x \oplus V_x^{\perp}$ there are points (v_0, ω_0) (with $v_0 \neq 0$) at which the rank of $\exp_V$ drops down (see Chapter 3, Section 2). One can claim only that the points at which the rank of $\exp_V$ decreases are in some sense scarce. It is more convenient to formulate the corresponding result not for the mapping $\exp_V$ itself but rather for its restriction to $S_\varepsilon \times V^{\perp}$. The image $A_\varepsilon^V = \exp_V(S_\varepsilon \times V^{\perp})$ is called the *nonholonomic wave front*. In other words, the wave front of radius ε is the set of endpoints of the length ε geodesics with origin x. In the holonomic case the wave front is, for small ε, simply the ε-sphere. In the nonholonomic case only the following inclusion holds.

Proposition 1. *For ε sufficiently small we have*

$$S_\varepsilon^V(x) \subsetneqq A_\varepsilon^V(x).$$

Remark 1. The specific features of the nonholonomic case are revealed in numerous "non-classical" effects. Here is one of them: There exists an ε such that for any $\delta_1, \delta_2 < \varepsilon$

$$S_{\delta_1}^V(x) \cap S_{\delta_2}^V(x) = \varnothing \quad \text{but} \quad A_{\delta_1}^V(x) \cap A_{\delta_2}^V(x) \neq \varnothing.$$

Consider the subset $\Sigma_i \subset S_\varepsilon \times V_x^{\perp}$ consisting of points at which the corank of $\exp_V$ is equal to i. The set of points at which the rank of $\exp_V$ drops down is small in the following sense.

Proposition 2.

$$\operatorname{codim} \Sigma_1 = 1, \quad \operatorname{codim}(\exp_V(\Sigma_1)) = 2.$$

The sets Σ_1 and $\exp_V \Sigma_1$ may have a fairly complicated structure. In particular, in general they are not manifolds. The assertion that Σ_1 has codimension 1 means that Σ_1 is contained in the union of a finite number of hypersurfaces. The assertion on the codimension of $\exp_V \Sigma_1$ has a similar meaning.

Remark 2. For $i > 1$ the codimension of Σ_i may be smaller than i.

1.7. The Action Functional. So far we considered mainly the variational problems with nonholonomic constraints, the Lagrangian for which is the length functional. There is another important Lagrangian, namely, the action functional. In the holonomic case the associated variational problems have the same solutions. More precisely, if a curve γ is a solution of the problem on the minimum of length, its parametrization such that $|\dot{\gamma}| = \text{const}$ is a solution of the problem on the minimum of action. This result is easily extended to nonholonomic case.

Proposition. *Let X be a Riemannian manifold, V a totally nonholonomic distribution on X. Fix two points $x, y \in X$ such that $\rho_V(x, y) = l$ and a number $T > 0$. Let γ be an admissible curve of length l which connects x and y. Then the parametrization of γ such that $\dot{\gamma} = l/T$ realizes the minimum of the action functional*

$$r_T(x, y) = \inf \int \langle \dot{\gamma}, \dot{\gamma} \rangle: \quad \dot{\gamma} \in V, \quad \gamma(0) = x, \quad \gamma(l) = y,$$

and the value of this functional on γ is l^2/T.

Remark. If V is a differential system, there may be no admissible curves of minimal length connecting x and y. However, the relation

$$r_T(x, y) = \frac{\rho_V^2(x, y)}{T}$$

is still valid and there is a sequence of curves γ_n such that $\dot{\gamma}_n \in V$, $l(\gamma_n) \to \rho_V(x, y)$ and the values of the action functional on γ_n converge to $r_t(x, y)$.

§2. Estimates of the Accessibility Set

As observed in Section 1, a totally nonholonomic distribution V determines on a smooth manifold M with metric ρ a new metric ρ_V. In this section we present two-sided estimates of this metric in terms of the original one and of the growth of V. All estimates are based on the normal forms of jets of distributions and of sets of vector fields described in Section 2, Chapter 1.

In Section 2.1 we present an estimate of ρ_V (the parallelotope theorem) which is exact by the order of magnitude and applies to regular distributions on a Riemannian manifold. This estimate is based on the Lemma on quasitriangular form (see Section 1.4, Chapter 1). In Section 2.2 this estimate is extended to the case of metrics associated with regular symmetric polysystems.

In Section 2.3 we present a much sharper result for generic distributions, an estimate of the "leading term" of the nonholonomic metric.

In Section 2.4 local estimates of Sections 2.1, 2.3 and results on generic distributions on compact manifolds (Section 2.3, Chapter 1) are used to derive "global" estimates of the nonholonomic metrics for generic distributions on compact Riemannian manifolds.

In Section 2.5 the estimates of the growth of the volume of balls in nonholonomic metrics are used to compute the Hausdorff dimension of a nonholonomic Riemannian manifold. These estimates are related to the study of the growth of the number of words in groups.

2.1. The Parallelotope Theorem. Let X be a Riemannian manifold, $\dim X = n$, V a regular totally nonholonomic distribution on X. Let us describe the shape of an ε-ball D_ε^V of the nonholonomic metric ρ_V on X for small ε. The nonholonomic ball $D_\varepsilon^V(x)$ is the set of points $y \in X$ which may be reached from x along an admissible curve of length $\leqslant \varepsilon$. As we know (see Section 1), any two points of X may be connected by an admissible curve. The question on the shape of $D_\varepsilon^V(x)$ amounts to that of connecting x by an admissible curve with a close point y which lies in an "inadmissible" direction. Let ξ_1, ξ_2 be admissible vector fields. The circuit of the parallelogram with the sides $\varepsilon\xi_1, \varepsilon\xi_2, -\varepsilon\xi_1, -\varepsilon\xi_2$ amounts to the shift of order ε^2 in the direction $\xi = [\xi_1, \xi_2] \in V_2$ (see e.g. Helgason [1978]). A generalization of this construction allows to shift by a distance of order ε^i in the direction $\xi \in V_i$ along an admissible curve of length $\sim \varepsilon$.

On the other hand, the lemma on quasitriangular form allows to show that the way to shift along a vector field $\xi \in V_i$ described above is the best possible by the order of magnitude. Thus the ball $D_\varepsilon^V(x)$ has a natural description in terms of the flag of V, or, more precisely, of its growth vector $(n_1^V, \ldots, n_k^V) = n$.

Define the function $\varphi = \varphi_V: \{1, \ldots., n\} \to \{1, \ldots, k\}$ by setting $\varphi(i) = j$ if $n_{j-1}^V \leqslant i \leqslant n_j^V$. Given a coordinate system $\{x_i\}$ in the neighborhood of x, define the parallelotope $\Pi_{C,\varepsilon}(x)$ in the following way (here C is a positive constant):

$$\Pi_{C,\varepsilon}(x) = \{y \in X; |x_i(y)| \leqslant C\varepsilon^{\varphi(i)}\}.$$

The Parallelotope Theorem (Gershkovich [1984a]). *Let X be a Riemannian manifold, $x \in X$, V a regular totally nonholonomic distribution on X. There exists a coordinate system $\{x^i\}$ in the neighborhood of x and such constants $c, C > 0$ that*

$$\Pi_{c,\varepsilon}(x) \subset D_\varepsilon^V(x) \subset \Pi_{C,\varepsilon}(x).$$

Remarks. 1. The first inclusion implies that there exist $\varepsilon_0 > 0$, $C > 0$ such that

$$\forall \varepsilon < \varepsilon_0 \quad D_{C\varepsilon^k}(x) \subset D_\varepsilon^V(x).$$

This result was independently proved by several authors (see references in Gershkovich [1984a]).

2. For a generic distribution the function φ which is used in the definition of the parallelotope may be described precisely in terms of the growth of the number of words in the free Lie algebra F with $n_1 = \dim V$ generators:

$$\Pi_{C,\varepsilon}(x) = \{y \in X; |x_i(y)| \leqslant C\varepsilon^l \text{ for } \tilde{n}_{l-1} < i \leqslant \tilde{n}_l\},$$

where $\tilde{n}_l$ is the number of words of length $\leqslant l$ in F (exact estimates for $\tilde{n}_l$ were given in Section 2, Chapter 1).

Thus the asymptotics of an ε-ball of the nonholonomic metric for a generic distribution depends only on the pair (n, n_1), i.e. on the dimensions of the space and of the distribution. For distributions of maximal growth this result will be sharpened in Section 2.3.

2.2. Polysystems and Finslerian Metrics. The estimate of a nonholonomic ε-ball $D_\varepsilon^V(x)$ obtained in Section 2.1 includes arbitrary positive constants c, C and hence remains valid if the Riemannian metric ρ is replaced by an arbitrary equivalent metric $\tilde{\rho}$. (Two metrics ρ, $\tilde{\rho}$ are called equivalent if there exist c_1, $C_1 > 0$ such that

$$\left.\forall x, y \in X \quad c_1 < \frac{\rho(x, y)}{\tilde{\rho}(x, y)} < C_1.\right)$$

In particular, the statement of the theorem remains valid for Finslerian metrics. (Recall that a Finslerian metric on a smooth manifold X is determined by the structure of a Banach space on tangent spaces $T_x X$ which depends smoothly on x.)

One of the objects that naturally give rise to Finslerian metrics are symmetric polysystems which are the traditional subject in control theory (see Agrachev et al. [1983], Jurdjevith and Sussmann [1972], Lobry [1973]) Recall that the germ of a symmetric polysystem on a smooth manifold X is determined by a set of germs of vector fields $\pm\xi_1, \ldots, \pm\xi_m$. A piecewise smooth curve $\gamma\colon \mathbb{R} \to X$ is said to be admissible with respect to the polysystem $\{\xi_i\}$ if there is a piecewise constant function $j(t)$ with values in $\{1, \ldots, m\}$ and a piecewise constant function $i(t)$ with values in $\{1, -1\}$ such that $\dot{\gamma}(t) = i(t)\xi_{j(t)}$.

Remark. Many properties of distributions are naturally extended to polysystems. A polysystem $\xi_1, \ldots, \xi_m$ on a smooth manifold X is called *regular* if $\xi_1, \ldots, \xi_m$ generate a regular distribution. In a similar way, one can speak of polysystems (or germs and jets of polysystems) of maximal growth. The results of Section 1, Chapter 1, show that for $m < \dim X$ the germ of a generic polysystem has maximal growth and is regular.

A symmetric polysystem $\{\xi_i\}$ determines on X a *nonholonomic Finslerian metric* $\tilde{\rho} = \rho_{\{\xi\}}$ which may be defined in two equivalent ways:

(1) $\tilde{\rho}(x, y)$ is the minimal time needed to reach y from x along an admissible curve.

(2) The metric $\tilde{\rho}$ may be obtained as ρ_V^K where ρ^K is the Finslerian metric defined by an arbitrary field of convex solid polyhedra K_X satisfying the condition $\operatorname{Conv}(\{\pm\xi_i\}) = K_X \cap V_x$.[2]

Proposition. *The two metrics defined above coincide*: $\rho_V^K = \rho$.

Since $\tilde{\rho}$ is equivalent to a Riemannian metric (cf. above) we get the parallelotope theorem for symmetric polysystems.

Theorem. *Let* $\{\xi_i\}_{i=1}^m$ *be a regular polysystem on a Riemannian manifold* X. *Let* $D_\varepsilon^{\{\xi_i\}}(x)$ *be the* ε*-ball centered at* x *with respect to the metric* $\tilde{\rho}$ *associated with the polysystem. There exist* $c > 0$, $C > 0$, $\varepsilon_0 > 0$ *such that for all* $\varepsilon < \varepsilon_0$

$$\Pi_{c,\varepsilon}(x) \subset D_\varepsilon^{\{\xi_i\}}(x) \subset \Pi_{C,\varepsilon}(x).$$

The parallelotope $\Pi_{C,\varepsilon}(x)$ *is associated with the distribution* $V = V(\xi_1, \ldots, \xi_m)$ *(see Section 2.1).*

2.3. Theorem on the Leading Term. Sharper estimates of $D_\varepsilon^V(x)$ for maximal growth distributions are based on the canonical form of the Euler-Lagrange equations (see Section 1.4). Let us first derive them for distributions with zero projection map (see Section 1.5). Then we shall prove that for an arbitrary maximal growth distribution there exists a distribution $\tilde{V}$ with zero projection map such that the balls $D_\varepsilon^{\tilde{V}}(x)$ and $D_\varepsilon^V(x)$ are sufficiently close.

Recall that a rigged distribution V has zero projection map if it satisfies the following conditions:

(1) For any $jc_{il}^j = 0$ if $1 \leqslant i \leqslant n_1$ and $n_{k-1} < l$,

(2) $c_{il}^j = 0$ if $1 \leqslant i \leqslant n_1$, $n_{k-2} < l$ and $j \leqslant n_{k-1}$.

Here $c_{ij}^l = \langle [\xi_i, \xi_j], \xi_l \rangle$.

Proposition 1. *Let* V *be the germ of a maximal growth distribution with zero projection map on a manifold* X. *Let* $\{\xi_i\}_{i=1}^{n_1}$ *be a basis of* V, $\{x_i\}$ *a system of coordinates in a neighborhood of* $x \in X$ *compatible with the basis. Let us use the coordinates* $\{x_i\}$ *to identify the* ε*-neighborhood of* x *with a neighborhood of the origin in* $\mathbb{R}^n$. *Then the* ε*-sphere* $S_\varepsilon^V(x)$ *of the nonholonomic metric* ρ_V *is quasihomogeneous, i.e. there exists an* $\varepsilon_0 > 0$ *such that for all* $\varepsilon > 0$, $\varepsilon \leqslant \varepsilon_0$ *we have*

$$S_\varepsilon^V(x) = \psi_\delta(S_{\varepsilon_0}^V(x)), \quad D_\varepsilon^V(x) = \psi_\delta(D_{\varepsilon_0}^V(x)),$$

where $\delta = \dfrac{\varepsilon}{\varepsilon_0}$, $\psi_\delta(x_1, \ldots, x_n) = (x_1\delta^{\varphi(1)}, \ldots, x_n\delta^{\varphi(n)})$ *and the function* φ *is defined by* $\varphi(i) = j$ *for* $n_{j-1} < i \leqslant n_j$ *(cf. Section 2.1). The quasihomogeneity of* $D_\varepsilon^V(x)$ *immediately implies that* $D_\varepsilon^V(x)$ *is homeomorphic to the unit ball in* $\mathbb{R}^n$ *and* $S_\varepsilon^V(x)$ *is homeomorphic to the unit sphere.*

Fix the dimension of the manifold n and the dimension of the distribution n_1.

---

[2] Concerning the fields of convex polyhedra see also Vershik and Chernyakov [1982].

Proposition 2. *For any two distributions V_1, V_2 of dimension n_1 on an n-dimensional manifold which have zero projection map and for a sufficiently small ε the balls $D_\varepsilon^{V_1}(x)$ and $D_\varepsilon^{V_2}(x)$ are isometric.*

The results obtained for distributions with zero projection map allow to estimate the distance between ε-balls of any two distributions. Let V_1 be a maximal growth distribution on a Riemannian manifold X_1 and V_2 a maximal growth distribution on a Riemannian manifold X_2 such that $\dim X_1 = \dim X_2$ and $\dim V_1 = \dim V_2$.

Recall the definition of the distance between two metric spaces considered by several authors (Minkowski and others), see Gromov [1981a] for details.

Definition. Let Z_1, Z_2 be metric spaces. The *distance* between Z_1 and Z_2 is

$$\rho_G(Z_1, Z_2) = \inf_{\{Z, f_1, f_2\}} \rho(f_1 Z_1, f_2 Z_2).$$

The infimum is taken over the triples (Z, f_1, f_2), where Z is a metric space and f_i an isometric embedding of Z_i into Z, $i = 1, 2$.

Proposition 3. *Let X be a Riemannian manifold, $x \in X$. For any two germs of maximal growth distributions V_1, V_2 on X we have*

$$\rho_G(D_\varepsilon^{V_1}(x), D_\varepsilon^{V_2}(x)) = O(\varepsilon^{k+1}).$$

Remark. One can prove that there exists a germ S of a diffeomorphism of $\mathbb{R}^n$ and positive constants C, δ_0 such that $V_{1x} = SV_{2x}$ and for all $(v_0, \omega_0) \in V_x \times V_x^\perp$ and for all $\delta \leqslant \delta_0$ we have

$$\rho(\gamma_{V_1,\delta}(v_0, \omega_0), \gamma_{SV_2,\delta}(Sv_0, S\omega_0)) \leqslant C\delta^{k+1},$$

where ρ is the Riemannian metric on X and k is the nonholonomicity degree of V_1.

Propositions 1–3 give an estimate of the leading term for the ε-ball in the metric ρ_V. Let Y be a nonholonomic ball of a maximal growth distribution with zero projection map.

Theorem (on the leading term). *Let X be a Riemannian manifold. Let V be the germ of a rigged maximal growth distribution of dimension n_1 at $x \in X$. Then there exist $\varepsilon_0 > 0$, $C > 0$ such that for all $\varepsilon < \varepsilon_0$ we have*

$$\rho_G(D_\varepsilon^V(x), \Psi_\varepsilon(Y)) \leqslant C\varepsilon^{k+1},$$

where $D_\varepsilon^V(x)$ is the ε-ball of the metric ρ_V, Ψ_ε is a quasihomogeneous dilation in $\mathbb{R}^n$ which depends only on n_1 (cf. Proposition 1).

The theorem on the leading term implies the following result.

Proposition 4. *Let V be the germ of a maximal growth distribution with zero projection map on a Riemannian manifold X, $\dim X = n$. Then for sufficiently small ε the nonholonomic ball $D_\varepsilon^V(x)$ is homeomorphic to the unit ball in $\mathbb{R}^n$ and the nonholonomic sphere is homeomorphic to the unit sphere in $\mathbb{R}^n$.*

Remarks. 1. The nonholonomic sphere is always a surface with singularities. In the next chapter we shall give a thorough description of the differential structure of $S_\varepsilon^V(x)$ for 3-dimensional Lie groups.

2. If the projection maps of two distributions V_1, V_2 coincide, we have an even stronger inequality

$$\rho_G(D_\varepsilon^{V_1}(x), D_\varepsilon^{V_2}(x)) = O(\varepsilon^{k+2}).$$

2.4. Estimates of Generic Nonholonomic Metrics on Compact Manifolds. Let X be a compact manifold, V a totally nonholonomic distribution on X. Assume that the nonholonomicity degree of V does not exceed k^V anywhere on X. The parallelotope theorem allows to obtain an estimate of nonholonomic metrics on X.

Proposition 1. *There exists a positive constant C such that*

$$\rho(x, y) \leqslant \rho_V(x, y) \leqslant C \max(\rho(x, y), \rho^{1/k^V}(x, y)),$$

where k^V is the upper bound of the nonholonomicity degree of V on X.

Our next aim is to estimate the typical value of k^V on a compact manifold.

Denote by $k^V(x)$ the nonholonomicity degree of V at the point x, and by $k(n, n_1)$ that of the germ of a maximal growth distribution of the same dimension n_1 on X.

Theorem. *Let X be a compact manifold,* $\dim X = n$, $n_1 < n$. *There exists a positive number C such that:*

(1) *If* $(n, n_1) \neq (3, 2)$ & $(n, n_1) \neq (5, 2)$, *for generic differential systems of dimension n_1 on X we have*

$$k^V = \max_{x \in X} k_x^V \leqslant k(n, n_1) + 1,$$

$$\rho(x, y) \leqslant \rho_V(x, y) \leqslant C \max(\rho^{1/(k+1)}(x, y), \rho(x, y))$$

(here $k = k(n, n_1)$).

(2) *For* $(n, n_1) = (3, 2)$ *or* $(n, n_1) = (5, 2)$ *one can guarantee for generic differential systems only the weaker inequalities*

$$k^V = \max_{x \in X} k_x^V = k(n, n_1) + 2,$$

$$\rho(x, y) \leqslant \rho_V(x, y) \leqslant C \max(\rho^{1/(k+2)}(x, y), \rho(x, y)).$$

The proof is based on the results of Section 2, Chapter 1, which connect the codimensions of degeneracies of distributions (or of sets of vector fields) and their growth defects. Let us explain it in more details.

Fix the numbers $n = \dim X$, $n_1 = \dim V$. Denote by $v_X^{n_1}$ the set of differential systems of dimension n_1 on X. The estimates which connect the codimensions of degeneracies and the growth defects (see Section 2.3, Chapter 1) imply:

Proposition 2. *Let X be a compact manifold,* $\dim X = n$, $n_1 < n$. *Then* (1) *there exists an open dense subset $\mathring{v} \subset v$ such that all differential systems from $\mathring{v}$ satisfy* *the following condition*:

$$\operatorname{codim}\{x \in X \mid k_x^V > k(n, n_1) + 1\} = \tilde{n}_{k+1} - n + 1.$$

(2) *If* $n_{k+1} - n + 1 > n$, *then for almost all differential systems* V *of dimension* n_1 *on* X *we have*

$$\max_{x \in X} k_x^V = k(n, n_1) + 1.$$

(Recall that we denoted by $\tilde{n}_k$ the number of words of length $\leqslant k$ in the free Lie algebra with n_1 generators, see Section 2, Chapter 1.)

Our next aim is to list the pairs (n, n_1) for which the above inequality $\tilde{n}_{k+1} > 2n - 1$ is true.

Observe that since $n \leqslant \tilde{n}_k$, for regular dimensions $n = n_k$ the inequality is the sharpest possible. In this case the inequality is checked with the help of explicit formulae for $\tilde{n}_{i+1} - \tilde{n}_i$, i.e. for the number of length i words in the free Lie algebra (see Section 2, Chapter 1). In this way we get

Proposition 3. *For fixed* n_1 *we have*

(1) *If the dimension* $n > n_1$ *is not regular, the inequality*

$$\tilde{n}_{k+1} > 2n - 1$$

holds.

(2) *If the dimension* n *is regular*, $n = \tilde{n}_k$, *the inequality*

$$\tilde{n}_{k+1} > 2n - 1$$

holds for all pairs (n, n_1) *with the following exceptions*:

(a) $n = 3$, $n_1 = 2$ (i.e. $k = 2$),

(b) $n = 5$, $n_1 = 2$ $(k = 3)$.

2.5. Hausdorff Dimension of Nonholonomic Riemannian Manifolds. Let X be a Riemannian manifold with metric ρ, V a totally nonholonomic regular distribution on X; as already explained, V determines a metric ρ_V on X. Let us compute the Hausdorff dimension d_H of the metric space (X, ρ_V). (These results were obtained by Mitchel [1985], they easily follow from the estimates of ε-balls, cf. Section 2.1.) We present geometric statements on the volume of nonholonomic balls and on the Hausdorff dimension of nonholonomic manifolds and then briefly explain a connection of these results with some problems in algebra.

We begin with an estimate of the volume $\operatorname{Vol} D_\varepsilon^V(x)$ of a nonholonomic ε-ball. The estimates of Section 2.1 give the following assertion.

Proposition 1. *Let* X *be a Riemannian manifold*, $\dim X = n$, V *a regular totally nonholonomic distribution on* X, $(n_1, \ldots, n_k)$ *the growth vector of* V. *There exist* ε_0, c, $C > 0$ *such that for all* $\varepsilon < \varepsilon_0$

$$c\varepsilon^M < \operatorname{Vol} D_\varepsilon^V(x) < C\varepsilon^M,$$

where

$$M = \sum_{i=1}^{k} i(n_i - n_{i-1}).$$

The estimates of Sections 2.1, 2.2 allow to compute the Hausdorff dimension $d_H(X, \rho_V)$. Recall the definition of the Hausdorff dimension of a metric space Y (see Mitchel [1985]). Let $D_1 \subset Y$ be a unit ball in Y, N_ε the minimal number of ε-balls which cover D_1. *Then*

$$d_H(Y) \stackrel{\text{def}}{=} -\lim_{\varepsilon\to 0} \frac{\log N_\varepsilon}{\log \varepsilon}.$$

Proposition 2. *Let X be a Riemannian manifold,* $\dim X = n$, *V a totally nonholonomic distribution on X, and $(n_1, \ldots, n_k)$ the growth vector of V. Then*

$$d_H(X, \rho_V) = \sum_{i=1}^{k} i(n_i - n_{i-1}).$$

Remarks. 1. Let us give an explicit formula for the dimension of a nonholonomic manifold $d_H(X, \rho_V)$ when V is a generic distribution. Denote by n_1 the dimension of V and by k its nonholonomicity degree. If the dimension $n = \dim X$ is regular, i.e. $n = \tilde{n}_k$ (cf. Section 2.4), then

$$d_H(X, \rho_V) = \sum_{i=1}^{k} \sum_{j|i} \mu(j) n_1^{i/j} = \frac{n_1^{k-1} - 1}{n_1 - 1} + \sum_{i=1}^{k} \sum_{\substack{j|i\\ j>1}} \mu(j) n_1^{i/j}.$$

Formulae for the growth of the number of words in the free Lie algebra (see Section 2, Chapter 1) yield

$$\dim X = n = d_H(X, \rho) = \sum_{i=1}^{k} \frac{1}{i} \sum_{j|i} \mu(j) n_1.$$

Thus the ratio of the dimensions satisfies the inequalities

$$k - 1 < \frac{d_H(X, \rho_V)}{d_H(X, \rho)} < k$$

(here $k = k(n, n_1)$).

2. For a maximal growth distribution the estimate of growth of the volume of a nonholonomic ε-ball may be sharpened:

$$\operatorname{Vol} D_\varepsilon^V(x) = \varepsilon^{d_H(X, \rho_V)} \cdot (\operatorname{Vol}_0 + \varepsilon \operatorname{Vol}_1 + O(\varepsilon^2)),$$

where Vol_0 does not depend on V (and depends only on the dimensions n, n_1) and Vol_1 is determined by the projection map of V at the point x.

2.6. The Nonholonomic Ball in the Heisenberg Group as the Limit of Powers of a Riemannian Ball. Let N be the Heisenberg group, $D_r = D_r(e)$ the ball of radius r with center $e \in N$ with respect to a left invariant Riemannian metric on N, V the unique (up to an automorphism) nonholonomic two-dimensional distribution on N (see Section 1.7, Chapter 1). Let $\psi_n\colon N \to N$ be a quasihomogeneous dilation which is linear (with coefficient n) in admissible directions and quadratic (with coefficient n^2) in the direction of the center of N (cf. Section 2.3).

Theorem 1.

$$\lim_{n\to\infty} \psi_n^{-1} D_r^n = D_r^V, \tag{10}$$

where the limit has the same meaning as in Section 2.3, D_r^V is the nonholonomic ball and D_r^n is the n-th power of the ball D_r in the group N.

Remarks. 1. The structure of D^V is described in Section 1, Chapter 3.

2. There is a remarkable relation for the nonholonomic ball D_r^V:

$$\psi_2^{-1}([D_r^V]^2) = D_r^V, \tag{11}$$

and, moreover, D_r^V is the only set satisfying this relation among all sets having the same intersection with the orthogonal complement of the center of N.

3. Relation (11) admits an optical interpretation in terms of propagation of wave fronts. In Section 1, Chapter 3 we show that the boundary of D_ε^V (the nonholonomic sphere) is the exterior part of the nonholonomic wave front.

4. A slight modification of the above result in terms of measures on N and their convolutions relates it to such topics as the law of large numbers and limit theorems on groups.

5. Both the theorem and the remarks above may be extended to the case of arbitrary nilpotent groups, the distribution V being replaced by the distribution from Example e), Section 1.5, Chapter 1, which determines a canonical quasihomogeneous structure on the group. For other groups the situation is more complicated, since on these groups there are no canonical quasihomogeneous structures. Still it is possible to introduce a locally almost quasihomogeneous structure (cf. Section 2.3), but we shall not dwell on that.

One can introduce a discrete version of the objects considered above. It is connected with an algebraic problem on the growth of the number of words in groups and leads to the study of certain nonholonomic Finslerian metrics. Let S be a finite subset of N, $S = \{s_1, \ldots, s_n\}$. Put $\tilde{S} = S \cup S^{-1}$, $S^{-1} = \{s_1^{-1}, \ldots, s_n^{-1}\}$. $\tilde{S}$ generates a symmetric polysystem $\{\pm\xi_1, \ldots, \pm\xi_n\}$ on N, where ξ_i is the left-invariant vector field on N such that $\xi_i(e) = \log s_i$. Recall that a polysystem determines a Finslerian metrics on N (see Section 2.2).

Theorem 2. *The sequence of sets $\psi_n^{-1}(\tilde{S}^n)$ approximates $D_1^{\{\xi_i\}}$, i.e.*

$$\lim_{n\to\infty} \rho(\psi_n^{-1}(\tilde{S}^n), D_1^{\{\xi_i\}}) = 0,$$

where $D_1^{\{\xi_i\}}$ *is the unit ball of the Finslerian metric associated with the polysystem* $\{\xi_i\}$.

Remark. Theorem 2 shows that the set of limits of normalized powers of discrete sets in N coincides with the set of unit balls of Finslerian metrics on N associated with polysystems, or, which is the same, with fields of convex symmetric polyhedra (cf. Section 2.2).

Chapter 3
Nonholonomic Variational Problems on Three-Dimensional Lie Groups

In this chapter we consider the simplest nonholonomic variational problems. We study three-dimensional nonholonomic Lie groups, i.e. groups with a left-invariant nonholonomic distribution. Our main subject is the study of the nonholonomic geodesic flow (NG-flow), more precisely, of the nonholonomic sphere, of the wave front (Section 1), and of the general dynamical properties of the flow (Section 2). The mixed bundle for Lie groups is the direct product $G \times (V \oplus V^{\perp})$. In Section 1.1 we show that the NG-flow on the mixed bundle is the semidirect product with base $V \oplus V^{\perp}$ and fiber G. In Section 1.2 we describe left-invariant metric tensors on Lie algebras; in Section 1.3 the normal forms for the equations of nonholonomic geodesics are obtained. In Section 1.4 we study the reduced flow on $V \oplus V^{\perp}$. In the subsequent Sections (1.5–1.7) we describe local properties of the flow on the fiber; in Section 1.5 we describe the ε-wave front of the NG-flow and the ε-sphere of the nonholonomic metrics which appear to be manifolds with singularities, the same for all three-dimensional nonholonomic Lie groups. In Section 1.6 we describe their topology and in Section 1.7 their metric structure.

§ 1. The Nonholonomic ε-Sphere and the Wave Front

1.1. Reduction of the Nonholonomic Geodesic Flow. Let G be a Lie group with a left-invariant metric ρ, $\mathfrak{g}$ its Lie algebra, V a totally nonholonomic left-invariant distribution on G. This distribution is determined by a linear subspace $V_e \subset \mathfrak{g}$ (see Section 1.5, Chapter 1). The mixed bundle associated with V (Section 1.3, Chapter 1) is the direct product $G \times (V_e \oplus V_e^{\perp})$[1]. It is convenient to write the equations of a nonholonomic geodesic γ on G in terms of the coefficients of the expansion of $\dot{\gamma}$ with respect to an orthonormal basis of $\mathfrak{g}$. Let $\{n_i\}$ be the growth vector of V, $\{\xi_i\}$ an orthonormal basis of $\mathfrak{g}$ such that $\{\xi_1, \ldots, \xi_{n_1}\}$ is a basis of V, $\ldots$, $\{\xi_{n_i+1}, \ldots, \xi_{n_{i+1}}\}$ is a relative basis of V_{i+1} modulo V_i, etc. The equations of nonholonomic geodesics have the following form (cf. Section 1.5, Chapter 2):

$$\begin{cases} \dot{\gamma} = \sum\limits_{i=1}^{n_1} v_i \xi_i, \\ \dot{v}_i = \sum\limits_{j<n_1, l>n_1} c_{ij}^l v_j \lambda_l - \sum\limits_{m,l<n_1} \Gamma_{ml}^i v_l v_m, \\ \dot{\lambda}_i = -\sum\limits_{j\leqslant n_1<l} c_{ij}^l v_j \lambda_l + \sum\limits_{m,l} \Gamma_{ml}^i v_l v_m, \end{cases}$$

[1] The index e will be sometimes omitted.

where c_{ij}^l are the structure constants of $\mathfrak{g}$, $c_{ij}^l = \langle [\xi_i, \xi_j], \xi_l \rangle$, and Γ_{ij}^l are the Christoffel symbols, $\Gamma_{ij}^l = \langle \nabla_{\xi_j} \xi_i, \xi_l \rangle$. Recall (see Dubrovin et al. [1979]) that the Christoffel symbols are expressed through the structure constants in the following way: $\Gamma_{ij}^l = \frac{1}{2}(c_{ji}^l + c_{li}^j + c_{lj}^i)$. The last two groups of differential equations on $V \oplus V^\perp$ have constant coefficients and determine a flow on $V \oplus V^\perp$; if we fix an initial point in $V \oplus V^\perp$, the first equation determines a flow on G.

Recall the definition of the semidirect product of dynamical systems (Cornfeld et al. [1980]). Let M be the direct product of smooth manifolds, $M = M_1 \times M_2$. Suppose that there is a dynamical system $T_1(t)$: $M_1 \to M_1$ on M_1 and a family of dynamical systems $T_2(x_1)(t)$: $M_2 \to M_2$ on M_2 which depends smoothly on $x_1 \in M_1$. The semidirect product of T_1 and T_2 is a dynamical system T on $M_1 \times M_2$ such that $T(t)(x_1, x_2) = (T_1(t)x_1, T_2(x_1)(t)(x_2))$. The space M_1 is the base and M_2 is the fiber of the semidirect product.

Proposition. *The nonholonomic geodesic flow regarded as a dynamical system on the mixed bundle of G is the semidirect product with base $V \oplus V^\perp$ and fiber G.*

This proposition is similar to the corresponding statement on ordinary left-invariant geodesic flows on Lie groups. The decomposition described above is called *reduction.* By contrast with symplectic reduction, further decomposition is, in general, impossible, since a left-invariant symplectic structure is lacking.

The structure of the semidirect product will be thoroughly studied in Sections 1.5, 1.6.

1.2. Metric Tensors on Three-Dimensional Nonholonomic Lie Algebras. In Section 1 of Chapter 1 we have listed all three-dimensional nonholonomic Lie algebras up to an isomorphism. Let us classify the metrics ρ_V on these algebras.

Observe that a metric ρ_V is determined by the restriction of a metric tensor g defined on the Lie algebra $\mathfrak{g}$ to the plane V: $\tilde{g} = g|_V$. Since the dilation of time $t \mapsto \lambda t$ is equivalent to the dilation of g, we shall determine $\tilde{g}$ only up to a constant factor. Finally, we are interested only in the equivalence classes of such metrics. In other words, we are dealing with the orbits of the group of automorphisms of a nonholonomic Lie algebra on the projective space whose points correspond to the restrictions of metric tensors to the plane V. The next assertion describes the set $\mathscr{C}\ell_{\mathfrak{g}}^{\rho}$ of these equivalence classes for all nonholonomic three-dimensional Lie algebras.

Proposition. *The number of classes of equivalent metrics for three-dimensional nonholonomic Lie groups is equal to*

(1) 1 *for the Heisenberg group,*
(2) 1 *for solvable* 3*-dimensional groups,*
(3) *a* 1*-parameter family for each simple nonholonomic Lie group*

$$(G, V) = (\mathrm{SO}(3), V), \quad (G, V) = (\mathrm{SL}_2\,\mathbb{R}, V_1), \quad (G, V) = (\mathrm{SL}_2\,\mathbb{R}, V_2).$$

In the latter case the set $\mathscr{C}\ell^p_g$ of equivalence classes is naturally isomorphic to $\mathbb{R}_+$; representatives of the equivalence classes are given by $g = \begin{pmatrix} 1 & 0 \\ 0 & m \end{pmatrix}$ in the basis of V described in Proposition 1 of Section 1.1, Chapter 1.

1.3. Structure Constants of Three-Dimensional Nonholonomic Lie Algebras. The set of structure constants c_{ij}^k, i, j, $k = 1, 2, 3$, of three-dimensional nonholonomic Lie algebras is an algebraic submanifold $S \subset \mathbb{R}^{27}$ determined by the following equations:

$$\begin{cases} c_{ij}^k = -c_{ji}^k \quad i, j, k = 1, 2, 3, \\ \sum_{r=1}^{3} c_{ij}^r c_{rk}^l + c_{jk}^r c_{ri}^l + c_{ki}^r c_{rj}^l = 0, \\ c_{12}^3 = 1, \quad c_{12}^1 = c_{12}^2 = 0. \end{cases} \tag{1}$$

The first group of equations corresponds to the skew symmetry of the Lie bracket, the second one to the Jacobi identity, and the third one reflects the nonholonomicity of V. Equations (1) imply

$$c_{23}^2 = -c_{13}^1, \quad c_{23}^3 = 0. \tag{2}$$

The structure of the algebraic manifold S may easily be described.

Proposition. *The algebraic manifold S is the union of two affine spaces, $S = P_3 \cup P_2$, where P_3 is a three-dimensional affine space, $P_3 \backslash P_2$ corresponds to simple Lie algebras, P_2 is a two-dimensional affine plane which corresponds to solvable Lie algebras; the line $P_3 \cap P_2$ corresponds to nilpotent Lie algebras. The space P_3 is singled out by the condition $c_{13}^3 = 0$ and the plane P_2 by the condition $c_{23}^1 = c_{23}^2 = 0$.*

1.4. Normal Forms of Equations of Nonholonomic Geodesics on Three-Dimensional Lie Groups. Consider a nonholonomic geodesic $\gamma\colon \mathbb{R} \to G$. The curve γ is a solution of the conditional variational problem on the minimum of length on the group G with constraint V. Recall (see Section 1.4, Chapter 2) that a nonholonomic geodesic satisfies the following equations:

$$\begin{cases} \nabla_{\dot\gamma}\dot\gamma = \dot\lambda\omega + \lambda\dot\gamma \lrcorner\, d\omega, \\ \langle \dot\gamma, \omega \rangle = 0, \end{cases} \tag{3}$$

where ∇ is the covariant derivative which corresponds to the Riemannian connection on G, ω is a 1-form annihilating V and λ is the Lagrange multiplier.

The nonholonomic geodesic flow on G is defined on the mixed bundle $G \times (V \oplus \mathbb{R})$, where in classical terms $\mathbb{R}$ is the space of Lagrange multipliers and in terms of Section 1, Chapter 1, it is the annihilator of V in $\mathfrak{g}^*$.

As the nonholonomic geodesic flow decomposes into semidirect product (see Section 1.1), we begin with the study of the flow on its base $V \oplus V^\perp$.

Choose an orthonormal basis ξ_1, ξ_2 in V. Replacing, if necessary, the metric tensor g with another one which has the same restriction to V, we may assume that $\xi_3 = [\xi_1, \xi_2] \perp V$, $|\xi_3| = 1$. Put $\dot{\gamma} = a_1\xi_1 + a_2\xi_2$. Denote by c_{ij}^k the structure constants of g: $c_{ij}^k = \langle [\xi_i, \xi_j], \xi_k \rangle$. Using the expression of the Christoffel symbols $\Gamma_{ij}^k = \langle \nabla_{\xi_j}\xi_i, \xi_l \rangle$ in terms of the structure constants (see Dubrovin et al. [1979] on the vierbein formalism),

$$\Gamma_{ij}^k = \tfrac{1}{2}(c_{ji}^k + c_{ki}^j + c_{kj}^i),$$

we reduce the system (3) to the following form:

$$\begin{cases} \dot{\gamma} = a_1\xi_1 + a_2\xi_2, \\ \dot{a}_1 = -\lambda a_2, \\ \dot{a}_2 = \lambda a_1, \\ \dot{\lambda} = a_1a_2(c_{31}^2 + c_{32}^1) + a_1^2c_{31}^1 + a_2^2c_{32}^2 + \lambda(c_{13}^3a_1 + c_{23}^3a_2). \end{cases}$$

In the sequel we shall consider only such geodesics γ for which $|\dot{\gamma}| = 1$, i.e. $a_1^2 + a_2^2 = 1$. Thus the nonholonomic geodesic flow is restricted to the subbundle $G \times (S^1 \times \mathbb{R})$ of the mixed bundle with fiber $\{(\dot{\gamma}, \lambda)\} = S^1 \times \mathbb{R}$, $S^1 \subset V$. Putting $a_1 = \sin\varphi$, $a_2 = \cos\varphi$ and taking in account (2), we get

$$\begin{cases} \dot{\gamma} = \sin\varphi\cdot\xi_1 + \cos\varphi\cdot\xi_2, \\ \dot{\varphi} = \lambda, \\ \dot{\lambda} = \tfrac{1}{2}\sin 2\varphi(c_{31}^2 + c_{32}^1) + c_{13}^1\cos 2\varphi + \lambda c_{13}^3\sin\varphi. \end{cases} \tag{4}$$

The proposition of Section 1.3 allows to obtain the normal form of the equation of nonholonomic geodesics for each of the three-dimensional nonholonomic Lie algebras (we write down only the last of the equations (4), since the first two are standard).

Theorem (on normal forms of equations of nonholonomic geodesics). *The equations of nonholonomic geodesics have the following form*:

N. *For the Heisenberg group* N: $\dot{\lambda} = 0$.

S. *For solvable groups* R_α, $\alpha \in \mathrm{SL}_2\mathbb{R}$:

$$\dot{\lambda} = \tfrac{1}{2}\det\alpha\cdot\sin 2\varphi - \lambda\,\mathrm{Sp}\,\alpha\cdot\sin\varphi.$$

SS. *For simple Lie groups*

$$\dot{\lambda} = f(m)\sin 2\varphi,$$

where $m \in \mathbb{R}$ *is the parameter of the metric tensor* (see Section 1.2) *and*

$$f(m) = \begin{cases} (m^{-1} - 1)/2 & \text{for } \mathrm{SO}(3), \\ 2(1 - m^{-1}) & \text{for } (\mathrm{SL}_2\mathbb{R}, V_1), \\ 2(m+1) & \text{for } (\mathrm{SL}_2\mathbb{R}, V_2). \end{cases}$$

1.5. The Flow on the Base $V \oplus V^{\perp}$ of the Semidirect Product. Consider the phase portrait of system (4) on the cylinder $S^1 \times \mathbb{R}$. The theorem above implies that for all algebras the normal forms of the equations of nonholonomic geodesics have the following form:

$$\begin{cases} \dot{\varphi} = \lambda, \\ \dot{\lambda} = \mu_1 \sin 2\varphi + \mu_2 \lambda \sin \varphi, \end{cases} \tag{5}$$

where $\mu_1 \leqslant 0$. Let us study this system.

Proposition 1. *Consider system* (5) *on the cylinder* $(\varphi, \lambda) \in S^1 \times \mathbb{R}$. *There are two alternative cases*:

(1) *For* $\mu_1 \neq 0$ *there exist four fixed points* $\lambda = 0$, $\varphi = \pi k/2$, $k = 0, 1, 2, 3$, *connected by four separatrices. All trajectories, with the exception of the separatrices, are closed. The separatrices connect the points* $\pi/2$ *and* $3\pi/2$.

(2) *For* $\mu_1 = 0$ *the fixed points form a circle determined by the equation* $\lambda = 0$. *All trajectories* (*for* $\lambda \neq 0$) *are closed.*

The phase portrait of system (5) for $\mu_1 \neq 0$ is given in Figure 1. Each integral curve of system (5) intersects the line $\varphi = 0$ or the line $\varphi = \pi$. Denote by α_t the trajectory which intersects the line $\varphi = 0$ at the point t and by τ_t its period. Let $\pm t_\infty$ be the intersection point of a separatrix with the line $\varphi = 0$. When $\mu_1 = 0$, we assume that $t_\infty = 0$. The phase portrait on the cylinder for $\mu_1 = 0$ is given in Figure 2.

Proposition 2. (1) *For* $t_\infty < t$ *the function* τ_t *is strictly decreasing and* $\tau_t \to 0$.

(2) *For* $0 < t < t_\infty$ *the function* τ_t *is strictly increasing and* $\lim \tau_t = T_0$ *for some* $T_0 > 0$.

Remarks. 1. The function τ_t is even.

2. When $\mu_1 = 0$, we assume that $T_0 = \infty$.

3. $t\tau_t \to 1$.

The plot of the period τ as a function of λ is represented on Figure 3 for $\mu_1 \neq 0$.

1.6. Wave Front of Nonholonomic Geodesic Flow, Nonholonomic ε-Sphere and Their Singularities. The propositions of the previous section completely describe the structure of the reduced nonholonomic geodesic flow on the base $V \oplus V^{\perp}$. Let us now describe the geodesics themselves regarded as curves on G. The projection of each integral curve of the nonholonomic geodesic flow in $G \times (V \oplus V^{\perp})$ onto $V \oplus V^{\perp}$ is a curve on $S^1 \times V^{\perp} \simeq S^1 \times \mathbb{R}$. In order to obtain the geodesic on the group itself, we must solve the equation

$$\gamma^{-1}\dot{\gamma} = \xi_1 \cos \varphi(t) + \xi_2 \sin \varphi(t)$$

($\gamma^{-1}\dot{\gamma}$ is the Cartan development of the curve on G).

Fix a point $x \in G$ and a positive number ε and denote by $NE(v_0, \lambda) = \exp_{V,\varepsilon}(v_0, \lambda)$ the endpoint of the geodesic of length ε with the origin x and initial data (v_0, λ). Since the metric and the distribution are left-invariant, we may assume that x is the unit element of G. We begin with local questions, such as the

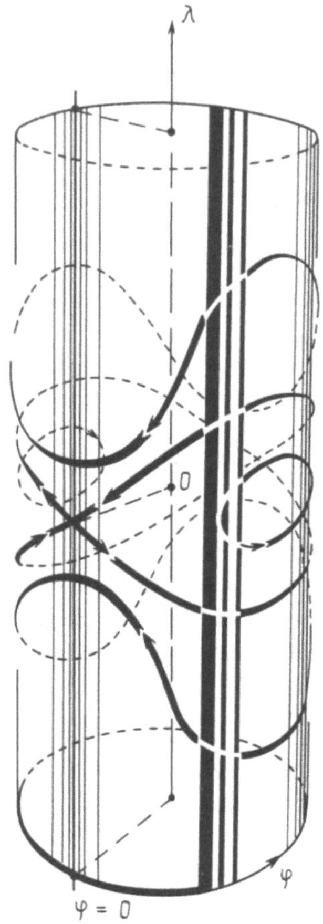

Fig. 1

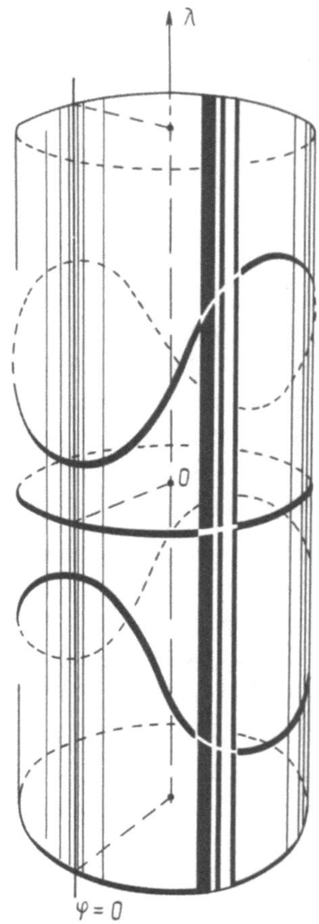

Fig. 2

study of the nonholonomic ε-sphere and ε-ball. As noted earlier (see Section 2, Chapter 2), $S_\varepsilon^V(x) \subset A_\varepsilon^V(x)$, where $A_\varepsilon^V(x)$ is the nonholonomic wave front, i.e. the set of endpoints of length ε geodesics. Thus A_ε^V is the image of the cylinder $S_\varepsilon \times \mathbb{R} \subset V \oplus V^\perp$, $A_\varepsilon^V = NE(S \times \mathbb{R})$. We shall use the Floquet theorem.

Let G be a Lie group, $\mathfrak{g}$ its Lie algebra, $f\colon \mathbb{R} \to \mathfrak{g}$ a periodic curve in $\mathfrak{g}$, $f(t + \tau) = f(t)$. Fix a point $x \in G$ and consider the family of curves γ_r^x, $r \in \mathbb{R}$, on G satisfying the same differential equation

$$(\gamma_r^x)^{-1}\dot{\gamma}_r^x = f(t), \tag{6}$$

with initial data $\gamma_r^x(r) = x$.

Theorem 1 (Floquet, cf. Hartman [1964], Lefshetz [1957]). *For any r_1, r_2 the integral curves $\gamma_{r_1}^x$, $\gamma_{r_1}^x$ of the differential equation* (6) *with initial data described*

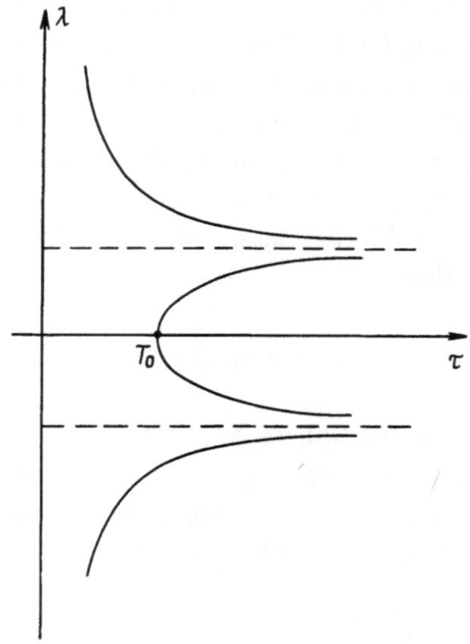

Fig. 3

above satisfy

$$\gamma_{r_1}^x(r_1+\tau)=\gamma_{r_2}^x(r_2+\tau).$$

The mapping $s\colon G\to G$ defined by $x\mapsto\gamma_r^x(r+\tau)$ is called the *monodromy map*; s is the right translation by

$$z_\tau=\gamma_r^x(r+\tau)(\gamma_r^x(r))^{-1}=\gamma_r^x(r+\tau)x^{-1}.$$

Denote by α_n, $n\in\mathbb{N}$, the curve (φ,λ) of period ε/n with $\lambda>0$ on $S^1\times\mathbb{R}$ which satisfies the system of equation (5), let α_{-n}, $n\in\mathbb{N}$, be the curve $-\alpha_n$. The Floquet theorem implies

Proposition 1. *For each natural n the mapping* $\exp_{V,\varepsilon}$ *shrinks the curve* $\tilde{\alpha}_n$ *into a point.*

Put

$$\tilde{z}_n^\varepsilon=\exp_{V,\varepsilon}(\tilde{\alpha}_n).$$

Proposition 2. (1) *The mapping* $\kappa\colon\mathbb{R}_+\to G$, *such that* $\kappa(\tau)=\tilde{z}_1^\tau$, *is smooth.*

(2) *We have* $\rho(e,\kappa(\delta))\delta^{-2}\xrightarrow[\delta\to 0]{}C_G$, *where* C_G *is a positive constant (which depends only on G).*

The complement of the union of curves $\Sigma=(S^1\times\mathbb{R}^1)\setminus\bigcup_{n\in(\mathbb{N}\cup-\mathbb{N})}\tilde{\alpha}_n$ is the union of a countable set of connected components. Let C_n be the connected component bounded by the curves $\tilde{\alpha}_n$ and $\tilde{\alpha}_{n+1}$, and C_{-n} the component bounded by $\tilde{\alpha}_{-n+1}$ and $\tilde{\alpha}_{-n}$; finally, denote by C_0 the components bounded by the curves $\tilde{\alpha}_{-1}$, $\tilde{\alpha}_1$.

Theorem 2 (*on the wave front*). *Let G be a three-dimensional nonholonomic Lie group. For a sufficiently small ε the following assertions are valid*:

(1) *The mapping $\exp_{V,\varepsilon}$ is nondegenerate, i.e. it has rank 2 at all points $x \in \Sigma$.*

(2) *The mapping $\exp_{V,\varepsilon}$ shrinks the curve $\tilde{\alpha}_n$ into the point z_n^ε. The germ of the wave front A_ε^V in the neighborhood of any of these points is diffeomorphic to the germ of the cone $K = \{(x, y, z); z = x^2 + y^2\}$ at the origin.*

(3) *For each $n \in \mathbb{N} \cup -\mathbb{N}$ the image of $\overline{C}_i$ under the mapping $\exp_{V,\varepsilon}$ is diffeomorphic to the 2-sphere.*

(4) $S_\varepsilon^V = \exp_{V,\varepsilon}(\overline{C}_0)$.

(5) $\exp_{V,\varepsilon}(C \setminus \overline{C}_0)$ *lies in the interior of the non-holonomic ε-ball.*

1.7. Metric Structure of the Sphere S_ε^V. Let us describe the shape of S_ε^V for small ε. Choose a coordinate system x_1, x_2, x_3 in the neighborhood of the unit element of G in such a way that the basic vector fields of the distribution V introduced in Section 1.4 have the following form:

$$\xi_1 = \frac{\partial}{\partial x_1}, \quad \xi_2 = \frac{\partial}{\partial x_2} + x_1 \frac{\partial}{\partial x_3} + \cdots.$$

Put $\tilde{\xi}_2 = \dfrac{\partial}{\partial x_2} + x_1 \dfrac{\partial}{\partial x_3}$, $\tilde{\xi}_1 = \xi_1$ and consider the following two systems of differential equations:

(1) the system of equations for nonholonomic geodesics in G

$$\begin{cases} \dot{\gamma}_1 = \cos \varphi_1 \xi_1 + \sin \varphi_1 \xi_2, \\ \dot{\varphi}_1 = \lambda_1 \\ \dot{\lambda}_1 = \mu_1 \sin 2\varphi_1 + \mu_2 \lambda_1 \sin \varphi_1; \end{cases} \tag{7}$$

(2) the system

$$\begin{cases} \dot{\gamma}_2 = \cos \varphi_2 \xi_1 + \sin \varphi_2 \xi_2, \\ \dot{\varphi}_2 = \lambda_2, \\ \dot{\lambda}_2 = 0, \end{cases} \tag{8}$$

System (8) is the system of equations for nonholonomic geodesics in the Heisenberg group.

The results of Section 2, Chapter 2, imply the following

Proposition. *Fix $(\varphi, \lambda) \in S^1 \times \mathbb{R}^1$ and let γ_1, γ_2 be the solutions of systems* (7) *and* (8), *respectively, with the initial data (φ, λ) and $\gamma_1(0) = \gamma_2(0)$. Then $\rho(\gamma_1(\varepsilon), \gamma_2(\varepsilon)) = O(\varepsilon^3)$.*

Thus, the ε-wave front A_ε^V and the ε-sphere for G coincide, up to terms of order ε^3, with the corresponding objects for the Heisenberg group N. The nonholonomic geodesics for the Heisenberg group may be found explicitly: system (8) is easily solved (see Vershik and Gershkovich, [1986a]). Denote the coordinates of its solution $\gamma_{\lambda,\varphi}(\varepsilon)$ by $x_{\varphi,\lambda}, y_{\varphi,\lambda}, z_{\varphi,\lambda}$. They are given by the following formulae:

$$x_{\varphi,\lambda} = \varepsilon\frac{\sin(\theta+\varphi)-\sin\varphi}{\theta},$$

$$y_{\varphi,\lambda} = \varepsilon\frac{\cos\varphi-\cos(\theta+\varphi)}{\theta}, \tag{9}$$

$$z_{\varphi,\lambda} = \varepsilon^2\frac{2\theta-(\sin(2\theta+2\varphi)-\sin 2\varphi)+4\sin\varphi(\cos(\theta+\varphi)-\cos\varphi)}{4\theta^2},$$

where $\theta = \lambda\varepsilon$.

Formulae (9) imply that (cf. also Section 2):

(1) A_ε^V and S_ε^V are quasihomogeneous surfaces: the change $\varepsilon \mapsto \varepsilon n$ amounts to an n-fold dilation in the directions x and y and to an n^2-fold dilation in the direction z.

(2) A_ε^V and S_ε^V have the symmetry group $\Gamma = \mathbb{Z}/2\mathbb{Z} \oplus \mathbb{Z}/2\mathbb{Z}$ consisting of transformations which change sign of any two coordinates in $\mathbb{R}^3$.

Figure 4 gives a rough picture of the wave front for the Heisenberg group (in this case the curves α are simply the horizontal circles $\lambda = 2\pi k/\varepsilon$). All "beads" represent the images of C_i, $i \neq 0$, and are mapped into the interior of the first

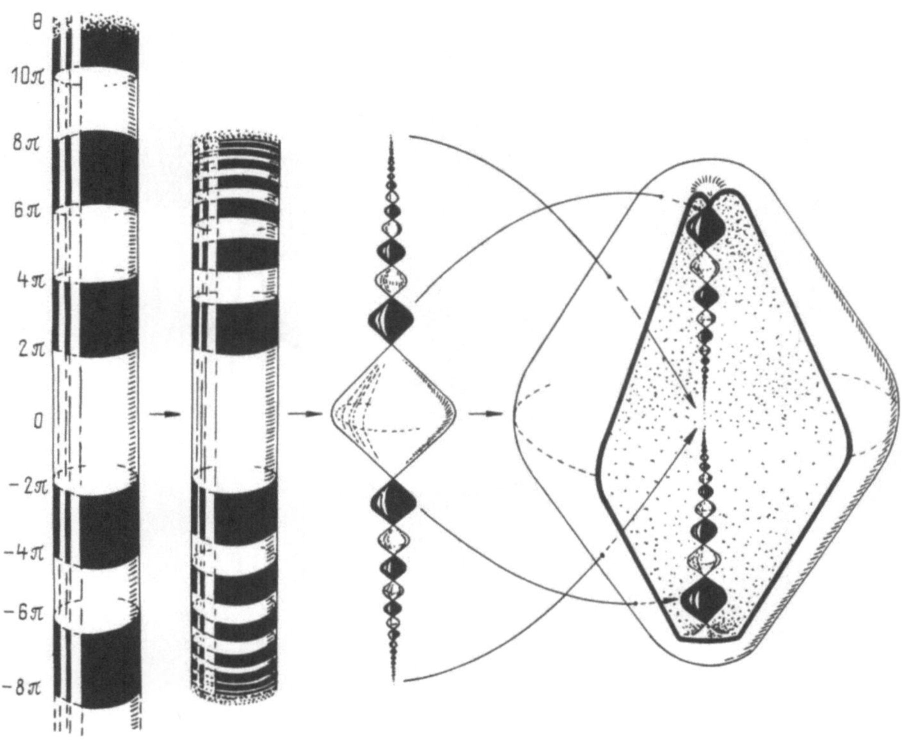

Fig. 4

one which is the image of C_0; the singular points lie on the z axis and have the coordinates z_n, $n \in \mathbb{N} \cup -\mathbb{N}$,

$$z_k = \frac{\varepsilon^2}{4k\pi} \xrightarrow[k\to\infty]{} 0.$$

The metric structure of the ε-sphere and the wave front is rather complicated. Their more or less exact shape is shown in Figure 5. Figure 4 represents the nonholonomic exponential mapping of the cylinder $S^1_\varepsilon \times V^\perp$ onto the nonholonomic wave front.

Since the quasihomogeneity degree for the Heisenberg group is (1, 1, 2), Proposition 1 assures that for small ε the nonholonomic ε-sphere for any three dimensional nonholonomic Lie group has a "similar" shape.

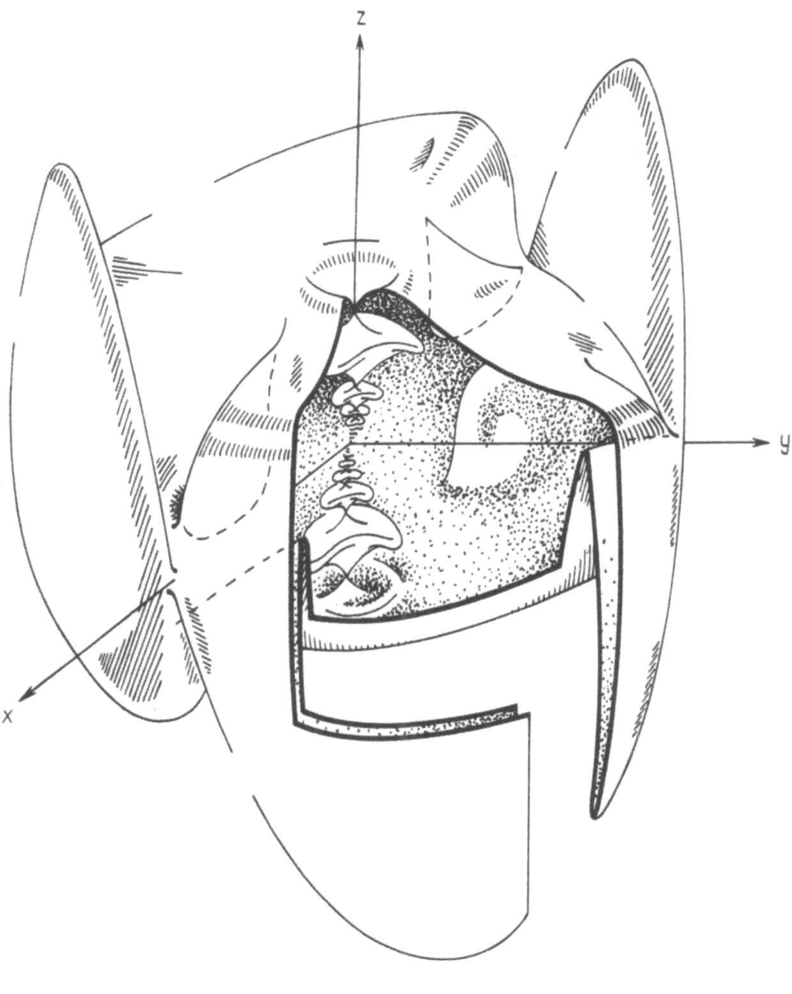

Fig. 5

Remark. Theorem 2, Section 1.6, shows that an ε-sphere for any three-dimensional nonholonomic Lie group is diffeomorphic to a rotation body. For the Heisenberg group the corresponding coordinates in $\mathbb{R}^3$ may be found explicitly. Namely, identify the group N with $\mathbb{R}^3$ by means of the standard coordinates

$$\begin{bmatrix} 1 & \alpha & \gamma \\ 0 & 1 & \beta \\ 0 & 0 & 1 \end{bmatrix} \mapsto (\alpha, \beta, \gamma + \alpha\beta).$$

Under this identification the nonholonomic ε-sphere becomes a rotation body.

§ 2. Nonholonomic Geodesic Flow on Three-Dimensional Lie Groups

In this section we study the dynamical properties of the nonholonomic geodesic flow on three-dimensional Lie groups and their compact homogeneous spaces. Since all trajectories of the reduced dynamical system on the base $V \oplus V^{\perp}$ (see Section 1.5) for NG-flows are closed, we get the monodromy maps on the fiber which we describe in Section 2.1; they allow to study the nonholonomic geodesic flows on Lie groups and their homogeneous spaces (Section 2.2). In Section 2.3 we describe the properties of NG-flows on SO(3) and on compact homogeneous spaces of the Heisenberg group. The flow on compact homogeneous spaces of $SL_2 \mathbb{R}$ which proves to be the most interesting one is described in Section 2.4. Finally, in Section 2.5 we describe the flow on a nilmanifold which is a compact homogeneous space of a grade 2 nilpotent group with m generators.

2.1. The Monodromy Maps. The results of Section 1.5 show that for $\mu_1 \neq 0$ all trajectories on the base $S^1 \times \mathbb{R}$ (with the exception of four separatrices and four fixed points) break up into four families of closed curves:

1, 2) α_τ^+ (respectively, α_τ^-), the trajectory of period τ in the upper (resp. lower) part of the cylinder, $\tau \in (0, \infty)$.

3, 4) β_τ^+ (respectively, β_τ^-), the trajectory of period τ that lies between the separatrices in the right (resp. left) part of the cylinder (see Figure 1).

For $\mu_1 = 0$, there are two families of periodic trajectories described above and a circle $\lambda = 0$ consisting of fixed points (see Figure 2). The Floquet theory (see Section 2.5) shows that in time τ all geodesics $\gamma_{V,t}(v_0, \lambda_0)$ starting from x with initial data $(v_0, \lambda_0) \in \alpha_\tau$ (respectively, $(v_0, \lambda_0) \in \beta_\tau$) will reach the same point xz_τ (resp., xy_τ). Thus we obtain

Proposition 1. *A nonholonomic geodesic* $\gamma = \gamma_{V,t}(v_0, \lambda_0)$ *with* $(v_0, \lambda_0) \in \alpha_\tau$ *(or* $(v_0, \lambda_0) \in \beta_\tau$*) satisfies the relation* $\gamma(t) = \gamma(t - n\tau) \cdot z_\tau^n$ *(respectively,* $\gamma(t) = \gamma(t - n\tau) y_\tau^n$*).*

Proposition 1 readily allows to describe the nonholonomic geodesic flow for all three-dimensional simply connected Lie groups, with the exception of the two-fold covering of SO(3), i.e. of the group $SU(2) \simeq S^3$.

Recall (see Bourbaki [1970]) that the Heisenberg group, the simply connected solvable groups and the universal covering of $SL_2 \mathbb{R}$ are homeomorphic (as manifolds) to the three-dimensional Euclidean space $\mathbb{R}^3$. The universal covering of SO(3) is homeomorphic to the three-dimensional sphere S^3.

Proposition 2. *Let G be a three-dimensional simply connected noncompact Lie group, V a left-invariant two-dimensional distribution on G, ρ a left-invariant metric on G. Then each nonholonomic geodesic goes off to infinity and has no limit points.*

Fix $(v_0, \lambda) \in \alpha_\tau$ (respectively, $(v_0, \lambda) \in \beta_\tau$; for any $x \in G$ denote by $\gamma^x_{v_0,\lambda}(t)$ the nonholonomic geodesic starting from x with initial data (v_0, λ). Consider the mapping $\mu_\tau: G \to G$, $\mu_\tau(x) = \gamma^x_{v_0,\lambda}(\tau)$. We shall call μ_τ the period τ monodromy map. Clearly, for $(v_0, \lambda) \in \alpha_\tau$ (respectively, for $(v_0, \lambda) \in \beta_\tau$) μ_τ acts as right multiplication by the element $z_\tau \in G$ (respectively, by $y_\tau \in G$): $\mu_\tau x = xz_\tau$ (resp., $\mu_\tau x = xy_\tau$).

Definition. The set $\{z_\tau; \tau \in (-\infty, \infty)\}$ (respectively, the set $\{y_\tau; \pm\tau \in (T_0, \infty)\}$) is called the *monodromy submanifold.*

The next assertion characterizes the points of the monodromy submanifold.

Proposition 3. *For each point of the monodromy submanifold there exists a 1-parameter family of the shortest geodesics (of equal lengths) connecting this point with the unit element.*

Thus each point of the monodromy submanifold is conjugate to the unit element.

In the classical case the set of points of a geodesic conjugate to a given one is discrete. Existence of submanifolds consisting of conjugate points is another nonholonomic feature. It is interesting to describe their structure. One can show that for the nilpotent group and for all simple three-dimensional nonholonomic groups the monodromy submanifold coincides with the 1-parameter subgroup z_τ if the coefficient μ_1 is zero; if $\mu_1 \neq 0$, the monodromy submanifold $\mathfrak{M}$ consists of the 1-parameter subgroup z_τ which corresponds to the α-family of closed curves and of a subset of the 1-parameter subgroup y_τ which cooresponds to the β-family. For solvable groups the monodromy submanifold does not admit such a simple description. The next assertion describes the monodromy submanifold $\mathfrak{M}$ for some of the three-dimensional nonholonomic Lie groups.

Proposition 4.

(1) *For the Heisenberg group* $\mathfrak{M} = \{z_\tau\}_{\tau \in \mathbb{R}}$ *is a 1-parameter subgroup of N (isomorphic to* $\mathbb{R}^1$*) with a generator from* $V^\perp$.

(2a) *For* SO(3) *with a bi-invariant metric* $\mathfrak{M}$ *is a 1-parameter subgroup (isomorphic to* S^1*) with a generator from* $V^\perp$.

(2b) *For* SO(3) *with a left-invariant metric (which is not bi-invariant)* $\mathfrak{M}$ *is the union of two 1-parameter subgroups,* $\mathfrak{M} = \{z_\tau\} \cup \{y_\tau\}$, *where* $\{z_\tau\}$ *is a 1-parameter subgroup of* SO(3) *with a generator from* $V^\perp$ *and* $\{y_\tau\}$ *is a* 1*-parameter subgroup whose generator corresponds to the center, i.e. to the fixed point on the cylinder* (see Figure 1). *Thus each of the components of* $\mathfrak{M}$ *is a projective line in* $SO(3) \simeq P_3\mathbb{R}$ *and these lines intersect at the unit element* $e \in SO(3)$.

(3a) *For the nonholonomic Lie group* $(SL_2\mathbb{R}, V_1)$ *with metric defined by the matrix* $\begin{pmatrix} 1 & 0 \\ 0 & 1 \end{pmatrix}$ *in the standard basis* $\xi_1 = \begin{pmatrix} 0 & 1 \\ 0 & 0 \end{pmatrix}$, $\xi_2 = \begin{pmatrix} 0 & 0 \\ 1 & 0 \end{pmatrix}$, $\xi_3 = \begin{pmatrix} 1 & 0 \\ 0 & -1 \end{pmatrix}$ *of the Lie algebra* $\mathfrak{sl}_2\mathbb{R}$ $\mathfrak{m} = \{z_\tau\}$ *is a* 1*-parameter subgroup of* $SL_2\mathbb{R}$ *with generator* ξ_3.

(3b) *Consider the nonholonomic Lie group* $(SL_2\mathbb{R}, V_1)$ *with metric defined in the standard basis by the matrix* $g = \begin{pmatrix} 1 & 0 \\ 0 & m \end{pmatrix}$, $m \neq 1$ (see Section 1.2). *Then the monodromy submanifold* $\mathfrak{M}$ *does not depend on m and is the union of three disjoint components*:

(a) *the* 1*-parameter subgroup* $\{z_\tau\}$ *with generator* $\xi_3 \in V_1^\perp$;

(b) *the ray of the* 1*-parameter subgroup*

$$\{\exp t\xi; t > T_0\}$$

with generator $\xi \in V_1$;

(c) *the ray* y_τ^- *of the same subgroup*,

$$\{\exp t\xi; t < -T_0\}.$$

The generator $\xi = \begin{pmatrix} 0 & 0 \\ 1 & 0 \end{pmatrix} \in V_1$ *for* $m < 1$ *and* $\xi = \begin{pmatrix} 0 & 1 \\ 0 & 0 \end{pmatrix}$ *for* $m > 1$.

Each of the components of $\mathfrak{M}$ *is homeomorphic to* $\mathbb{R}$.

(4) *For the nonholonomic Lie group* $(SL_2\mathbb{R}, V_2)$ *the submanifold* $\mathfrak{M}$ *consists of three components*, $\mathfrak{M} = \alpha \cup \beta^+ \cup \beta^-$, *where*

(a) $\alpha = \{\alpha_\tau\}_{\tau \in \mathbb{R}}$

is the 1*-parameter subgroup with generator* $\begin{pmatrix} 0 & \gamma \\ -\gamma & 0 \end{pmatrix}$ *which is homeomorphic to* S^1;

(b) $\beta^+ = \{\beta_\tau\}_{\tau \in (T_0, \infty)}$ *is the ray of the* 1*-parameter subgroup with generator* $\xi = \begin{pmatrix} 1 & 0 \\ 0 & -1 \end{pmatrix}$;

(c) $\beta^- = \{\beta_\tau\}_{\tau \in (-\infty, -T_0)}$ *is the ray of the same subgroup* $\{\exp t\xi\}$ *(the two last components are homeomorphic to* $\mathbb{R}^1$).

Remark. The metrics given by (3b) are called generic metrics on $(SL_2\mathbb{R}, V_1)$; the metric (3a) is called special.

2.2. Nonholonomic Geodesic Flow on SO(3). In the previous Section we described the nonholonomic geodesic flow on all simply connected three-dimensional Lie groups with the exception of the universal covering of SO(3), i.e. S^3. Let us describe the flow on S^3 beginning with the case of bi-invariant metrics. In this case Proposition 4, Section 2.1, shows that the monodromy submanifold is the circle S^1 in S^3 whose tangent vector is orthogonal to V and the monodromy map is the rotation along this circle by angle φ_t. Let ξ_1, ξ_2 be an orthonormal basis of V, $\xi_3 = [\xi_1, \xi_2] \perp V$. The next theorem describes the nonholonomic geodesic flow on SO(3) (and on S^3) with bi-invariant metrics.

Theorem 1 (on NG-flow for bi-invariant metrics). *Consider the group* SO(3) *with a bi-invariant metric and with a two-dimensional left-invariant nonholonomic distribution V. Let γ be the geodesic with initial data $(v_0, \lambda) \in S^1 \times V^\perp$. Then*

(1) *For $\lambda = 0$, γ is a 1-parameter subgroup $(\simeq S^1)$ of* SO(3) *with an admissible generator.*

(2) *If $\lambda \neq 0$ and φ_τ is commensurable with π, i.e. $\varphi_\tau = \dfrac{2\pi k}{l}$, $k, l \in \mathbb{Z}$, $l \neq 0$, then γ is a periodic trajectory. The mapping $\gamma: \mathbb{R} \to$* SO(3) *is given by*

$$\gamma(t) = \exp \frac{2k\pi}{l\tau} t\xi_3 \exp\left(\xi_1 \cos \frac{2\pi}{\tau} t + \xi_2 \sin \frac{2\pi}{\tau} t\right);$$

the geodesic γ lies on a two-dimensional topological torus in SO(3) *determined by the mapping*

$$\kappa: (\varphi, \psi) \in S^1 \times S^1 \mapsto \exp(\varphi\xi_3) \exp(\xi_1 \cos \psi t + \xi_2 \sin \psi t).$$

(3) *If φ_τ is incommensurable with π, the trajectory is everywhere dense on the torus $\kappa(S^1 \times S^1) \subset$* SO(3).

Assume now that the metric is left-invariant (but is not bi-invariant). In this case the monodromy submanifold consists of points z_τ each of which lies on its own 1-parameter subgroup $\Psi_\tau \simeq S^1_\tau$ and defines a rotation of this subgroup by angle φ_τ. We have the following obvious result.

Assertion. *The angle φ_τ is a continuous function of τ.*

The nonholonomic geodesic flow associated with a left-invariant metric on SO(3) is described as follows.

Theorem 2 (on NG-flow for left-invariant metrics on SO(3)). *Consider the group* SO(3) *with a left-invariant metric (which is not bi-invariant) and with a two-dimensional left-invariant nonholonomic distribution V. Let γ be the geodesic with initial data $(\varphi_0, \lambda) \in S^1 \times V^\perp$. Then*

(1) *If $(\varphi_0, \lambda) = \left(\dfrac{\pi k}{2}, 0\right)$, $k = 0, 1, 2, 3$, is a fixed point of system* (5) *on $S^1 \times \mathbb{R}$, the nonholonomic geodesic is a 1-parameter subgroup of* SO(3).

(2) *If the initial data of γ lie on a period τ curve, i.e. $(\varphi_0, \lambda) \in \alpha_\tau$, or $(\varphi_0, \lambda) \in \beta_\tau$, and the angle φ_τ is commensurable with π, γ is a periodic trajectory on* SO(3).

(3) *If $(\varphi_0, \lambda) \in \alpha_\tau$, or $(\varphi_0, \lambda) \in \beta_\tau$, and φ_τ is incommensurable with π, γ is an everywhere dense trajectory on a two-dimensional (topological) torus in* SO(3).

(4) *If (φ_0, λ) lies on a separatrix, the closure of the positive half-trajectory $\{\gamma_{\varphi_0,\lambda}(t), t > 0\}$ contains a left translate of a 1-parameter subgroup in* SO(3), *and the closure of the negative half-trajectory contains another left translate of the same 1-parameter subgroup.*

These translates of a 1-parameter subgroup are limit cycles ω_+, ω_- of the nonholonomic geodesic γ which approaches to them at an exponential rate.

We described the motion on the group SO(3), i.e. on the fiber of the mixed bundle $(V \oplus V^{\perp}) \times \mathrm{SO}(3)$. The motion in the mixed bundle is described by the following assertion.

Theorem 3. *Consider an arbitrary left-invariant metric on* SO(3). *The 4-dimensional manifolds* $\alpha_\tau \times \mathrm{SO}(3)$, $\beta_\tau \times \mathrm{SO}(3)$ *are invariant sets of the NG-flow. Each of these 4-dimensional components breaks down into the union of 2-dimensional and 1-dimensional invariant tori. The type of a trajectory is determined in the following way*:

(1) *If* γ *is a periodic nonholonomic geodesic on* SO(3), *then* $(\gamma, \dot{\gamma}, \lambda)$ *is a periodic trajectory in the mixed bundle.*

(2) *If the closure of* γ *is a two-dimensional topological torus, then the closure of* $(\gamma, \dot{\gamma}, \lambda)$ *is a 2-dimensional topological torus in* $\mathrm{SO}(3) \times (V \oplus V^{\perp})$.

In Section 2.1 we have shown that the nonholonomic geodesic flows on simply connected three-dimensional noncompact Lie groups are trivial. Geodesics escape to infinity and have no limit points. As in the classical (holonomic) case, the question on the nonholonomic geodesic flows on compact homogeneous spaces of Lie groups is much more interesting.

2.3. NG-Flow on Compact Homogeneous Spaces of the Heisenberg Group. The Heisenberg group N is the group of upper triangular matrices $\left\{ \begin{pmatrix} 1 & \alpha & \beta \\ 0 & 1 & \gamma \\ 0 & 0 & 1 \end{pmatrix}, \alpha, \beta, \gamma \in \mathbb{R}^1 \right\}$, its Lie algebra $\mathfrak{N}$ is the algebra of strictly upper triangular matrices $\begin{pmatrix} 0 & a & b \\ 0 & 0 & c \\ 0 & 0 & 0 \end{pmatrix}$ with basis

$$\xi_1 = \begin{pmatrix} 0 & 1 & 0 \\ 0 & 0 & 0 \\ 0 & 0 & 0 \end{pmatrix}, \quad \xi_2 = \begin{pmatrix} 0 & 0 & 0 \\ 0 & 0 & 1 \\ 0 & 0 & 0 \end{pmatrix}, \quad \xi_3 = \begin{pmatrix} 0 & 0 & 1 \\ 0 & 0 & 0 \\ 0 & 0 & 0 \end{pmatrix}.$$

The algebra $\mathfrak{N}$ may be identified with the subalgebra of the Lie algebra of vector fields in $\mathbb{R}^3$; this identification is specified by the correspondence

$$\xi_1 \mapsto \frac{\partial}{\partial x_1}, \quad \xi_2 \mapsto \frac{\partial}{\partial x_2} + x_1 \frac{\partial}{\partial x_3}, \quad \xi_3 \mapsto \frac{\partial}{\partial x_3}.$$

The Heisenberg group may be identified with $\mathbb{R}^3$:

$$\begin{pmatrix} 1 & \alpha & \beta \\ 0 & 1 & \gamma \\ 0 & 0 & 1 \end{pmatrix} \mapsto (\alpha, \gamma, \beta + \alpha\gamma) \in \mathbb{R}^3.$$

Under this identification, a left-invariant metric on N corresponds to a metric on $\mathbb{R}^3$ defined in the coordinates (x, y, z) by the tensor

$$g = \begin{bmatrix} 1 & 0 & 0 \\ 0 & (1 + x_1^2)^{-1/2} & 0 \\ 0 & 0 & 1 \end{bmatrix}.$$

Recall that a uniform subgroup of a Lie group G is a discrete subgroup such that the coset space G/D is compact. Uniform subgroups of the Heisenberg group form a 1-parameter series, the elements of which are labelled by positive integers $k \in \mathbb{N}$. The subgroup D_k is generated by three matrices

$$D_k = \left\{ \begin{bmatrix} 1 & 1 & 0 \\ 0 & 1 & 0 \\ 0 & 0 & 1 \end{bmatrix}, \begin{bmatrix} 1 & 0 & 0 \\ 0 & 1 & 1 \\ 0 & 0 & 1 \end{bmatrix}, \begin{bmatrix} 1 & 0 & \frac{1}{k} \\ 0 & 1 & 0 \\ 0 & 0 & 1 \end{bmatrix} \right\}$$

(see Auslender et al. [1963]).

Recall that up to an automorphism there exists only one non-holonomic distribution in the Heisenberg group, $V = \mathrm{Lin}(\xi_1, \xi_2)$. The monodromy submanifold for the Heisenberg group is a 1-parameter subgroup with generator ξ_3. The monodromy map for a geodesic γ with initial data (φ_0, λ) (and period $2\pi/\lambda$) is the right multiplication by $\begin{bmatrix} 1 & 0 & 4\pi/\lambda^2 \\ 0 & 1 & 0 \\ 0 & 0 & 1 \end{bmatrix}$ or, which amounts to the same, the shift by $4\pi/\lambda^2$ along the z-axis.

The nonholonomic geodesic flow on N is described by the following theorems.

Theorem 1 (on the motion in the fiber). *Let D_k be a uniform subgroup of the Heisenberg group N, γ a nonholonomic geodesic on N/D_k with initial data (φ_0, λ). Then*

(1) *If $\lambda = 0$, γ is the projection to N/D_k of a 1-parameter subgroup homeomorphic to S^1.*

If $\lambda \neq 0$ and λ^2 is commensurable with π, then γ is a periodic curve in N/D_k.

(3) *If $\lambda \neq 0$ and λ^2 is incommensurable with π, then γ is an everywhere dense trajectory on a two-dimensional (topological) torus in N/D_k.*

We described the NG-flow on the Heisenberg group. The flow on the mixed bundle $N/D_k \times (V \oplus V^\perp)$ is described by the following statement which is similar to the proposition of Section 2.2.

Proposition. *Let D_k be a uniform subgroup of N, and γ a nonholonomic geodesic on N/D_k with initial data (φ_0, λ). Then*

(1) *If $\lambda = 0$, the trajectory in $N/D_k \times (V \oplus V^\perp)$ is periodic (homeomorphic to S^1).*

(2) *If $\lambda \neq 0$, and λ^2 is commensurable with π, then $(\gamma, \dot{\gamma}, \lambda)$ is a periodic curve.*

(3) *If $\lambda \neq 0$ and λ^2 is incommensurable with π, the closure of the trajectory $(\gamma, \dot{\gamma}, \lambda)$ is a two-dimensional torus in $N/D_k \times (V \oplus V^\perp)$.*

Let us describe the dynamics in the mixed bundle.

The trajectory α_τ for the Heisenberg group is a circle S^1_λ determined by the condition $\lambda = 2\pi/\tau$. A complete description of the dynamics in the mixed bundle is given in the following theorem.

Theorem 2. *Let N be the Heisenberg group, V a two-dimensional left-invariant nonholonomic distribution on N, $D_k \subset N$ uniform subgroup. Then the four-dimensional manifolds $M_\lambda = N/D_k \times S_\lambda$ are invariant with respect to the NG-flow on the mixed bundle $N/D_k \times (V \oplus V^\perp)$. Other invariant sets are the projections to N/D_k of 1-parameter subgroups in N with admissible generators. The dynamics on the invariant manifolds is described as follows:*

(1) *If λ^2 is commensurable with π, the manifold M_λ consists of invariant closed trajectories.*

(2) *If $\lambda \neq 0$ and λ^2 is incommensurable with π, the manifold M_λ is the union of two-dimensional tori. On each of these tori the geodesics are uniform irrational windings (the winding angle is fixed on each torus).*

2.4. Nonholonomic Geodesic Flows on Compact Homogeneous Spaces of $\mathrm{SL}_2\mathbb{R}$. In Section 1.7, Chapter 1, we showed that the set of two-dimensional nonholonomic left-invariant distributions on $\mathrm{SL}_2\mathbb{R}$ consists of two orbits of the group of automorphisms of $\mathrm{SL}_2\mathbb{R}$. Fix a basis in $\mathfrak{sl}_2\mathbb{R}$:

$$\xi_1 = \begin{pmatrix} 0 & 1 \\ 0 & 0 \end{pmatrix}, \quad \xi_2 = \begin{pmatrix} 0 & 0 \\ 1 & 0 \end{pmatrix}, \quad \xi_3 = \begin{pmatrix} 1 & 0 \\ 0 & -1 \end{pmatrix}.$$

A representative of the first orbit is the distribution $V_1 = \mathrm{Lin}(\xi_1, \xi_2)$, a representative of the second is $V_2 = \mathrm{Lin}(\xi_1 + \xi_2, \xi_3)$. Recall that an element $a \in \mathfrak{sl}_2\mathbb{R}$ is called hyperbolic if a has real eigenvalues λ_1, λ_2 of opposite signs; a is called parabolic, if $\lambda_1 = \lambda_2 = 0$ and elliptic if λ_1, λ_2 are not real.

The elements of V_1 have the following type:

$$\begin{pmatrix} 0 & \alpha \\ \beta & 0 \end{pmatrix} \text{ is } \begin{cases} \text{hyperbolic for } \alpha\beta > 0, \\ \text{parabolic for } \alpha\beta = 0, \\ \text{elliptic for } \alpha\beta < 0. \end{cases}$$

$V_1^\perp = \left\{\begin{pmatrix} \gamma & 0 \\ 0 & -\gamma \end{pmatrix}\right\}$ consists of hyperbolic matrices. Nonzero matrices from V_2 are hyperbolic.

$V_2^\perp = \left\{\begin{pmatrix} 0 & \gamma \\ -\gamma & 0 \end{pmatrix}\right\}$ consists of elliptic matrices.

The nonholonomic geodesic flow on $(\mathrm{SL}_2\mathbb{R}, V)$ is rather closely related to the ordinary geodesic flow on the Lobatchevsky plane, as explained below. If D is a lattice in $\mathrm{SL}_2\mathbb{R}$ and $\mathrm{SL}_2\mathbb{R}/D$ is compact, this relation extends to the flow on $(\mathrm{SL}_2\mathbb{R}/D, V)$ and the geodesic flow on $\mathrm{SL}_2\mathbb{R}/D$.

Fix $a \in \mathfrak{sl}_2 \mathbb{R}$. The 1-parameter subgroup with generator a gives rise to a dynamical system on $\mathrm{SL}_2 \mathbb{R}$:

$$T_a^t\colon \mathrm{SL}_2 \mathbb{R} \to \mathrm{SL}_2 \mathbb{R}, \quad T_a^t x = x \exp(ta).$$

Recall that (see, for instance, Dubrovin et al. [1979], Cornfeld et al. [1982]) this dynamical system is naturally isomorphic either to the geodesic flow on the Lobachevsky plane if a is hyperbolic, or to the horocyclic flow if a is parabolic, or to the cyclic flow if a is elliptic.

Fix a Haar measure on $\mathrm{SL}_2 \mathbb{R}$. The properties of the quotient flows on $\mathrm{SL}_2 \mathbb{R}$ are described by the following theorem.

Theorem 1. (Cornfeld et al. [1982]) (1) *The geodesic flow on a compact surface* $X = S^1 \backslash \mathrm{Sl}_2 \mathbb{R}/D$ *is ergodic, mixing and, moreover, is a K-system.*

(2) *The horocyclic flow on X is ergodic.*

The nonholonomic geodesic flow on the mixed bundle $\mathrm{SL}_2 \mathbb{R}/D \times (V \oplus V^{\perp})$ is the semidirect product with base $S^1 \times V^{\perp}$ and fiber $\mathrm{SL}_2 \mathbb{R}/D$. The flow on the base is periodic (cf. Section 2.4). Fix a point $(\varphi_0, \lambda) \in S^1 \times V^{\perp}$ and consider the following dynamical system on $\mathrm{SL}_2 \mathbb{R}$:

$$S^t_{(\varphi_0, \lambda)}\colon \mathrm{SL}_2 \mathbb{R} \to \mathrm{SL}_2 \mathbb{R}\colon x \mapsto x\gamma_{(\varphi_0, \lambda)}(t)$$

where $\gamma_{(\varphi_0, \lambda)}$ is the nonholonomic geodesic with initial data (φ_0, λ). Observe that the Haar measure on $\mathrm{SL}_2 \mathbb{R}$ is invariant with respect to S^t. Assume that (φ_0, λ) does not lie on a separatrix (cf. Section 2.5). Then (φ_0, λ) lies on a periodic trajectory; let $\tau = \tau(\varphi_0, \lambda)$ be its period and z_τ its monodromy map.

Proposition 1. *Fix a point* $(\varphi_0, \lambda) \in S^1 \times V^{\perp}$, $\lambda \neq 0$, *which does not lie on a separatrix. The monodromy map is the right multiplication by an element* $z_\tau \in \mathrm{SL}_2 \mathbb{R}$ *which lies on the monodromy submanifold. Thus*

$$S^{n\tau}_{(\varphi_0, \lambda)} \simeq T^n_{z_\tau}$$

Recall that a cascade is a dynamical system with discrete time. Proposition 1 shows that the dynamical systems $S^{t\tau}_{(\varphi_0, \lambda)}$, T_{z_τ} have a common cascade.

The behaviour of the NG-flow S^t is to a large extent determined by the properties of the cascade generated by the monodromy S^τ and coinciding with the cascade T^n embedded into the flow T^t whose properties are determined by the type of the monodromy element z_τ. Combining Theorem 1 and Proposition 1 we get

Proposition 2. *Let D be a uniform subgroup in* $\mathrm{SL}_2 \mathbb{R}$. *Consider the flow* S^t *on* $\mathrm{SL}_2 \mathbb{R}/D$. *Let* (φ_0, λ) *be a point on a trajectory of period* τ *in* $S^1 \times V^{\perp}$; *let* z_τ *be the corresponding monodromy element. Then*

(1) *If* $\log z_\tau$ *is hyperbolic, the flow* S^t *is ergodic on* $\mathrm{SL}_2 \mathbb{R}/D$.

(2) *If* $\log z_\tau$ *is parabolic, the flow* S^t *is also ergodic.*

(3) *If* $\log z_\tau$ *is elliptic,* $\mathrm{SL}_2 \mathbb{R}/D$ *breaks up into the union of 2-dimensional and 1-dimensional invariant tori.*

Consider now the nonholonomic geodesic flow on the mixed bundle $SL_2\mathbb{R}/D \times (S^1 \times V^{\perp})$. The four-dimensional manifolds $SL_2\mathbb{R}/D \times \alpha_\tau$ and $SL_2\mathbb{R}/D \times \beta_\tau$ are invariant with respect to the flow. Denote by $\tilde{S}^t$ the restriction of the flow to any of these sets. Each of the sets $SL_2\mathbb{R}/D \times \alpha_\tau$, $SL_2\mathbb{R}/D \times \beta_\tau$ possesses an invariant measure $\frac{d\varphi}{\lambda} \times \mu$ (where μ is the Haar measure on $SL_2\mathbb{R}$) with respect to the flow $\tilde{S}^t$. The flow $\tilde{S}^t$ is the semidirect product of the flow on the circle α_τ, or β_τ, (base) and the flow on the homogeneous space $SL_2\mathbb{R}/D$ (fiber). The properties of $\tilde{S}^t$ are as follows.

Proposition 3. (1) *If the flow $S^t_{(\varphi,\lambda)}$ is ergodic on the fiber, then the flow $\tilde{S}^t$ is ergodic on* $SL_2\mathbb{R}/D \times S^1$.

(2) *If $SL_2\mathbb{R}/D$ breaks up into the union of 2-dimensional and 1-dimensional tori which are invariant with respect to $S^t_{(\varphi,\lambda)}$, then $SL_2\mathbb{R}/D \times S^1$ also breaks up into the union of 2-dimensional and 1-dimensional tori which are invariant with respect to $\tilde{S}^t$.*

Propositions 2, 3 describe the features of the nonholonomic geodesic (NG) flow on four-dimensional invariant submanifolds $SL_2\mathbb{R}/D \times \alpha_\tau$ and $SL_2\mathbb{R}/D \times \beta_\tau$ of the mixed bundle depending on the type of $\log z_\tau$ and $\log y_\tau$. The type of these elements for nonholonomic Lie groups $(SL_2\mathbb{R}, V_1)$, $(SL_2\mathbb{R}, V_2)$ may be extracted from the properties of V_1, V_2 described at the beginning of Section 2.4. Thus we get the following description of the NG-flow on these groups.

Theorem 2 (on the NG-flow on $(SL_2\mathbb{R}, V_1)$). *Fix a genric left-invariant metric on $(SL_2\mathbb{R}, V_1)$. Then the invariant sets of the NG-flow are*

(1) $(SL_2\mathbb{R}/D) \times \alpha_\tau$, $\tau \in (-\infty, \infty)$,

(2) $(SL_2\mathbb{R}/D) \times \beta_\tau$, $\tau \in (-\infty, -T_0) \cup (T_3, \infty)$,

(3) $(SL_2\mathbb{R}/D) \times \mathrm{sep}_i$, $i = 1, 2, 3, 4$,

where sep_i, $i = 1, \ldots, 4$, are the four separatrices of systems (5) *on the cylinder* (see Section 1.5 and Figure 1).

(4) *1-parameter subgroups which correspond to fixed points on the base with generators $\xi_1 = \begin{pmatrix} 0 & 1 \\ 0 & 0 \end{pmatrix}$, $\xi_2 = \begin{pmatrix} 0 & 0 \\ 1 & 0 \end{pmatrix}$. On each of the sets of types* (1), (2) *the flow is ergodic. Each of the trajectories of types* (3), (4) *is everywhere dense on* $SL_2\mathbb{R}/D$.

The next assertion describes the NG-flow on $(SL_2\mathbb{R}, V_1)$ with special metrics.

Proposition 4. *Fix a special metric on $(SL_2\mathbb{R}, V_1)$. Then there are two types of invariant sets for the NG-flow*:

(1) $SL_2\mathbb{R}/D \times \alpha_\tau$; *on this set the flow is ergodic.*

(2) *1-parameter subgroup with an arbitrary admissible generator $\begin{pmatrix} 0 & \alpha \\ \beta & 0 \end{pmatrix}$; this subgroup is everywhere dense on $SL_2\mathbb{R}/D$ if $\alpha\beta \geqslant 0$ and is periodic if $\alpha\beta < 0$.*

Let us now describe th NG-flows on compact homogeneous spaces of the nonholonomic group $(SL_2\mathbb{R}, V_2)$.

Theorem 3 (on the NG-flow on $(\mathrm{SL}_2\mathbb{R}, V_2)$). *Let D be a uniform subgroup of $\mathrm{SL}_2\mathbb{R}$. Then the structure of the NG-flow on invariant subsets described in Theorem 2 is as follows:*

(1) *$\mathrm{SL}_2\mathbb{R}/D \times \alpha_\tau$ breaks up into the union of 2-dimensional and 1-dimensional invariant tori.*

(2) *NG-flow on $\mathrm{SL}_2\mathbb{R}/D \times \beta_\tau$ is ergodic and is the semidirect product, with a flow on the base having a discrete spectrum and that on the fiber being a K-system.*

(3) *Each of the trajectories of types* (3), (4) *is everywhere dense on $\mathrm{SL}_2\mathbb{R}/D$.*

2.5. Nonholonomic Geodesic Flow on Some Special Multidimensional Nilmanifolds. Fix an integer $m \geqslant 0$. Define the maximal nilpotent algebra $\mathfrak{N}_m^2$ of degree 2 with m generators in the following way. As a linear space it is generated by the set of elements

$$\xi_1, \ldots, \xi_m, \{\xi_{i,j}\}, \quad 1 \leqslant i < j \leqslant m. \tag{10}$$

The Lie brackets are defined by

$$\begin{aligned} [\xi_i, \xi_j] &= \xi_{i,j} && \text{for } 1 \leqslant i < j \leqslant m, \\ [\xi_i, \xi_{j,k}] &= 0 && \text{for } 1 \leqslant i \leqslant m, 1 \leqslant j < k \leqslant m, \\ [\xi_{i,j}, \xi_{k,l}] &= 0 && \text{for } 1 \leqslant i < j \leqslant m, 1 \leqslant k < l \leqslant m. \end{aligned}$$

Denote by N_m^2 the corresponding Lie group. The group N_m^2 is maximal in the sense that any other nilpotent Lie group with m generators of degree 2 is its image under an appropriate homomorphism. Define on N_m^2 a left-invariant nonholonomic m-dimensional distribution $V = \mathrm{Lin}(\xi_1, \ldots, \xi_m)$.

Proposition 1. *The distribution V has the nonholonomicity degree 2 and any m-dimensional nonholonomic left-invariant distribution $\tilde{V}$ on N_m^2 such that $\tilde{V} + [\tilde{V}, \tilde{V}] = \mathfrak{N}_m^2$ may be transformed into V by an automorphism of N_m^2.*

It is also easy to show that all metric tensors on V lie on the same orbit of the group of automorphisms of (N_m^2, V). Let us choose a metric tensor in such a way that ξ_i, $i = 1, \ldots, m$, form an orthonormal basis of V.

The equations of nonholonomic geodesics in the present case are particularly simple. It is natural to denote the coordinates in the $m(m-1)/2$-dimensional space $V^\perp$ of Lagrange multipliers by λ_{ij}, $1 \leqslant i < j \leqslant m$. Put $\lambda_{ij} = -\lambda_{ji}$ for $i > j$, $\Lambda = (\lambda_{ij})$. Then

$$\begin{cases} \dot{\gamma} = \sum_{i=1}^{m} v_i \xi_i, \\ \dot{v}_i = \sum_{j=1}^{m} \lambda_{ij} v_j, \\ \dot{\lambda}_{ij} = 0. \end{cases} \tag{11}$$

System (11) may be conveniently rewritten in matrix form. Denote by A the m-dimensional vector with coordinates v_i, $i = 1, \ldots, m$. The last two groups of equations which determine the reduced flow on the base (see Section 2.1) take the form

$$\begin{cases} \dot{A} = \Lambda A, \\ \dot{\Lambda} = 0. \end{cases} \tag{12}$$

By an orthogonal transformation $\alpha \in \mathrm{SO}(V) \simeq \mathrm{SO}(m)$ the skew-symmetric matrix Λ may be transformed to the canonical form:

$$\Lambda = \begin{bmatrix} B_1 & & \\ & \ddots & \\ & & B_m \end{bmatrix}, \quad \text{where} \quad B_i = \begin{pmatrix} 0 & \omega_i \\ -\omega_i & 0 \end{pmatrix},$$

and $\{\pm i\omega_j\}$ is the set of eigenvalues (frequencies) of Λ. In the chosen basis the motion on V is described by the equation

$$\dot{\gamma} = \sum_{i=1}^{[m/2]} a_i \cos \omega_i(t + \tau_i)\xi_{2i-1} + a_i \sin \omega_i(t + \tau_i)\xi_{2i}. \tag{13}$$

From (13) it is clear that the behaviour of the trajectory γ on the base depends on commensurability of the frequencies ω_j. It is described by the following simple assertion.

Proposition 2. (1) *For a generic skew-symmetric matrix Λ the frequencies ω_j are mutually incommensurable. The trajectory A on V is an everywhere dense* (*and equidistributed*) *winding on a $[m/2]$-dimensional torus.*

(2) *If the frequencies ω_j break up into l groups of mutually commensurable numbers* (*while the numbers from different groups are incommensurable*), *the trajectory A is an everywhere dense winding on an l-dimensional torus.* (In particular, in the most degenerate case when all frequencies are commensurable the trajectory is periodic.)

Consider the motion on the fiber. The group N_m^2 is homeomorphic to $\mathbb{R}^M$, $M = m(m+1)/2$. The structure of the nonholonomic geodesic flow on N_m^2 is quite simple: all geodesics go off to infinity. It is very interesting to study the NG-flow on compact homogeneous spaces N_m^2/D, where D is a uniform subgroup (Auslander et al. [1963]). Let us use the identification $N_m^2 \simeq \mathbb{R}^M$ to get explicit formulae for nonholonomic geodesics which allow to study the motion on the fiber N_m^2/D.

Assume that the basis in V is so chosen that the matrix Λ has the canonical (block diagonal) form and that its frequencies are arranged in decreasing order $\omega_1 > \cdots > \omega_{[m/2]}$. Let us define a homomorphism $\varphi\colon \mathfrak{N}_m^2 \to \operatorname{Vect} \mathbb{R}^M$, $M = m + \dfrac{m(m-1)}{2}$. To this end fix a basis in $\mathbb{R}^M$ and label the coordinates in $\mathbb{R}^M$ in the same way as the basis in $\mathfrak{N}_m^2$:

$$\{x_i\}, \quad 1 \leqslant i \leqslant m, \quad \{x_{i,j}\}, \quad 1 \leqslant i < j \leqslant m,$$

We define φ on the basis in $\mathfrak{N}_m^2$ by

$$\varphi(\xi_i) = \frac{\partial}{\partial x_i} + \sum_{j=i+1}^{m} x_j \frac{\partial}{\partial x_{i,j}}, \quad 1 \leqslant i \leqslant m;$$

$$\varphi(\xi_{i,j}) = \frac{\partial}{\partial x_{i,j}}, \quad 1 \leqslant i < j \leqslant m.$$

The homomorphism φ allows to identify N_m^2 with $\mathbb{R}^M$. Denote by $\gamma_{A,\Lambda}$ the nonholonomic geodesic with the origin $0 \in \mathbb{R}^M$ and initial data (A, Λ). The curve $\gamma_{A,\Lambda}$ is determined by equation (13), and we may assume that $\tau_i = 0$ for $i = 1, \ldots, m$. (This may be achieved by an appropriate orthogonal transformation of V.) Equation (13) may be integrated explicitly.

Proposition 3. *Let Λ be a skew-symmetric matrix with nonzero frequencies $\omega_1, \ldots, \omega_m$ which are mutually incommensurable. Then the coordinates $x_i(\gamma_{A,\Lambda}(t))$ are given by the following formulae:*

(1) $x_{2i-1}(\gamma_{A,\Lambda}(t)) = \dfrac{a_i}{\omega_i} \sin \omega_i t,$

(2) $x_{2i}(\gamma_{A,\Lambda}(t)) = \dfrac{a_i}{\omega_i}(1 - \cos \omega_i t),$

(3) $x_{2i-1,2j-1}(\gamma_{A,\Lambda}(t)) = \dfrac{a_i a_j}{2\omega_i(\omega_i + \omega_j)}(1 - \cos(\omega_i + \omega_j)t) + \dfrac{a_i a_j}{2\omega_i(\omega_i - \omega_j)}(1 - \cos(\omega_i - \omega_j)t),$

(4) $x_{2i-1,2i}(\gamma_{A,\Lambda}(t) = \dfrac{a_i^2}{4\omega_i^2}(2\omega_i t - \sin 2\omega_i t),$

(5) $x_{2i-1,2j}(\gamma_{A,\Lambda}(t)) = \dfrac{a_i a_j}{\omega_i(\omega_j - \omega_i)} \sin(\omega_j - \omega_i)t - \dfrac{a_i a_j}{\omega_i(\omega_i + \omega_j)} \times \sin(\omega_i + \omega_j)t, \; i \neq j,$

(6) $x_{2i,2j}(\gamma_{A,\Lambda}(t)) = \dfrac{a_i a_j}{\omega_i(\omega_i - \omega_j)} \sin(\omega_i - \omega_j)t - \dfrac{a_i a_j}{\omega_i(\omega_i + \omega_j)} \times \sin(\omega_i + \omega_j)t.$

The subalgebra of $\mathfrak{N}_m^2$ spanned by $\xi_{2i-1}, \xi_{2i}, \xi_{2i-1,2i}$ is isomorphic to the Heisenberg algebra; denote it by $\mathfrak{N}_{(i)}$ and let $N_{(i)} \subset N_m^2$ be the corresponding Lie subgroup. The orthogonal projection $P_i: \mathbb{R}^M \to \mathbb{R}^3 \simeq N_{(i)}$ maps the nonholonomic geodesic flow on N_m^2 onto that on the Heisenberg group. The subgroups $N_{(i)}$ mutually commute. Denote their sum by $\tilde{N}$.

Proposition 4. *The nonholonomic geodesic flow on N_m^2 is the semidirect product of a flow T_1 on the base $\tilde{N}$ and a flow T_2 on the fiber $\mathbb{R}^{M-m-[m/2]}$. The flow T_1 is the direct product of nonholonomic geodesic flows on the Heisenberg groups $N_{(i)}$.*

References*

Agrachev, A.A., Vakhrameev, S.A., Gamkrelidze, R.V. [1983]: Differential geometry and group-theoretical methods in optimal control theory. Itogi Nauki Tekh., Ser. Probl. Geom., *14*, 3–56. English transl.: J. Sov. Math. *28*, 145–182 (1985), Zbl.542.93045

Akhiezer, N.I. [1981]: The Calculus of Variations. Vishcha Shkola: Kharkov. English transl: Harwood Academic Publishers: London, 1988, Zbl.507.49001

Aleksandrov, A.D. [1947a, b]: Geometry and topology in the USSR. I–II. Usp. Mat. Nauk *2*, No. 4, 3–58; *2*, No. 5, 9–92 (Russian), Zbl.36,379

Aleksandrov, A.D. [1958, 1959a, b]: On the principle of maximum. I–III. Izv. Vyssh. Uchebn. Zaved., Mat., 1958, No. 5, 126–157, Zbl.123,71; 1959, No. 3, 3–12; 195, No. 5, 16–32 (Russian), Zbl.125,59

Appell, P. [1953]: Traité de Mécanique Rationnelle, T.1, 2. Gauthier-Villars: Paris, Zbl.67,132

Arnol'd, V.I. [1974]: Mathematical Methods of Classical Mechanics. Nauka: Moscow. English transl.: Graduate Texts in Math. *60*. Springer-Verlag: New York-Berlin-Heidelberg, 1978, Zbl.386.70001

Arnol'd, V.I. [1983]: Singularities in the calculus of variations. Itogi Nauki Tekh., Ser. Sovrem. Probl. Mat. *22*, 3–55. English transl.: J. Sov. Math. *27*, 2679–2713 (1984), Zbl.537.58012

Arnol'd, V.I. [1988]: Geometrical Methods in the Theory of Ordinary Differential Equations. 2nd ed., Die Grundlehren der Math. Wissenschaften *250*. Springer-Verlag: New York-Berlin-Heidelberg, Zbl.507.34003

Arnol'd, V.I., Givental, A.B. [1985]: Symplectic geometry. In: Dynamical Systems IV. Itogi Nauki Tekh. Ser. Sovrem. Probl. Mat., Fundam. Napravleniya *4*, 7–139. English transl. in: Encyclopaedia of Math. Sciences *4*, 1–136. Springer-Verlag: Berlin-Heidelberg-New York, 1988, Zbl.592.58038

Arnol'd, V.I., Kozlov, V.V., Nejshtadt, A.I. [1985]: Mathematical aspects of classical and celestial mechanics. In: Dynamical Systems III. Itogi Nauki Tekh., Ser. Sovrem. Probl. Mat., Fundam. Napravleniya *3*, 5–304. English transl.: Encyclopaedia of Math. Sciences *3*. Springer-Verlag: Berlin-Heidelberg-New York, 1988, Zbl.612.70002

Auslander, L., Green, L., Hahn, F., Markus, L. and Massey, W. [1963]: Flows on Homogeneous Spaces. Ann. Math. Stud. *53*. Princeton University Press: Princeton, Zbl.106,368

Bennequin, D. [1983]: Entrelacement et équations de Pfaff. Astérisque *107/108*, 87–161, Zbl.573.58022

Berezin, F.A. [1974]: Hamiltonian formalism in the general Lagrange problem. Usp. Mat. Nauk *29*, No. 3, 183–184 (Russian), Zbl.307.70020

Birkhoff, G. [1967]: Lattice Theory. Am. Math. Soc. Colloq. Publ. *25*. Am. Math. Soc.: Providence, Zbl.153,25

Bishop, R.L., Crittenden, R.J. [1964]: Geometry of Manifolds. Academic Press: New York, Zbl.132,160

Boothby, W.M., Wang, H.C. [1958]: On contact manifolds. Ann. Math., II. Ser. *68*, 721–734, Zbl.84,392

Bourbaki, N. [1972]: Groupes et Algèbres de Lie, Ch. 1–3. Hermann: Paris. English transl.: Lie Groups and Lie Algebras. Chap. 1–3, Springer-Verlag: Berlin-Heidelberg-New York, 1981, Zbl.244.22007

Brendelev, V.N. [1978]: On commutation relations in nonholonomic mechanics. Vestn. Mosk. Univ., Ser. I, 1978, No. 6, 47–54. English transl.: Mosc. Univ. Mech. Bull. *33*, No. 5–6, 30–35 (1978), Zbl.401.70017

*For the convenience of the reader, references to reviews in Zentralblatt für Mathematik (Zbl.), compiled using the MATH database, and Jahrbuch über die Fortschritte der Mathematik (Jbuch) have, as far as possible, been included in this bibliography.

Bröcker, Th. [1975]: Differential Germs and Catastrophes. Cambridge University Press: Cambridge, Zbl.302.58006
Brockett, R. [1982]: Control theory and singular Riemannian geometry. In: New Directions in Applied Mathematics, Pap. present. 1980 on the ocass. of the Case centen. Celebr., (Hilton, P.J., Young, C.S. (Eds.)), 11–27, Zbl.483.49035
Brockett, R.W., Millman, R.S., and Sussmann, H.J. (eds) [1982]: Differential Geometric Control Theory. Progress in Math. *27*, Birkhäuser: Boston-Basel-Stuttgart, Zbl.503.00014
Carathéodory, C. [1909]: Untersuchungen über die Grundlagen der Thermodynamik. Math. Ann. *67*, 355–386
Cartan E. [1926]: Les groupes d'holonomie des espaces généralisés. Acta Math. *48*, 1–42, Jbuch 52, 723
Cartan, E. [1935]: La Méthode du Repère Mobile, la Théorie des Groupes Continus et les Espaces Généralisés. Hermann: Paris, Zbl.10,395
Chaplygin S.A. [1949]: Papers on Dynamics of Nonholonomic Systems. GITTL: Moscow-Leningrad (Russian)
Chow, W.L. [1939]: Systeme von linearen partiellen Differential-gleichungen erster Ordnung. Math. Ann. *117*, 98–105, Zbl.22,23
Cornfel'd, I.P., Fomin, S.V., Sinai, Ya.G. [1980]: Ergodic Theory. Nauka: Moscow. English transl.: Die Grundlehren der Mathematischen Wissenschaften *245*. Springer-Verlag: New York-Berlin-Heidelberg 1982, Zbl.493.28007
Davydov, A.A. [1985]: The attainability boundary is quasi-Hölderian. Tr. Semin. Vectorn. Tensorn. Anal., Prilozh. Geom. Mekh. Fiz. *22*, 25–30. English transl.: Sel. Math. Sov. *9*, No. 3, 229–234 (1990), Zbl.593.49028
Dirac, P.A.M. [1964]: Lectures on Quantum Mechanics. Yeshiva University: New York
Dobronravov, V.V. [1970]: Elements of Mechanics of Nonholonomic Systems. Vyssh. Shkola: Moscow (Russian), Zbl.216,248
Dubrovin, B.A., Fomenko, A.T., Novikov, S.P. [1979]: Modern Geometry. Methods and Applications I. Nauka: Moscow. English transl.: Graduate Texts in Math. *93*. Springer-Verlag: New York-Berlin-Heidelberg, 1984, Zbl.433.53001
Fajbusovich, L. [1988]: A Hamiltonian formalism for nonholonomic dynamical systems connected with generalized Toda flows. In: Nov. Global'nom Anal. 1988, 167–171 (Russian), Zbl.706.58058
Fefferman, C., Phong, D.H. [1979]: On the lowest eigenvalue of pseudodifferential operators. Proc. Natl. Acad. Sci. USA *76*, 6055–6056, Zbl.434.35071
Fefferman, C., Sanchez-Calle, A. [1986]: Fundamental solution for second order subelliptic operators. Ann. Math., II. Ser. *124*, 247–272, Zbl.613.35002
Filippov, A.F. [1959]: On certain questions of the optimal control theory. Vestn. Mosk. Univ., Ser. Mat. Mech. Astron. Fiz. Chim. *14*, No. 2, 25–32 English transl.: J. Soc. Ind. Appl. Math., Ser. A: Control 1 (1962), 76–84 (1964), Zbl.90,69
Folland, G.E. [1973]: A fundamental solution for a subelliptic operator. Bull. Am. Math. Soc. *79*, 373–376, Zbl.256.35020
Folland, G.E. [1975]: Subelliptic estimates and function spaces on nilpotent Lie groups. Ark. Mat. *13*, 161–207, Zbl.312.35026
Folland, G.E., Stein, E.M. [1974]: Estimates for $\bar{\partial}$-complex and analysis on the Heisenberg group. Commun. Pure Appl. Math. *27*, 429–522, Zbl.293.35012
Franklin, P., Moore, C.L.E. [1931]: Geodesics of Pfaffians. J. Math. Phys. *10*, 157–190, Zbl.2,412
Gardner, R.B. [1967]: Invariants of Pfaffian systems. Trans. Am. Math. Soc. *126*, 514–533, Zbl.161,413
Gardner, R.B. [1983]: Differential geometric methods interfacing control theory. In: Differential Geometric Control Theory. Proc. Conf. Mich. Technol. Univ. 1982, Prog. Math. *27*, 117–180, Zbl.522.49028
Gaveau, B. [1976]: Systèmes dynamiques associés à certains opérateurs hypoelliptiques. Bull. Sci. Math. II. Ser. *102*, 203–229, Zbl.391.35019
Gaveau, B. [1977]: Principe de moindre action, propagation de la chaleur et estimées sous elliptiques sur certains groupes nilpotents. Acta Math. *139*, 95–153, Zbl.366.22010

Gaveau, B., Greiner, P., Vantier, J. [1985]: Intégrales de Fourier quadratiques et calcul symbolique exact sur le groupe d'Heisenberg. J. Funt. Anal. *68*, 248–272, Zbl.609.43007

Gaveau, B., and Vauthier, J. [1986]: The Dirichlet problem on the Heisenberg group III. Can. J. Math. *38*, No. 3, 666–671, Zbl.603.22003

Gel'fand, I.M., Fomin, S.V. [1961]: Variational Calculus. Fizmatgiz: Moscow. English transl.: Prentice Hall: Englewood Cliffs, N.J., 1965, 3rd ed., Zbl.127,54

Gershkovich, V.Ya. [1984a]: Two-sided estimates of metrics generated by absolutely non-holonomic distributions on Riemannian manifolds. Dokl. Akad. Nauk SSSR *278*, No. 5, 1040–1044. English transl.: Sov. Math., Dokl. *30*, 506–510 (1984), Zbl.591.53033

Gershkovich, V.Ya. [1984b]: A variational problem with nonholonomic constraints on SO(3). Geometriya i Topologiya v Global'nykh Nelinejnykh Zadachakh, Nov. Global'nom Anal. 1984, 149–152, Zbl.554.58021

Gershkovich, V.Ya. [1985]: The method of penalty metrics in problems with nonholonomic constraints. In: Applications of Topology in Modern Analysis, 138–145. Voronezh University: Voronezh (Russian)

Gershkovich, V. [1988]: On normal form of distribution jets. In: Topology and Geometry, Rokhlin Semin. 1984–1986, Lect. Notes Math. *1346* (Viro, O.Ya. (Ed.)), 77–98. Zbl.649.58004

Gershkovich, V. [1989]: Short time hypoelliptic diffusion. Proc. of the 5th Internat. Conf. on Probability Theory, Vilnius, Vol. 3, 130–131

Gershkovich, V. [1990]: Estimates of nonholonomic ε-balls. In: Lect. Notes Math. *1543* (Borisovich, Yu.G. Gliklikh, Yu.E. (Eds.))

Gershkovich, V., Vershik, A. [1988]: Nonholonomic manifolds and nilpotent analysis. J. Geom. Phys. *5*, No. 3, 407–452, Zbl.693.53006

Gibbs, J.W. [1873]: Graphical methods in the thermodynamics of fluids. Trans. Connec. Acad. *2*, 309–342, Jbuch 5, 585

Godbillon, C. [1969]: Géométrie Différentielle et Mécanique Analytique. Collection Méthodes. Hermann: Paris, Zbl.174,246

Golubitsky, M., Guillmin, V. [1973]: Stable Mappings and their Singularities. Graduate Texts in Mathematics *14*. Springer-Verlag: New York, Berlin-Heidelberg Zbl.294.58004

Gorbatenko, E.M. [1985]: Differential geometry of nonholonomic manifolds (after V.V. Vagner). Geom. Sbornik, 31–43, Tomsk University: Tomsk

Goursat, E. [1922]: Leçons sur le problème de Pfaff. Hermann: Paris, Jbuch 48, 538

Gray, J.W. [1959]: Some global properties of contact structures. Ann. Math., II. Ser. *69*, No. 2, 421–450, Zbl.92,393

Griffiths, P.A. [1983]: Exterior Differential Systems and the Calculus of Variations. Progress in Math. *25*. Birkhäuser: Boston-Basel-Berlin, Zbl.512.49003

Grigoryan, A.T., Fradlin, B.N. [1977]: Mechanics in USSR. Nauka: Moscow (Russian)

Grigoryan, A.T., Fradlin, B.N. [1982]: A History of Mechanics of the Rigid Body. Nauka: Moscow (Russian)

Gromoll, D., Klingenberg, W., Meyer, W. [1968]: Riemannsche Geometrie im Grossen. Lect. Notes Math. *55*, Zbl.155,307

Gromov, M. [1981a]: Groups of polynomial growth and expanding maps. Publ. Math., Inst. Hautes Etud. Sci. *53*, 53–78, Zbl.474.20018

Gromov, M. [1981b]: Structures métriques pour les variétés Riemanniennes. Redigé par J. Lafontaine et P. Pansu. Textes Mathématiques, 1. Cedic/Nathan. VII: Paris, Zbl.509.53034

Günter, N.M. [1941]: Lectures on the Calculus of Variations. ONTI: Leningrad-Moscow (Russian)

Hartman, P. [1964]: Ordinary Differential Equations. J. Wiley & Sons: New York-London-Sydney, Zbl.125,321

Haynes, G.W., Hermes, H. [1970]: Nonlinear controllability via Lie theory. SIAM J. Control Optimization, *8*, 450–460, Zbl.229.93012

Helffer, B., Nourrigat, J. [1985]: Hypoellipticité Maximale pour des Opérateurs Polynômes de Champs de Vecteurs. Progress in Math. *58*. Birkhäuser: Basel-Boston-Stuttgart, Zbl.568.35003

Helgason, S. [1978]: Differential Geometry, Lie Groups, and Symmetric Spaces. Academic Press: New York-London, Zbl.451.53038

Hertz, H.R. [1894] Die Prinzipien der Mechanik. Gesammelte Werke, Bd. 2. J.A. Barth: Leipzig, Jbuch 25, 1310
Hörmander, L. [1967]: Hypoelliptic second order differential equations. Acta Math. *119*, 147–171, Zbl.156,107
Hörmander, L., Melin, A. [1978]: Free systems of vector fields. Ark. Mat. *16*, 83–88, Zbl.383.35013
Hueber, H., Müller, D. [1989]: Asymptotics for some Green kernels of the Heisenberg group and the Martin boundary. Math. Ann. *283*, 97–119, Zbl.639.31005
Hulanicki, A. [1976]: The distribution of energy in the Brownian motion in Gaussian field and analytic hypoellipticity of certain subelliptic operators on the Heisenberg group. Stud. Math. *56*, 165–173, Zbl.336.22007
Jakubczyk, B., Przytycki, F. [1984]: Singularities of k-tuples of vector fields. Diss. Math. *213*, Zbl.565.58007
Judjevic V., Sussmann, H.J. [1972]: Controllability of nonlinear systems. J. Differ. Equations *12*, 95–116, Zbl.242.49040
Kannai, Y. [1977]: Off diagonal short time asymptotics for fundamental solutions of diffusion equations. Commun. Partial Differ. Equations *2*, 781–830, Zbl.381.35039
Karapetyan, A.V. [1981]: On realization of nonholonomic constraints by viscous friction forces and celtic stones stability. Prikl. Mat. Mekh. *45*, 542–51. English transl.: J. Appl. Math. Mech. *45*, 30–36 (1982), Zbl.493.70008
Kobyashi, S., Nomizu, K. [1963, 1969]: Foundations of Differential Geometry, Vol. 1, 2. Interscience Publ.: New York-London, Zbl.119,375, Zbl.175,485
Kozlov, V.V. [1982a]: Dynamics of systems with non-integrable constraints. I. Vestn. Mosk. Univ., Ser. I, No. 3, 92–100. English transl., Mosc. Univ. Mech. Bull. *37*, No. 3–4, 27–34 (1982), Zbl.501.70016
Kozlov, V.V. [1982b]: Dynamics of systems with non-integrable constraints. II. Vestn. Mosk. Univ., Ser. I, No. 4, 70–76. English transl. Mosc. Univ. Mech. Bull. *37*, No. 3–4, 74–80 (1982), Zbl.508.70012
Kozlov, V.V. [1983]: Dynamics of systems with constraints. III. Vestn. Mosk. Univ., Ser. I, No. 3, 102–111. English transl.: Mosc. Univ. Mech. Bull. *38*, No. 3, 40–51 (1983), Zbl.516.70017
Kumpera, A., Ruisz, C. [1982]: Sur l'èquivalence locale des systèmes de Pfaff en drapeau. In: Monge-Ampère Equations and Related Topics. Proc. Semin., Firenze, 1980, 201–248, Zbl.516.58004
Lanczos, C. [1949]: The Variational Principles of Mechanics. Univ. Toronto Press: Toronto, Zbl.37,399
Laptev, G.F. [1953]: Differential geometry of embedded manifolds. Tr. Mosk. Mat. O-va *2*, 275–382 (Russian), Zbl.53,428
Lefshetz, S. [1957]: Differential Equations. Geometric Theory. Interscience Publishers: New York-London-Sydney, Zbl.80,64
Levitt, N., Sussmann, H.J. [1975]: On controllability by means of two vector fields. SIAM J. Control Optimization, *13*, No. 6, 1271–1281, Zbl.313.93006
Lobry, C. [1973]: Dynamical polysystems and control theory. In: Geom. Methods Syst. Theory, Proc. NATO Adv. Study Inst. London, 1–42, Zbl.279.93012
Lumiste Yu.G. [1966]: Connections in homogeneous bundles. Mat. Sb., Nov. Ser. *69* (111), 434–469. English transl.: Am. Math. Soc., Transl., II. Ser. *92*, 231–274 (1970), Zbl.143,446
Lutz, R. [1974]: Structures de contact en codimension quelconque. In: Géométrie Différentielle, Colloque, Santiago de compostella, Lect. Notes Math. *392* (Vidal, E. (Ed.)), 23–29, Zbl.294.5303
Manin, Yu.I. [1984]: Gauge Field Theory and Complex Geometry. Nauka: Moscow. English transl.: Die Grundlehren der Mathematischen Wissenschaften *289*. Springer-Verlag: New York-Berlin-Heidelberg, 1988, Zbl.576.53002
Martinet, I. [1971]: Formes de contact sur les variétés de dimension 3. In: Proc. Liverpool Singularities-Symp. II, Lect. Notes Math. *209* (Wall, C.T.C. (Ed.)), 142–163, Zbl.215,230
Melin, A. [1983]: Parametrix constructions for right invariant differential operators on nilpotent groups. Ann. Global Anal. Geom. *1*, No. 1, 79–130, Zbl.524.58044
Menikoff, A., Sjøstrand, J. [1978]: On the eigenvalues of a class of hypoelliptic operators. Math. Ann. *235*, 55–85, Zbl.375.35014

Métivier, G. [1976]: Fonction spectrale et valeurs propres d'opérateurs non elliptiques. C. R. Acad. Sci. Paris, Ser. *A283*, 453–456, Zbl.337.35057

Milnor, J. [1963]: Morse Theory. Ann. Math. Studies *51*. Princeton Univ. Press: Princeton N. J., Zbl.108,104

Mitchell, J. [1985]: On Carnot-Caratheodory metrics. J. Differ. Geom. *21*, 35–45, Zbl.554.53023

Mormul, P. [1988]: Singularities of vector fields in $\mathbb{R}^4$. Institute of Math. of the Polish Academy of Sciences, Preprint No. 420: Warszawa

Nachman, A. [1982]: The wave equation on the Heisenberg group. Commun. Partial Differ. Equations 7, 675–714, Zbl.524.35065

Nagano, T. [1966]: Linear differential systems with singularities and an application to transitive Lie algebras. J. Math. Soc. Japan. *18*, No. 4, 398–404, Zbl.147,235

Nagel, A., Stein, E.M., Wainger, S. [1985]: Balls and metrics defined by vector field. I. Acta Math. *155*, No. 1–2, 103–147, Zbl.578.32044

Nejmark, Yu.I., Fufaev, N.A. [1967]: Dynamics of Nonholonomic Systems. Nauka: Moscow (Russian), Zbl.171,455

Pansu, P. [1983]: Croissance des boules et des géodésiques fermées dans les nilvariétés. Ergodic Theory Dyn. Syst. *3*, 415–445, Zbl.509.53040

Rashevsky, P.K. [1938]: Any two points of a totally nonholonomic space may be connected by an admissible line. Uch. Zap. Ped. Inst. im. Liebknechta. Ser. Phys. Math., Vol. 2, 83–94 (Russian)

Rashevsky, P.K. [1947]: A Geometric Theory of Partial Differential Equations. ONTI: Moscow-Leningrad (Russian), Zbl. 36,64

Rashevsky, P.K. [1954]: Linear differential geometrical objects. Dokl. Akad. Nauk SSSR *97*, 606–611 (Russian), Zbl.57,134

Rotschild, L., Stein, E.M. [1976]: Hypoelliptic differential operators and nilpotent groups. Acta Math. *137*, 247–320, Zbl.346.35030

Routh, E.G. [1905]: Dynamics of a System of Rigid Bodies, Vol. 1. McMillan & Co: London, Jbuch. 36, 749

Rund, H. [1959]: Differential Geometry of Finsler Spaces. Die Grundlehren der Math. Wissenschaften *101*. Springer-Verlag: New York-Berlin-Heidelberg, Zbl.87,366

Sanchez-Calle, A. [1984]: Fundamental solutions and geometry of the sum of squares of vector fields. Invent. Math. *78*, 143–160, Zbl.582.58004

Shcherbakov, R.N. [1980]: Twenty issues of Geometricheski Sbornik (a bibliography). In: Geom. Sb. *21*, 79–90 (Russian)

Schoenberg, I. [1931]: Notes on geodesics in higher spaces. Bull. Am. Math. Soc. *37*, No. 5, 345, Jbuch 57, 967

Schouten, J.A. [1928]: On nonholonomic connections. Proc. Amsterdam, Nederl. Akad. Wetensch., Ser. A, *31*, No. 3, 291–299, Jbuch 54, 758

Schouten, J.A., van Kampen, E.R. [1930]: Zur Einbettung und Krümmungstheorie nichtholonomer Gebilde. Math. Ann. *103*, 752–783, Jbuch 56, 635

Schouten, J.A., Kulk, V.D. [1949]: Pfaff's Problem and its Generalization. Clarendon Press: Oxford, Zbl.33,369

Sidorov, A.F., Shapeev, V.P., Yanenko, N.N. [1984]: Method of Differential Constraints and its Applications to Gas Dynamics. Nauka: Moscow (Russian), Zbl.604.76062

Sintsov, D.M. [1972]: Papers on Nonholonomic Geometry. Vischa Shkola: Kiev (Russian)

Smirnov, V.I. [1957]: Lectures on Higher Mathematics, Vol. 4 (3rd edition). GITTL: Moscow. German transl.: Berlin (1958), Zbl.44,320

Solov'ev, A. [1982]: Second fundamental form of a distribution. Math. Zametki *31*, No. 1, 139–146. English transl.: Math. Notes *31*, 71–75 (1982), Zbl.492.53028

Solov'ev, A. [1984]: Curvature of a distribution. Mat. Zametki *35*, 111–124. English transl.: Math. Notes *35*, 61–68 (1984), Zbl.553.53017

Solov'ev, A. [1988]: On Pontryagin classes of completely geodesic subbundle. Sib. Mat. Zh. *29*, No. 3, 216–219. English transl.: Sib. Math. J. *29*, 510–512 (1988), Zbl.654.57012

Stein E.M. [1987]: Problems in harmonic analysis related to curvature and oscillatory integrals. Proc. Int. Congr. Math., Berkeley, USA 1986, Vol. 1, 196–221, Zbl.718.42012

Sternberg, S. [1964]: Lectures on Differential Geometry. Prentice Hall: Englewood Cliffs, Zbl.129,131
Strichartz, R.S. [1978]: Sub-Riemannian geometry. J. Differ. Geom. *24*, 221–263, Zbl.609.53021
Suslov, D.K. [1946]: Theoretical Mechanics. ONTI: Moscow-Leningrad (Russian)
Sussmann, H.J. [1973]: Orbits of families of vector fields and integrability of distributions. Trans. Am. Math. Soc. *180*, 171–188, Zbl.274.58002
Sussmann, H.J. [1983]: Lie brackets and local controllability: a sufficient condition for scalar-input systems. SIAM J. Control, Optimization *21*, 686–713, Zbl.523.49026
Synge, J.L. [1927]: On geometry of dynamics. Philos. Trans. R. Soc. Lond., Ser. A, *226*, No. 2, 31–107
Synge, J.L. [1928]: Geodesics in non-holonomic geometry. Math. Ann. *99*, 738–751, Jbuch 54, 758
Synge, J.L. [1936]: Tensorial Methods in Dynamics. Toronto University Press: Toronto, Zbl.17,41
Taylor, M. [1984]: Noncommutative Microlocal Analysis. Part I. Mem. Am. Math. Soc. No. 313. Am. Math. Soc.: Providence, Zbl.554.35025
Taylor, M. [1986]: Noncommutative Harmonic Analysis. Mathematical Surveys Monographs, No. 22. Am. Math. Soc.: Providence, Zbl.604.43001
Taylor, T.J.S. [1989a]: Some aspects of differential geometry associated with hypoelliptic second order operators. Pac. J. Math. *136*, No. 2, 355–378, Zbl.698.35041
Taylor, T.J.S. [1989b]: Off diagonal asymptotics of hypoelliptic diffusion equations and singular Riemannian geometry. Pac. J. Math. *136* (2), 379–399, Zbl.692.35011
Treves, F. [1980]: Introduction to Pseudodifferential and Fourier Integral Operators. Plenum Press: New York-London, Zbl.453.47027
Ulam, S.M. [1960]: A Collection of Mathematical Problems. Interscience Publ.: New York-London, Zbl.86,241
Vagner, V.V. [1935]: Differential geometry of nonholonomic manifolds. Tr. Semin. Vectorn. Tenzorn. Anal. *213*, 269–314 (Russian), Zbl.12,318
Vagner, V.V. [1940]: Differential geometry of nonholonomic manifolds. In: The VIII-th International Competition for the N.I. Lobatschewski Prize. Report. The Kazan Physico-Mathematical Society: Kazan
Vagner, V.V. [1941]: A geometric interpretation of nonholonomic dynamical systems. Tr. Semin. Vectorn. Tenzorn. Anal. *5*, 301–327
Vagner, V.V. [1965]: Geometria del calcolo delle variazioni, Vol. 2. C.I.M.E.: Roma
Varchenko, A.N. [1981]: On obstructions to local equivalence of distributions. Mat. Zametki *29*, No. 6, 939–947. English transl.: Math. Notes *29*, 479–484 (1981), Zbl.471.58004
Varopoulos, N.Th. [1986]: Analysis on nilpotent Lie groups. J. Funct. Anal. *66*, No. 3, 406–431, Zbl.595.22008
Varopoulos, N.Th. [1986]: Analysis on Lie groups. J. Funct. Anal. *76*, No. 2, 346–410, Zbl.634.22008
Varopoulos, N.Th. [1988]: Estimations du noyau de la chaleur sur une variété. C. R. Acad. Sci., Paris, Ser. I *307*, 527–529, Zbl.649.58032
Varopoulos, N.T. [1988]: Estimations gaussiennes du noyau de la chaleur sur les variétés. C. R. Acad. Sci., Paris, Ser. I *307*, 861–863, Zbl.671.58039
Vasil'ev, A.M., Efimov, N.N., Rashevsky, P.K. [1967]: The studies of differential geometry in Moscow University in the Soviet period. Vestn. Mosk. Univ., Ser I *22*, No. 5, 12–23 (Russian), Zbl.155,9
Vershik, A.M. [1984]: Classical and nonclassical dynamics with constraints. In: Geometriya: Topologiya v Global'nykh Nelinejnykh Zadachakh, Nov. Global'nom Anal. 1984, 23–48. English transl.: Lect. Notes Math. *1108*, 278–301 (1984), Zbl.538.58012
Vershik, A., Berestouski, V. [1992]: Manifolds with intrinsic metrics and non-holonomic spaces. Advances in Soviet Math. *9*, 255–267
Vershik, A.M., Chernyakov, A.G. [1982]: Fields of convex polyhedra and the Pareto-Smale optimum. Optimizatsiya *28* (45), 112–146 (Russian), Zbl.498.90065
Vershik, A.M., Faddeev, L.D. [1972]: Differential geometry and Lagrangian mechanics with constraints. Dokl. Akad Nauk SSSR *202*, No. 3, 555–557. English transl.: Sov. Phys. Dokl. *17*, 34–36 (1972), Zbl.243.70014

Vershik, A.M., Faddeev, L.D. [1975]: Lagrangian mechanics in invariant form. In: Probl. Theor. Phys., Vol. II, 129–141. English transl.: Sel. Math. Sov. *1*, 339–350, Zbl.518.58015

Vershik, A.M., Gershkovich, V.Ya. [1986a]: Nonholonomic problems and geometry of distributions. In: Griffiths, P.A. Exterior differential systems and the calculus of variations (Russian translation), 317–349. MIR: Moscow. English transl.: Acta Appl. Math. *12*, No. 2, 181–209, 1988, Zbl.666.58004

Vershik, A.M., Gershkovich, V.Ya. [1986b]: Geodesic flows on $SL_2\mathbb{R}$ with nonholonomic constraints. Zap. Nauchn. Semin. Leningr., Otd. Mat. Inst. Steklova *155*, 7–17. English transl.: J. Sov. Math. *41*, No. 2, 891–898 (1988), Zbl.673.58034

Vershik, A., Gershkovich, V. [1987]: The geometry of the nonholonomic sphere of three-dimensional Lie groups. In: Geometriya i Teoriya Osobennosej v Nelinnejnykh Uravneniyakh, Nov. Global'nom Anal. 1987, 61–75. English transl.: Lect. Notes Math. *1334* (Borisovich, Yu.G., Gliklikh, Yu.E. (Eds.)), 309–331 (1988), Abl.637.53061

Vershik, A., Gershkovich, V. [1988]. Determination of the functional dimension of the orbit space of generic distributions. Mat. Zametki *44*, 596–603. English transl.: Math. Notes *44*, 806–810 (1988), Zbl.694.58005

Vershik, A., Gershkovich, V. [1989]: A nilpotent Lie algebra bundle over a nonholonomic manifold (Nilpotentization). Zap. Nauchn. Semin. Leningr., Otd. Mat. Inst. Steklova *172*, 21–40 (Russian), Zbl.705.58004

Vershik, A., Granichina, O. [1991]: Reduction of nonholonomic variational problems to the isoperimetric problems and connections in principal bundles. Mat. Zametki *49*, 37–44

Veselov, A.P., Veselova, L.E. [1986]: Flows on Lie groups with nonholonomic constraints and integrable non-Hamiltonian systems Funkts. Anal. Prilozh. *20*, No. 4, 65–66. English transl.: Funct. Anal. Appl. *20*, 308–309, Zbl.621.58024

Veselov, A., Veselova, L. [1988]: Integrable nonholonomic systems on Lie groups. Math. Zametki *44*, No. 5, 604–619. English transl.: Math. Notes *44*, No. 5, 810–819 (1988), Zbl.662.58020

Vinogradov, A.M., Krasil'shchik, I.S., Lychagin, V.V. [1981]: Introduction to the Geometry of Nonlinear Differential Equations. Nauka: Moscow. English transl.: Geometry of Jet Spaces and Nonlinear Partial Differential Equations. Gordon and Breach Publ.: New York, 1986, Zbl.592.35002, Zbl.722.35001

Vosilyus, R.V. [1983]: A contravariant theory of differential prolongation in a model of a space with connection. Itogi Nauki Tekh. Ser. Probl. Geom. *14*, 101–176. English transl.: J. Sov. Math. *28*, 208–256 (1985), Zbl.528.58039

Vranceanu, G. [1931]: Parallèlisme et courbure dans une variété non holonome. Atti del Congresso Internaz. dei Matematici, Bologna 1928, No. 5., 63–75. Zanichelli: Bologna

Whittaker, E.T. [1937]: A Treatise on the Analytical Dynamics of Rigid Bodies. Cambridge Univ. Press: Cambridge, Zbl.61,418

Young, L.C. [1969]: Lectures on the Calculus of Variations and Optimal Control Theory. W.B. Saunders Co: Philadelphia-London-Toronto, Zbl.177,378

Zhitomirsky, M. (= Zhitomirskij, M.) [1990]: Normal forms of two-dimensional distribution germs in $\mathbb{R}^4$. Mat. Zametki *48* (Russian)

Additional Bibliographical Notes

In the present survey we have left aside a number of topics which are closely connected with the subjects considered. Below we list certain of them along with some additional references (making no claims of completeness). This list was compiled specially for the English edition.

1. Hypoelliptic operators and their connections with nonholonomic geometry. These connections were acknowledged starting with the classical theorem of Hörmander [1967] on the sum of squares and especially after Stein [1986] had noticed that a hypoelliptic operator defines the structure of a nilpotent Lie algebra on tangent spaces to the manifold. A thorough exposition of these connections is given by Gershkovich and Vershik [1989]. At present these questions are the subject of an ever increasing number of works which bear on the following main topics.

(a) *The spectrum of the nonholonomic Laplace operator*, including its asymptotics, estimates for its minimal eigenvalue and explicit calculation of the spectrum for certain Lie groups:

M. Taylor [1986], Gershkovich and Vershik [1988], Vershik and Gershkovich [1989], Metivier [1976], Fefferman and Phong [1979], Menikoff and Sjstrand [1978].

(b) *Geometrical aspects of hypoelliptic diffusion*, including estimates of diffusion for small times, estimates for long times, the Harnack inequality, diagonal asymptotics for transition probabilities of the corresponding Markov process, explicit formulae for the fundamental solution for a number of Lie groups: Hulanicki [1976], Varopoulos [1986a, 1986b, 1988], Gaveau [1976, 1977], Kannai [1977], Gershkovich [1989], T. Taylor [1989].

(c) *Fundamental solutions of hypoelliptic operators, Green's formula and the Cauchy problem*: Folland [1973, 1975], Hueber and Muller [1989], Fefferman and Sanchez-Calle [1986], Sanchez-Calle [1984], Gaveau and Vantier [1986].

(d) *Propagation of singularities and wave fronts of hypoelliptic operators*: Gaveau [1976, 1977], Vershik and Gershkovich [1988], Baryshnikov [1990], Nachman [1982].

(e) *Symbolic calculus of hypoelliptic operators.* The papers on this subject are numerous. Most of them deal, however, with the simplest nonholonomic manifolds: the Heisenberg group, or CR-manifolds. See, for instance, Melin [1983], M. Taylor [1986], Gaveau, Greiner and Vantier [1986].

(f) *Harmonic analysis and hypoelliptic operators on nilpotent Lie groups.* This topic has ample intersections wtih the five preceding. See, for instance, M. Taylor [1986], Folland and Stein [1974], Folland [1975], Melin [1983], Helfer and Nourrigat [1985], Gaveau and Vantier [1986], Rotschild and Stein [1976], Varopoulos [1986], Nachman [1982], Hulanicki [1976], Gaveau, Greiner and Vantier [1986].

2. As already noticed in the Introduction, in this survey we deal only with nonholonomic variational problems and completely leave aside nonholonomic mechanics and the related geometry of the straightest lines. Papers on the geometry of the straightests may be divided into the following three groups:

(a) *Papers of the end of the XIX-th and the beginning of the XX-th century*, in which the contact structure in $\mathbb{R}^3$ was studied, by analogy with the theory of two-dimensional surfaces. The references starting with the papers of abbé Issaly, may be found in Sintsov [1972].

(b) *A series of papers of the* 20s *and* 30s in which the curvature tensor of a nonholomonic manifold was defined, and some of its modern extensions. The milestones here are the papers of Vranceanu [1931], Synge [1936], Schouten [1928, 1930], Cartan [1926, 1935], Vagner [1935, 1940, 1941, 1965]. See also Solov'ev [1984, 1988].

(c) *The general theory of connections in a fibre bundle*; see, for instance, Laptev [1953], Loomiste [1966]. A good list of references is given by Aleksandrov [1947], and by Vasil'ev, Efimov and Rashevsky [1967].

3. Nonholonomic mechanics. The list of papers, starting from XVIII-th century, is enormous. A good review is given by Grigoryan and Fradlin [1982]. Let us single out the following subjects.

(a) *The study of particular nonholonomic systems* which is covered in the classical textbooks: Dobronravov [1970], Suslov [1946], Chaplygin [1949], Appell [1953], Lanczos [1949], Routh [1905], Whittaker [1937].

(b) *The Hamiltonian formalism of nonholonomic mechanics*, see Gaveau [1977], Baryshnikov [1990], Veselov and Veselova [1986].

(c) *The Lagrangian formalism*, see Arnol'd, Kozlov and Nejshtadt [1985], Brendelev [11], Vershik and Faddeev [1972, 1975], Karapetyan [1981], Kozlov [1982, 1983], Schouten [1928], Synge [1927, 1928].

(d) *Integrability of nonholonomic systems*, see Arnol'd, Kozlov and Nejshtadt [1985], Fajbusovich [1988], Veselov and Veselova [1986, 1988].

(e) *Stability of the free movement of nonholonomic systems*, see Nejmark and Fufaer [1967], Synge [1927, 1928, 1936].

4. Problems of existence of distributions with given properties on a manifold, i.e. of regular distributions with a given growth vector, apparently are very difficult. Most papers (see, for instance, Bennequin [1983], Martinet [1971], Bouthby and Wang [1958], Gray [1959], Lutz [1972, 1982], Solov'ev [1982]) deal with the simplest case, that of the contact structure. Even in this case the situation is not completely clear. A special case of a distribution which is the horizontal space of a connection is considered by Chernyakov [1988]. We also mention a few papers which deal with the related problem of existence of sets of vector fields on a manifold with specific properties: Kumpera and Ruisz [1982], Mormul [1988], Zhitomirsky [1990], Jacubczyk and Przytycki [1984].

5. Fields of cones and polysystems. We mention here the following subjects which are related to nonholonomic geometry:

(a) *The study of the accessibility sets for fields of cones and for polysystems on a manifold*, see Gershkovich [1984], Davydov [1985].

(b) *Control by a set of vector fields.* This subject is a part of an extremely vast domain of control theory; we list only several papers which are more close to our approach: Brockett, Millman and Sussmann [1983], Jurdjevith and Sussmann [1972], Levit and Sussmann [1975], Lobry [1973], Sussmann [1973], Agrachev, Vakhrameev and Gamkrelidze [1983], Sussmann [1983].

6. Penalty metrics. In many questons it is useful to approximate a nonholonomic metrics by a sequence of Riemannian "penalty metrics". This idea probably goes back to Gray [1959] and was used by several authors, see Strichartz [1978], T. Taylor [1989], Gershkovich [1985].

7. Normal forms of germs and jets of sets of vector fields and distributions and estimates of the number of orbits. See Kumpera and Ruisz [1982], Vershik and Gershkovich [1988], Zhitomirsky [1990], Varchenko [1981].

II. Integrable Systems II

Contents

Introduction

The three chapters that follow are conceived as independent surveys dealing with various group-theoretical constructions of finite-dimensional integrable systems. The major part of Chapter 1 is devoted to the so-called Calogero-Sutherland systems, which historically are among the first examples of this kind. A somewhat wider range of applications is provided by the Kostant-Adler scheme and its generalization known as the r-matrix construction; these are discussed in Chapter 2 with an emphasis on concrete examples. A natural class of Lie algebras where the r-matrix construction gives interesting results includes semi-simple Lie algebras and loop algebras (or affine Lie algebras); the latter appear already in Chapter 1 in connection with periodic Toda lattices and are exploited more profoundly in Chapter 2. The r-matrix approach applied to loop algebras gives a natural passage to algebraic-geometric methods; explanation of these links is another major theme of Chapter 2. (We also recommend it to the reader to consult the survey "Integrable systems. 1" by B. Dubrovin, I. Krichever and S. Novikov, EMS vol. 4, Springer-Verlag 1990.) Finally, Chapter 3 is concerned entirely with the quantization problem for a particular, though very interesting, family of integrable systems, the nonperiodic Toda lattices.

We hope that some overlaps of the first two chapters will not annoy the reader but may rather lead to a better understanding of the various aspects of the subject.

Chapter 1
Integrable Systems and Finite-Dimensional Lie Algebras

M.A. Olshanetsky, A.M. Perelomov

Translated from the Russian
by A.G. Reyman

In this survey we consider integrable systems whose construction makes use of root systems of simple (usually finite-dimensional) Lie algebras.

Some special cases related to the Lie algebra SU(n) have been widely discussed in the physics literature starting from the 1960s (see, for instance Calogero [1971] and Sutherland [1971]). Further progress in this area was made possible by the advances of the inverse scattering method as applied to finite-dimensional systems. The key tool of this approach is the representation of Hamilton's equations of motion in the form of a Lax equation

$$\frac{d}{dt}L = [L, M]. \tag{1}$$

Here L and M are matrices which depend on the dynamical variables. In general, these matrices also depend on the auxiliary (spectral) parameter, which allows one to analyze these systems and produce their solutions by means of algebraic-geometric methods. This topic is discussed in detail, for instance, in Dubrovin et al. [1985] (see also the survey by A. Reyman and M. Semenov-Tian-Shansky in this volume).

Here we shall be concerned with the simpler case where there is no dependence on the spectral parameter. The construction of solutions in this case also simplifies and reduces to operations of linear algebra which have an invariant group-theoretical meaning. The only exceptions are systems with potentials given by the Weierstrass $\mathcal{P}$-function and the generalized periodic Toda lattices.

For more details regarding the problems treated in this chapter the reader may consult Calogero [1978], Moser [1975], [1980], Olshanetsky and Perelomov [1981], Perelomov [1990].

§ 1. Hamiltonian Systems on Coadjoint Orbits of Lie Groups

The study of a vast "experimental material" has led to the following remarkable fact: for most known integrable systems, the phase space $\mathcal{M}$ may be identified with the dual space $\mathfrak{g}^*$ of some Lie algebra $\mathfrak{g}$, equipped with the standard

Lie-Poisson structure

$$\{X_j, X_k\} = C^i_{jk} X_i$$

(see Arnol'd [1974], Berezin [1967], Kirillov [1972], Weinstein [1983]).

Thus, let G be a Lie group, $\mathfrak{g}$ its Lie algebra, $\mathfrak{g}^*$ the dual space of $\mathfrak{g}$, i.e. the space of linear functionals on $\mathfrak{g}$. Consider the action of G on itself by conjugation:

$$\varphi_g \colon h \to ghg^{-1}.$$

The identity element e of G is fixed by this action: $\varphi_g \colon e \to e$. Therefore the linearization of φ_g at $h = e$ gives rise to the action of G on the Lie algebra $\mathfrak{g}$. This is the *adjoint representation* of G; $\mathrm{Ad}(g)\colon \mathfrak{g} \to \mathfrak{g}$. The adjoint action induces the dual action $\mathrm{Ad}^*(g)$ of G on $\mathfrak{g}^*$ called the *coadjoint representation* of G.

Let f be some point in $\mathfrak{g}^*$. Acting on it by coadjoint transformations $\mathrm{Ad}^*(g)$, where g runs over the entire group, we get the *coadjoint orbit* $\mathcal{O}_f$ passing through f.

The Lie-Berezin Poisson structure on this orbit is nondegenerate and determines a closed nondegenerate 2 form ω called the *Kirillov form.*

The manifold $(\mathcal{O}, \omega)$ is a homogeneous symplectic space: the group G acts transitively on $\mathcal{O}_f$ leaving invariant the form ω. Thus, to every element ξ of the Lie algebra $\mathfrak{g}$ there corresponds a Hamiltonian vector field X_ξ on $\mathfrak{g}^*$ induced by the associated Hamiltonian function H_ξ. The correspondence $\xi \to H_\xi$ can be arranged in such a way that

$$H_{[\xi,\eta]} = \{H_\xi, H_\eta\}, \tag{2}$$

i.e. this is a homomorphism from the Lie algebra $\mathfrak{g}$ into the Lie algebra of functions with Poisson bracket. A group action with this property is said to be *Hamiltonian* (Arnol'd [1974] uses the term "Poisson action").

The class of homogeneous symplectic spaces essentially coincides with the class of coadjoint orbits of Lie groups. More precisely, we have the following theorem.

Theorem (Kirillov [1972]). *Every homogeneous symplectic manifold whose group of motions is a connected Lie group G is locally isomorphic to a coadjoint orbit of the group G or of its central extension by means of $\mathbb{R}$.*

Examples.

1. $G = \mathrm{SO}(3)$, the rotation group of three-dimensional Euclidean space, $\mathfrak{g}^* \sim \mathfrak{g} = \{x\colon x = (x_1, x_2, x_3)\}$. The orbits are two-dimensional spheres $S^2 = \{x\colon x^2 = \rho^2\}$. The origin is an orbit, too.

2. $G = \mathrm{SU}(m + n)$, $\mathfrak{g}^* \sim \mathfrak{g} = \{x\}$, $x^+ = -x$. Suppose that the eigenvalues of the matrix f fall into two sets containing m and n elements, respectively, so that within each set the eigenvalues coincide. Then the orbit through f has the form

$$\mathcal{O} = \mathrm{SU}(m + n)/\mathrm{SU}(m) \times \mathrm{SU}(n) \times U(1) \tag{3}$$

and is isomorphic to the so-called complex Grassmann manifold G_{mn}.

3. Let G be the group of real upper triangular matrices with $\det g = 1$. The Lie algebra $\mathfrak{g}$ consists of real upper triangular matrices with trace zero. By using the inner product $(x, y) = \operatorname{tr}(xy)$ in $\mathrm{sl}(n, \mathbb{R})$, the dual space $\mathfrak{g}^*$ can be identified with the lower triangular matrices $\{x\}$, $\operatorname{tr} x = 0$. The coadjoint action of G on $\mathfrak{g}^*$ is given by

$$\mathrm{Ad}^*(g)\colon x \to (gxg^{-1})_-, \tag{4}$$

where the minus subscript shows that the elements above the principal diagonal of the matrix in question are replaced by zeros. For instance, the G- orbit in $\mathfrak{g}^*$ passing through the element

$$f = \begin{bmatrix} 0 & & 0 \\ 1 & 0 & \\ & \ddots & \ddots \\ 0 & 1 & 0 \end{bmatrix}, \tag{5}$$

consists of matrices x of the form

$$x = \begin{bmatrix} a_1 & & & 0 \\ b_1 & a_2 & & \\ & \ddots & \ddots & \\ 0 & & b_{n-1} & a_n \end{bmatrix}, \quad \sum a_j = 0. \tag{6}$$

Associated with this simplest orbit of the group of upper triangular matrices is a dynamical system called the Toda lattice (to be described later). For systems related to other orbits of this group see Kamalin and Perelomov [1985] and Goodman and Wallach [1984].

4. All coadjoint orbits of a compact Lie group G are simply connected and admit a G-invariant Kähler structure (Borel [1954]).

§ 2. The Moment Map

Let $\mathscr{M}$ be a manifold endowed with a Poisson structure $\{\ ,\ \}$ and suppose that we are given a Hamiltonian action of a Lie group G on $\mathscr{M}$. Since the Hamiltonian function $H_\xi(x)$ corresponding to an element ξ of the Lie algebra $\mathfrak{g}$ is linear in ξ, we can write

$$H_\xi(x) = \langle \mu(x), \xi \rangle. \tag{7}$$

Here $\mu(x)$ lies in the space $\mathfrak{g}^*$ dual to $\mathfrak{g}$ and $\langle \mu, \xi \rangle$ is the value of μ on $\xi \in \mathfrak{g}$.

In this way, a Hamiltonian action of G on $\mathscr{M}$ gives rise to the mapping

$$\mu\colon \mathscr{M} \to \mathfrak{g}^* \tag{8}$$

which is called the *moment map*.

Examples.

1. Let $\mathscr{M} = \mathbb{R}^{2n}$ with its standard Poisson bracket and let $G = \mathrm{Sp}(2n, \mathbb{R})$ be the real symplectic group. The dual space $\mathfrak{g}^*$ can be identified with the Lie algebra $\mathfrak{g}$ which consists of $2n \times 2n$ matrices $\mathscr{A}$ such that

$$\mathscr{A}I + I\mathscr{A}' = 0, \quad \mathscr{A} = \begin{pmatrix} A & B \\ C & D \end{pmatrix}, \quad I = \begin{pmatrix} 0 & -I \\ I & 0 \end{pmatrix}. \tag{9}$$

It follows that $\mathscr{A}$ has the form

$$\mathscr{A} = \begin{pmatrix} A & B \\ C & -A' \end{pmatrix}, \quad B' = B, \quad C' = C. \tag{10}$$

The corresponding Hamiltonian is

$$H_{\mathscr{A}} = \langle \mathscr{A}x, x \rangle = (\mathscr{A}x, Ix) = (p, Bp) + (q, Cq) + (p, Aq). \tag{11}$$

The moment map μ: $\mathbb{R}^{2n} \to \mathbb{R}^N$, $N = 2n^2 + n$, takes $x = (p, q)$ to the matrix $\mathscr{A}$ with

$$A = p \otimes q, \quad B = p \otimes p, \quad C = q \otimes q. \tag{12}$$

2. Let $\mathscr{M} = T^*X = \{x, y\}$, where X is the space of $n \times n$ Hermitian matrices. The symplectic structure of $\mathscr{M}$ is defined by

$$\omega = \mathrm{tr}(dy \wedge dx). \tag{13}$$

This symplectic structure determines a Poisson structure on $\mathscr{M}$.

The group $G = \mathrm{U}(n)$ of $n \times n$ unitary matrices acts on $\mathscr{M}$ as follows:

$$g\colon x \to gxg^+, \quad y \to gyg^+, \quad gg^+ = I. \tag{14}$$

The moment map for this action is

$$\mu\colon (x, y) \to [x, y] = xy - yx. \tag{15}$$

Suppose we have a Hamiltonian system with symmetry group G, i.e. a quadruple

$$(\mathscr{M}, \{\ ,\ \}, H, G) \tag{16}$$

such that the action of G is Hamiltonian and H is invariant under G. Then the functions H_ξ for $\xi \in \mathfrak{g}$, and hence the entire moment map, are integrals of motion for the Hamiltonian system in question. To give examples of Hamiltonian systems with symmetries it is convenient to fix the action of G on $\mathscr{M}$ and then look for a G-invariant Hamiltonian H.

Example. Let H be the Hamiltonian of the n-dimensional oscillator:

$$H = \frac{1}{2}(p^2 + q^2), \tag{17}$$

where $\mathscr{M} = \mathbb{R}^{2n}$ with the standard Poisson bracket. This Hamiltonian is $\mathrm{SO}(n)$ invariant; moreover, it is invariant under the larger group $\mathrm{SO}(2n)$. The transformations of SO(2n) do not in general preserve the symplectic form $\omega = dp_j \wedge dq_j$. We know, however, that ω is invariant under the group Sp(2n, $\mathbb{R}$). Therefore the (Hamiltonian) symmetry group of (17) is $G = \mathrm{Sp}(2n, \mathbb{R}) \cap \mathrm{SO}(2n)$ which is a maximal compact subgroup of Sp(2n, $\mathbb{R}$) and is isomorphic to U(n). The Lie algebra of G consists of matrices of the form

$$\mathscr{A} = \begin{pmatrix} A & B \\ -B & A \end{pmatrix}, \quad A' = -A, \quad B' = B. \tag{18}$$

The integrals of motion are

$$\begin{gathered} L_{jk} = q_j p_k - q_k p_j, \quad L_{jk} = -L_{kj}, \\ F_{jk} = q_j q_k + p_j p_k. \end{gathered} \tag{19}$$

§ 3. The Projection Method

A Hamiltonian system with a symmetry group can be reduced to a system with fewer degrees of freedom. As a rule, motion on the reduced phase space becomes more complicated and the manifest symmetry of the problem disappears or becomes "hidden".

If we are interested in the explicit integration of the equations of motion of a completely integrable Hamiltonian system, we cannot be satisfied with Liouville's theorem since the underlying integration procedure is usually too difficult to be carried out. In some cases, however, the problem can be solved by the so-called *projection method*. This method was introduced by Olshanetsky and Perelomov [1976b] and was used to solve several concrete problems in Olshanetsky and Perelomov [1976c], [1979] and Olshanetsky and Rogov [1978].

The basic idea of the projection method is reverse to the idea of reduction. The projection method consists of constructing a new dynamical system in a phase space $\tilde{\mathscr{M}}$ of higher dimension in such a way that

a) the original system $(\mathscr{M}, \omega, H)$ is obtained from $(\tilde{\mathscr{M}}, \tilde{\omega}, \tilde{H})$ by projection;

b) the equations of motion of the system $(\tilde{\mathscr{M}}, \tilde{\omega}, \tilde{H})$ can be integrated in closed form;

c) the projection π is given by explicit formulae.

Let us give an abstract formulation of the method. Suppose we have a dynamical system on a manifold $\mathscr{M} = \{x\}$:

$$h_t \colon \mathscr{M} \to \mathscr{M}, \quad x_t = h_t \cdot x_0. \tag{20}$$

Suppose further that there exists another dynamical system on a manifold $\tilde{\mathscr{M}} = \{y\}$:

$$\tilde{h}_t \colon \tilde{\mathscr{M}} \to \tilde{\mathscr{M}}, \quad y_t = \tilde{h}_t \cdot y_0 \tag{21}$$

and a projection $\pi \colon \tilde{\mathscr{M}} \to \mathscr{M}$ which sends the flow $\tilde{h}_t$ into the flow h_t:

$$x_t = \pi y_t = \pi \tilde{h}_t y_0. \tag{22}$$

For this formula to make sense the following consistency condition must be satisfied: if $\pi y_1 = \pi y_2$ for $t = 0$ $(y_1, y_2 \in \tilde{\mathcal{M}})$, then $\pi \tilde{h}_t y_1 = \pi \tilde{h}_t y_2$ for all t. This condition is fulfilled if, for instance, a Lie group $G = \{g\}$ acts on $\tilde{\mathcal{M}}$ commuting with the phase flow $\tilde{h}_t$ and $\mathcal{M} = \tilde{\mathcal{M}}/G$, the quotient space by the action of G.

Note that if there exists a mapping ρ: $\mathcal{M} \to \tilde{\mathcal{M}}$ such that $\pi \cdot \rho = \mathrm{Id}$ (Id is the identity map of $\mathcal{M}$), then the dynamics on $\mathcal{M}$ can be expressed as

$$x_t = \pi \tilde{h}_t \rho x_0. \tag{23}$$

Let us now illustrate the method by some simple examples.

1. Consider the free motion of a particle of unit mass on the plane (q_1, q_2). In this case

$$\begin{gathered} \tilde{\mathcal{M}} = \mathbb{R}^4 = \{(\mathbf{p}, \mathbf{q})\}, \quad \mathbf{p} = (p_1, p_2), \quad \mathbf{q} = (q_1, q_2), \\ \tilde{\omega} = dp_1 \wedge dq_1 + dp_2 \wedge dq_2, \\ \tilde{H} = \frac{1}{2}\mathbf{p}^2 = \frac{1}{2}(p_1^2 + p_2^2). \end{gathered} \tag{24}$$

We take π to be the standard *radial projection*:

$$\begin{gathered} \pi(\mathbf{p}, \mathbf{q}) = (p, q), \\ q = r = \sqrt{q_1^2 + q_2^2}, \quad p = p_r = \frac{\mathbf{pq}}{q}. \end{gathered} \tag{25}$$

Then upon projecting we come down to a system $(\mathcal{M}, \omega, H)$ with

$$\begin{gathered} \mathcal{M} = \{(p, q): q > 0\}, \quad \omega = dp \wedge dq, \\ H = H_l = \frac{1}{2}\left(p^2 + \frac{l^2}{q^2}\right). \end{gathered} \tag{26}$$

Here $l^2 = (q_1 p_2 - q_2 p_1)^2$ is an integral of motion for the system $(\tilde{\mathcal{M}}, \tilde{\omega}, \tilde{H})$.

It is not hard to verify that the consistency condition for the system (26) is satisfied. (This is a consequence of the invariance of the system $(\tilde{\mathcal{M}}, \tilde{\omega}, \tilde{H})$ under the group $G = \mathrm{SO}(2)$ of rotations of the plane.)

The covering equations of motion on $\tilde{\mathcal{M}}$ are easily integrated. We have

$$\mathbf{q}(t) = \mathbf{a} + \mathbf{b}t \tag{27}$$

and without loss of generality we can assume $(\mathbf{a}, \mathbf{b}) = 0$. This leads immediately to a solution of the equations of motion on $\mathcal{M}$:

$$q(t) = \sqrt{a^2 + b^2 t^2}, \quad p(t) = \frac{b^2 t}{\sqrt{a^2 + b^2 t^2}}. \tag{28}$$

For a generalization of this construction to higher dimensions see Section 7.

2. On the same plane (q_1, q_2) we consider a two-dimensional isotropic oscillator with frequency ν. Here $\tilde{\mathcal{M}}$ and $\tilde{\omega}$ are the same as above, and the Hamil-

tonian $\tilde{H}$ is

$$\tilde{H} = \frac{1}{2}(\mathbf{p}^2 + \nu^2\mathbf{q}^2). \tag{29}$$

With the same projection we get a system $(\mathcal{M}, \omega, H)$ with Hamiltonian

$$H = \frac{1}{2}\left(p^2 + \frac{l^2}{q^2} + \nu^2 q^2\right). \tag{30}$$

We then have

$$\mathbf{q}(t) = \mathbf{a} \cos \nu t + \frac{\mathbf{b}}{\nu} \sin \nu t, \quad (\mathbf{a}, \mathbf{b}) = 0 \tag{31}$$

and so

$$q(t) = \sqrt{a^2 \cos^2 \nu t + \frac{b^2}{\nu^2} \sin^2 \nu t}. \tag{32}$$

For generalizations see Olshanetsky and Perelomov [1976b].

3. We consider the free motion (i.e. the geodesic motion with respect to the SO(2, 1)-invariant Riemannian metric) on the upper sheet of a hyperboloid of two sheets:

$$\mathbb{H}^2 = \{\mathbf{x}\colon \mathbf{x}^2 = x_0^2 - x_1^2 - x_2^2 = 1, x_0 > 0\}. \tag{33}$$

The projection π is defined by

$$q = \pi \cdot \mathbf{x} = \operatorname{Ar} \cosh x_0. \tag{34}$$

Projecting then gives a system $(\mathcal{M}, \omega, H)$ with Hamiltonian

$$H = \frac{p^2}{2} + g^2 \sinh^{-2} q, \tag{35}$$

where $g^2 = l^2/2 = \frac{1}{2}(x_1\dot{x}_2 - x_2\dot{x}_1)^2$.

The equations of the geodesic flow on the hyperboloid are easily integrated:

$$\mathbf{x}(t) = \mathbf{a} \cosh kt + \mathbf{b} \sinh kt \tag{36}$$

with

$$\mathbf{a}^2 = a_0^2 - a_1^2 - a_2^2 = 1, \quad \mathbf{b}^2 = -1, \quad (\mathbf{a}, \mathbf{b}) = 0, \tag{37}$$

where $k^2/2$ is the total energy of the system. This yields an explicit expression for $q(t)$:

$$q(t) = \operatorname{Ar} \cosh(\cosh q_0 \cdot \cosh kt), \tag{38}$$

where $k = \dfrac{\sqrt{2}g}{\sinh q_0}$. Extensions of this result to the case of several degrees of freedom are given by Olshanetsky and Perelomov [1976c].

4. Suppose that a particle of unit mass moves along a geodesic on a two-dimensional sphere

$$S^2 = \{\mathbf{x} : \mathbf{x}^2 = x_0^2 + x_1^2 + x_2^2 = 1\}. \tag{39}$$

We define the projection π by

$$q = \pi \cdot \mathbf{x} = \arccos x_0. \tag{40}$$

In this case projecting gives a system $(\mathcal{M}, \omega, H)$ with Hamiltonian

$$H = \frac{p^2}{2} + g^2 \sin^{-2} q, \quad 0 < q < \pi. \tag{41}$$

By integrating the equations of geodesics on the sphere we find

$$\mathbf{x}(t) = \mathbf{a} \cos kt + \mathbf{b} \sin kt, \tag{42}$$

with

$$\mathbf{a}^2 = a_1^2 + a_2^2 + a_3^2 = 1, \quad \mathbf{b}^2 = 1, \quad (\mathbf{a}, \mathbf{b}) = 0.$$

This yields

$$q(t) = \arccos(\cos q_0 \cdot \cos kt), \tag{43}$$

where $k = \dfrac{\sqrt{2g}}{\sin q_0}$. For generalizations see Olshanetsky and Perelomov [1976c].

5. Consider the geodesic motion on a hyperboloid of one sheet

$$\mathbb{H}^2 = \{\mathbf{x} : \mathbf{x}^2 = x_0^2 - x_1^2 - x_2^2 = -1\}. \tag{44}$$

We define the projection π by

$$q = \pi \cdot \mathbf{x} = \operatorname{Ar} \sinh x_0. \tag{45}$$

By projecting we obtain a system $(\mathcal{M}, \omega, H)$ with Hamiltonian

$$H = \frac{p^2}{2} - g^2 \cosh^{-2} q. \tag{46}$$

We notice that in contrast to the previous cases there are three different types of geodesics on the one-sheeted hyperboloid. Accordingly we shall obtain three different expressions for $q(t)$.

a) Let the initial conditions be such that

$$\begin{aligned} \mathbf{x}(t) &= \mathbf{a} \cosh kt + \mathbf{b} \sinh kt, \\ \mathbf{a}^2 &= a_0^2 - a_1^2 - a_2^2 = -1, \quad \mathbf{b}^2 = 1, \quad (\mathbf{a}, \mathbf{b}) = 0. \end{aligned} \tag{47}$$

Then

$$q(t) = \operatorname{Ar} \sinh(\alpha \sinh kt), \quad k = \frac{\sqrt{2g}}{\sqrt{\alpha^2 - 1}}. \tag{48}$$

b) The second case is

$$\begin{aligned} \mathbf{x}(t) &= \mathbf{a} \cos kt + \mathbf{b} \sin kt, \\ \mathbf{a}^2 &= -1, \quad \mathbf{b}^2 = -1, \quad (\mathbf{a}, \mathbf{b}) = 0. \end{aligned} \tag{49}$$

Then

$$q(t) = \operatorname{Ar\,sinh}(\alpha \cos kt), \quad k = \frac{\sqrt{2}g}{\sqrt{\alpha^2 + 1}}. \tag{50}$$

c) Finally, the last case is

$$\begin{gathered} \mathbf{x}(t) = \mathbf{a} + \mathbf{b}t, \\ \mathbf{a}^2 = -1, \quad \mathbf{b}^2 = 0, \quad (\mathbf{a}, \mathbf{b}) = 0, \end{gathered} \tag{51}$$

and then

$$q(t) = \operatorname{Ar\,sinh}(\alpha t), \quad \alpha = \sqrt{2}g. \tag{52}$$

Generalizations of these formulae to several degrees of freedom are given by Olshanetsky and Rogov [1978].

6. As in Example 3 let us consider the free motion on the upper sheet of the two-sheeted hyperboloid, but instead of the radial projection we take the so-called *horospherical projection*:

$$q = \pi \cdot \mathbf{x} = \log(x_0 - x_1). \tag{53}$$

By projecting we get a system with Hamiltonian

$$H = \frac{p^2}{2} + g^2 \exp(-2q) \tag{54}$$

which is equivalent to the so-called Toda lattice of two particles.

Using the formula (36) for the geodesic motion

$$\mathbf{x}(t) = \mathbf{a} \cosh kt + \mathbf{b} \sinh kt, \tag{55}$$

where we set for convenience

$$\mathbf{a} = (a_0, a_1, 0), \quad \mathbf{b} = (0, 0, 1), \quad a_0^2 - a_1^2 = 1, \tag{56}$$

we come down to an explicit expression for $q(t)$:

$$q(t) = \log(\alpha \cosh kt), \quad k\alpha = \sqrt{2}g. \tag{57}$$

Generalizations of this formula to the Toda lattice of any number of particles and to modified Toda lattices are given in Kostant [1979] and Olshanetsky and Perelomov [1979], [1980].

§4. Description of the Calogero-Sutherland Type Systems

Systems of Calogero-Sutherland type describe one-dimensional n-particle dynamics. They are defined by the Hamiltonian

$$H = \frac{1}{2} \sum_{j=1}^{n} p_j^2 + U(q_1, \ldots, q_n), \tag{58}$$

where the potential U has the form

$$U(q_1, \ldots, q_n) = g^2 \sum_{j<k}^{n} v(q_j - q_k). \tag{59}$$

Originally these systems were studied in the quantum case for special choices of the function $v(\xi)$ in $U(q)$:

$$v(\xi) = \xi^{-2} + \frac{\omega^2}{2g^2}\xi^2 \quad \text{(Calogero [1971])} \tag{60}$$

and

$$v(\xi) = \sin^{-2}\xi \quad \text{(Sutherland [1971])}. \tag{61}$$

Later it turned out that systems with the interaction given by (60) and (61) also determine the evolution of the singularities in the solutions of some well-known equations of mathematical physics, such as the Kadomtsev-Petviashvili equation (Krichever [1980]) or the Benjamin-Ono equation (Case [1976]). These systems and their generalizations provide an important nontrivial illustration to the general theory of Section 3.

It turns out that integrable cases are not exhausted by the interactions defined above but embrace a wider class of systems. We shall describe some of them which are natural generalizations of (60) and (61). For that purpose we first extend the class of functions v and consider the following five types:

$$\begin{aligned}
&\text{I} \quad v(\xi) = \xi^{-2},\\
&\text{II} \quad v(\xi) = \sinh^{-2}\xi,\\
&\text{III} \quad v(\xi) = \sin^{-2}\xi,\\
&\text{IV} \quad v(\xi) = \mathcal{P}(\xi),\\
&\text{V} \quad v(\xi) = \xi^{-2} + \frac{\omega^2}{2g^2}\xi^2,
\end{aligned} \tag{62}$$

where $\mathcal{P}(\xi)$ is the Weierstrass function. Note that case IV includes the previous ones as limiting cases where the periods of the Weierstrass function go to infinity and it degenerates into a trigonometric or a rational function. Case V differs from case I by an additional oscillatory term.

Further generalizations involve a change in the form of the arguments of v and are related to hidden symmetries of these systems. As is well known (Helgason [1978]) every real simple Lie group G or, more precisely, every Riemannian symmetric space $X = G/K$, where K is a maximal compact subgroup, has an associated irreducible restricted root system R in a corresponding maximal Abelian subspace $\mathfrak{H}$. This root system may be of type A_n, B_n, C_n, D_n, E_6, E_7, E_8, F_4, G_2 or may be nonreduced of type BC_n. In addition, in cases I and V one can also consider the noncrystallographic systems $I_2(n)$, H_3, H_4 (Bourbaki [1968]). Let $R = \{\alpha\}$ be an irreducible root system in the configuration space $\mathfrak{H}$ and let R_+ be the set of positive roots with respect to some ordering in $\mathfrak{H}$. Let g_α^2 be some constants which are the same for the whole equivalence class of the root α under the action of the Weyl group W. Let $q_\alpha = (q, \alpha)$ be the scalar product of the position vector q and the root α. Define the potential $U(q)$ by

$$U(q) = \sum_{\alpha \in R_+} g_\alpha^2 v(q_\alpha), \tag{63}$$

where $v(q_\alpha)$ is given by (62). In the simplest cases where $v(q_\alpha)$ is of type I, II, or V and the root system R is of type A_{n-1}, the potential $U(q)$ coincides with the Calogero-Sutherland potentials (60), (62). In what follows the systems with interaction (63) will be labelled by the combination of a Roman index which numbers the function $v(q_\alpha)$ (62) and the standard code of the root system. Thus, the Sutherland system (61) is denoted by III A_{n-1}.

We shall write explicitly the potentials corresponding to the classical root systems. To do so we consider the nonreduced root system BC_n which contains other classical root systems as subsystems. The potentials of type I BC_n – V BC_n have the form

$$U(q) = g^2 \sum_{k<l}^{n} [v(q_k - q_l) + \varepsilon v(q_k + q_l)] + g_1^2 \sum_{k=1}^{n} v(q_k) + g_2^2 \sum_{k=1} v(2q_k). \tag{64}$$

Other potentials are obtained by taking special values of the interaction constants:

$$\text{I } A_{n-1} - \text{V } A_{n-1}\colon \varepsilon = g_1^2 = g_2^2 = 0,$$

$$\text{I } B_n - \text{V } B_n\colon \varepsilon = 1, \quad g_2^2 = 0,$$

$$\text{I } C_n - \text{V } C_n\colon \varepsilon = 1, \quad g_1^2 = 0,$$

$$\text{I } D_n - \text{V } D_n\colon \varepsilon = 1, \quad g_2^2 = g_1^2 = 0.$$

Trajectories of the system with potential (63) are subject to some constraints arising from the singularities of the functions $v(q_\alpha)$ (62). For instance, for the system of type A_{n-1} (59) the ordering of the particles remains the same in the course of evolution. This shows that the configuration space is effectively only part of $\mathfrak{H}$. For systems of type I, II, and V the configuration space is the Weyl chamber:

$$\Lambda = \{q \in \mathfrak{H} | q_\alpha > 0, \alpha \in R_+\}, \tag{65}$$

whereas for systems of type III and IV it is the Weyl alcove:

$$\Lambda_\alpha = \Lambda \cap \{q \in \mathfrak{H} | (q, \omega) < d\}, \tag{66}$$

where ω is the maximal root of R and d is the real period of $v(\xi)$.

§ 5. The Toda Lattice

Another family of integrable systems to be discussed in this survey consists of the Toda lattice and its group-theoretic generalizations.

The periodic Toda lattice is defined by a Hamiltonian of the form (58) but unlike previous examples only the nearest neighbours interact, and the potential is given by

$$U(q) = \sum_{j=1}^{n} g_j^2 \exp 2(q_j - q_{j+1}), \tag{67}$$

where $q_{n+1} = q_1$ and the factor of two in the exponent is taken for convenience. If $g_n^2 = 0$, we are dealing with the nonperiodic (open) Toda lattice. In this case, by a suitable shift in the positions q_j all the constants g_j^2 $(j \neq n)$ can be reduced to 1 which is assumed in what follows.

This system allows the following generalization, an important part of which was first proposed by Bogoyavlensky [1976]. The construction is based on the so-called affine Lie algebras which are a subclass of Kac-Moody algebras (Kac [1968], Moody [1968], Helgason [1978]). We shall first give the main definitions and discuss some relevant concepts concerning affine Lie algebras.

A *Kac-Moody algebra* $\mathfrak{g}(A)$ is defined in terms of $3n + 3$ generators $\{h_j, e_j, f_j\}$, $j = 0, \ldots, n$. The defining commutation relations are formulated by means of an integral $(n + 1) \times (n + 1)$ matrix $A = (a_{ij})$ called a *Cartan matrix*:

$$\begin{gathered} [e_j, f_k] = \delta_{jk} h_k, \quad [h_j, h_k] = 0, \\ [h_i, e_j] = a_{ij} e_j, \quad [h_j, f_k] = -a_{jk} f_k, \\ (\operatorname{ad} e_j)^{1-a_{ij}} e_i = 0, \quad (\operatorname{ad} f_j)^{1-a_{ji}} f_i = 0. \end{gathered} \tag{68}$$

The Cartan matrix must obey the following conditions: $a_{ii} = 2$, $a_{ij} \leqslant 0$ for $i \neq j$, and $a_{ij} = 0$ implies $a_{ji} = 0$. A Kac-Moody algebra $\mathfrak{g}(A)$ is finite-dimensional if and only if all principal minors of A are positive; these Lie algebras are semi-simple. If all proper principal minors of A are positive but $\det A = 0$ (hence $\operatorname{rank} A = n$), the associated Kac-Moody algebra is called an affine Lie algebra. In this case $\mathfrak{g}(A)$ is infinite-dimensional and has a one-dimensional center formed by the linear combinations

$$c = \sum_{j=0}^{n} \gamma_j h_j, \tag{69}$$

where $\gamma = (\gamma_0, \ldots, \gamma_n)$ is such that $\gamma A = 0$. The quotients of affine Lie algebras by their centers can be described alternatively as follows.

Consider a simple Lie algebra L. Let $\operatorname{Aut} L$ be the group of all automorphisms of L and $\operatorname{Aut}^0 L$ the subgroup of inner automorphisms; recall that the quotient $\operatorname{Aut} L/\operatorname{Aut}^0 L$ can be identified with the symmetry group of the Dynkin diagram of L. An automorphism C of L is called a *Coxeter* automorphism if the subalgebra L^C of fixed points of C is Abelian and C has the minimal order among all automorphisms with this property in the coset $\operatorname{Aut} L/\operatorname{Aut}^0 L$. Such a C exists in each coset and is unique up to conjugation by inner automorphisms. Let h denote the order of C.

Consider now the Lie algebra $\tilde{L}(C)$ of Laurent polynomials in the variable λ with coefficients in L

$$g(\lambda) = \sum g_j \lambda^j \tag{70}$$

such that

$$Cg_j = \exp\left(\frac{2\pi i}{h} j\right) g_j.$$

Obviously, $\tilde{L}(c)$ is graded by powers of λ; this is called the *principal gradation.*

To establish a link with the definition (68) consider three subspaces in $\tilde{L}(C)$: $L_0 = L^C, L_{\pm 1} = \left\{ g\lambda^{\pm 1}: Cg = \exp\left(\pm\frac{2\pi i}{h}\right)g \right\}$. In these subspaces there exist a set of vectors $\{h_0, \ldots, h_n\} \in L_0$ which span L_0, and two bases $\{e_0, \ldots, e_n\} \in L_1$, $\{f_0, \ldots, f_n\} \in L_{-1}$ such that the commutation relations (68) hold with an affine Cartan matrix A.

The order of C in the quotient group Aut L/Aut^0 L is called the height of $\tilde{L}(C)$, which can be 1, 2, or 3. The type of the associated affine Lie algebra is denoted by $X_l^{(k)}$ where X_l is the type of L in the standard classification and k is the height. Affine Lie algebras of height 1 are defined by an inner automorphism and are in one-to-one correspondence with simple Lie algebras. There are also affine algebras of type $A_l^{(2)}$, $D_l^{(2)}$, $E_6^{(2)}$, and $D_4^{(3)}$.

We shall now define generalized Toda lattices associated with Kac-Moody algebras. Let $\mathfrak{H}$ be the linear span of the generators $h_0, \ldots, h_n$. A simple root α_i is a linear functional on $\mathfrak{H}$ defined by means of the Cartan matrix:

$$\alpha_i(h_j) = a_{ji}. \tag{71}$$

Let $\Pi = \{\alpha_0, \ldots, \alpha_n\}$ be the set of simple roots of $\mathfrak{g}(A)$. For instance, the set of simple roots of an affine Lie algebra $\tilde{L}(C)$ of height 1 consists of the simple roots of L together with the minimal root. The generalized Toda potential associated with Π is defined on the configuration space $\mathfrak{H}$ by the formula

$$U(q) = \sum_{\alpha \in \Pi} \exp 2q_\alpha, \quad (q_\alpha = \alpha(q)). \tag{72}$$

The potential (67) corresponds to the affine Lie algebra $A_{n-1}^{(1)}$. Finite-dimensional simple Lie algebras give rise to nonperiodic Toda lattices; affine Lie algebras give rise to generalized periodic lattices. Although the definition (72) makes sense for any Cartan matrix, one can move far enough only in the affine (or finite-dimensional) case.

We shall list the potentials associated with classical algebras of height 1. Set

$$V_k = \sum_{j=1}^{k-1} \exp 2(q_j - q_{j+1}).$$

Then we have the correspondence:

$$
\begin{aligned}
B_n^{(1)}&: U(q) = V_n(q) + \exp 2q_n + \exp(-2q_1 - 2q_2),\\
C_n^{(1)}&: U(q) = V_n(q) + \exp 4q_n + \exp(-4q_1),\\
D_n^{(1)}&: U(q) = V_n(q) + \exp 2(q_{n-1} + q_n) + \exp 2(-q_1 - q_2).
\end{aligned} \tag{73}
$$

Deleting the last terms in these formulae (as in (67)) we get the nonperiodic lattices associated with the Lie algebras of type $B_n(\mathrm{so}(2n+1))$, $C_n(\mathrm{sp}(n))$ and $D_n(\mathrm{so}(2n))$.

Associated to the simple algebra of type G_2 is the nonperiodic lattice with potential

$$U(q) = \exp 2(q_3 - q_2) + \exp 2(q_1 + q_2 - 2q_3) \quad (q_1 + q_2 + q_3 = 0) \tag{74}$$

and its periodic counterpart $G_2^{(1)}$ which differs from (74) by the additional term

$$\exp 2(q_2 + q_3 - 2q_1). \tag{75}$$

The periodic lattice of type $D_4^{(3)}$ related to the Lie algebra so(8) has a similar form: its potential is the sum of (74) and

$$\exp 2(q_2 - q_1). \tag{76}$$

§ 6. Lax Representation. Proof of Complete Integrability

As was mentioned earlier, to prove complete integrability it is convenient to use a Lax representaion (see (1)). We shall first exhibit Lax pairs for Calogero-Sutherland systems of type I–IV and for generalized Toda lattices, which will allow us to analyze the integrability of these systems. Next we shall consider systems of type V.

Lax pairs for system of type I A_{n-1}–IV A_{n-1} were found by Calogero [1975] and Moser [1975]. They are given by the matrices

$$
\begin{aligned}
L_{jk} &= p_j\delta_{jk} + ig(1 - \delta_{jk})x(q_j - q_k),\\
M_{jk} &= g\left[\delta_{jk}\sum_{l\neq j} z(q_j - q_l) - (1 - \delta_{jk})y(q_j - q_k)\right].
\end{aligned} \tag{77}
$$

Substituting these matrices into the Lax equation $\dot{L} = [L, M]$ and requiring this equation to be equivalent to the original Hamilton's equations we find

$$y(\xi) = x'(\xi) \tag{78}$$

and a functional relation for $x(\xi)$ and $z(\xi)$:

$$x'(\xi)x(\eta) - x'(\eta)x(\xi) = x(\xi + \eta)[z(\xi) - z(\eta)]. \tag{79}$$

The function $v(\xi)$ which occurs in the potential (63) is given by

$$v(\xi) = x(\xi)x(-\xi) + \text{const}. \tag{80}$$

The functional equation (79) was solved in several papers, e.g. Calogero [1976], Olshanetsky and Perelomov [1976a]. It turns out that

$$z(\xi) = -x''(\xi)/2x(\xi). \tag{81}$$

All odd solutions $x(\xi)$ have the form

$$x(\xi) = \begin{cases} \xi^{-1} & \text{I} \\ a \coth a\xi, \quad a \sinh^{-1} a\xi & \text{II} \\ a \cot a\xi, \quad a \sin^{-1} a\xi & \text{III} \\ a\dfrac{\operatorname{cn} a\xi}{\operatorname{sn} a\xi}, \quad a\dfrac{\operatorname{dn} a\xi}{\operatorname{sn} a\xi}, \quad \dfrac{a}{\operatorname{sn} a\xi} & \text{IV} \end{cases} \tag{82}$$

Here cn, sn and dn are the Jacobi elliptic functions.

It is easily verified that the potentials of type I–IV (62) result from these expressions by using (80). Substituting ia for a takes the function $x(\xi)$ of type II to the function $x(\xi)$ of type III. In the limit as $a \to 0$ the function $x(\xi)$ of type II or III goes over into the function $x(\xi)$ of type I. The same relationship holds for the matrices L (77).

Lax representations for systems of type I–IV related to other classical root systems (see (64)) were found by Olshanetsky and Perelomov [1976a]. These Lax pairs are valid only if the interaction constants obey the following relations:

$$\begin{gathered} g_1^2 + \sqrt{2}g_1 g_2 - 2g_2^2 = 0 \quad \text{if} \quad g_1 \neq 0, \\ g, g_2 \quad \text{arbitrary if} \quad g_1 = 0. \end{gathered} \tag{83}$$

In the general case, the matrices L and M have dimension $2n + 1$ and can be written in the form

$$L = P + X, \quad M = -D + Y, \tag{84}$$

where

$$P = \operatorname{diag}(p_1, \ldots, p_n, 0, -p_n, \ldots, -p_1); \tag{85}$$

$$X = \begin{vmatrix} A_1 & C_1^T & B_1 \\ \overline{C}_1 & 0 & -\overline{C}_1 \\ -B_1 & -C_1^T & -A_1 \end{vmatrix}, \quad Y = \begin{vmatrix} A_2 & C_2^T & B_2 \\ -\overline{C}_2 & 0 & -\overline{C}_2 \\ B_2 & C_2^T & A_2 \end{vmatrix},$$

$$D = \operatorname{diag}(d_1, d_2, \ldots, d_n, d_0, d_n, \ldots, d_2, d_1), \tag{86}$$

where A_1, B_1, A_2, B_2 are $n \times n$ matrices:

$$\{A_1\}_{kl} = i(1 - \delta_{kl})gx(q_k - q_l),$$

$$\{B_1\}_{kl} = ig_2\sqrt{2}\delta_{kl}x(2q_k) + ig(1 - \delta_{kl})x(q_k + q_l),$$

$$\{A_2\}_{kl} = i(1 - \delta_{kl})gy(q_k - q_l),$$

$$\{B_2\}_{kl} = ig_2\sqrt{2}\delta_{kl}y(2q_k) + ig(1 - \delta_{kl})y(q_k + q_l),$$

$$\{C_1\}_k = ig_1 x(q_k),$$
$$\{C_2\}_k = ig_1 y(q_k),$$
$$d_k = i(e_k - c),$$
$$e_k = \sum_{r=1}^{n} g[z(q_k - q_r) + z(q_k + q_r)] + g_1^2 g^{-1} z(q_k) + \sqrt{2} g_2 z(2q_k),$$
$$e_0 = 2g \sum_{k=1}^{n} z(q_k), \quad c = \frac{2}{2n+1}\left[\sum_{k=1}^{n} e_k + \frac{1}{2} e_0\right].$$

As before, the functions $x(\xi)$, $y(\xi)$ and $z(\xi)$ satisfy (78) and (79) so that the interaction potential has the form (62).

For systems of Calogero-Sutherland type related to exceptional root systems no Lax pairs have been found until now. These systems are shown to be completely integrable only for certain values of the interaction constants g_α^2 by using the integrability of the associated quantum system, see (Olshanetsky and Perelomov [1981]).

We shall now proceed to the Lax pairs for Toda lattices. A Lax pair for the ordinary lattice (67) was constructed by Manakov [1974] and Flaschka [1974a, b]. We shall write it in the form

$$\begin{aligned} L_{jk} &= p_j \delta_{jk} + \delta_{j,k+1} + \delta_{j,k-1} \exp 2(q_j - q_{j+1}), \\ M_{jk} &= 2 \exp 2(q_j - q_{j+1}) \delta_{j,k-1}. \end{aligned} \tag{87}$$

It is not hard to see that the Lax equation coincides with the equations of motion for the Toda lattice.

A generalization of this formula which uses extended root systems was given by Bogoyavlensky [1976]; actually, this generalization works for all affine Lie algebras. The configuration space $\mathfrak{H}$ in this case is spanned by the generators $\{h_j\}$. Let e_j and f_j be the raising and lowering generators associated to the simple root α_j and let λ be the parameter of the standard grading of the affine algebra, see (70). The Lax pair is then given by

$$\begin{aligned} L &= p + \lambda \sum_{\alpha_j \in \Pi} e_j + \lambda^{-1} \sum_{\alpha_j \in \Pi} \exp(2q_{\alpha_j}) \cdot f_j, \\ M &= \lambda^{-1} \sum_{\alpha_j \in \Pi} \exp(2q_{\alpha_j}) \cdot f_j. \end{aligned} \tag{88}$$

The equivalence of the Lax equation for the pair (88) and the system of Hamilton's equations follows directly from the commutation relations (68).

To show that the above systems are completely integrable consider the invariants of L of the form

$$I_k = \frac{1}{k} \operatorname{tr} L^k \tag{89}$$

which obviously are integrals of the motion. The matrix structure of L (see (77), (84) and (88)) implies that these integrals have the form

$$I_k = \frac{1}{k} \sum p_j^k + \text{terms of lower degree in the momenta.}$$

Hence the functional independence of the I_k follows from the functional independence of the quantities

$$s_k = \sum p_j^k. \tag{90}$$

Note that the polynomials s_k are invariants of the corresponding simple (or affine) Lie algebra. The degrees of the basic invariants (which generate the whole algebra of invariants) are well known. For the classical root systems they are

$$A_{n-1}: k = 2, 3, \dots, n,$$

$$B_n, C_n, BC_n: k = 2, 4, \dots, 2n,$$

$$D_n: k = 2, 4, \dots, 2(n-1), n.$$

The involutivity of the integrals for systems of type I–III and for nonperiodic Toda lattices can easily be proved by using the argument of Calogero [1975]. For that purpose consider first a system of type II. Since the potential $U(q)$ is repulsive it is not hard to show that $q_\alpha(t) \to \infty$ as $t \to \infty$. (For systems of type A_{n-1} this means that the particles move away infinitely far from one another.) It is clear from the structure of L that in this case the integral I_k (89) tends to s_k (90), hence $\{I_k, I_j\} \to \{s_k, s_j\} = 0$. Since the Poisson bracket of any two integrals is also an integral, this implies $\{I_k, I_j\} = 0$. Thus the systems of type II A_{n-1} and II BC_n are completely integrable.

Systems of type I and III are obtained from systems of type II by taking the limit as $a \to 0$ or by substituting $a \mapsto ia$, respectively, which implies their integrability.

For nonperiodic Toda lattices, one can show that $q_\alpha(t) \to -\infty$ as $t \to \infty$ (see (72)). In particular, for the ordinary lattice (67) this means that $q_{j+1}(t) - q_j(t) \to \infty$. As before, it follows that $I_k(t) \to s_k$ and the previous reasoning remains valid.

It is more difficult to prove the involutivity of the integrals of motion for systems of type IV. In this case it is convenient to take for the basic integrals the eigenvalues of L (Perelomov [1976]). The involutivity of the integrals for periodic Toda lattices is shown for instance, in Reyman [1980], Reyman et al. [1979] (see also the survey by A. Reyman and M. Semenov-Tian-Shansky in this volume).

For systems of type I A_{n-1} – IV A_{n-1} (59) and for the ordinary Toda lattice (67), in addition to the integrals I_k (89) it is convenient to consider another set of integrals J_k which are the coefficients of the characteristic polynomial of L:

$$\det(L - \lambda E) = \lambda^n + \sum_{k=1}^{n} J_k \lambda^{n-k}.$$

The integrals I_k and J_k are related by Newton's formulae:

$$I_k = I_{k-1} J_1 - I_{k-2} J_2 + \cdots - (-1)^{k-1} I_1 J_{k-1} - (-1)^k k J_k.$$

For systems of type I A_{n-1} – IV A_{n-1} the integral J_n has a simple representation (Sawada and Kotera [1975], Wojciechowski [1977]):

$$J_n = \exp\left\{-\frac{g^2}{2}\sum_{k,l} v(q_k - q_l)\frac{\partial}{\partial p_k}\frac{\partial}{\partial p_l}\right\}\prod_{j=1}^{n} p_j.$$

A similar expression for the Toda lattice has the form

$$J_n = \exp\left\{-\frac{1}{2}\sum_{j=1}^{n} \exp\{2(q_j - q_{j+1})\}\frac{\partial}{\partial p_j}\frac{\partial}{\partial p_{j+1}}\right\}\prod_{l=1}^{n} p_l.$$

Other integrals result from the simple recursion relation

$$J_{k-1} = -\frac{1}{n-k+1}\{\sum q_l, J_k\} = \frac{1}{n-k+1}\left(\sum_{l=1}^{n}\frac{\partial}{\partial p_l}\right)J_k.$$

A slight modification of the Lax equation allows one to treat also systems of type V. It turns out that the equations of motion in this case are equivalent to the matrix equations

$$\dot{L}^{\pm} = [L^{\pm}, M] \pm i\omega L^{\pm}, \tag{91}$$

where M is the same as in the Lax pair for the system of type I and

$$L^{\pm} = L \pm i\omega q \tag{92}$$

with

$$q = \operatorname{diag}(q_1, \ldots, q_n) \tag{93}$$

for systems of type V A_{n-1} and

$$q = \operatorname{diag}(q_1, \ldots., q_n, 0, -q_n, \ldots, -q_1) \tag{94}$$

for systems of type V BC_n. To show the equivalence one uses the relation

$$[q, M] = X, \tag{95}$$

where X is the off-diagonal part of L.

From (91) it follows that the quantities

$$B_k^{\pm} = \frac{1}{k}\operatorname{Tr}(L^{\pm})^k$$

have a simple time dependence:

$$B_k^{\pm}(t) = B_k^{\pm}(0)\exp(\mp ik\omega t).$$

Using (91) one can also find the integrals of motion. For this purpose consider the matrices

$$N_1 = L^+L^-, \quad N_2 = L^-L^+.$$

They satisfy the usual Lax equation

$$\dot{N}_j = [N_j, M], \quad j = 1, 2.$$

Therefore traces of powers of these matrices are integrals of motion.

§ 7. Explicit Integration of the Equations of Motion

The equations of motion for the systems of type I, II, III, V and for the generalized nonperiodic Toda lattices can be solved explicitly by means of the projection method described in Section 3. The examples treated in Section 3 are extended in this section to include systems with more than one degree of freedom.

A) Systems of Calogero Type. We consider here the systems of type I and V, see (62) (Olshanetsky and Perelomov [1976b]). The construction makes use of the space $X_n^0 = \{x\}$ of $n \times n$ traceless complex Hermitian matrices, which can be viewed as Euclidean space with metric given by

$$ds^2 = \operatorname{tr}(dx)^2. \tag{96}$$

The equations of geodesics in X_n^0 can be written in Hamiltonian form: the phase space is $T^*X_n^0$, the symplectic structure is determined by the 2-form ω (13), and the Hamiltonian is $H = \frac{1}{2}\operatorname{tr}(y^2)$ where y is the momentum variable. The phase space $T^*X_n^0$ plays the role of $\tilde{\mathcal{M}}$ (see Section 3). The equation of a geodesic is

$$\ddot{x} = 0 \tag{97}$$

and its solutions are straight lines

$$x(t) = at + b. \tag{98}$$

Clearly, the metric (96) and the form ω (13) are invariant under unitary transformations (14). Let us reduce the Hermitian matrix $x(t)$ to diagonal form $q(t)$:

$$x(t) = u(t)q(t)u^{-1}(t), \tag{99}$$

where

$$q(t) = \operatorname{diag}(q_1, \ldots, q_n). \tag{100}$$

We can assume that $q_1 > q_2 > \cdots > q_n$. The reduction of x to q is called the *spherical projection.*

We shall derive the equations of motion for $q(t)$ and $p(t) = \dot{q}(t)$ given that $x(t)$ satisfies (97). Differentiating (99) with respect to time we find

$$u(t)L(t)u^{-1}(t) = a, \tag{101}$$

with

$$L(t) = p(t) + [q, M], \tag{102}$$

$$M = -u^{-1}(t)\dot{u}(t), \tag{103}$$

where L is a Hermitian and M a skew-Hermitian matrix. Differentiating (101) again we have

$$\dot{L} = [L, M].$$

Thus the pair of matrices (102) and (103) which describes the geodesic flow (97) satisfies a Lax equation.

On the other hand, it is easily verified that the Lax matrices (77) for systems of type I A_{n-1} satisfy (102). This establishes the equivalence of these two systems.

We must note, however, that not every geodesic is mapped by the spherical projection into a trajectory of a system of type I A_{n-1}. Indeed, consider the angular momentum (15) of the geodesic (98):

$$\mu = [x, \dot{x}] = [b, a] = u[q, L]u^{-1}. \tag{104}$$

It is not hard to see that the commutator $[q, L]$ has a rather special structure:

$$[q, L] = g(v^T \otimes v - I), \quad (v = 1, 1, \ldots, 1), \tag{105}$$

i.e., it has $n - 1$ coinciding eigenvalues. (The symmetry group G which determines the momentum map μ is $\mathrm{SU}(n)$; the subgroup G_μ which leaves invariant the momentum μ (104) consists of those matrices in $\mathrm{SU}(n)$ which commute with $[q, L]$.)

It is now easy to obtain a solution of the Cauchy problem for systems of type I A_{n-1}. We set $u(0) = \mathrm{I}$ and reconstruct the parameters of the geodesic from the initial data:

$$a = L(0), \quad b = q(0).$$

Using (99) we conclude that *the coordinates $q_j(t)$ are the eigenvalues of the matrix*

$$q(0) + L(0)t. \tag{106}$$

To determine solutions of the equations of type V A_{n-1} we take the same space X_n^0 but instead of the free motion we consider the equation of a harmonic oscillator:

$$\ddot{x} + \omega^2 x = 0. \tag{107}$$

Its solutions are

$$x(t) = a\omega^{-1} \sin \omega t + b \cos \omega t, \quad a, b \in X_n^0.$$

As before, taking the spherical decomposition (99) of $x(t)$ and differentiating twice with respect to t we are able to write the dynamical system (107) in the form (91). This equivalence implies that *the coordinates $q_j(t)$ of the trajectories of the system of type* V A_{n-1} *are the eigenvalues of the matrix*

$$q(0) \cos \omega t + \omega^{-1} L(0) \sin \omega t. \tag{108}$$

These results can be used to derive several useful corollaries. Let us first consider the scattering process for systems of type I A_{n-1}.

Since the potential decays as $|q_j - q_k| \to \infty$, we have

$$q_j(t) \sim p_j^{\pm} t + q_j^{\pm}, \quad t \to \pm\infty.$$

If the particles are numbered in such a way that $q_1 < q_2 < \cdots < q_n$, then $p_1^- > p_2^- > \cdots > p_n^-$ and $p_1^+ < p_2^+ < \cdots < p_n^+$. The explicit formulae show that the asymptotic values $p_j^\pm$ and $q_j^\pm$ differ only by a permutation:

$$p_1^- = p_n^+, \quad p_2^- = p_{n-1}^+, \ldots, p_n^- = p_1^+, \tag{109}$$

$$q_1^- = q_n^+, \quad q_2^- = q_{n-1}^+, \ldots, q_n^- = q_1^+. \tag{110}$$

A converse statement, due to Ya. Sinai, is also true: If in a one-dimensional n-body problem with a repulsive pair potential which decays as some power of the distance the asymptotic momenta are related by (109), then the potential has the form I (62).

The explicit formulae for solutions show a simple relationship between trajectories in the systems of type I and V (Perelomov [1978]). Let $q_j(t)$ be a solution for a system of type I and $\tilde{q}_j(t)$ a solution for a system of type V. Then

$$\tilde{q}_j(t) = q_j(\omega^{-1}\tan(\omega t))\cdot\cos\omega t. \tag{111}$$

Finally, consider the symmetric polynomials of degree k in the coordinates $q_j(t)$. It follows that these are polynomials of degree k in t if $\omega = 0$ or in $\sin\omega t$ and $\cos\omega t$ if $\omega \neq 0$.

All the above results extend to other systems of type I and V which have a Lax representation, i.e. to the general systems I BC_n and V BC_n. For this purpose one has to consider the configuration space of $(2n+1)\times(2n+1)$ traceless Hermitian matrices satisfying

$$xf + fx^+ = 0, \tag{112}$$

where

$$f = \begin{pmatrix} 0 & 0 & J \\ 0 & 1 & 0 \\ J & 0 & 0 \end{pmatrix}$$

and J is an $n \times n$ matrix with units on the opposite diagonal. Formulae (106) and (108) now involve the matrix $L(0)$ (84) and

$$q = \mathrm{diag}(q_1, \ldots, q_n, 0, -q_n, \ldots, -q_1).$$

B) Systems of Sutherland Type. We shall construct solutions of the equations of motion for systems of type II and III.

We shall first discuss systems of type II A_{n-1}. Consider the space X_n^- of $n \times n$ positive definite Hermitian matrices with determinant 1. This is a homogeneous space: $X_n^- = \mathrm{SL}(n, \mathbb{C})/\mathrm{SU}(n)$,

$$x \to gxg^+, \quad x \in X_n^-. \tag{113}$$

In particular, every point x can be represented as

$$x = gg^+.$$

The space X_n^- has a metric ds^2 invariant under the transformations of the form (113):

$$ds^2 = \operatorname{tr}(x^{-1}\,dx)^2. \tag{114}$$

The curvature of X_n^- relative to this metric is nonpositive, which explains the minus supercript in our notation. One easily derives the equation of geodesics:

$$\frac{d}{dt}\left(x^{-1}(t)\frac{d}{dt}x(t)\right) = 0. \tag{115}$$

Let us regard X_n^- as a submanifold of $\mathrm{SL}(n, \mathbb{C})$. The matrices $x^{-1}(t)\dot{x}(t)$ and $\dot{x}(t)x^{-1}(t)$ are tangent vectors to $\mathrm{SL}(n, \mathbb{C})$ at the identity but are not in general tangent to X_n^-. Their sum, however, is tangent to X_n^-. Therefore instead of (115) we shall consider the equation

$$\frac{d}{dt}\frac{x^{-1}\dot{x} + \dot{x}x^{-1}}{2} = 0. \tag{116}$$

Every geodesic in X_n^- can be written as

$$x(t) = b\exp\{2\mathfrak{a}t\}b^+, \tag{117}$$

where $b \in \mathrm{SL}(n, \mathbb{C})$, $\mathfrak{a}^+ = \mathfrak{a}$, $\operatorname{tr}\mathfrak{a} = 0$.

From linear algebra one knows that a Hermitian matrix can be reduced to diagonal form by a unitary transformation. For any curve $x(t)$ in X_n^- we then have

$$x(t) = u(t)\exp\{2aq(t)\}u^+(t), \tag{118}$$

where $u \in \mathrm{SU}(n)$, a is a parameter and

$$q = \operatorname{diag}(q_1, \ldots, q_n). \tag{119}$$

This is the so-called spherical coordinate system on X_n^-. With respect to the parametrization (118) of $x(t)$ we have

$$\frac{1}{2}(x^{-1}\dot{x} + \dot{x}x^{-1}) = 2auLu^+, \tag{120}$$

where

$$L = p + \frac{1}{4a}[e^{-2aq}Me^{2aq} - e^{2aq}Me^{-2aq}], \tag{121}$$

$$M = u^{-1}\dot{u} \quad \text{is the angular velocity}, \tag{122}$$

and $p = \dot{q}$ is the relative velocity in the coordinate system (118). Differentiating (120) with respect to time we find

$$\frac{d}{dt}\frac{x^{-1}\dot{x} + \dot{x}x^{-1}}{2} = 2au(\dot{L} - [L, M])u^+.$$

If $x(t)$ is a geodesic, the left-hand side of this equation vanishes. It follows that the geodesic flow on X_n^- has a Lax representation with the matrices L and M given by (121) and (122).

It can be verified directly that the Lax pair (77) for systems of type II A_{n-1} satisfies (121) provided that the function $x(\xi)$ entering in L is $x(\xi) = a \coth a\xi$ (see (82)).

Thus if the geodesic $x(t)$ is chosen in such a way that under the spherical parametrization (118) its angular velocity (122) has the specific form (77) then the evolution of the radial part $q(t)$ (119) has a Lax representation which, as was shown in the previous section, is equivalent to the original Hamiltonian system.

The special choice of geodesics has a simple mechanical interpretation. Consider the angular momentum $\mu = \frac{1}{2}(x^{-1}\dot{x} - \dot{x}x^{-1})$ of the curve $x(t)$ relative to the Hamiltonian action of the group SU(n) on the phase space $\tilde{\mathcal{M}} = T^*X_n^-$. It turns out that if the angular velocity M has the desired form (77), the angular momentum is time-independent and is given by

$$\mu = -2ia^2 g(v^T \otimes v - I), \quad v = (1, \ldots, 1). \tag{123}$$

On the other hand, the momentum of the geodesic (117) is

$$\mu = -(b\mathfrak{a}b^{-1} - b^{+^{-1}}\mathfrak{a}b^{+}). \tag{124}$$

By equating these two expressions we obtain constraints for the parameters $(\mathfrak{a}, b)$ of the geodesics which are projected to the trajectories of systems of type II A_{n-1}. Thus the geodesic flow (116) is reduced by the spherical projection at the fixed momentum (123) to the Sutherland system.

The above construction immediately yields an explicit expression for the trajectories. Let us indicate a solution of the Cauchy problem. Setting $u(0) = I$ we find from (117) and (118)

$$\begin{gathered} b = \exp\{aq(0)\}, \\ \{\mathfrak{a}\}_{jk} = ap_j(0)\delta_{jk} + ia^2 g \sinh^{-1} a(q_j(0) - q_k(0))(1 - \delta_{jk}). \end{gathered} \tag{125}$$

We then have the following result: *the quantities* $2aq_j(t)$ *(which are solutions of the equations of motion for systems of type* II A_{n-1}*) are the logarithms of the eigenvalues of the matrix*

$$x(t) = b\exp\{2\mathfrak{a}t\}b^{+},$$

where the matrices $\mathfrak{a}$ and b are given by (125).

The passage to solutions for systems of type III A_{n-1} is performed by formally substituting ia for the parameter a in the formulae above. Geometrically, this substitution means a passage from the space X_n^- of positive definite Hermitian matrices to the space $X_n^+ = \mathrm{SU}(n)$ of unitary matrices. The group SU(n) is a Riemannian symmetric space of nonnegative curvature with respect to the invariant metric (114). The relationship between X_n^- and X_n^+ is a special case of duality established by E. Cartan for a large class of symmetric spaces, which generalizes the well-known relationship between hyperbolic and spherical geometries.

In a similar way, systems of type III BC_n result from geodesic flows on the complex Grassmann manifolds $\mathrm{SU}(2n+1)/\mathrm{SU}(n) \oplus \mathrm{U}(n+1)$, while systems of type II BC_n are related to the dual noncompact symmetric spaces $SU(n+1, n)/SU(n) \oplus U(n+1)$. In the later case solutions are also obtained by diagonalizing a matrix of the form (117) where $b \in SU(n+1, n)$ and $\mathfrak{a}$ satisfies (112).

C) Systems with Two Types of Particles. A simple change of variables proposed by Calogero [1975] gives a generalization of systems of type II A_{n-1}. Consider the substitution

$$q_j \to q_j + i\frac{\pi}{2a}, \quad 0 < n_1 < j \leqslant n. \tag{126}$$

Then the potential

$$U(q) = g^2 \sum_{k<j} a^2 \sinh^{-2} a(q_j - q_k) \tag{127}$$

goes over into

$$U(q) = g^2 \sum_{\substack{j<i\leqslant n_1 \\ n_1<j<i\leqslant n}} a^2 \sinh^{-2} a(q_i - q_j) - g^2 \sum_{\substack{i>n_1 \\ j\leqslant n_1}} a^2 \cosh^{-2} a(q_i - q_j). \tag{128}$$

This system contains n_1 particles of the same sign and $n_2 = n - n_1$ particles of the opposite sign. We shall assume for definiteness that $n_1 \geqslant n_2$. Obviously, the substitution (126) does not affect the main results of the previous section. In particular, the quantities $\exp 2aq_j(t)$ are the eigenvalues of the matrix $x(t)$. However, this prescription does not tell us whether or not the system with potential (128) has bound states. Using the geometric approach one can show that bound states always exist but are unstable except in the case $n_1 = n_2 = 1$ (Olshanetsky and Rogov [1978]).

To show this let us note that the substitution (126) amounts to passing from the space X_n^- of positive definite Hermitian matrices to the space X_{n_1,n_2} of Hermitian matrices of signature (n_1, n_2). The latter is a symmetric pseudo-Riemannian space with respect to the metric (114). It can be shown that every geodesic in X_{n_1,n_2} has the form $x(t) = b\exp\{2\mathfrak{a}t\}\sigma b^+$ where $b \in SL(n, \mathbb{C})$, $\sigma = \mathrm{diag}(E_{n_1}, -E_{n_2})$, and the matrix $\exp\{2\mathfrak{a}t\}$ can be reduced by conjugation in $SU(n_1, n_2)$ to the form

$$\exp\{2\mathfrak{a}t\} = \left[\begin{array}{ccccccc}
e^{2\alpha_1 t} & & & & & & \\
 & \ddots & & & & 0 & \\
 & & e^{2\alpha_r t} & & & & \\
 & & & e^{2\beta_1 t}\cos 2\varphi_1 t & \cdots & ie^{2\beta_1}\sin 2\varphi_1 t & \\
 & & & \vdots & \ddots & \ddots & \\
 & & & & e^{2\beta_k t}\cos 2\varphi_k t \ldots & ie^{2\beta_k t}\sin 2\varphi_k t & \\
 & & & ie^{2\beta_1 t}\sin 2\varphi_1 t & \vdots & e^{2\beta_1 t}\cos 2\varphi_k t & \vdots \\
 & & & & \ddots & \ddots & \\
 & 0 & & & ie^{2\beta_k t}\sin 2\varphi_k t \ldots & e^{2\beta_k t}\cos 2\varphi_k t &
\end{array}\right] \tag{129}$$

where

$$n_1 - n_2 \leqslant r \leqslant n, \quad k \leqslant n_2, \quad r + 2k = n,$$
$$\sum_{j=1}^{r} \alpha_j + 2 \sum_{j=1}^{k} \beta_j = 0, \tag{130}$$

It follows that the geodesics break up into $n_2 + 1$ classes. Each class is characterized by k "compact" parameters φ_j and $r + k$ "noncompact" parameters (α_j, β_l). We observe that the geodesic associated with (129) lies in a bounded region of X_{n_1,n_2} if and only if $\alpha_j = 0$, $\beta_l = 0$, $j = 1, \ldots, r$, $l = 1, \ldots, k$. These are precisely the geodesics that are mapped by (118) into bound states of the system which correspond to finite trajectories. Such a geodesic depends on k parameters whereas a generic geodesic depends on $n - 1$ parameters. If $n \geqslant 3$, a bounded geodesic is not generic and hence is unstable; if $n = 2$, we have a system with one degree of freedom (46) where there exist stable bounded trajectories (50).

D) The Nonperiodic Toda Lattice. We shall first consider the ordinary nonperiodic Toda lattice with potential

$$U(q) = \sum_{j=1}^{n-1} \exp 2(q_j - q_{j+1}). \tag{131}$$

According to the general ideas of Section 3 we must indicate the extended phase space $\tilde{\mathcal{M}}$ whose projection would be the phase space of the dynamical system with potential (131). In our case this is the phase space of the geodesic flow on the space of symmetric positive definite unimodular matrices, $X_n = \mathrm{SL}(n, \mathbb{R})/\mathrm{SO}(n)$. The Riemannian metric and geodesics are given by the same formulae as in subsection B. (One should only remember that the matrices are real.)

We shall now define the projection and analyze the reduced phase space. Let Z be the subgroup of $\mathrm{SL}(n, \mathbb{R})$ consisting of upper triangular matrices with units on the principal diagonal and let H be subgroup of diagonal matrices. Each matrix x in X_n can be written uniquely as

$$x = z(x)h^2(x)z^T(x), \quad z(x) \in Z, \quad h(x) \in H. \tag{132}$$

The pair $(h(x), z(x))$ is called the *horospherical coordinates* in the space X_n. For $n = 2$, X_2 is the Lobachevsky plane and (h, z) are the true horospherical coordinates. The diagonal matrix $h(x) = \mathrm{diag}(h_1, \ldots, h_n)$ is called the horospherical projection. Let $\Delta_j(x)$ be the j-th lower principal minor of x. The horospherical projection is then given by

$$h_1 = \frac{1}{\Delta_{n-1}}, \quad h_2 = \frac{\Delta_{n-1}}{\Delta_{n-2}}, \ldots, h_{n-1} = \frac{\Delta_2}{\Delta_1}, \quad h_n = \Delta_1. \tag{133}$$

We have the following two propositions.

1) Under the condition

$$2(bab^{-1})_{j,j-k} = \begin{cases} 1, & k = 1, \\ 0, & k > 1, \end{cases} \tag{134}$$

the horospherical projection (132) maps the geodesic flow (115) given by (117) into motion defined by the potential (131).

2) Let $(q^0, p^0) = (q_1^0, \ldots, q_n^0, p_1^0, \ldots, p_n^0)$ be the initial data for the nonperiodic Toda lattice. Define the $n \times n$ matrix $\mathfrak{a}$ by

$$\mathfrak{a}_{jk} = p_j^0 \delta_{jk} + e^{q_{j-1}^0 - q_j^0} \delta_{j,k+1} + e^{q_j^0 - q_{j+1}^0} \delta_{j,k-1}, \tag{135}$$

and let Δ_j be the j-th lower principal minor of $\exp\{2\mathfrak{a}t\}$. Then the solution of the Cauchy problem is given by

$$q_j(t) = q_j^0 + \log \frac{\Delta_{n+1-j}}{\Delta_{n-j}}. \tag{136}$$

For the proof consider the horospherical decomposition (132) of the geodesic $x(t)$ (117):

$$x(t) = z(x) e^{2q(t)} z^T(x). \tag{137}$$

Then

$$\dot{x}x^{-1} = z\{2p + e^{2q} M^T e^{-2q} + M\} z^{-1},$$

where $p = \dot{q}$ and

$$M = z^{-1}\dot{z}. \tag{138}$$

Let us denote

$$L = p + \frac{1}{2} M + \frac{1}{2} e^{2q} M^T e^{-2q}. \tag{139}$$

Then

$$\dot{x}x^{-1} = 2zhz^{-1}. \tag{140}$$

Differentiating this again we find

$$\frac{d}{dt} \dot{x}x^{-1} = 2z(\dot{L} - [L, M])z^{-1}. \tag{141}$$

Since $x(t)$ is a geodesic it satisfies (115). Therefore the left-hand side of (141) vanishes and thus the geodesic flow has a Lax representation with matrices (138), (139). Note that M lies in the Lie algebra of Z.

On the other hand, the Lax pair of the Toda lattice also obeys the relation (139). Moreover, the matrix M is triangular and has the special form (87). This condition singles out a particular class of geodesics. It can be cast in a more natural form by using the moment map μ for the action of Z:

$$\mu = (\dot{x}x^{-1})_-, \tag{142}$$

where the minus subscript means that all entries of the matrix in brackets lying on and above the main diagonal are replaced by zeros. Equation (139) and the matrix structure of L (87) show that the momentum of the geodesics related to the Toda lattice is

$$\{\mu\}_{jk} = \delta_{j,k+1}. \tag{143}$$

Calculating the momentum (142) of the geodesic given by (117) and using (143) we obtain (134), which proves our first proposition.

To prove the second proposition compare the expression for the geodesics (117) and the horospherical decomposition (137). Setting $z(0) = I$ we find

$$b = \exp q(0) = \operatorname{diag}(e^{\overset{0}{q}_1}, \ldots, e^{\overset{0}{q}_n}). \tag{144}$$

Equation (140) and the explicit form of L (87) imply the expression (135) for a in terms of the initial data. From (117) and (137) we find

$$e^{2q(t)} = z^{-1} b \exp\{2at\} b^T (z^T)^{-1}. \tag{145}$$

Since b is a diagonal matrix, it normalizes the subgroup of triangular matrices, so that $z^{-1}b = bz_1$ where z_1 is some element of Z. Therefore (145) can be written as

$$\exp 2(q(t) - q(0)) = z_1 \exp\{2at\} z_1^T.$$

Combined with the definition of the horospherical projection (see (132) and (133)) this yields (136).

Let us briefly indicate how these two propositions are modified for the generalized nonperiodic Toda lattices (72). One must deal with those symmetric spaces for which the Lie algebra of the corresponding isometry group is the so-called normal form (see Helgason [1978]). Such spaces are uniquely determined by the associated root systems. Each space of this type has a horospherical coordinate system and a horospherical projection $h(x)$ with values in a Cartan subgroup of the isometry group. The main formula (136) in the general case has the form

$$q(t) = q(0) + \frac{1}{2} \log h(\exp\{2at\}). \tag{146}$$

As (136), it is obtained by projecting the geodesic flow. One can also write a constraint analogous to (134) which specifies the geodesics projected into the trajectories of the associated Toda lattice.

To conclude, we note that in each concrete case the calculations required by (146) reduce to taking a matrix exponential and a subsequent computation of minors, i.e. to purely algebraic operations.

It can be shown that the phase space of the generalized Toda lattice coincides with the coadjoint orbit of the Borel subgroup $B = ZH$ passing through $\sum_{j \in \Pi} f_j$. The dimension of this orbit is twice the rank of the symmetric space and its Lie-Poisson structure is given by the canonical form $dp \wedge dq$.

As regards periodic Toda lattices, explicit solutions were obtained by Krichever [1978a] for the affine algebra of type $A_{n-1}^{(1)}$. Their interpretation in terms of Hamiltonian reduction was given by Reyman and Semenov-Tian-Shansky [1979].

§ 8. Bibliographical Notes

In classical mechanics the moment map has been used in simple situations since long ago. For general systems see, for instance, Marsden and Weinstein [1974].

The projection method was first used by Olshanetsky and Perelomov [1976b] to integrate the equations of motion of Hamiltonian systems of Calogero type, and later was applied to other systems (Olshanetsky and Rogov [1978]). The relationship between the moment map, the projection method, and Lax representations is discussed in Kazhdan et al. [1978].

Hamiltonian systems defined by (58), (59), (60) were first investigated by Calogero [1971] in the quantum case; systems defined by (58), (59), (68) were studied by Sutherland [1971]. Generalizations of these systems associated with root systems of Lie algebras and symmetric spaces were considered by Olshanetsky and Perelomov [1976a].

The potential $U(q)$ of the form (67) was first studied by Toda [1967]. A generalization in terms of root systems and simple Lie algebras was given by Bogoyavlensky [1976].

A Lax representation for the Calogero-Sutherland systems was found by Moser [1975], Calogero [1975] and Olshanetsky and Perelomov [1976a]. A Lax pair for Toda lattices was obtained by Flaschka [1974a, b], Manakov [1974] and Bogoyavlensky [1976].

Explicit integration of the equations of motion for the Calogero-Sutherland systems was carried out by Olshanetsky and Perelomov [1976c]; for non-periodic Toda lattices this was done by Olshanetsky and Perelomov [1979] and Kostant [1979].

References*

Arnol'd, V.I. [1974]: Mathematical Methods of Classical Mechanics. Nauka: Moscow, 431 pp. English transl.: Graduate Texts in Mathematics *60*, Springer-Verlag: New York – Berlin-Heidelberg 1978, 462 pp., 1978, Zbl.386.70001

Berezin, F.A. [1967]: Some remarks on the universal enveloping algebra of a Lie algebra. Funkts. Anal. Prilozh. *1*, No. 2, 1–14. English transl.: Funct. Anal. Appl. *1*, 91–102, Zbl.227.22020

Bogoyavlensky, O.I. (= Bogoyavlenskij, O.I.) [1976]: On perturbations of the periodic Toda lattice. Commun. Math. Phys. *51*, 201–209

Borel, A. [1954]: Kählerian coset spaces of semisimple Lie groups. Proc. Natl. Acad. Sci. USA *40*, 1147–1151, Zbl.58,160

Bourbaki, N. [1968]: Groupes et Algèbres de Lie. Ch. IV. Hermann: Paris, 288 pp, Zbl.186,330

Calogero, F. [1971]: Solution of the one-dimensional *n*-body problem with quadratic and/or inversely quadratic pair potential. J. Math. Phys. *12*, 419–436

Calogero, F. [1975]: Exactly solvable one-dimensional many body-problems. Lett. Nuovo Cim. *13*, 411–427

* For the convenience of the reader, references to reviews in Zentralblatt für Mathematik (Zbl.), compiled using the MATH database, have, as far as possible, been included in this bibliography.

Calogero, F. [1976]: On a functional equation connected with many-body problem. Lett. Nuovo Cim. *16*, 77–80

Calogero, F. [1978]: Integrable many-body problems. In: Nonlinear equations in physics and mathematics. Proc. NATO adv. Study Inst., Istanbul 1977 (Barut, A.O. (Ed.)), 3–53, Zbl.399.70022

Case, K.M. [1976]: Meromorphic solutions of the Benjamin-Ono equation. Physica A *96*, 173–182

Dubrovin, B.A., Krichever, I.M., Novikov, S.P. [1985]: Dynamical Systems IV: Integrable systems. In: Itogi Nauki Tekh., Ser. Sovrem. Probl. Mat., Fundam. Napravleniya *4*, 179–285. English transl.: Encyclopaedia of Mathematical Sciences *4*, 173–280. Springer-Verlag: Berlin-Heidelberg-New York 1990, Zbl.591.58013

Flaschka, H. [1974a]: The Toda lattice. I. Existence of integrals. Phys. Rev. B *9*, 1924–1925

Flaschka, H. [1974b]: On the Toda lattice. II. Inverse scattering solution. Progr. Theor. Phys. *51*, 703–716

Goodman, R., Wallach, N.R. [1984]: Classical and quantum mechanical systems of Toda lattice type II. Commun. Math. Phys. *94*, 177–217, Zbl.592,58028

Helgason, S. [1978]: Differential Geometry, Lie Groups and Symmetric Spaces. Academic Press: New York, 643 pp., Zbl.451.53088

Kac, V.G. [1968]: Simple irreducible graded Lie algebras of finite growth. Izv. Akad. Nauk SSSR, Ser. Mat. 32, No. 6, 1323–1367. English transl.: Math. USSR, Izv. *2*, 1271–1311 (1968), Zbl.222.17007

Kamalin, S.A., Perelomov, A.M. [1985]: Construction of canonical coordinates on polarized coadjoint orbits of graded Lie groups. Commun. Math. Phys. *97*, 553–568, Zbl.578.58016

Kazhdan, D., Kostant, B., Sternberg, S. [1978]: Hamiltonian group actions and dynamical systems of Calogero type. Commun. Pure Appl. Math. *31*, 491–508, Zbl.368.58008

Kirillov, A.A. [1972]: Elements of the Theory of Representations. Nauka: Moscow, 336 pp. English transl.: Grundlehren der Mathematischen Wissenschaften *220*. Springer-Verlag: New York–Berlin–Heidelberg 1976, 315 pp., Zbl.264.22011

Kostant, B. [1979]: The solution to a generalized Toda lattice and representation theory. Adv. Math. *34*, 195–338, Zbl.433.22008

Krichever, I.M. [1978a]: Algebraic curves and nonlinear difference equations. Usp. Mat. Nauk *33*, No. 4, 215–216. English transl.: Russ. Math. Surv. *33*, No. 4, 255–256 (1978), Zbl.382.39003

Krichever, I.M. [1980]: Elliptic solutions of the Kadomtsev-Petviashvili equation and integrable systems of particles. Funkts. Anal. Prilozh. *14*, No. 4, 45–54. English transl.: Funct. Anal. Appl. *14*, 282–290 (1981), Zbl.462.35080

Manakov, S.V. [1974]: Complete integrability and stochastization of discrete dynamical systems. Zh. Exp. Teor. Fiz. *40*, 269–274. English transl.: Sov. Phys. ZETP *40*, 269–274 (1975)

Marsden, J., Weinstein, A. [1974]: Reduction of symplectic manifolds with symmetry. Rep. Math. Phys. *5*, 121–130, Zbl.327.58005

Moody, R.V. [1968]: A new class of Lie algebras. J. Algebra *10*, 221–230, Zbl.191,30

Moser, J. [1975]: Three integrable Hamiltonian systems connected with isospectral deformations. Adv. Math. *16*, 197–220, Zbl.303.34019

Moser, J. [1980]: Various aspect of integrable Hamiltonian systems. In: Dynamical Systems, C.I.M.E. Lect. Bressanone 1978. Prog. Math. *8*, 233–289, Zbl.468.58011

Olshanetsky, M.A. (= Olshanetskij, M.A.), Perelomov, A.M. [1976a]: Completely integrable Hamiltonian systems connected with semisimple Lie algebras. Invent. Math. *37*, 93–108, Zbl.342.58017

Olshanetsky, M.A., Perelomov, A.M. [1976b]: Explicit solutions of the Calogero model in the classical case and geodesic flows on symmetric spaces of zero curvature. Lett. Nuovo Cim. *16*, 333–339

Olshanetsky, M.A., Perelomov, A.M. [1976c]: Explicit solutions of some completely integrable systems. Lett. Nuovo Cim. *17*, 97–101

Olshanetsky, M.A., Perelomov, A.M. [1979]: Explicit solutions of the classical generalized Toda models. Invent. Math. *54*, 261–269, Zbl.419.58008

Olshanetsky, M.A., Perelomov, A.M. [1980]: The Toda lattice as a reduced system. Teor. Mat. Fiz. *45*, No. 1, 3–18. English transl.: Theor. Math. Phys. *45*, 843–854 (1981)

Olshanetsky, M.A., Perelomov, A.M. [1981]: Classical integrable finite-dimensional systems related to Lie algebras. Phys. Rep. *71C*, No. 5, 313–400

Olshanetsky, M.A., Perelomov, A.M. [1983]: Quantum integrable systems related to Lie algebras. Phys. Rep. *94*, 313–404

Olshanetsky, M.A., Rogov, V.B. [1978]: Bound states in completely integrable systems with two types of particles. Ann. Inst. Henri Poincare, Sect. A 29, 169–177, Zbl.416.58014

Perelomov, A.M. [1976]: Completely integrable classical systems connected with semisimple Lie algebras. III. Lett. Math. Phys. *1*, 531–534

Perelomov, A.M. [1978]: The simple relation between certain dynamical systems. Commun. Math. Phys. *63*, 9–11, Zbl.435.70013

Perelomov, A.M. [1983a]: Systems with constraints. Preprint ITEP No. 111, Moscow, 26 pp.

Perelomov, A.M. [1983b]: Motion of a rigid body about a fixed point. Preprint ITEP No. 147, Moscow, 42 pp.

Perelomov, A.M. [1990]: Integrable Systems of Classical Mechanics and Lie Algebras. Nauka: Moscow, 240 pp., English transl.: Birkhäuser: Basel – Boston – Berlin, 307 pp., Zbl.699.70003, Zbl.717.70003

Reyman, A.G. [1980]: Integrable Hamiltonian systems connected with graded Lie algebras. Zap. Nauchn. Semin. Leningr. Otd. Mat. Inst. Steklova *95*, 3–54. English transl.: J. Sov. Math. *19*, No. 5, 1507–1545 (1982), Zbl.488.70013

Reyman, A.G., Semenov-Tian-Shansky, M.A. [1979]: Reduction of Hamiltonian system, affine Lie algebras and Lax equations. I. Invent. Math. *54*, 81–100, Zbl.403.58004

Reyman, A.G., Semenov-Tian-Shansky, M.A. [1981]: Reduction of Hamiltonian systems, affine Lie algebras and Lax equations. II. Invent. Math. *63*, 423–432, Zbl.442.58016

Reyman, A.G., Semenov-Tian-Shansky, M.A., Frenkel, I.B. [1979]: Graded Lie algebras and completely integrable systems. Dokl. Akad. Nauk SSSR *247*, No. 4, 802–805. English transl.: Sov. Math., Dokl. *20*, 811–814 (1979), Zbl.437.58008

Sawada, K., Kotera, T. [1975]: Integrability and a solution for the one-dimensional n-particle system with inversely quadratic pair potential. J. Phys. Soc. Japan *39*, 1614–1618

Sutherland, B. [1971]: Exact results for a quantum many-body problem in one dimension. Phys. Rev. A *4*, 2019–2021

Toda, M. [1967]: Wave propagation in anharmonic lattices. J. Phys. Soc. Japan *23*, 501–506

Weinstein, A. [1983]: The local structure of Poisson manifolds. J. Differ. Geom. *18*, 523–557, Zbl.524.58011

Wojciechowski, S. [1977]: Involutive set of integrals for completely integrable many body problems with pair interactions. Lett. Nuovo Cim. *18*, 103–107

Chapter 2
Group-Theoretical Methods in the Theory of Finite-Dimensional Integrable Systems

A.G. Reyman, M.A. Semenov-Tian-Shansky

Translated by the authors

Introduction

The present survey is devoted to a general group-theoretic scheme which allows to construct integrable Hamiltonian systems and their solutions in a systematic way. This scheme originates from the works of Kostant [1979a] and Adler [1979] where some special but very instructive examples were studied. Some years later a link was established between this scheme and the so-called classical R-matrix method (Faddeev [1984], Semenov-Tian-Shansky [1983]). One of the advantages of this approach is that it unveils the intimate relationship between the Hamiltonian structure of an integrable system and the specific Riemann problem (or, more generally, factorization problem) that is used to find its solutions. This shows, in particular, that the Hamiltonian structure is completely determined by the Riemann problem. The simplest system which may be studied in this way is the open Toda lattice already described in Chapter 1 by Olshanetsky and Perelomov. (The Toda lattices will be considered here again in a more general framework.) However, the most interesting examples are related to infinite-dimensional Lie algebras. In fact, it can be shown that the solutions of Hamiltonian systems associated with finite-dimensional Lie algebras have a too simple time dependence (roughly speaking, like trigonometric polinomials). By contrast, genuine mechanical problems often lead to more sophisticated (e.g. elliptic or abelian) functions.

An appropriate class of Lie algebras which give rise to integrable systems of this type are affine Lie algebras, or loop algebras. Their role in the theory of integrable systems was first indicated by Reyman, Semenov-Tian-Shansky and Frenkel [1979]. Our survey is aimed to include a comprehensive list of mechanical systems that may be realized within the general scheme. Along with the Toda lattices and their various generalizations we discuss numerous cases of the motion of the n-dimensional rigid body: free tops with a fixed point, heavy tops in a linear potential (the generalized Lagrange and Kowalewski cases), or in an arbitrary quadratic potential. We also discuss the motion of rigid bodies in ideal fluid (the Clebsch and Steklov cases) and a new family of integrable left-invariant geodesic flows on SO(4) related to the exceptional Lie algebra G_2. Other examples include the motion of a point on a sphere in a quadratic potential (C. Neumann's problem), various types of anharmonic oscillators and the Garnier system.

To describe the phase spaces of integrable systems we frequently use the Hamiltonian reduction technique. Its brief description is included into the survey. The version of the reduction that we shall use goes back to Lie [1893] and is based on the theory of Poisson manifolds. We take the opportunity to advertize for this approach (cf. Weinstein [1983]). One of its advantages is that it allows to extend the reduction technique to non-Hamiltonian group actions. (Semenov-Tian-Shansky [1986]). This generalization is crucial for applications to difference Lax equations (see Section 12).

The second major theme covered in this survey is the connection with algebraic geometry and the finite-band integration technique. Using the general factorization theorem we are able to give short qualitative proofs of the complete integrability and linearization theorems and to express solutions of the Lax equations by means of Riemann's theta functions.

In the major part of Chapter 2 we use the simplest R-matrix which is connected with the decomposition of an affine Lie algebra into a linear sum of its graded subalgebras. A more general decomposition which leads to the so-called elliptic R-matrices is studied in Section 11. Among the integrable systems related to elliptic R-matrices is the Steklov case of the motion of a rigid body in ideal fluid.

Other examples of R-matrices seem less important at present, due mainly to the lack of interesting applications. In the main body of the paper we preferred to consider only special types of R-matrices. The general theory is briefly outlined in Section 2. One of the principal motivations for the definition of R-matrices was the study of a new type of Poisson brackets related to integrable systems on a lattice and to difference Lax equations. V. Drinfel'd [1983] was the first to realize that the theory of R-matrices leads to new geometrical concepts, namely, the theory of Poisson Lie groups. An extension of the general scheme to this broader context involves much of this new geometry in a rather nontrivial way. We decided to add an extra section which deals with these subjects. The topics discussed there include the Poisson reduction method which generalizes the ordinary Hamiltonian reduction, the duality theory for Poisson Lie groups, and the theory of dressing transformations (which may be regarded as a non-linear counterpart of the coadjoint representation). We also discuss the tensor formalism which is very convenient in the study of lattice systems and present a number of examples.

§1. Poisson Manifolds

A *Poisson bracket* on a smooth manifold $\mathcal{M}$ is a Lie bracket on the space $C^\infty(\mathcal{M})$ of smooth functions on $\mathcal{M}$ which satisfies the Leibniz rule

$$\{\varphi_1, \varphi_2\varphi_3\} = \{\varphi_1, \varphi_2\}\varphi_3 + \{\varphi_1, \varphi_3\}\varphi_2.$$

In local coordinates $\{x_i\}$ the Poisson bracket is given by

$$\{\varphi, \psi\}(x) = \sum_{ij} c_{ij}(x)\frac{\partial\varphi}{\partial x_i}\frac{\partial\psi}{\partial x_j},$$

where the tensor $c_{ij}(x)$ is skew-symmetric and satisfies the first order differential equation which follows from the Jacobi identity:

$$\sum_l \left(c_{lj}\frac{\partial c_{ik}}{\partial x_l} + c_{lk}\frac{\partial c_{ji}}{\partial x_l} + c_{li}\frac{\partial c_{kj}}{\partial x_l}\right) = 0.$$

A manifold $\mathcal{M}$ equipped with a Poisson bracket is called a *Poisson manifold*. Each function φ on $\mathcal{M}$ determines a Hamiltonian vector field Ξ_φ defined by the formula $\Xi_\varphi \cdot \psi = \{\varphi, \psi\}$, or, in coordinate form,

$$\Xi_\varphi = \sum_{ij} c_{ij}(x)\frac{\partial\varphi}{\partial x_i}\frac{\partial}{\partial x_j}. \tag{1.1}$$

The function φ is called the Hamiltonian of Ξ_φ.

A mapping $f: \mathcal{M} \to \mathcal{N}$ of Poisson manifolds is called a *Poisson mapping* if it preserves the Poisson brackets, i.e.

$$\{\varphi, \psi\}_{\mathcal{N}} \circ f = \{\varphi \circ f, \psi \circ f\}_{\mathcal{M}}$$

for any $\varphi, \psi \in C^\infty(\mathcal{N})$. In this case f maps the flow of the Hamiltonian $\varphi \circ f$ on $\mathcal{M}$ onto the flow of φ on $\mathcal{N}$.

A submanifold $\mathcal{M}$ of a Poisson manifold $\mathcal{N}$ is a *Poisson submanifold* if there is a Poisson structure on $\mathcal{M}$ such that the embedding $\mathcal{M} \hookrightarrow \mathcal{N}$ is a Poisson mapping (such a structure is unique if it exists). It is easy to check that $\mathcal{M}$ is a Poisson submanifold if and only if any Hamiltonian vector field in $\mathcal{N}$ is tangent to $\mathcal{M}$.

Obviously, symplectic manifolds carry a natural Poisson structure. The main theorem of the theory of Poisson manifolds which goes back to Lie [1893] asserts that an arbitrary Poisson manifold admits a stratification whose strata are symplectic manifolds (they may also be characterized as the *minimal Poisson submanifolds*). We shall refer to these strata as the *symplectic leaves*. A function φ is called a *Casimir function* (of a given Poisson structure on $\mathcal{M}$) if $\{\varphi, \psi\} = 0$ for all $\psi \in C^\infty(\mathcal{M})$. (In other words, Casimir functions are those functions which give rise to trivial equations of motion.) A function φ is a Casimir function if and only if it is constant on each symplectic leaf in $\mathcal{M}$.

The simplest example of a Poisson manifold which is not symplectic is the dual space of a Lie algebra equipped with the *Lie-Poisson bracket* (or, the *Kirillov bracket*) defined by the condition that for linear functions on $\mathfrak{g}^*$ it coincides with the Lie bracket on $\mathfrak{g}$:

$$\{X, Y\}(\xi) = \xi([X, Y]), \quad X, Y \in \mathfrak{g}, \quad \xi \in \mathfrak{g}^*. \tag{1.2}$$

Choose a basis $\{e_i\}$ in $\mathfrak{g}$. The tensor $c_{ij}(\xi)$ for the Lie-Poisson bracket has the form $c_{ij}(\xi) = \sum c_{ij}^k \xi_k$ where $\xi_k = \xi(e_k)$ and c_{ij}^k are the structure constants of $\mathfrak{g}$. The Hamiltonian equations (1.1) have the form

$$\frac{d\xi}{dt} = -\mathrm{ad}^*\, d\varphi(\xi) \cdot \xi, \tag{1.3}$$

where ad^* denotes the coadjoint representation of $\mathfrak{g}$ in $\mathfrak{g}^*$: the operator $\mathrm{ad}^* X$ in $\mathfrak{g}^*$ is the adjoint of $-\mathrm{ad}\, X$. This shows that $X \in \mathfrak{g}$ regarded as a linear function on $\mathfrak{g}^*$ is the Hamiltonian of the Hamiltonian action of the 1-parameter subgroup $\exp(-tX)$. Hence, the Casimir functions on $\mathfrak{g}^*$ are precisely the functions invariant under the coadjoint action. It is easy to check that symplectic leaves in $\mathfrak{g}^*$ are the orbits of this action (cf. Section 5).

Bibliographical Notes. After Lie [1893] the Lie-Poisson bracket and symplectic structure on coadjoint orbits were rediscovered by many authors (Berezin [1967], Kirillov [1972], Kostant [1970], Souriau [1970]). The global version of the theorem on symplectic leaves was given by Kirillov [1976]. A modern exposition of the theory is given by Weinstein [1983].

§2. The R-Matrix Method and the Main Theorem

The algebraic construction described in this section allows to unify the following main features of the inverse scattering method:

(a) The solution of the equations of motion reduces to the Riemann-Hilbert problem;

(b) The equations of motion are Hamiltonian with respect to a natural Poisson bracket;

(c) Integrals of the motion are the spectral invariants of an auxiliary linear operator (Lax operator). These integrals are in involution with each other with respect to the Poisson bracket mentioned above.

This construction is based on the interplay of two different structures of a Lie algebra (or two Lie-Poisson brackets) on the same space.

2.1. The Involutivity Theorem. Let $\mathfrak{g}$ be a Lie algebra, $R \in \operatorname{End} \mathfrak{g}$ a linear operator. We shall say that R is an *R-matrix* and defines on $\mathfrak{g}$ the structure of a *double Lie algebra* if the bracket

$$[X, Y]_R = \frac{1}{2}([RX, Y] + [X, RY]) \tag{2.1}$$

is a Lie bracket, i.e. if it satisfies the Jacobi identity (the skew symmetry of (2.1) is obvious for any R). We denote the Lie algebra with the bracket (2.1) by $\mathfrak{g}_R$. In the dual space $\mathfrak{g}^* \simeq \mathfrak{g}_R^*$ there are therefore two Poisson brackets: the Lie-Poisson brackets of $\mathfrak{g}$ and $\mathfrak{g}_R$.

The Lie-Poisson bracket of $\mathfrak{g}_R$ will be referred to as the *R-bracket*, for short.

A class of double Lie algebras most important for applications is constructed as follows. Assume that there is a vector space decomposition of $\mathfrak{g}$ into a direct sum of two Lie subalgebras, $\mathfrak{g} = \mathfrak{g}_+ \dotplus \mathfrak{g}_-$. Let $P_\pm$ be the projection operator onto $\mathfrak{g}_\pm$ parallel to the complementary subalgebra and put

$$R = P_+ - P_-. \tag{2.2}$$

In this case the bracket (2.1) is given by

$$\begin{gathered} [X, Y]_R = [X_+, Y_+] - [X_-, Y_-], \\ X_\pm = P_\pm X, \quad Y_\pm = P_\pm Y, \end{gathered} \tag{2.3}$$

i.e. is the difference of Lie brackets in $\mathfrak{g}_+$ and $\mathfrak{g}_-$. We shall briefly write $\mathfrak{g}_R = \mathfrak{g}_+ \ominus \mathfrak{g}_-$. The Jacobi identity for $[X, Y]_R$ is obvious from (2.3). Other examples and the general theory of R-matrices will be discussed at the end of this section (cf. Belavin and Drinfel'd [1982], Semenov-Tian-Shansky [1983]).

The definition of double Lie algebras is motivated by the following basic theorem. Let $I(\mathfrak{g})$ be the ring of Casimir functions on $\mathfrak{g}^*$ (i.e. the ring of its coadjoint invariants).

Theorem 2.1. (i) *Functions from $I(\mathfrak{g})$ are in involution with respect to the R-bracket on $\mathfrak{g}^*$.*

(ii) *The equations of motion induced by a function $\varphi \in I(\mathfrak{g})$ with respect to the R-bracket have the form*

$$\frac{dL}{dt} = -\mathrm{ad}_{\mathfrak{g}}^* M \cdot L, \quad L \in \mathfrak{g}^*, \quad M = \frac{1}{2} R(d\varphi(L)). \tag{2.4}$$

If $\mathfrak{g}$ admits a nondegenerate invariant bilinear form, so that $\mathfrak{g}^* = \mathfrak{g}$ and $\mathrm{ad}^* = \mathrm{ad}$, equations (2.4) take the Lax form

$$\frac{dL}{dt} = [L, M]. \tag{2.5}$$

It is convenient to rewrite (2.5) in the equivalent form

$$\frac{dL}{dt} = [L, M_\pm], \quad M_\pm = \frac{1}{2}(R \pm 1)(d\varphi(L)). \tag{2.6}$$

Proof. The R-bracket of two functions φ, ψ on $\mathfrak{g}^*$ is given by

$$\{\varphi, \psi\}_R(L) = \frac{1}{2}\langle L, [R(d\varphi(L)), d\psi(L)] + [d\varphi(L), R(d\psi(L))]\rangle.$$

The invariance of φ is equivalent to the relation $\langle L, [d\varphi(L), X]\rangle = 0$ for all $X \in \mathfrak{g}$. This proves (i). The equations of motion $\dot{\psi} = \{\varphi, \psi\}_R$ with an invariant Hamiltonian φ take the form

$$\dot{\psi}(L) = \frac{1}{2}\langle L, [R(d\varphi(L)), d\psi(L)]\rangle = -\frac{1}{2}\langle \mathrm{ad}^* R(d\varphi(L)) \cdot L, d\psi(L)\rangle,$$

whence $\dot{L} = -\frac{1}{2}\mathrm{ad}^* R(d\varphi(L)) \cdot L$. □

Warning. One should not confuse the approach based on double Lie algebras with the so called Magri-Lenard scheme (Magri [1978]) where the same equations are Hamiltonian with respect to different Poisson brackets. In general, the equations of motion induced by Casimir functions with respect to the R-bracket are not Hamiltonian with respect to the Lie-Poisson bracket of $\mathfrak{g}$.[1]

Theorem 2.1 has a transparent geometrical meaning: it shows that the trajectories of the dynamical systems with Hamiltonians $\varphi \in I(\mathfrak{g})$ lie in the intersection of two families of orbits in $\mathfrak{g}^*$, the coadjoint orbits of $\mathfrak{g}$ and $\mathfrak{g}_R$. Indeed, the coadjoint orbits of $\mathfrak{g}_R$ are preserved by all Hamiltonian flows in $\mathfrak{g}_R^*$. On the other hand, due to (2.4) the flow is always tangent to $\mathfrak{g}$-orbits in $\mathfrak{g}^*$. We shall see below that in many cases the intersections of orbits are the "Liouville tori" for our dynamical systems (cf. Section 9).

2.2. Factorization Theorem. The scheme outlined so far incorporates only two of the three main features of the inverse scattering method: the Poisson

[1] The hierarchies of compatible Poisson brackets associated with R-matrices are described in Section 4.

brackets and the Lax form of the equations of motion. As it happens, the most important feature of this method, the reduction of the equations of motion to the Riemann problem, is already implicit in our scheme. An abstract version of the Riemann problem is provided by the *factorization problem* in Lie groups.

We shall state the factorization theorem (which is a global version of Theorem 2.1) for the simplest R-matrices of the form (2.2). Let G be a connected Lie group with the Lie algebra $\mathfrak{g}$, and let $G_\pm$ be its subgroups corresponding to $\mathfrak{g}_\pm$.

Theorem 2.2. *Let $\varphi \in I(\mathfrak{g})$, $X = d\varphi(L)$. Let $g_\pm(t)$ be the smooth curves in $G_\pm$ which solve the factorization problem*

$$\exp tX = g_+(t)^{-1} g_-(t), \tag{2.7}$$

$g_\pm(0) = e$. Then the integral curve $L(t)$ of equation (2.4) *with $L(0) = L$ is given by*

$$L(t) = \mathrm{Ad}_G^* g_\pm(t) \cdot L. \tag{2.8}$$

Proof. Differentiating (2.8) with respect to t we get

$$\frac{d}{dt} L(t) = \mathrm{ad}^*(\dot{g}_+(t) g_+(t)^{-1}) \cdot L(t) = \mathrm{ad}^*(\dot{g}_-(t) g_-(t)^{-1}) \cdot L(t).$$

It remains to check that $\dot{g}_\pm(t) g_\pm(t)^{-1} = -M_\pm(t)$ (recall that $M_\pm(t) = \pm P_\pm X(t)$ where $X(t) = d\varphi(L(t))$). The Ad_G^*-invariance of φ implies that $X(t) = \mathrm{Ad}\, g_\pm(t) \cdot X$. Writing (2.7) in the form $g_+(t) \exp tX = g_-(t)$ and differentiating with respect to t we get $\dot{g}_+(t) g_+(t)^{-1} + \mathrm{Ad}\, g_-(t) \cdot X = \dot{g}_-(t) g_-(t)^{-1}$. Since $\dot{g}_\pm(t) g_\pm(t)^{-1} \in \mathfrak{g}_\pm$, this implies $\dot{g}_\pm(t) g_\pm(t)^{-1} = \mp P_\pm X(t)$. $\square$

We shall also give another, more geometric, proof of Theorem 2.2 which shows that Lax equations are obtained by a change of variables from the simplest G-invariant Hamiltonian systems on T^*G admitting explicit solutions.

Let us identify the cotangent bundle T^*G with $G \times \mathfrak{g}^*$ by means of left translations; then left-invariant functions on T^*G are identified with functions on $\mathfrak{g}^*$. Since the canonical Poisson bracket of left-invariant functions is again left-invariant, this induces a Poisson structure on $\mathfrak{g}^*$ which coincides with the Lie-Poisson bracket.

The Lie algebra $\mathfrak{g}_R$ corresponds to the Lie group $G_R = G_+ \times G'_-$ where G'_- is the group G_- with the opposite multiplication $g \cdot h = hg$. Define the mapping $\sigma\colon G_R \to G$ by $\sigma(g_+, g_-) = g_+ g_-$. Then at the unit element $e \in G_R$ we have $d\sigma = \mathrm{id}$ (the identity map $\mathfrak{g}_R \to \mathfrak{g}$). At an arbitrary point $(g_+, g_-) \in G_R$ the differential of σ is given by $d\sigma(g_+, g_-) = \mathrm{Ad}\, g_-^{-1} \circ \mathrm{id}$; this follows from the identity $\sigma(h_+ g_+, h_- \circ g_-) = h_+ \sigma(g_+, g_-) h_-$. Hence σ is an immersion, and we may extend it to a symplectic mapping $\sigma_*\colon T^*G_R \to T^*G$. The same identity also implies that if $\varphi \in C^\infty(T^*G)$ is bi-invariant, then its pullback $\varphi \circ \sigma_*$ is a left-invariant function on T^*G_R. Now, bi-invariant functions on T^*G may be canonically identified with Casimir functions on $\mathfrak{g}^*$. Since σ_* is symplectic, we reproduce the first assertion of Theorem 2.1: Casimir functions are in involution with respect to the R-bracket.

Next observe that for $\varphi \in I(\mathfrak{g})$ the corresponding equations of motion of T^*G have the form $d/dt(g, \xi) = (d\varphi(\xi), 0)$ and are easily solved:

$$\xi(t) = \text{const}, \quad g(t) = g \exp(t\, d\varphi(\xi)).$$

Consider the integral curves with $g = e$, $\xi = L$. The mapping σ_*^{-1} takes these curves into $(\sigma^{-1}(\exp tX), d\sigma^*(L))$, $X = d\varphi(L)$. Now, if $\exp tX = g_+(t)^{-1} g_-(t)$, then recalling that $d\sigma(g_+, g_-) = \mathrm{Ad}_G\, g_-^{-1} \circ \sigma$ we get

$$(\sigma^{-1}(\exp tX), d\sigma^*(L)) = ((g_+(t)^{-1}, g_-(t)), \mathrm{Ad}^*\, g_-(t) \cdot L).$$

Hence the integral curve of the Hamiltonian φ on $\mathfrak{g}^*$ with respect to the R-bracket is given by

$$L(t) = \mathrm{Ad}^*\, g_-(t) \cdot L = \mathrm{Ad}^*\, g_+(t) \cdot L$$

(since $\mathrm{Ad}_G^* \exp tX \cdot L = L$), as desired.

2.3. Classical Yang-Baxter Identity and the General Theory of Classical *R*-Matrices. Let us now discuss the conditions imposed on an R-matrix by the Jacobi identity for the R-bracket. Recall that the R-bracket associated with $R \in \mathrm{End}\, \mathfrak{g}$ is given by

$$[X, Y]_R = \frac{1}{2}([RX, Y] + [X, RY]).$$

Put

$$B_R(X, Y) = [RX, RY] - R([RX, Y] + [X, RY]). \tag{2.9}$$

Proposition 2.3. *The R-bracket* (2.1) *satisfies the Jacobi identity if and only if for any $X, Y, Z \in \mathfrak{g}$ we have*

$$[B_R(X, Y), Z] + [B_R(Y, Z), X] + [B_R(Z, X), Y] = 0. \tag{2.10}$$

The necessary and sufficient condition (2.10) is usually replaced by sufficient conditions which are bilinear rather than trilinear. The simplest sufficient condition is the so called *classical Yang-Baxter identity* (CYBE)

$$B_R(X, Y) = 0. \tag{2.11}$$

Another important sufficient condition is the *modified classical Yang-Baxter identity* (mCYBE)

$$B_R(X, Y) = -c^2[X, Y], \quad c = \text{const}. \tag{2.12}$$

By rescaling we may always assume that $c = 1$. Note that the R-matrices (2.2) satisfy mCYBE with $c = 1$.

The reason for the study of classical R-matrices satisfying the modified classical Yang-Baxter identity is that although they do not in general have the simple form (2.2), one can still associate with them a factorization problem. By contrast, the ordinary classical Yang-Baxter identity (2.11) represents a degenerate case and does not lead to a factorization problem.

Let us briefly describe the corresponding construction. Given an R-matrix which satisfies mCYBE with $c = 1$, put

$$R_{\pm} = \frac{1}{2}(R \pm I). \tag{2.13}$$

Proposition 2.4. *We have*

$$[R_{\pm}X, R_{\pm}Y] = R_{\pm}([X, Y]_R), \tag{2.14}$$

i.e. $R_{\pm}\colon \mathfrak{g}_R \to \mathfrak{g}$ *is a Lie algebra homomorphism.*

Put $\mathfrak{g}_{\pm} = \operatorname{Im} R_{\pm}$, $\mathfrak{k}_{\pm} = \ker R_{\mp}$.

Proposition 2.5. (i) $\mathfrak{g}_{\pm} \subset \mathfrak{g}$ *is a Lie subalgebra;*
(ii) $\mathfrak{k}_{\pm} \subset \mathfrak{g}_{\pm}$ *is an ideal.*

Define the mapping $\theta_R\colon \mathfrak{g}_+/\mathfrak{k}_+ \to \mathfrak{g}_-/\mathfrak{k}_-$ by setting $\theta_R\colon (R + I)X \mapsto (R - I)X$. Note that θ_R is well defined since $X \in \mathfrak{k}_{\pm}$ implies $(R \mp I)X = 0$.

Proposition 2.6. (i) θ_R *is a Lie algebra isomorphism.*
(ii) *Consider the combined mapping* $i_R = R_+ \oplus R_-$:

$$\mathfrak{g}_R \xrightarrow{i_R} \mathfrak{g} \oplus \mathfrak{g}.$$

This is a Lie algebra embedding and its image consists of those pairs $(X, Y) \in \mathfrak{g}_+ \oplus \mathfrak{g}_-$ *for which* $\theta_R(\bar{X}) = \bar{Y}$ *(here* $\bar{X}$, $\bar{Y}$ *is the residue class of* X, Y *in* $\mathfrak{g}_+/\mathfrak{k}_+$, $\mathfrak{g}_-/\mathfrak{k}_-$, *respectively).*
(iii) *Each* $X \in \mathfrak{g}$ *has a unique decomposition*

$$X = X_+ - X_-$$

with $(X_+, X_-) \in \operatorname{Im} i_R$.

The operator θ_R is called the *Cayley transform* of R.
Now let G, G_R be local Lie groups which correspond to $\mathfrak{g}$, $\mathfrak{g}_R$. The homomorphisms $R_{\pm}$ give rise to Lie group homomorphisms which we denote by the same letters. We put $G_{\pm} = R_{\pm}(G_R)$, $K_{\pm} = \operatorname{Ker} R_{\mp}$ and extend the Cayley transform θ_R to a Lie group isomorphism $\theta_R\colon G_+/K_+ \to G_-/K_-$. The combined mapping $G_R \xrightarrow{R_+ \times R_-} G \times G$ is a Lie group embedding and its image consists of those pairs $(x, y) \in G_+ \times G_-$ for which $\theta_R(\bar{x}) = \bar{y}$. Consider the map $m\colon G \times G \to G\colon (x, y) \to xy^{-1}$. The composition map $f\colon G_R \xrightarrow{m \circ (R_+ \times R_-)} G$ is a local homeomorphism and hence an arbitrary element $x \in G$ which is sufficiently close to the identity admits a unique representation

$$x = x_+ x_-^{-1} \tag{2.15}$$

with $(x_+, x_-) \in \operatorname{Im}(R_+ \times R_-)$.
The proof of Theorem 2.2 extends to the present setting with only minor changes.
Thus, in order to construct integrable systems we must indicate a Lie algebra $\mathfrak{g}$ with an R-matrix, determine its invariants, and describe suitable Poisson subspaces in $\mathfrak{g}_R^*$. This task becomes easier when $\mathfrak{g}$ is graded, see Section 3.

2.4. Examples. We end this section with some examples of classical R-matrices satisfying mCYBE which do not have the simplest form (2.2). Let $\mathfrak{g}$ be a complex semisimple Lie algebra, $\mathfrak{h}$ its Cartan subalgebra, Δ its root system, $\Delta_+ \subset \Delta$ the set of positive roots (with respect to some fixed order). For $\alpha \in \Delta$, let $\mathfrak{g}_\alpha \subset \mathfrak{g}$ be the corresponding root space. Put $\mathfrak{n}_\pm = \bigoplus_{\alpha \in \Delta_+} \mathfrak{g}_{\pm\alpha}$, $\mathfrak{b}_\pm = \mathfrak{h} + \mathfrak{n}_\pm$. Clearly, we have

$$\mathfrak{g} = \mathfrak{n}_+ + \mathfrak{h} + \mathfrak{n}_-. \tag{2.16}$$

Fix a nondegenerate linear operator $\theta \in \operatorname{End} \mathfrak{h}$ such that $\det(I - \theta) \neq 0$ and define $R_\theta \in \operatorname{End} \mathfrak{g}$ by the formulae

$$\begin{aligned} X &= X_+ + X_- + (I - \theta)X_0, \\ R_\theta X &= X_+ - X_- + (I + \theta)X_0. \end{aligned} \tag{2.17}$$

It is easy to see that R_θ satisfies mCYBE, $\operatorname{Im}(R_\theta \pm I) = \mathfrak{b}_\pm$, $\operatorname{Ker}(R_\theta \mp I) = \mathfrak{n}_\pm$ and its Cayley transform coincides with θ under the natural identification $\mathfrak{b}_\pm/\mathfrak{n}_\pm \simeq \mathfrak{h}$. The operator R_θ is skew with respect to the Killing form on $\mathfrak{g}$ if and only if θ is orthogonal. Note that by setting formally $\theta = 0$ in (2.17) we get an operator which has the form (2.2), i.e. is the difference of two complementary projectors. In a more abstract way, consider a decomposition of an arbitrary Lie algebra $\mathfrak{g}$ into a linear sum of three Lie subalgebras

$$\mathfrak{g} = \mathfrak{g}_+ + \mathfrak{g}_0 + \mathfrak{g}_-$$

such that $\mathfrak{g}_0$ normalizes both $\mathfrak{g}_+$ and $\mathfrak{g}_-$. Let P_+, P_-, P_0 be the projection operators onto $\mathfrak{g}_+, \mathfrak{g}_-, \mathfrak{g}_0$ defined by this decomposition.

Proposition 2.7. *Let* $R_0 \in \operatorname{End} \mathfrak{g}_0$ *be a linear operator satisfying the modified Yang-Baxter equation* (2.12). *Then*

$$R = P_+ - P_- + R_0 P_0 \tag{2.18}$$

also satisfies (2.12).

The next example shows, however, that the structure of general R-matrices satisfying mCYBE is more complicated than suggested by (2.18). Let $\mathfrak{g}$ be a semisimple algebra. Recall that its parabolic subalgebra $\mathfrak{p}$ is any subalgebra which contains a Borel subalgebra. Let $\mathfrak{a}$ be the split component of $\mathfrak{p}$ (the maximal split subalgebra lying in the center of $\mathfrak{p}$) and $\mathfrak{n}$ the nilpotent radical of $\mathfrak{p}$. Two parabolic subalgebras $\mathfrak{p}_1$, $\mathfrak{p}_2$ are said to be associated if their split components $\mathfrak{a}_1, \mathfrak{a}_2$ are conjugate in $\mathfrak{g}$, i.e. $\mathfrak{a}_1 = \operatorname{Ad} x \cdot \mathfrak{a}_2$ for some x in the adjoint group of $\mathfrak{g}$. Fix a Borel subalgebra $\mathfrak{b} \subset \mathfrak{g}$ and let $\bar{\mathfrak{b}}$ be the opposite Borel subalgebra. Let τ be an automorphism of $\mathfrak{g}$ that sends $\mathfrak{b}$ into $\bar{\mathfrak{b}}$. Choose two parabolic subalgebras $\mathfrak{p}_+ \supset \mathfrak{b}$ and $\mathfrak{p}_- \supset \bar{\mathfrak{b}}$.

Proposition 2.8. *An R-matrix* $R \in \operatorname{End} \mathfrak{g}$ *for which* $\mathfrak{g}_\pm = \mathfrak{p}_\pm$ *exists only if* $\mathfrak{p}_+$ *and* $\mathfrak{p}_-$ *are associated.*

Put $\mathfrak{l}_\pm = \mathfrak{p}_\pm \cap \tau\mathfrak{p}_\pm$ and let $\mathfrak{m}_\pm$ be the semisimple component of $\mathfrak{l}_\pm$. We have the decomposition

$$\mathfrak{p}_\pm = \mathfrak{l}_\pm + \mathfrak{n}_\pm = \mathfrak{m}_\pm + \mathfrak{a}_\pm + \mathfrak{n}_\pm$$

and hence $\mathfrak{g}_\pm/\mathfrak{k}_\pm \simeq \mathfrak{l}_\pm$. An operator Ad x such that $\mathfrak{a}_- = \operatorname{Ad} x\mathfrak{a}_+$ induces a linear isomorphism of $\mathfrak{m}_+$ onto $\mathfrak{m}_-$.

Proposition 2.9. (i) *Assume that* $\operatorname{Ad} x|_{\mathfrak{m}_+}$ *has no fixed vectors. Then there exists an R-matrix on* $\mathfrak{g}$ *such that* $\mathfrak{g}_\pm = \mathfrak{p}_\pm$, $\mathfrak{k}_\pm = \mathfrak{n}_\pm$ *and the restriction of its Cayley transform* $\theta \in \operatorname{Hom}(\mathfrak{l}_+, \mathfrak{l}_-)$ *to* $\mathfrak{m}_+$ *coincides with* $\operatorname{Ad} x|_{\mathfrak{m}_+}$. (ii) *All such R-matrices are parametrized by linear operators* $\theta_0 \in \operatorname{Hom}(\mathfrak{a}_+, \mathfrak{a}_-)$ (ii) *If* θ_0 *is an isometry, the corresponding R-matrix on* $\mathfrak{g}$ *is skew* (*with respect to the Killing form*).

One can prove that all skew R-matrices on $\mathfrak{g}$ satisfying mCYBE are constructed in this way. The condition that Ad x has no fixed points in $\mathfrak{m}_+$ is unavoidable. In particular, if $\mathfrak{p}_+ = \tau(\mathfrak{p}_-)$, there are no R-matrices with $\mathfrak{g}_\pm = \mathfrak{p}_\pm$ unless $\mathfrak{p}_+ = \mathfrak{b}$, $\mathfrak{p}_- = \bar{\mathfrak{b}}$. Indeed, in this case $\mathfrak{m}_+ = \mathfrak{m}_-$, and hence $\operatorname{Ad} x|_{\mathfrak{m}_+}$ is an automorphism of a semi-simple Lie algebra $\mathfrak{m}_+$. Such an automorphism always has fixed vectors. We refer the reader to Belavin and Drinfel'd [1984] where a similar construction is carried out for affine Lie algebras.

2.5. Bibliographical Notes. The first version of the general scheme exposed in Section 2 appeared in papers of Kostant [1979a] and Adler [1979] and was then developed by a number of authors (Symes [1980a,b], Reyman, Semenov-Tian-Shansky and Frenkel [1979]). The paper of Semenov-Tian-Shansky [1983] established a link between the Kostant-Adler scheme and the classical R-matrix method which was developed in connection with the quantum inverse scattering method (Sklyanin [1979], Faddeev [1980]). The factorization theorem was established by the authors (Reyman and Semenov-Tian-Shansky [1979]) and was subsequently many times rediscovered. The R-bracket of the form (2.3) was already used by van Moerbeke and Mumford [1979]. The general theory of classical R-matrices, in particular, the theory of the Cayley transform is due to Belavin and Drinfel'd [1982]; the role of the modified Yang-Baxter equation in connection with the factorization problems was discovered by Semenov-Tian-Shansky [1983]. Examples in Section 2.4 are basically due to Belavin and Drinfel'd [1982, 1984]; our exposition follows Semenov-Tian-Shansky [1983].

§ 3. Gradings and Orbits. Toda Lattices

3.1. Graded Algebras. A *grading* (or, more precisely, a $\mathbb{Z}$-*grading*) of a Lie algebra $\mathfrak{g}$ is a decomposition of $\mathfrak{g}$ into a linear sum of subspaces

$$\mathfrak{g} = \bigoplus_{i \in \mathbb{Z}} \mathfrak{g}_i$$

such that $[\mathfrak{g}_i, \mathfrak{g}_j] \subset \mathfrak{g}_{i+j}$.

A graded Lie algebra[2] admits an obvious decomposition into a sum of its subalgebras, $\mathfrak{g} = \mathfrak{g}_+ + \mathfrak{g}_-$, where

$$\mathfrak{g}_+ = \bigoplus_{i=0}^{\infty} \mathfrak{g}_i, \quad \mathfrak{g}_- = \bigoplus_{i=1}^{\infty} \mathfrak{g}_{-i}. \tag{3.1}$$

Let $\mathfrak{g}_j^* = \{f \in \mathfrak{g}^*; f|_{\mathfrak{g}_i} = 0 \text{ for } i \neq j\}$. It is natural to define the full algebraic dual $\hat{\mathfrak{g}}^*$ of $\mathfrak{g}$ as the projective limit of the spaces $\mathfrak{g}_{M,N}^* = \bigoplus_{j=-M}^{N} \mathfrak{g}_j^*$. In other words, the elements of $\hat{\mathfrak{g}}^*$ are formal series $f = \sum_j f_j$, $f_j \in \mathfrak{g}_j^*$. However, for our purposes we need only the restricted dual $\mathfrak{g}^*$ which consists of finite sums. Below the dual space of a graded Lie algebra always means the restricted dual. We have the decomposition

$$\mathfrak{g}^* = \mathfrak{g}_+^* + \mathfrak{g}_-^*, \quad \mathfrak{g}_+^* = \bigoplus_{i \geqslant 0} \mathfrak{g}_i^*, \quad \mathfrak{g}_-^* = \bigoplus_{i<0} \mathfrak{g}_i^*.$$

Let $P_\pm$ be the projection operators onto $\mathfrak{g}_\pm$ and let $R = P_+ - P_-$ be the operator associated with the decomposition (3.1). We have $\mathfrak{g}_R = \mathfrak{g}_+ \ominus \mathfrak{g}_-$. It is easy to check the following formula for the coadjoint representation of $\mathfrak{g}_R$. For $\xi \in \mathfrak{g}^*$ denote by $\xi_\pm$ the components of ξ in $\mathfrak{g}_\pm^*$: $\xi = \xi_+ + \xi_-$. Then

$$\operatorname{ad}^*_{\mathfrak{g}_R} X \cdot \xi = (\operatorname{ad}^* X_+ \cdot \xi)_+ - (\operatorname{ad}^* X_- \cdot \xi)_-, \quad X_\pm = P_\pm X. \tag{3.2}$$

The orbits of $\mathfrak{g}_R$ in $\mathfrak{g}^*$ are direct products of orbits of $\mathfrak{g}_+$ and $\mathfrak{g}_-$.

Lemma 3.1. *Fix $m, n \geqslant 0$ and $f \in \mathfrak{g}^*_{-m-1}$. The subspace $f + \bigoplus_{i=-m}^{n-1} \mathfrak{g}_i^*$ is a Poisson subspace in $\mathfrak{g}_R^*$.*

Proof. One has to show that this subspace is invariant with respect to $\operatorname{ad}^* \mathfrak{g}_R$. This follows from (3.2) and the inclusion $\operatorname{ad}^* \mathfrak{g}_i \cdot \mathfrak{g}_j^* \subset \mathfrak{g}_{j-i}^*$. □

The space $\bigoplus_{i=0}^{n} \mathfrak{g}_i^*$ is the dual of the quotient algebra $\mathfrak{g}_+/J_n$ where $J_n = \bigoplus_{i>n} \mathfrak{g}_i$, and the Poisson bracket in it coincides with the Lie-Poisson bracket of $\mathfrak{g}_+/J_n$. The same is true for the algebra $\mathfrak{g}_-$. In particular, the Poisson subspace $\mathfrak{g}_0^* + \mathfrak{g}_1^*$ is the dual of the Lie algebra $\mathfrak{g}_+/\bigoplus_{i>1} \mathfrak{g}_i = \mathfrak{g}_0 \ltimes \mathfrak{g}_1$, the semi-direct product of the Lie algebra $\mathfrak{g}_0$ and an abelian subspace $\mathfrak{g}_1$. The orbits of semi-direct products will be described in Section 5 in terms of Hamiltonian reduction.

In the sequel the main examples of integrable systems will be associated with the subspace $\mathfrak{g}_{-1}^* + \mathfrak{g}_0^* + \mathfrak{g}_1^*$.

Lemma 3.2. *Every orbit of the R-bracket in the subspace $\mathfrak{g}_{-1}^* + \mathfrak{g}_0^* + \mathfrak{g}_1^*$ is the sum of a one-point orbit of $\mathfrak{g}_-$ in $\mathfrak{g}_{-1}^*$ and an orbit of the semi-direct product $\mathfrak{g}_0 \ltimes \mathfrak{g}_1$ in $\mathfrak{g}_0^* + \mathfrak{g}_1^*$. In particular, the orbits lying in $\mathfrak{g}_0^*$ coincide with the coadjoint orbits of $\mathfrak{g}_0$.*

[2] This term has nothing to do with $\mathbb{Z}_2$-graded algebras, or Lie superalgebras, which satisfy a different set of axioms.

3.2. First Examples: Toda Lattices. We are now in a position to describe the classical Toda lattice and its generalizations.

Let $\mathfrak{g}$ be a real split simple Lie algebra, $\mathfrak{a}$ its split Cartan subalgebra, Δ its root system, Δ_+ a set of positive roots, $P \subset \Delta_+$ the system of simple roots. For $\alpha \in \Delta$ let $\mathfrak{g}_\alpha$ be the corresponding root space and $e_\alpha \in \mathfrak{g}_\alpha$ a root vector.

The root space decomposition $\mathfrak{g} = \mathfrak{a} + \bigoplus_{\alpha \in \Delta} \mathfrak{g}_\alpha$ gives rise to a natural grading on $\mathfrak{g}$ (the so called *principal grading*). By definition, the *height* $d(\alpha)$ of a root α is an additive function which is equal to 1 for simple roots. In other words, if $\beta = \sum_{\alpha \in P} k_\alpha \cdot \alpha$ then $d(\beta) = \sum k_\alpha$. Put

$$\mathfrak{g}_0 = \mathfrak{a}, \quad \mathfrak{g}_i = \bigoplus_{d(\alpha)=i} \mathfrak{g}_\alpha, \quad i \neq 0.$$

Then $[\mathfrak{g}_i, \mathfrak{g}_j] \subset \mathfrak{g}_{i+j}$. Let us now apply the construction of Section 2 to the standard decomposition (3.1) of $\mathfrak{g}$. (Observe that $\mathfrak{g}_+ = \bigoplus_{i \geqslant 0} \mathfrak{g}_i$ is a Borel subalgebra of $\mathfrak{g}$ and $\mathfrak{g}_-$ is the opposite nilpotent subalgebra, so that $\mathfrak{g} = \mathfrak{g}_+ + \mathfrak{g}_-$ is the usual *Bruhat decomposition*.) The Killing form $(\ ,\ )$ allows us to identify $\mathfrak{g}_i^*$ with $\mathfrak{g}_{-i}$, so that

$$\mathfrak{g}_+^* = \bigoplus_{i \leqslant 0} \mathfrak{g}_i, \quad \mathfrak{g}_-^* = \bigoplus_{i > 0} \mathfrak{g}_i.$$

Let us take as a Hamiltonian the simplest Casimir function

$$H(X) = \frac{1}{2}(X, X) \tag{3.3}$$

and consider the simplest non-trivial orbits. Namely, put $f = \sum_{\alpha \in P} e_{-\alpha}$ and let $\mathcal{O}_f$ be the $\mathfrak{g}_+$-orbit of f in $\mathfrak{a} + \mathfrak{g}_{-1}$.

Proposition 3.3. *The points of $\mathcal{O}_f$ have the form*

$$\xi = p + \sum_{\alpha \in P} c_\alpha e_{-\alpha}, \quad p \in \mathfrak{a}, \tag{3.4}$$

where $c_\alpha > 0$. The Lie-Poisson brackets of $p_\beta = (p, \beta)$ and c_α are

$$\{p_\beta, c_\alpha\} = (\alpha, \beta) c_\alpha, \quad \{c_\alpha, c_\beta\} = \{p_\alpha, p_\beta\} = 0. \tag{3.5}$$

It is convenient to introduce the variable $q \in \mathfrak{a}$ such that $c_\alpha = \exp \alpha(q)$. Fix a basis $\{h_i\}$ in $\mathfrak{a}$ and let $\{g_i\}$ be the dual basis with respect to the Killing form: $(h_i, g_j) = \delta_{ij}$. Put

$$q_i = (q, h_i), \quad p_i = (p, g_i).$$

Then

$$\{p_i, p_j\} = \{q_i, q_j\} = 0, \quad \{p_i, q_j\} = \delta_{ij},$$

i.e. the variables p_i, q_i are canonical. The orbit $\mathcal{O}_f$ is parametrized by the canonical variables as follows:

$$\xi = \sum_{i=1}^{l} p_i h_i + \sum_{\alpha \in P} \exp\left(\sum_{i=1}^{l} q_i (\alpha, g_i)\right) \cdot e_{-\alpha}. \tag{3.6}$$

The orbit which gives rise to the *generalized Toda lattice* has the form $\mathcal{O} = \mathcal{O}_f + e$ where $e = \sum_{\alpha \in P} e_\alpha$ is a one-point orbit in $\mathfrak{g}_-^*$. For $L = \xi + e$, the Hamiltonian (3.3) becomes

$$H(L) = \frac{1}{2}(p, p) + \sum_{\alpha \in P} d_\alpha e^{(\alpha, q)}, \quad d_\alpha = (e_\alpha, e_{-\alpha}). \tag{3.7}$$

The behaviour of the dynamical system with Hamiltonian (3.7) depends crucially on the signs of the coefficients d_α: the flow of (3.7) is complete if and only if all $d_\alpha \geqslant 0$ (cf. Section 3.3). In the sequel we normalize the root vectors e_α in such a way that $d_\alpha = 1$.

There are several useful forms of the Hamiltonian (3.7) depending on the choice of $\{h_i\}$. Let $\alpha_1, \ldots, \alpha_l$ be the simple roots; taking $h_i = \alpha_i$, we find

$$H = \frac{1}{2}\sum c_{ij} p_i p_j + \sum_i e^{q_i}, \quad c_{ij} = (\alpha_i, \alpha_j). \tag{3.8}$$

The coefficients c_{ij} are simply related to the coefficients of the *Cartan matrix* $n_{ij} = 2(\alpha_i, \alpha_j)/(\alpha_j, \alpha_j)$ (and coincide with them for simple Lie algebras of *ADE* type). In these coordinates Newton's equations take the form

$$\ddot{q}_i = -\sum_j c_{ij} e^{q_j}. \tag{3.9}$$

Upon a linear change of variables, $q_i = \sum_j c_{ij}\varphi_j$, they become

$$\ddot{\varphi}_i = -\exp \sum_j c_{ij}\varphi_j \tag{3.10}$$

and in this form are usually encountered in the generalized Liouville field theory.

In orthogonal coordinates, where $\{h_i\}$ is an orthonormal basis in $\mathfrak{a}$, the Toda Hamiltonian (3.7) takes the form

$$H(L) = \frac{1}{2}\sum p_i^2 + \sum_{\alpha \in P} \exp\left(\sum_i q_i(\alpha, e_i)\right). \tag{3.11}$$

Since $dH(L) = L$, the Lax representation for the Toda lattice has the form

$$\frac{dL}{dt} = [L, M_\pm], \tag{3.12}$$

where

$$L = p + \sum_{\alpha \in P} e^{(\alpha, q)} e_{-\alpha} + e, \quad M_+ = p + e, \quad M_- = -\sum_{\alpha \in P} e^{(\alpha, q)} e_{-\alpha}. \tag{3.13}$$

There are $l = \operatorname{rank} \mathfrak{g}$ functionally independent invariant polynomials on $\mathfrak{g}$. Restricted to the orbit $\mathcal{O}$ they remain functionally independent. (This follows from their functional independence on $\mathfrak{a}$ by considering points $\xi + e$ with $e^{(\alpha, q)}$ small enough.) By Theorem 2.1, these invariants Poisson commute on $\mathcal{O}$. Hence the generalized Toda Hamiltonians are completely integrable in the Liouville sense.

Table 1

A_l ($l \geqslant 1$)	$\frac{1}{2}\sum_{i=1}^{l+1} p_i^2 + \sum_{i=1}^{l} e^{q_i - q_{i+1}}, \quad \sum_{i=1}^{l+1} p_i = \sum_{i=1}^{l+1} q_i = 0$
B_l ($l \geqslant 2$)	$\frac{1}{2}\sum_{i=1}^{l} p_i^2 + \sum_{i=1}^{l-1} e^{q_i - q_{i+1}} + e^{q_l}$
C_l ($l \geqslant 2$)	$\frac{1}{2}\sum_{i=1}^{l} p_i^2 + \sum_{i=1}^{l-1} e^{q_i - q_{i+1}} + e^{2q_l}$
D_l ($l \geqslant 3$)	$\frac{1}{2}\sum_{i=1}^{l} p_i^2 + \sum_{i=1}^{l-1} e^{q_i - q_{i+1}} + e^{q_{l-1}+q_l}$
E_6	$\frac{1}{2}\sum_{i=1}^{8} p_i^2 + \sum_{i=1}^{5} e^{q_i - q_{i+1}} + e^{(-q_1-q_2-q_3+q_4+\cdots+q_8)/2}, \quad \sum_{i=1}^{6} p_i = \sum_{i=1}^{6} q_i = p_7 - p_8 = q_7 - q_8 = 0$
E_7	$\frac{1}{2}\sum_{i=1}^{8} p_i^2 + \sum_{i=1}^{6} e^{q_i - q_{i+1}} + e^{(-q_1-q_2-q_3-q_4+q_5+q_6+q_7+q_8)/2}, \quad \sum_{i=1}^{8} p_i = \sum_{i=1}^{8} q_i = 0$
E_8	$\frac{1}{2}\sum_{i=1}^{9} p_i^2 + \sum_{i=1}^{7} e^{q_i - q_{i+1}} + e^{(-q_1-\cdots-q_5+2q_6+2q_7+2q_8-q_9)/3}, \quad \sum_{i=1}^{9} p_i = \sum_{i=1}^{9} q_i = 0$
F_4	$\frac{1}{2}\sum_{i=1}^{4} p_i^2 + e^{1/2(q_1-q_2-q_3-q_4)} + e^{q_4} + e^{q_3-q_4} + e^{q_2-q_3}$
G_2	$\frac{1}{2}\sum_{i=1}^{3} p_i^2 + e^{q_1-q_2} + e^{q_2+q_3-2q_1}, \quad \sum_{i=1}^{3} p_i = \sum_{i=1}^{3} q_i = 0$

In Table 1 we give a list of Toda Hamiltonians (in an orthogonal basis) for all simple Lie algebras. Since the root systems of type A_n, E_6, E_7, E_8 and G_2 in $\mathfrak{a} = \mathbb{R}^l$ are easier to describe in terms of the standard basis in a larger space ($\mathbb{R}^{l+1}$ or $\mathbb{R}^{l+2}$) we write the corresponding Hamiltonians with more than l position variables and indicate the associated linear constraints. (Of course, the non-constrained system is also completely integrable.)

There is another realization of Toda lattices based on the Iwasawa decomposition of $\mathfrak{g}$. Let θ be a Cartan involution in $\mathfrak{g}$ which leaves $\mathfrak{a}$ invariant, and let $\mathfrak{k}$ be the fixed subalgebra of θ. The *Iwasawa decomposition* is $\mathfrak{g} = \mathfrak{b} + \mathfrak{k}$ where $\mathfrak{b} = \mathfrak{g}_+ = \mathfrak{a} + \bigoplus_{\alpha \in \Delta_+} \mathfrak{g}_\alpha$ is the same Borel subalgebra as before. The maximal compact subalgebra $\mathfrak{k}$ is spanned by the vectors $e_\alpha - e_{-\alpha}$, $\alpha \in \Delta_+$, where $e_{-\alpha} = -\theta e_\alpha$. The dual of $\mathfrak{b}$ is identified with the "symmetric" space $\mathfrak{p} = \mathfrak{k}^\perp$ spanned by $\mathfrak{a}$ and $e_\alpha + e_{-\alpha}$, $\alpha \in \Delta_+$. The element $f \in \mathfrak{b}^*$ which determines the Toda orbit is now represented by $f' = \sum_{\alpha \in P} (e_\alpha + e_{-\alpha})$. The $\mathfrak{b}$-orbit through f' in $\mathfrak{p}$ consists of the points

$$L' = 2p + \sum_{\alpha \in P} e^{(\alpha, q)/2}(e_\alpha + e_{-\alpha}), \tag{3.14}$$

where $q \in \mathfrak{a}$, $p \in \mathfrak{a}^*$ are canonically conjugate variables. (This choice of parametrization makes a comparison with (3.7) more convenient.) The Toda Hamiltonian $1/8(L', L')$ becomes

$$H = \frac{1}{2}(p, p) + \sum_{\alpha \in P} (e_\alpha, e_{-\alpha}) e^{(\alpha, q)} \tag{3.15}$$

which is the same as (3.7). Since $e_{-\alpha} = -\theta e_\alpha$, $(e_\alpha, e_{-\alpha})$ is positive. Thus the Iwasawa decomposition yields only Toda lattices with positive potentials. Without loss of generality we may assume that $(e_\alpha, e_{-\alpha}) = 1$. The corresponding Lax equation is

$$\frac{d}{dt} L' = [L', M'], \quad M' = \frac{1}{4} \sum_{\alpha \in P} e^{(\alpha, q)/2}(e_\alpha - e_{-\alpha}). \tag{3.16}$$

This Lax pair is gauge equivalent to the previous Lax pair (3.13):

$$\begin{gathered} L = \operatorname{Ad} e^{q/2} \cdot L', \\ \frac{1}{2}(M_+ + M_-) = \operatorname{Ad} e^{q/2} \cdot M' + e^{-q/2} \frac{d}{dt}(e^{q/2}). \end{gathered} \tag{3.17}$$

In particular, both Lax pairs give rise to the same set of integrals of motion for the Toda lattice. The "symmetric" realization (3.14) is often more convenient because the Iwasawa factorization on the Lie group G is globally defined. (This is not the case for the Bruhat factorization.)

Remark. If the compact subalgebra $\mathfrak{k}$ has nontrivial characters, we may consider the shifted Lax matrix $\hat{L} = L' + \chi$ where $\chi \in \mathfrak{k}^* \simeq \mathfrak{b}^\perp$ is a character (one-point orbit) of $\mathfrak{k}$. This happens precisely when $\mathfrak{g}$ is of type C_n, i.e. $\mathfrak{g} = \mathfrak{sp}(2n, \mathbb{R})$ and $\mathfrak{k} = \mathfrak{u}(n)$. By calculating $1/8 \cdot (\hat{L}, \hat{L})$ we find the potential

$$V(q) = \sum_{i=1}^{n-1} e^{q_i - q_{i+1}} + e^{2q_n} + c e^{q_n}, \tag{3.18}$$

where c is arbitrary. It can be viewed as the ordinary $\mathfrak{sl}(n)$-lattice with a Morse-type potential $e^{2q_n} + c e^{q_n}$ attached at the endpoint. In particular, for $n = 1$, $c = -1$ we have the Hamiltonian

$$H = \frac{1}{2} p^2 + e^{2q} - e^q$$

which describes a particle in the *Morse potential.*

3.3. Solutions and Scattering for Toda Lattices. The Factorization Theorem 2.2. allows us to determine the trajectories of the Toda lattice associated with $\mathfrak{g}$. Let G be a real split simple Lie group with Lie algebra $\mathfrak{g}$, let $\mathfrak{b} = \mathfrak{a} + \mathfrak{n}$ be its Borel subalgebra, $\bar{\mathfrak{n}}$ the opposite nilpotent subalgebra, $\mathfrak{k}$ the maximal compact subalgebra, and let $B, A, N, \bar{N}, K$ be the corresponding subgroups of G.

Consider the Bruhat factorization $e^{tL} = b(t)\bar{n}(t)$ of e^{tL}, where $b(t) \in B$, $\bar{n}(t) \in \bar{N}$ and L is given by (3.13). Then the Toda trajectory is

$$L(t) = \operatorname{Ad}_G b(t)^{-1} \cdot L = \operatorname{Ad}_G n(t) \cdot L. \tag{3.19}$$

To prove that this decomposition exists for all t we replace the Lax pair (3.13) by the gauge equivalent one (3.14), (3.16). The associated factorization problem is now the Iwasawa decomposition which exists globally. Namely, let $e^{tL'} = b(t)k(t)$ where $b(t) \in B$, $k(t) \in K$. Then

$$L'(t) = \mathrm{Ad}\, b(t)^{-1} \cdot L' = \mathrm{Ad}\, k(t)^{-1} \cdot L'. \tag{3.20}$$

Note that these factorization recipes for solving the Toda lattice are essentially the same as the one discussed in the last section of the survey by Olshanetsky and Perelomov in this volume. In particular, writing $b(t) = n(t)e^{c(t)}$ where $n(t) \in N$, $c(t) \in \mathfrak{a}$, we have from (3.20)

$$q(t) = q(0) + c(t). \tag{3.21}$$

The Cartan part $e^{c(t)}$ of the Iwasawa decomposition $e^{tL'} = n(t)e^{c(t)}k(t)$ may be analyzed by using representation theory (Kostant [1979b]). We shall briefly outline this argument.

Let π_λ be an irreducible finite-dimensional representation of G with highest weight λ, and let v_λ be the corresponding highest weight vector. We fix a K-invariant scalar product in the representation space. Then

$$\|\pi_\lambda(e^{-tL'})v_\lambda\| = \|\pi_\lambda(k(t)^{-1}e^{-c(t)}n^{-1}(t)v_\lambda\| = e^{-\lambda(c(t))}\|v_\lambda\|.$$

Let $\mathfrak{a}_+$ be the positive Weyl chamber in $\mathfrak{a}$. The Lax matrix L' is conjugate under K to a unique element, H, in $\bar{\mathfrak{a}}_+$. It can be shown that H is regular, i.e. $\alpha(H) > 0$ for any positive root α. We have $L' = \mathrm{Ad}\, k^{-1} \cdot H$ for some $k \in K$, and hence $e^{tL'} = k^{-1}e^{tH}k$. Thus

$$\|\pi_\lambda(e^{-tL'})v_\lambda\| = \|\pi_\lambda(e^{-tH}k)v_\lambda\|.$$

We shall assume that $\|v_\lambda\| = 1$. Decomposing the vector $\pi_\lambda(e^{-tH}k)v_\lambda$ with respect to the weight basis $\{v_\mu\}$ we find

$$e^{-2\lambda(c(t))} = \sum_\mu |(\pi\lambda(e^{-tH})\pi_\lambda(k)v_\lambda, v_\mu)|^2 = \sum_\mu |(\pi_\lambda(k)v_\lambda, v_\mu)|^2 e^{-2\mu(H)t}.$$

Since H is regular, the leading exponential terms in the above expressions are $e^{-2\lambda(H)t}$ as $t \to -\infty$ and $e^{-2\lambda^{w_0}(H)t}$ as $t \to +\infty$, where w_0 is the complete reflection in the Weyl group (w_0 maps $\mathfrak{a}_+$ to $-\mathfrak{a}_+$), so that λ^{w_0} is the lowest weight of π_λ. Assuming that $(\pi_\lambda(k)v_\lambda, v_\lambda) \neq 0$ and $(\pi_\lambda(k)v_\lambda, v_{\lambda^{w_0}}) \neq 0$ we find the asymptotics

$$\lambda(c(t)) = \lambda(H)t - \log|(\pi_\lambda(k)v_\lambda, v_\lambda)|, \quad t \to -\infty,$$

$$\lambda(c(t)) = \lambda(H^{w_0})t - \log|(\pi_\lambda(k)v_\lambda, v_{\lambda^{w_0}})|, \quad t \to +\infty.$$

This implies the following proposition.

Proposition 3.4. *The Toda lattice motion is asymptotically free*:

$$p(t) \sim H, \quad q(t) \sim Ht + q_- \quad \text{as} \quad t \to -\infty,$$

$$p(t) \sim H^{w_0}, \quad q(t) \sim H^{w_0}t + q_+ \quad \text{as} \quad t \to +\infty.$$

The scattering transformation which relates the asymptotic momenta as $t \to \pm\infty$ is the complete reflection w_0 in the Weyl group. The asymptotic behaviour of the Lax matrix is $L(t) \sim H$ as $t \to -\infty$, and $L(t) \sim H^{w_0}$ as $t \to +\infty$.

3.4. Bibliographical Notes. A Lax representation for Toda lattices was first obtained by Manakov [1974] and Flaschka [1974a,b]. The generalized Toda lattices associated with root systems were constructed by Bogoyavlensky [1976]. An interpretation of Toda lattices in terms of group theory given by Kostant [1979a] was at the origin of the general scheme exposed in this survey. The scattering problem for the ordinary A_n Toda lattice was analyzed by Moser [1975] and later studied in the general case by Kostant [1979b]. There is an interesting relationship between the factorization recipe for solving the Toda lattice and some fast computational algorithms for diagonalizing symmetric matrices (see Deift and Li [1989]).

§ 4. Affine Lie Algebras and Lax Equations with a Spectral Parameter

The most interesting examples of Lax equations are those where the Lax matrices depend on a spectral parameter. They are connected with the so-called affine Lie algebras (loop algebras).

4.1. Construction of Loop Algebras. Let $\mathfrak{g}$ be a Lie algebra. Its *loop algebra* $\mathfrak{L}(\mathfrak{g})$ is the Lie algebra of Laurent polynomials in the variable λ with coefficients in $\mathfrak{g}$:

$$\mathfrak{L}(\mathfrak{g}) = \mathfrak{g}[\lambda, \lambda^{-1}] = \left\{X(\lambda) = \sum_i x_i\lambda^i;\ x_i \in \mathfrak{g}\right\}.$$

The commutator in $\mathfrak{L}(\mathfrak{g})$ is given by $[x\lambda^i, y\lambda^j] = [x, y]\lambda^{i+j}$, or $[X, Y](\lambda) = [X(\lambda), Y(\lambda)]$.

Let σ be an automorphism of $\mathfrak{g}$ of order n. The *twisted loop algebra* $\mathfrak{L}(\mathfrak{g}, \sigma)$ is the subalgebra of $\mathfrak{L}(\mathfrak{g})$ defined by

$$\mathfrak{L}(\mathfrak{g}, \sigma) = \{X \in \mathfrak{L}(\mathfrak{g}) \colon X(\varepsilon\lambda) = \sigma X(\lambda)\},$$

where $\varepsilon = \exp 2\pi i/n$. (If $n > 2$, it is assumed that $\mathfrak{g}$ is complex.) In other words,

$$\mathfrak{L}(\mathfrak{g}, \sigma) = \left\{\sum_i x_i\lambda^i \colon \sigma(x_i) = \varepsilon^i x_i\right\}. \tag{4.1}$$

The most interesting case for applications is when σ is an involution, i.e. $n = 2$. In this case we may assume that $\mathfrak{g}$ and $\mathfrak{L}(\mathfrak{g})$ are real.

The algebra $\mathfrak{L}(\mathfrak{g}, \sigma)$ has a natural grading by powers of λ:

$$\mathfrak{L}(\mathfrak{g}, \sigma) = \bigoplus_{i \in \mathbb{Z}} \mathfrak{g}_i\lambda^i, \tag{4.2}$$

where $\mathfrak{g}_i = \{x \in \mathfrak{g}: \sigma(x) = \varepsilon^i x\}$ are the eigenspaces of σ. Below we shall also consider other gradings.

If $(\ ,\ )$ is a nondegenerate invariant bilinear form on $\mathfrak{g}$, we may define a nondegnerate invariant form on $\mathfrak{L}(\mathfrak{g})$ by

$$(X, Y) = \operatorname{Res}_{\lambda=0} \lambda^{-1}(X(\lambda), Y(\lambda))\, d\lambda. \tag{4.3}$$

The restriction of this form to $\mathfrak{L}(\mathfrak{g}, \sigma)$ also remains nondegenerate.

The algebraic dual of $\mathfrak{L}(\mathfrak{g}, \sigma)$ consists of formal Laurent series. However, we shall always consider the "restricted" dual space $\mathfrak{L}(\mathfrak{g}, \sigma)^* = \bigoplus_i \mathfrak{g}_i^* \lambda^i$ which consists of Laurent polynomials (cf. the beginning of Section 3.1). For $\sigma = \mathrm{id}$ we have

$$\mathfrak{L}(\mathfrak{g})^* = \mathfrak{g}^*[\lambda, \lambda^{-1}].$$

The nondegenerate invariant bilinear form (4.3) allows to identify $\mathfrak{L}(\mathfrak{g}, \sigma)^*$ with $\mathfrak{L}(\mathfrak{g}, \sigma)$ so that the coadjoint representation of $\mathfrak{L}(\mathfrak{g}, \sigma)$ is identified with the adjoint representation. In the sequel we shall usually assume that $\mathfrak{g}$ admits a nondegenerate invariant bilinear form, so that $\mathfrak{L}(\mathfrak{g}, \sigma)^*$ and $\mathfrak{L}(\mathfrak{g}, \sigma)$ are identified.

Twisted loop algebras $\mathfrak{L}(\mathfrak{g}, \sigma)$ where $\mathfrak{g}$ is semi-simple are also called *affine Lie algebras*.

Following the general scheme of Section 2, to construct Lax equations associated with $\mathfrak{L}(\mathfrak{g}, \sigma)$ we must indicate:

(1) a decomposition of $\mathfrak{L}(\mathfrak{g}, \sigma)$ into two subalgebras;
(2) the invariants of $\mathfrak{L}(\mathfrak{g}, \sigma)$;
(3) the orbits or suitable Poisson subspaces of the R-bracket.

This may easily be done. A decomposition into two subalgebras is defined by the grading (4.2). In agreement with (3.1) we put

$$\mathfrak{L}(\mathfrak{g}, \sigma)_+ = \bigoplus_{i \geqslant 0} \mathfrak{g}_i \lambda^i, \quad \mathfrak{L}(\mathfrak{g}, \sigma)_- = \bigoplus_{i<0} \mathfrak{g}_i \lambda^i. \tag{4.4}$$

The pairing $\langle u\lambda^i, x\lambda^j \rangle = u(x)\delta_{i,-j}$ between $\mathfrak{L}(\mathfrak{g})^*$ and $\mathfrak{L}(\mathfrak{g})$ allows us to identify the duals $\mathfrak{L}(\mathfrak{g}, \sigma)_\pm^*$ with the subspaces

$$\mathfrak{L}(\mathfrak{g}, \sigma)_+^* = \bigoplus_{i \leqslant 0} \mathfrak{g}_i^* \lambda^i, \quad \mathfrak{L}(\mathfrak{g}, \sigma)_-^* = \bigoplus_{i>0} \mathfrak{g}_i^* \lambda^i. \tag{4.5}$$

To give a coordinate expression for the R-bracket in $\mathfrak{L}(\mathfrak{g}, \sigma)^*$ defined by the decomposition (4.4), choose a basis e^a in $\mathfrak{g}$ and let C_c^{ab} be the structure constants of $\mathfrak{g}$: $[e^a, e^b] = \sum C_c^{ab} e^c$. Let $L(\lambda) = \sum u_i \lambda^i$, $u_i \in \mathfrak{g}^*$, and put $u_i^a = u_i(e^a)$. Then

$$\{u_i^a, u_j^b\} = \varepsilon_{ij} C_c^{ab} u_{i+j}^c \tag{4.6}$$

where $\varepsilon_{ij} = 1$ for $i, j \leqslant 0$, $\varepsilon_{i,j} = -1$ for $i, j > 0$, and $\varepsilon_{ij} = 0$ for $i \leqslant 0, j > 0$.

The invariants of $\mathfrak{L}(\mathfrak{g}, \sigma)$ are easily described.

Lemma 4.1. *Let φ be an invariant polynomial on $\mathfrak{g}^*$. For any integers m, n and for $L \in \mathfrak{L}(\mathfrak{g}, \sigma)^*$ set*

$$\varphi_{mn}[L] = \operatorname{Res}_{\lambda=0}(\lambda^{-n}\varphi(\lambda^m L(\lambda)))\, d\lambda.$$

Then φ_{mn} is an invariant polynomial on $\mathfrak{L}(\mathfrak{g}, \sigma)^*$.

By Lemma 3.1, the subspace $\bigoplus_{-m}^{n} \mathfrak{g}_i^* \lambda^i + f\lambda^{n+1}$ is a Poisson subspace of the R-bracket in $\mathfrak{L}(\mathfrak{g}, \sigma)^*$, provided $m \geqslant -1$, $n \geqslant 0$. (This is also seen from (4.6).) In other words, the subspaces $\bigoplus_{-m}^{0} \mathfrak{g}_i^* \lambda^i$ and $\bigoplus_{1}^{n} \mathfrak{g}_i^* \lambda^i + f\lambda^{n+1}$ are invariant under the coadjoint action of the subalgebras $\mathfrak{L}(\mathfrak{g}, \sigma)_+$ and $\mathfrak{L}(\mathfrak{g}, \sigma)_-$, respectively.

Proposition 4.2. *Let φ be an invariant polynomial on $\mathfrak{g}^*$ and set*

$$\varphi(\lambda^m L(\lambda)) = \sum \alpha_i[L]\lambda^i, \quad \varphi(\lambda^{-n-1} L(\lambda)) = \sum \beta_i[L]\lambda^{-i}.$$

Then $\alpha_1, \ldots, \alpha_m, \beta_1, \ldots, \beta_n$ are Casimir functions of the R-bracket in the subspace $\bigoplus_{-m}^{n} \mathfrak{g}_i \lambda^i \subset \mathfrak{L}(\mathfrak{g}, \sigma)^*$.

Proof. We will show that these polynomials are invariant under the corresponding coadjoint action. Consider, for instance, α_i and note that α_i depends only on the coefficients $u_{-m}, \ldots, u_{-m+i-1}$. Hence α_i is not affected by the action of $\mathfrak{L}(\mathfrak{g}, \sigma)_-$ which is trivial on $\mathfrak{L}(\mathfrak{g}, \sigma)_+^*$. The coadjoint action of an element $X(\lambda) \in \mathfrak{L}(\mathfrak{g}, \sigma)_+$ on L is

$$\mathrm{ad}_R^* X \cdot L = (\mathrm{ad}^* X \cdot L)_+,$$

where the $+$ subscript denotes projection to $\mathfrak{L}(\mathfrak{g}, \sigma)_+^*$ parallel to $\mathfrak{L}(\mathfrak{g}, \sigma)_-^*$. Since α_i depends only on the u_i, $i \leqslant 0$, the variation of α_i under the variation $\mathrm{ad}_R^* X \cdot L$ of L is the same as under the variation $\delta L = \mathrm{ad}^* X \cdot L$. But then $\delta\alpha_i = 0$ because α_i is an invariant polynomial on $\mathfrak{L}(\mathfrak{g}, \sigma)^*$ with respect to its original Lie algebra structure. □

Thus, in spite of the fact that the original algebra $\mathfrak{L}(\mathfrak{g}, \sigma)$ is infinite-dimensional, the orbits of the R-bracket in the polynomial dual of $\mathfrak{L}(\mathfrak{g}, \sigma)$ are finite-dimensional. Since the ring of invariants of $\mathfrak{L}(\mathfrak{g}, \sigma)$ has infinitely many independent generators, it is natural to expect that when restricted to an orbit of the R-bracket, these invariants become a complete Poisson-commuting family. In Section 9 this is shown for most orbits by means of algebro-geometric arguments.

4.2. Hierarchies of Poisson Structures for Lax Equations with a Spectral Parameter. There is an interesting phenomenon which is quite typical of Lax equations: usually they admit a whole hierarchy of compatible Poisson brackets. (Poisson brackets are called *compatible* if their linear combinations are again Poisson brackets.) These hierarchies naturally emerge in the R-matrix approach, and Lax equations with a spectral parameter provide a vast source of examples.

Let $\mathfrak{g}$ be a Lie algebra. A linear operator A in $\mathfrak{g}$ is called *intertwining* if it commutes with the adjoint representation of $\mathfrak{g}$, i.e. if

$$A[X, Y] = [AX, Y] = [X, AY] \tag{4.7}$$

for all $X, Y \in \mathfrak{g}$.

Proposition 4.3. *If R is an R-matrix and A an intertwining operator in $\mathfrak{g}$, then RA is also an R-matrix.*

Proof. Recall that the Jacobi identity for the R-bracket is equivalent to (2.10) where $B_R(X, Y)$ is defined by (2.9). Let $J_R(X, Y, Z)$ denote the left-hand side of (2.10). Clearly, if A is intertwining, $B_{RA}(X, Y) = B_R(AX, AY)$, and hence (2.10) implies $AJ_{RA}(X, Y, Z) = 0$. This concludes the proof if A is invertible; otherwise replace A by $A + \alpha I$ and let $\alpha \to 0$. Note that if R satisfies the modified Yang-Baxter equation, i.e. $B_R(X, Y) = -[X, Y]$, then $B_{RA}(X, Y) = -A^2[X, Y]$ and the equation $J_{RA}(X, Y, Z) = 0$ is obvious. □

Since the intertwining operators form a linear family, the Lie brackets $[\ ,\ ]_{RA}$ with R fixed also form a linear family. By duality we get a linear family of compatible Lie-Poisson brackets on $\mathfrak{g}^*$.

Observe that if A is invertible, the RA-bracket in $\mathfrak{g}^*$ is obtained from the R-bracket by the linear change of variables $L \mapsto (A^*)L$ in $\mathfrak{g}^*$.

The most interesting class of Lie algebras possessing a large number of intertwining operators are loop algebras $\mathfrak{L}(\mathfrak{g})$ (or, more, generally, algebras of meromorphic functions with values in $\mathfrak{g}$, see Section 4.5). Indeed, multiplication operators by scalar Laurent polynomials (respectively, by scalar meromorphic functions) are intertwining. Let R be the standard R-matrix (3.1) on $\mathfrak{L}(\mathfrak{g})$ and $\hat{q}$ the multiplication operator by $q \in \mathbb{C}[\lambda, \lambda^{-1}]$. Then $R \cdot \hat{q}$ is an R-matrix on $\mathfrak{L}(\mathfrak{g})$. We shall give explicit expressions for $R\hat{q}$-brackets in $\mathfrak{L}(\mathfrak{g})^*$ in terms of the coefficients of the Lax matrix $L(\lambda) = \sum u_i \lambda^i$, $u_i \in \mathfrak{g}^*$.

Let e^a be a basis in $\mathfrak{g}$ and C_c^{ab} the corresponding structure constants of $\mathfrak{g}$. Put $u_i^a = u_i(e^a)$. Denoting the $R\hat{\lambda}^k$-bracket by $\{\ ,\ \}_k$, we have

$$\{u_i^a, u_j^b\}_k = \varepsilon_{ij} C_c^{ab} u_{i+j-k}^c, \tag{4.8}$$

where

$$\varepsilon_{ij} = \begin{cases} 1, & i, j \leqslant k, \\ -1, & i, j > k, \\ 0, & i \leqslant k, \quad j > k. \end{cases}$$

The corresponding Hamiltonian equations in $\mathfrak{L}(\mathfrak{g}^*)$ have the form

$$\dot{u}_i = \sum_j \varepsilon_{ij} \mathscr{I}_{i+j-k}\left(\frac{\delta H}{\delta u_j}\right), \tag{4.9}$$

where the operator $\mathscr{I}_j$: $\mathfrak{g} \to \mathfrak{g}^*$ is given by

$$\mathscr{I}_j(X) = -\mathrm{ad}^* X \cdot u_j. \tag{4.10}$$

It is not hard to determine Poisson subspaces for the brackets $\{\ ,\ \}_k$, exactly as it was done for the basic bracket $\{\ ,\ \}_0$. In particular, for a fixed $a \in \mathfrak{g}^*$ and $m, n \in \mathbb{Z}$ $(m, n \geqslant 0)$ the subspace

$$\mathfrak{L}_{mn}^a = \left\{ \sum_{i=-m}^{n} u_i \lambda^i + a\lambda^{n+1} \right\} \tag{4.11}$$

is a common Poisson subspace for the brackets $\{\ ,\ \}_k$ with $-m-1 \leqslant k \leqslant n$, so that there is a $(n+m+2)$-parameter family of compatible brackets in $\mathfrak{L}^a_{mn}$. The corresponding Hamiltonian equations of motion in $\mathfrak{L}^a_{mn}$ can be written as

$$\frac{d}{dt}(u_{-m}, \ldots, u_n) = \mathscr{I}^{(k)}\left(\frac{\delta H}{\delta u_{-m}}, \ldots, \frac{\delta H}{\delta u_n}\right), \tag{4.12}$$

where $\mathscr{I}^{(k)}$ is an $(n+m+1) \times (n+m+1)$-operator-valued matrix:

$$\mathscr{I}^{(k)} = \left|\begin{array}{ccc|ccc} 0 & \cdots & \mathscr{I}_{-m} & & & \\ \vdots & \cdot^{\cdot^{\cdot}} & \vdots & & 0 & \\ \mathscr{I}_{-m} & \cdots & \mathscr{I}_k & & & \\ \hline & & & -\mathscr{I}_{k+2}, \ldots, & -\mathscr{I}_n, & -\mathscr{I}_{n+1} \\ & & & \vdots & \cdot^{\cdot^{\cdot}} & \\ & 0 & & -\mathscr{I}_n & \cdot^{\cdot^{\cdot}} & 0 \\ & & & -\mathscr{I}_{n+1} & & \end{array}\right|, \tag{4.13}$$

where $\mathscr{I}_j\,(-m \leqslant j \leqslant n)$ is given by (4.10) and $\mathscr{I}_{n+1}X = -\mathrm{ad}^*\, X \cdot a$. The invariant Hamiltonian

$$\varphi_f[L] = \mathrm{Res}_{\lambda=0}\,\{f(\lambda)\varphi(L(\lambda))\,d\lambda\} \tag{4.14}$$

with respect to the $R\hat{q}$-bracket gives rise to the Lax equation

$$\frac{dL}{dt} = [L, M_\pm], \tag{4.15}$$

where $M_\pm = \pm P_\pm(f(\lambda)q(\lambda)\nabla\varphi(L(\lambda))$. Clearly, these equations depend only on the product fq. In particular, the Hamiltonian $\varphi_{f\lambda^{-k}}$ produces the same equations of motion with respect to the bracket $\{\ ,\ \}_k$ for all k.

Obviously, the same construction applies to twisted loop algebras $\mathfrak{L}(\mathfrak{g}, \sigma)$. The only requirement is that the multiplier $q(\lambda)$ should satisfy $q(\varepsilon^d\lambda) = q(\lambda)$ where d is the order of σ.

4.3. Structure Theory of Affine Lie Algebras. This theory extends the structure theory of finite-dimensional simple Lie algebras. Here we shall need only some elementary properties of affine root systems which are outlined below without proof.

We begin by establishing an isomorphism between certain twisted algebras $\mathfrak{L}(\mathfrak{g}, \sigma)$. In this section $\mathfrak{g}$ is a complex simple Lie algebra.

Let Aut $\mathfrak{g}$ (respectively, Int $\mathfrak{g}$) be the group of all (respectively, inner) automorphisms of $\mathfrak{g}$. It is well known that the group of outer automorphisms Out $\mathfrak{g}$ = Aut $\mathfrak{g}$/Int $\mathfrak{g}$ coincides with the symmetry group of the Dynkin diagram of $\mathfrak{g}$.

Theorem 4.4. *Let* $\mathfrak{g}_1$ and $\mathfrak{g}_2$ *be complex simple Lie algebras. The Lie algebras* $\mathfrak{L}(\mathfrak{g}_1, \sigma_1)$ *and* $\mathfrak{L}(\mathfrak{g}_2, \sigma_2)$ *are isomorphic (as non-graded Lie algebras) if and only if* $\mathfrak{g}_1 \simeq \mathfrak{g}_2$ *and the images of* σ_1 *and* σ_2 *in the group* Out $\mathfrak{g}_1 \simeq$ Out $\mathfrak{g}_2$ *are conjugate.*

Thus the type of the algebra $\mathfrak{L}(\mathfrak{g}, \sigma)$ is determined by the conjugacy class of σ in Out $\mathfrak{g}$. The group Out $\mathfrak{g}$ is nontrivial only for Lie algebras of type A_n, D_n, E_6, and is equal to $\mathbb{Z}_2$ except for D_4. Accordingly, there are the following different types of affine Lie algebras: (i) the algebras $X_n^{(1)} \simeq \mathfrak{L}(\mathfrak{g})$ where $\mathfrak{g}$ is a simple Lie algebra of type X; (ii) the algebras $X_n^{(2)} = \mathfrak{L}(\mathfrak{g}, \sigma)$ where $\mathfrak{g}$ is of type $X_n = A_n$, D_n or E_6 and σ is an involution in $\mathfrak{g}$ which induces a nontrivial element of Out $\mathfrak{g}$; (iii) the algebra $\mathscr{D}_4^{(3)} = \mathfrak{L}(\mathfrak{g}, \sigma)$ where $\mathfrak{g}$ is of type D_4 and σ is an automorphism of order 3 which is nontrivial in Out $D_4 \simeq S_3$.

To illustrate Theorem 4.4 let us establish an isomorphism between $\mathfrak{L}(\mathfrak{g}, \sigma)$ and $\mathfrak{L}(\mathfrak{g})$ where σ is an inner automorphism of order n. Let $\sigma = \exp \operatorname{ad} X$ for $X \in \mathfrak{g}$, and let $\gamma(\lambda) = \exp\left(\frac{n}{2\pi i} \log \lambda \cdot \operatorname{ad} X\right)$. Since $\sigma^n = \mathrm{id}$, one can easily check that γ is a homomorphism from $\mathbb{C}^*$ into Int $\mathfrak{g}$. Moreover, $\gamma(\varepsilon) = \sigma$ and hence $\gamma(\varepsilon\lambda) = \sigma\gamma(\lambda)$. As the weights of γ: $\mathbb{C}^* \to \operatorname{End} \mathfrak{g}$ have the form λ^k, $k \in \mathbb{Z}$, it follows that $\gamma(\lambda)X(\lambda)$ is a Laurent polynomial for any $X \in \mathfrak{L}(\mathfrak{g})$. Hence the mapping $X(\lambda) \mapsto \gamma^{-1}(\lambda)X(\lambda)$ maps $\mathfrak{L}(\mathfrak{g}, \sigma)$ onto the subalgebra $\mathfrak{L}_n(\mathfrak{g}) = \{X \in \mathfrak{L}(\mathfrak{g}): X(\varepsilon\lambda) = X(\lambda)\}$. Obviously, $\mathfrak{L}_n(\mathfrak{g})$ is isomorphic to $\mathfrak{L}(\mathfrak{g})$.

We shall now describe the *root-space decomposition* of an affine Lie algebra. Consider first the loop algebra $\mathfrak{L}(\mathfrak{g})$. Let $\mathfrak{a} \subset \mathfrak{g}$ be a Cartan subalgebra, $\mathfrak{g} = \mathfrak{a} + \bigoplus_{\alpha \in \Delta} \mathfrak{g}_\alpha$ the root-space decomposition of $\mathfrak{g}$. Then

$$\mathfrak{L}(\mathfrak{g}) = \bigoplus_{\substack{\alpha \in \Delta \cup \{0\} \\ i \in \mathbb{Z}}} \mathfrak{g}_{\alpha, i}, \quad \mathfrak{g}_{\alpha, i} = \mathfrak{g}_\alpha \cdot \lambda^i, \quad \mathfrak{g}_{0, i} = \mathfrak{a} \cdot \lambda^i. \tag{4.16}$$

In order to distinguish between the subspaces $\mathfrak{g}_{\alpha, i}$ for different i let us adjoin to $\mathfrak{L}(\mathfrak{g})$ a new generator, the derivation d:

$$[d, X(\lambda)] = \lambda \frac{\partial}{\partial \lambda} X(\lambda), \tag{4.17}$$

or, equivalently,

$$[d, x \cdot \lambda^i] = i x \lambda^i, \quad x \in \mathfrak{g}.$$

Put $\hat{\mathfrak{g}} = \mathfrak{L}(\mathfrak{g}) + \mathbb{C}d$ and let $\hat{\mathfrak{a}} = \mathfrak{a} + \mathbb{C}d$ be the *extended Cartan subalgebra*. The root-space decomposition with respect to $\hat{\mathfrak{a}}$ may be written as

$$\hat{\mathfrak{g}} = \hat{\mathfrak{a}} + \bigoplus_{a \in \hat{\Delta}} \mathfrak{g}_a$$

where $\hat{\Delta}$ is the *affine root system* in $\hat{\mathfrak{a}}^*$,

$$\hat{\Delta} = \{a = (\alpha, i): \alpha \in \Delta \cup \{0\}, i \in \mathbb{Z}, a \neq (0, 0)\}. \tag{4.18}$$

If $\alpha \neq 0$, the subspaces $\mathfrak{g}_a$, $a = (\alpha, i)$, are one-dimensional. If $\alpha = 0$, $i \neq 0$, then $\mathfrak{g}_a = \mathfrak{a} \cdot \lambda^i$; such roots $a = (0, i)$ are called *imaginary*.

Let $\Delta_+ \subset \Delta$ be a subsystem of positive roots in Δ. We define the *positive roots in* $\hat{\Delta}$ by

$$\hat{\Delta}_+ = \{(\alpha, i)\text{: either } i > 0 \text{ or } \alpha \in \Delta_+, i = 0\}. \tag{4.19}$$

Clearly, $\hat{\Delta} = \hat{\Delta}_+ \cup -\hat{\Delta}_+$. A root $a \in \hat{\Delta}_+$ is called *simple* if it cannot be expressed as a nontrivial sum of positive roots.

Proposition 4.5. (i). *Every root is uniquely expressed as a sum of simple roots with integral coefficients of the same sign.*

(ii) *The system of simple roots for $\mathfrak{L}(\mathfrak{g})$ is*

$$\hat{P} = \{(\alpha_1, 0), \ldots, (\alpha_l, 0), (\omega, 1)\} \tag{4.20}$$

where $\alpha_1, \ldots, \alpha_l \in P$ are the simple roots of $\mathfrak{g}$ and ω is the minimal root (i.e. such that $\omega - \alpha$ is not a root for any $\alpha \in \Delta_+$).

Consider now the general case of a twisted loop algebra $\mathfrak{L}(\mathfrak{g}, \sigma)$. The subalgebra $\mathfrak{g}_0$ of fixed points of σ is reductive. Let $\mathfrak{h}_0 \subset \mathfrak{g}_0$ be its Cartan subalgebra. In complete analogy with (4.16) one defines the root-space decomposition of $\mathfrak{L}(\mathfrak{g}, \sigma)$ with respect to $\mathfrak{h}_0 + \mathbb{C}d$. The affine roots have the form $a = (\alpha, i)$ where α is a weight of $\mathfrak{h}_0$ in $\mathfrak{g}$ and $i \in \mathbb{Z}$. If $\alpha \neq 0$, i.e. if a is not imaginary, then $\dim \mathfrak{g}_a = 1$. The subsystem of positive roots is defined as in (4.19) with Δ_+ replaced by a subsystem of positive roots of $\mathfrak{g}_0$. Every root is an integral combination of simple roots. The number of simple roots is equal to rank $\mathfrak{g}_0 + 1$, hence their projections to $\mathfrak{h}_0$ satisfy a linear relation

$$\sum_{a \in \hat{P}} n_a \bar{a} = 0, \tag{4.21}$$

where $\bar{a}$ denotes the $\mathfrak{h}_0$-component of a. All n_a are positive integers.

As in the finite-dimensional cases, the affine root systems may be coded by Dynkin diagrams. Below in Table 2 we list the Dynkin diagrams for all irreducible affine root systems, and indicate the coefficients n_a for each simple root.

A concrete realization of an affine Lie algebra $\mathfrak{A}$ as a loop algebra $\mathfrak{L}(\mathfrak{g}, \sigma)$ induces a grading (4.2) on $\mathfrak{A}$. From (4.18) it is clear that a grading on $\mathfrak{A}$ is determined by an additive integer-valued function on the root system which is non-negative on $\hat{\Delta}_+$ and is completely fixed by its values on simple roots. In other words, a grading is determined by assigning an arbitrary non-negative weight $i(a)$ to each simple root $a \in \hat{P}$. There is one distinguished grading, called the *principal grading*, for which $i(a) = 1$ for each $a \in \hat{P}$. Equivalently, this is the grading by the height of a root: by definition, for $b = \sum_{a \in \hat{P}} k_a \cdot a \in \hat{\Delta}$ *height* $(b) = \sum k_a$.

Conversely, for each grading of an affine algebra $\mathfrak{A} = X_n^{(k)}$ there exists an automorphism σ of $\mathfrak{g} = X_n$ such that $\mathfrak{A}$ is realized as $\mathfrak{L}(\mathfrak{g}, \sigma)$ and the grading of $\mathfrak{A}$ is induced by the canonical grading of $\mathfrak{L}(\mathfrak{g}, \sigma)$ by powers of λ. The non-degenerate invariant inner product $(\ ,\ )$ induced on $\mathfrak{A}$ by (4.3) is independent, up to a constant factor, of the particular realization $\mathfrak{A} = \mathfrak{L}(\mathfrak{g}, \sigma)$. Note that $\mathfrak{A}_i$ is orthogonal to $\mathfrak{A}_j$ unless $i + j = 0$.

Table 2

$A_l^{(1)}$ ($l \geqslant 2$)	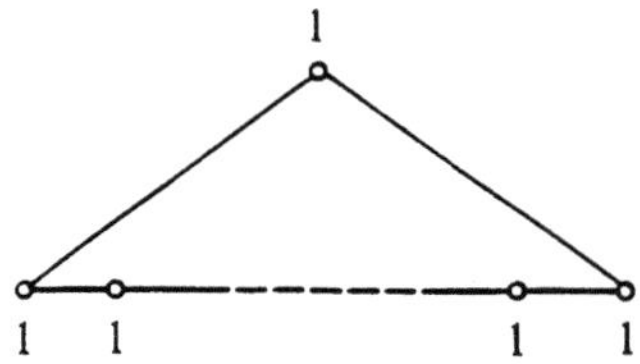	$E_6^{(1)}$	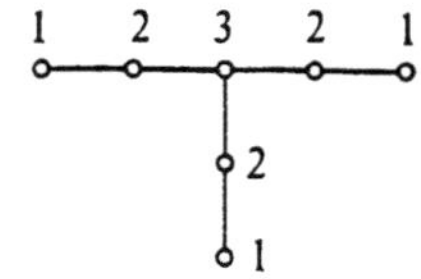
$A_1^{(1)}$		$E_7^{(1)}$	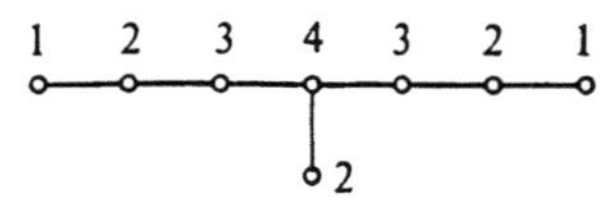
$B_l^{(1)}$ ($l \geqslant 3$)	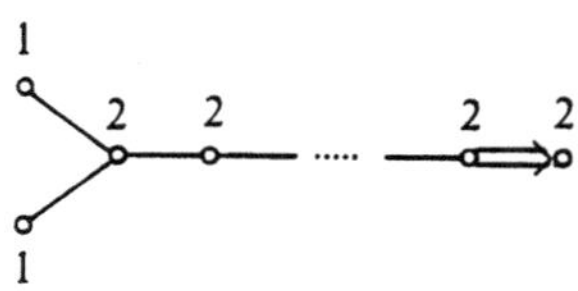	$E_8^{(1)}$	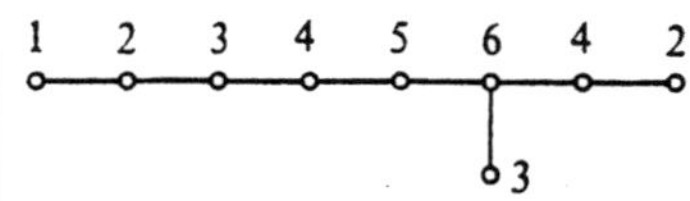
$C_l^{(1)}$ ($l \geqslant 2$)		$F_4^{(1)}$	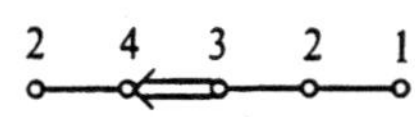
$D_l^{(1)}$ ($l \geqslant 4$)	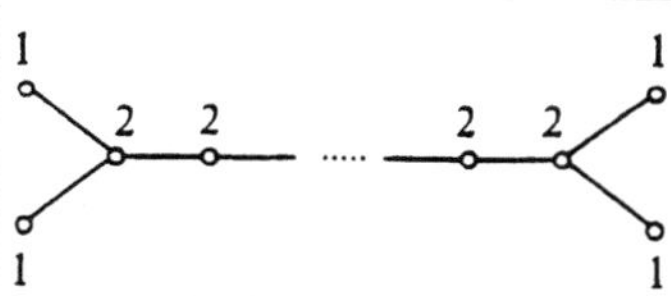	$G_2^{(1)}$	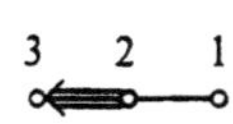
$A_{2l}^{(2)}$ ($l \geqslant 2$)	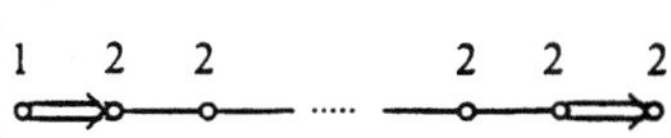	$D_{l+1}^{(2)}$ ($l \geqslant 2$)	
$A_2^{(2)}$		$E_6^{(2)}$	
$A_{2l-1}^{(2)}$ ($l \geqslant 3$)	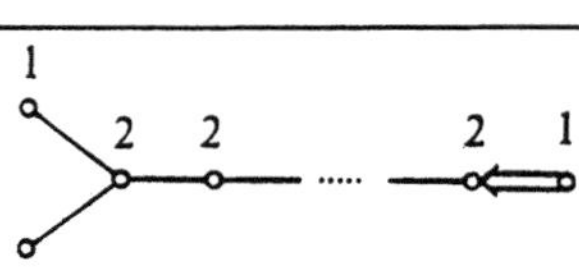	$D_4^{(3)}$	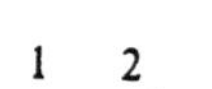

4.4. Periodic Toda Lattices. In complete analogy with Section 3.2, the root space decomposition of affine Lie algebras can be used to construct periodic Toda lattices.[3]

Let $\mathfrak{A}$ be a real split affine Lie algebra, so that the decomposition (4.16) is valid over the reals. Let $\mathfrak{A} = \bigoplus_i \mathfrak{A}_i$ be the principal grading, so that $\mathfrak{a} = \mathfrak{A}_0 \subset \mathfrak{A}$ is a Cartan subalgebra. Let $\mathfrak{A}_1 = \bigoplus_{a \in \hat{P}} \mathbb{R} \cdot e_a$ be the subspace spanned by the root vectors e_a corresponding to simple roots, $e_a \in \mathfrak{g}_a$. The orbit $\mathcal{O}_{\hat{f}}$ of the algebra $\mathfrak{A}_+$ which passes through $\hat{f} = \sum_{a \in \hat{P}} e_{-a}$, $e_{-a} \in \mathfrak{g}_{-a}$, consists of points of the form

$$\xi = p + \sum_{a \in \hat{P}} c_a e_{-a}, \tag{4.22}$$

where $p \in \mathfrak{a}$, $c_a > 0$, $\prod_a c_a^{n_a} = 1$ and n_a are the coefficients in (4.21). Let $\bar{a}$ denote the $\mathfrak{a}$-component of the root $a = (\alpha, i)$, i.e. $\bar{a} = \alpha$. The Lie-Poisson brackets of the variables c_a and $p_\beta = (p, \beta)$, $\beta \in \mathfrak{a}$, are $\{p_\beta, c_a\} = (\bar{a}, \beta) c_a$, $\{c_a, c_b\} = 0$, $\{p_\alpha, p_\beta\} = 0$.

As in Section 3.2 the orbit $\mathcal{O}_{\hat{f}}$ may be identified with $T^*\mathfrak{a} = \mathfrak{a} \oplus \mathfrak{a}^*$; the point ξ which corresponds to $(q, p) \in \mathfrak{a} \oplus \mathfrak{a}^*$ is $\xi = p + \sum_{a \in \hat{P}} \exp(\bar{a}, q) \cdot e_{-a}$. The orbit $\mathcal{O}$ associated with the periodic Toda lattice is the sum of $\mathcal{O}_{\hat{f}}$ and a one-point orbit $\hat{e} = \sum_{a \in \hat{P}} e_a$ of the subalgebra $\mathfrak{A}_-$:

$$\mathcal{O} = \left\{ L = \hat{e} + p + \sum_{a \in \hat{P}} e^{(\bar{a}, q)} e_{-a} \right\}. \tag{4.23}$$

The Hamiltonian $H = \frac{1}{2}(L, L)$ is

$$H = \frac{1}{2}(p, p) + \sum_{a \in \hat{P}} d_a e^{(\bar{a}, q)}, \tag{4.24}$$

where $d_a = (e_a, e_{-a})$. The shift $q \mapsto q + \gamma$ multiplies d_a by e^{γ_a} where $\sum_a \gamma_a n_a = 0$, so that effectively we have a one-parameter family of different Toda Hamiltonians associated with $\mathfrak{A}$. In orthogonal coordinates with basis $\{e_i\}$, H becomes

$$H = \frac{1}{2} \sum_i p_i^2 + \sum_{a \in \hat{P}} d_a e^{\sum_i (\bar{a}, e_i) q_i}.$$

The coefficients $(\bar{a}, e_i)$ can easily be computed from the *Cartan matrix*

$$A_{ab} = 2 \frac{(\bar{a}, \bar{b})}{(\bar{b}, \bar{b})}$$

[3] The term "periodic" is slightly misleading. It is due to the fact that the Toda system which corresponds to the algebra $A_{n-1}^{(1)}$ describes a system of n points on a lattice with the nearest neighbour interaction and with periodic boundary conditions. Systems associated with other affine Lie algebras are not periodic in this sense.

Table 3

$A_l^{(1)}$ $l \geqslant 1$	$H_{A_l} + e^{q_{l+1}-q_l}$, $\sum_{i=1}^{l+1} p_i = \sum_{i=1}^{l+1} q_i = 0$	$G_2^{(1)}$	$H_{G_2} + e^{q_1+q_2-2q_3}$, $p_1 + p_2 + p_3 = q_1 + q_2 + q_3 = 0$
$B_l^{(1)}$ $l \geqslant 2$	$H_{B_l} + e^{-q_1-q_2}$	$F_4^{(1)}$	$H_{F_4} + e^{-q_1-q_2}$
$C_l^{(1)}$ $l \geqslant 2$	$H_{C_l} + e^{-2q_1}$	$A_{2l}^{(2)}$ $l \geqslant 1$	$\frac{1}{2}\sum_{i=1}^{l} p_i^2 + \sum_{i=1}^{l-1} e^{q_i-q_{i+1}} + e^{q_l} + e^{-2q_1}$
$D_l^{(1)}$ $l \geqslant 4$	$H_{D_l} + e^{-q_1-q_2}$	$A_{2l-1}^{(2)}$ $l \geqslant 3$	$\frac{1}{2}\sum_{i=1}^{l} p_i^2 + \sum_{i=1}^{l-1} e^{q_i-q_{i+1}} + e^{2q_l} + e^{-q_1-q_2}$
$E_6^{(1)}$	$H_{E_6} + e^{-q_7-q_8}$, $\sum_{i=1}^{6} p_i = \sum_{i=1}^{6} q_i = p_7 - p_8 = q_7 - q_8 = 0$	$D_{l+1}^{(2)}$ $l \geqslant 2$	$\frac{1}{2}\sum_{i=1}^{l} p_i^2 + \sum_{i=1}^{l-1} e^{q_i-q_{i+1}} + e^{q_l} + e^{-q_1}$
$E_7^{(1)}$	$H_{E_7} + e^{q_7-q_8}$, $\sum_{i=1}^{8} p_i = \sum_{i=1}^{8} q_i = 0$	$E_6^{(2)}$	$\frac{1}{2}\sum_{i=1}^{4} p_i^2 + e^{q_2-q_3} + e^{q_3+q_4} + e^{-q_1} + e^{1/2(q_1-q_2-q_3-q_4)}$
$E_8^{(1)}$	$H_{E_8} + e^{q_9-q_1}$, $\sum_{i=1}^{9} p_i = \sum_{i=1}^{9} q_i = 0$	$D_4^{(3)}$	$\frac{1}{2}\sum_{i=1}^{3} p_i^2 + e^{q_1-q_2} + e^{q_2-q_3} + e^{2q_3-q_1-q_2}$, $\sum_{i=1}^{3} p_i = \sum_{i=1}^{3} q_i = 0$

of the affine Lie algebra. Finally, since $dH(L) = L$ the Lax representation for the Toda lattice is given by

$$\frac{d}{dt}L = [L, L_{\pm}], \quad L_+ = p + \hat{e}, \quad L_- = -\sum_{a \in \hat{P}} e^{(\bar{a}, q)} e_{-a}.$$

In Table 3 we list the potentials for all affine Lie algebras of type $X_n^{(k)}$, using the notation of Section 3.2, Table 1. As in Table 1, the potentials of type $A_n^{(1)}$, $E_n^{(1)}$ ($n = 6, 7, 8$), $G_2^{(1)}$ and $D_4^{(3)}$ are written with extra variables, and the corresponding linear constraints are indicated. We normalize the root vectors in such a way that $d_a = (e_a, e_{-a}) = 1$.

Remark. As in Section 3.2, one can define periodic lattices with Morse-type "impurities" based on the affine Lie algebras of type $C_n^{(1)}$. Let $\mathfrak{g} = \mathfrak{sp}(2n, \mathbb{R})$ and let $\mathfrak{g} = \mathfrak{n} + \mathfrak{a} + \mathfrak{k}$ be the Iwasawa decomposition of $\mathfrak{g}$ with $\mathfrak{k} = \mathfrak{u}(n)$. We extend this decomposition to the loop algebra $\mathfrak{L}(\mathfrak{g})$ by setting

$$\mathfrak{L}_+ = \mathfrak{n} + \mathfrak{a} + \bigoplus_{i>0} \mathfrak{g}\lambda^i,$$

$$\mathfrak{L}_- = \mathfrak{k} + \bigoplus_{i<0} \mathfrak{g}\lambda^i.$$

We have $\mathfrak{L}(\mathfrak{g}) = \mathfrak{L}_+ + \mathfrak{L}_-$. The dual $\mathfrak{L}_+^*$ is identified with $\mathfrak{p} + \bigoplus_{i<0} \mathfrak{g}\lambda^i$ where $\mathfrak{p} = \mathfrak{k}^\perp$. The Toda orbit in $\mathfrak{L}_+^*$ consists of the matrices

$$L = p + \sum_{\alpha \in P} e^{(\alpha, q)}(e_\alpha + e_{-\alpha}) + e^{(\omega, q)} e_{-\omega} \lambda^{-1},$$

where summation is over the simple roots of $\mathfrak{sp}(2n, \mathbb{R})$ and ω is the minimal root. The subalgebra $\mathfrak{L}_-$ has a two-dimensional family of characters which are represented in the dual $\mathfrak{L}_-^* \simeq \mathfrak{n} \oplus \bigoplus_{i>0} \mathfrak{g}\lambda^i$ by the one-point orbits

$$\chi = c_1 \begin{pmatrix} 0 & I \\ 0 & 0 \end{pmatrix} + c_2 \begin{pmatrix} 0 & I \\ -I & 0 \end{pmatrix} \lambda.$$

Taking the modified Lax matrix $\tilde{L} = L + \chi$ and calculating $H = \frac{1}{2} \operatorname{Res} \lambda^{-1} \operatorname{tr} \tilde{L}(\lambda)^2$ gives the Hamiltonian

$$H = \frac{1}{2} \sum p_i^2 + \sum e^{q_i - q_{i+1}} + e^{2q_n} + \alpha e^{q_n} + \beta e^{-q_1} \tag{4.25}$$

which differs from the $C_n^{(1)}$ Toda lattice described earlier.

4.5. Multi-pole Lax Equations. Lax matrices discussed so far are rational functions on the Riemann sphere with poles only at $\lambda = 0, \infty$. After an appropriate completion loop algebras admit different decompositons into the sum of two subalgebras which yield Lax equations with arbitrary poles. In Section 11 this construction will be generalized to include Lax equations with spectral parameter on an elliptic curve.

Let $\mathfrak{g}$ be a complex Lie algebra. Fix a finite set $D \subset \mathbb{C}P_1$ (we assume that $\infty \in D$). Let $R_D(\mathfrak{g})$ be the algebra of rational functions with values in $\mathfrak{g}$ that are regular outside D. For $v \in D$ let λ_v be the local parameter at v, i.e. $\lambda_v = \lambda - v$ for $v \neq \infty$, $\lambda_\infty = 1/\lambda$. Let $\mathfrak{L}(\mathfrak{g})_v = \mathfrak{g} \otimes \mathbb{C}((\lambda_v))$ be the algebra of formal Laurent series in the local parameter λ_v with coefficients in $\mathfrak{g}$; clearly, $\mathfrak{L}(\mathfrak{g})_v$ is a completion of the loop algebra $\mathfrak{L}(\mathfrak{g}) = \mathfrak{g} \otimes \mathbb{C}[\lambda, \lambda^{-1}]$. Let

$$\mathfrak{L}(\mathfrak{g})_v^+ = \mathfrak{g} \otimes \mathbb{C}[[\lambda_v]] \tag{4.26}$$

be its subalgebra consisting of formal power series. For $v = \infty$ we put

$$\mathfrak{L}(\mathfrak{g})_\infty^+ = \mathfrak{g} \otimes \lambda^{-1}\mathbb{C}[[\lambda^{-1}]] \tag{4.27}$$

(in other words, $\mathfrak{L}(\mathfrak{g})_\infty^+$ consists of formal power series in $\lambda_\infty = 1/\lambda$ without constant term).

Put

$$\mathfrak{L}(\mathfrak{g})_D = \bigoplus_{v \in D} \mathfrak{L}(\mathfrak{g})_v, \quad \mathfrak{L}(\mathfrak{g})_D^+ = \bigoplus_{v \in D} \mathfrak{L}(\mathfrak{g})_v^+. \tag{4.28}$$

There is a natural embedding $R_D(\mathfrak{g}) \to \mathfrak{L}(\mathfrak{g})_D$ which assigns to a rational function $X \in R_D(\mathfrak{g})$ the set of its Laurent expansions at the points $v \in D$.

Proposition 4.6. *We have*

$$\mathfrak{L}(\mathfrak{g})_D = R_D(\mathfrak{g}) \dotplus \mathfrak{L}(\mathfrak{g})_D^+. \tag{4.29}$$

Proof. Fix $X = \{X_\nu\}_{\nu\in D}$. For $\nu \in D$ let X_ν be the polynomial in λ_ν^{-1} such that $X_\nu - X_\nu^- \in \mathfrak{L}(\mathfrak{g})_\nu^+$. There exists a unique rational function $X^0 \in R_D(\mathfrak{g})$ whose principal parts at $\nu \in D$ coincide with X_ν^-. (Mind the normalization condition at $\lambda = \infty$.) Clearly, $X - X^0 \in \mathfrak{L}(\mathfrak{g})_D^+$. □

Let P_0 be the projection operator onto $R_D(\mathfrak{g})$ parallel to $\mathfrak{L}(\mathfrak{g})_D^+$. It is given by the following formula

$$(P_0X)(\lambda) = \sum_{\nu\in D} \operatorname{Res}_\nu(r(\lambda, \mu)X_\nu(\mu_\nu)\, d\mu), \tag{4.30}$$

where

$$r(\lambda, \mu) = \frac{I}{\lambda - \mu} \tag{4.31}$$

and I is the identity operator in $\mathfrak{g}$. The kernel (4.31) (which is essentially the Cauchy kernel) is called the *rational r-matrix*.

Let $K(\ ,\ )$ be a nondegenerate invariant form on $\mathfrak{g}$. We put

$$\langle X, Y\rangle = \sum_{\nu\in D} \operatorname{Res}_\nu K(X_\nu, Y_\nu)\, d\lambda. \tag{4.32}$$

Proposition 4.7. (i) *The bilinear form* (4.32) *is invariant and nondegenerate on* $\mathfrak{L}(\mathfrak{g})_D$. (ii) *The subalgebras* $R_D(\mathfrak{g})$, $\mathfrak{L}(\mathfrak{g})_D^+$ *are isotropic.*

Corollary. $(\mathfrak{L}(\mathfrak{g})_D^+)^* \simeq R_D(\mathfrak{g})$.

We shall apply Theorem 2.2 to the decomposition (4.29). Since we are interested in Lax equations with rational Lax matrices, we shall consider only orbits of $\mathfrak{L}(\mathfrak{g})_D^+$ in $R_D(\mathfrak{g})$. In order to understand the Lie-Poisson structure in the space $R_D(\mathfrak{g})$ observe first that there is another model of the dual space $(\mathfrak{L}(\mathfrak{g})_D^+)^*$ associated with the decomposition

$$\mathfrak{L}(\mathfrak{g})_D = \mathfrak{L}(\mathfrak{g})_D^+ \dotplus \mathfrak{L}(\mathfrak{g})_D^-, \tag{4.33}$$

where

$$\mathfrak{L}(\mathfrak{g})_D^- = \bigoplus_{\nu\in D} \mathfrak{L}(\mathfrak{g})_\nu^-,$$

$$\mathfrak{L}(\mathfrak{g})_\nu^- = \mathfrak{g}\otimes \lambda_\nu^{-1}\mathbb{C}[\lambda_\nu^{-1}] \quad (\nu \neq \infty), \quad \mathfrak{L}(\mathfrak{g})_\infty^- = \mathfrak{g}\otimes \mathbb{C}[\lambda].$$

Clearly, $(\mathfrak{L}(\mathfrak{g})_D^+)^* \simeq \mathfrak{L}(\mathfrak{g})_D^-$. An element $X = \left\{\sum_i c_{i\nu}\lambda_\nu^{-i}\right\}_{\nu\in D} \in \mathfrak{L}(\mathfrak{g})_D^-$ may be interpreted as a set of principal parts of a rational function on $\mathbb{C}P_1$ at $\nu \in D$. The two realizations of the dual are related by the mapping $\mathfrak{L}(\mathfrak{g})_D^- \to R_D(\mathfrak{g})$ which assigns to X the function $f = \sum_\nu \sum_i c_{i\nu}\lambda_\nu^{-i}$. The realization of $(\mathfrak{L}(\mathfrak{g})_D^+)^*$ as $\mathfrak{L}(\mathfrak{g})_D^-$ makes quite transparent the Lie-Poisson structure in the space of rational functions in terms of the decomposition of a function into partial fractions (or principal parts). The Poisson structure in $R_D(\mathfrak{g})$ is the direct product over $\nu \in D$ of the Poisson structures in the spaces of partial fractions $\left\{\sum_i c_{i\nu}\lambda_\nu^{-i}\right\}$ at each pole

ν, which in turn are the already familiar standard structures associated with the "local" algebras $\mathfrak{L}(\mathfrak{g})_\nu^+$. In particular, Poisson subspaces (orbits) in $R_D(\mathfrak{g})$ are direct products of Poisson subspaces (orbits) in $(\mathfrak{L}(\mathfrak{g})_\nu^+)^*$. For example, we know that $\left\{\sum_{i \leqslant N} c_i \lambda_\nu^{-i}\right\}$ is a Poisson subspace in $(\mathfrak{L}(\mathfrak{g})_\nu^+)^*$. Let $\mathfrak{D} = \sum_{\nu \in D} n_\nu \cdot \nu$ be a divisor with support in D and let

$$R(\mathfrak{g}; \mathfrak{D}) = \left\{\sum_\nu \sum_{i \leqslant n_\nu} c_{\nu i} \lambda_\nu^{-i}\right\}.$$

In other words, $R(\mathfrak{g}, \mathfrak{D})$ consists of such rational functions that the order of their pole at ν does not exceed n_ν. It follows that $R(\mathfrak{g}; \mathfrak{D})$ is a Poisson subspace in $R_D(\mathfrak{g})$.

The invariants of the algebra $\mathfrak{L}(\mathfrak{g})_D$ are generated by the functionals

$$\varphi_{f,\mu}\colon X = \{X_\nu\}_{\nu \in D} \mapsto \operatorname{Res}_\mu(f(\lambda_\mu)\varphi(X_\mu(\lambda_\mu))\, d\lambda), \tag{4.34}$$

where $\mu \in D$, $\varphi \in I(\mathfrak{g})$ and $f(\lambda_\mu) \in \mathbb{C}((\lambda_\mu))$ is a formal Laurent series. Since we are only interested in Lax equations in finite-dimensional subspaces $R(\mathfrak{g}; \mathfrak{D}) \subset R_D(\mathfrak{g})$, we may assume that f in (4.34) is a finite sum. (Indeed, the series f can be truncated without affecting the value of $\varphi_{f,\mu}$ on $R(\mathfrak{g}; \mathfrak{D})$). Let us denote the gradient of an invariant functional $\varphi[X]$ by $M = (M_\nu[X])_{\nu \in D}$. It is characterized by the property

$$\operatorname{ad} M_\nu[X] \cdot X_\nu = 0 \quad \text{for all} \quad \nu \in D.$$

If X_ν is the Laurent expansion of a rational function $X \in R_D(\mathfrak{g})$, the series $M_\nu[X]$ are locally convergent. (We assume here the all formal series $f(\lambda_\nu)$ in the definition of invariants are Laurent polynomials).

Proposition 4.9 (i) *The Hamiltonian equation of motion on $R_D(\mathfrak{g})$ generated by an invariant Hamiltonian φ has the Lax form*

$$\frac{dL}{dt} = [L, M_0], \quad L \in R_D(\mathfrak{g}), \quad M_0 = P_0(M[L]). \tag{4.35}$$

(ii) *For sufficiently small $t \in \mathbb{C}$ there exists a holomorphic function $g_0(\lambda, t)$ on $\mathbb{C}P_1 \backslash D$ with values in the Lie group G such that for all $\nu \in D$ the functions* $\exp tM_\nu[L] \cdot g_0(\lambda, t)^{-1}$ *are regular in the vicinity of ν. The integral curve of the Lax equation* (4.35) *starting from L is given by*

$$L(t) = \operatorname{Ad} g_0(\lambda, t) \cdot L. \tag{4.36}$$

The factorization problem described above is the so-called *matrix Cousin problem.*

Remark 1. The choice of the set D which enters the definition of our basic Lie algebra is, of course, arbitrary. It would be more natural to deal, instead of $\mathfrak{L}(\mathfrak{g})_D$, with the *algebra of adèles*

$$\mathfrak{L}(\mathfrak{g})_{\mathbb{A}} = \coprod_{\nu \in \mathbb{C}P_1} \mathfrak{g} \otimes \mathbb{C}((\lambda_\nu)).$$

[By definition, the elements of $\mathfrak{L}(\mathfrak{g})_{\mathbb{A}}$ are the sets $X = \{X_v\}_{v \in \mathbb{C}P_1}$ such that all X_v, except for a finite number of them, belong to $\mathfrak{L}(\mathfrak{g})_v^+$.] We preferred to deal with $\mathfrak{L}(\mathfrak{g})_D$ because its definition is simpler.

Remark 2. There is another approach to Lax equations with rational Lax matrices which avoids localization. Let us fix two disjoint sets of poles $D, D' \subset \mathbb{C}P_1$, $\infty \in D$, and let $R_{D \cup D'}(\mathfrak{g})$ be the algebra of rational functions on $\mathbb{C}P_1$ that are regular outside of $D \cup D'$. Then

$$R_{D \cup D'}(\mathfrak{g}) = R_D(\mathfrak{g}) \dot{+} {}^0R_{D'}(\mathfrak{g}), \tag{4.37}$$

where the functions in ${}^0R_{D'}$ are required to be zero at $\lambda = \infty$. Define the inner product on $R_{D \cup D'}(\mathfrak{g})$ by

$$\langle X, Y \rangle = \sum_{v \in D} \operatorname{Res}_v K(X_v, Y_v)\, d\lambda. \tag{4.38}$$

This pairing sets the subspaces $R_D(\mathfrak{g})$, ${}^0R_{D'}(\mathfrak{g})$ into duality, and we may equip $R_D(\mathfrak{g})$ with the Lie-Poisson bracket associated with ${}^0R_{D'}(\mathfrak{g})$. It is easy to see that this construction reduces to the proceding one. Namely, consider the embedding

$$i_D \colon R_{D \cup D'}(\mathfrak{g}) \to \mathfrak{L}(\mathfrak{g})_D = \bigoplus_{v \in D} \mathfrak{L}(\mathfrak{g})_v$$

which assigns to each function $X \in R_{D \cup D'}(\mathfrak{g})$ its Laurent expansions at $v \in D$. Clearly, i_D maps ${}^0R_{D'}$ into $\mathfrak{L}(\mathfrak{g})_D^+ = \bigoplus_{v \in D} \mathfrak{L}(\mathfrak{g})_v^+$ and the bilinear form (4.38) on $R_{D \cup D'}$ is compatible with the form (4.32) on $\mathfrak{L}(\mathfrak{g})_D$. Hence the subspaces $R(\mathfrak{g}; \mathfrak{D}) \subset R_D(\mathfrak{g})$ associated with divisors $\mathfrak{D} = \sum_{v \in D} n_v \cdot v$ are Poisson subspaces in $R_D(\mathfrak{g}) \subset ({}^0R_{D'}(\mathfrak{g}))^*$ and the Poisson structure in this subspace coincides with the one induced by $\mathfrak{L}(\mathfrak{g})_D^+$. It is also easy to establish a correspondence between Lax equations on $R_D(\mathfrak{g})$ that are constructed with the use of decompositions (4.29) and (4.37). The choice of D' in (4.37) is arbitrary and does not affect the Poisson structure on the subspaces $R(\mathfrak{g}; \mathfrak{D}) \subset R_D(\mathfrak{g})$ or the supply of invariant Hamiltonians.

Remark 3. As in Section 4.2, we may define a hierarchy of compatible Poisson brackets on $R_D(\mathfrak{g})$. On the subspace $R(\mathfrak{g}, \mathfrak{D})$ associated with a divisor $\mathfrak{D} = \sum_{v \in D} n_v \cdot v$ the family of compatible Poisson brackets is generated by multiplication operators $\hat{q}$, where $q \in \mathbb{C}(\lambda)$, $(q) \geqslant -\mathfrak{D}$ (i.e. q is regular outside D and the order of its pole at $v \in D$ does not exceed n_v). Clearly, this family depends on $N = \sum_v n_v + 1$ parameters. The proof is similar to the one given in Section 4.2 (it is simplified by the use of adelic language).

4.6. Bibliographical Notes. Affine Lie algebras are the most important subclass of Kac-Moody algebras, see Kac [1984] or Goddard and Olive [1988]. In partial violation of the standard terminology, we use the name of affine Lie algebras for the loop algebras $\mathfrak{L}(\mathfrak{g}, \sigma)$ where $\mathfrak{g}$ is simple; affine Lie algebras in the sense of Kac-Moody are their central extensions. Affine Lie algebras were suggested as a tool for studying Lax equations by Reyman, Semenov-Tian-Shansky and Frenkel [1979] and by Adler and van Moerbeke [1980a, b]. Further applications were given by Reyman [1980], Reyman and Semenov-Tian-Shansky [1979], [1981], Leznov and Saveli'ev [1979] and others. Compatible Poisson brackets for integrable equations were first introduced by Magri [1978]. The R-matrix construction of compatible brackets given in Section 4.2 is due to Reyman and Semenov-Tian-Shansky [1988]. Generalized Toda lattices corresponding to extended root systems were originally discovered by Bogoyavlensky [1976]. The involutivity of Lax integrals of motion for these systems can be proved entirely by finite-dimensional means (Kostant [1982]). The complexity of this proof is another argument in favour of the use of affine Lie algebras. The Toda lattice with Morse type impurities (4.25) was missed in the early days of the theory. Using a different technique Sklyanin [1987] has shown that the more general Hamiltonian

$$H = \frac{1}{2}\sum_{i=1}^{n} p_i^2 + \sum_{i=1}^{n-1} e^{q_i - q_{i+1}} + c_1 e^{2q_n} + c_2 e^{q_n} + c_3 e^{-2q_1} + c_4 e^{-q_1} \tag{4.39}$$

is also completely integrable. It is interesting to compare these systems with the classification theorem of Adler and van Moerbeke [1982]: the only algebraically completely integrable systems with Hamiltonian

$$H = \frac{1}{2}\sum_{i=1}^{n} p_i^2 + \sum_{j=1}^{n+1} \exp \sum_{i=1}^{n} N_{ij} q_i,$$

where N_{ij} is a matrix of rank n, are the Toda lattices associated with affine Lie algebras. It is important here that the potential contains at most $n + 1$ exponential terms. On the other hand, the potentials in (4.25) and (4.39) have $n + 2$ and $n + 3$ terms, respectively. The R-matrix interpretation of Lax equations with rational Lax matrices was proposed by Reyman and Semenov-Tian-Shansky [1986a]. The adelic formalism in the theory of classical R-matrices was introduced earlier by Cherednik [1983b] and independently by Drinfel'd (cf. Drinfel'd [1987]). This formalism also extends to Lax equations with a spectral parameter on an elliptic curve which are studied in Section 11. We do not consider here twisted algebras of rational functions on $\mathbb{C}P_1$ although they lead to some interesting possibilities (cf. Mikhailov [1981]).

§ 5. Hamiltonian Reduction and Orbits of Semi-direct Products

In this survey we shall frequently deal with phase spaces obtained by Hamiltonian reduction. For this reason we shall briefly describe the reduction procedure in the form suitable for our purposes.

5.1. Hamiltonian Reduction. Let $\mathcal{M}$ be a symplectic manifold and G a Lie group which acts on $\mathcal{M}$ by symplectic transformations. Assume that the space of orbits $\mathcal{M}/G$ is a smooth manifold. In this case there is a natural Poisson structure on $\mathcal{M}/G$ such that the canonical projection $\pi: \mathcal{M} \to \mathcal{M}/G$ is a Poisson mapping. Indeed, the space $C^\infty(\mathcal{M}/G)$ may be identified with the space of G-invariant functions on $\mathcal{M}$. Since the action of G preserves the symplectic form on $\mathcal{M}$, this latter space is a Lie subalgebra in $C^\infty(\mathcal{M})$ with respect to the Poisson bracket, and this induces a Poisson Bracket on $\mathcal{M}/G$. The space $\mathcal{M}/G$ is called the *reduced space* (obtained by reduction of M with respect to G).

The difficult part of reduction consists in describing the symplectic leaves in $\mathcal{M}/G$. In this section we shall discuss it under the assumption that the action of G is Hamiltonian; it is well known that an arbitrary symplectic action can be made Hamiltonian by replacing, if necessary, the group G by its central extension (Arnol'd [1974], Kostant [1970]).

Recall that the action of G on $\mathcal{M}$ is Hamiltonian if the action of each 1-parameter subgroup $\exp(-tX)$, $X \in \mathfrak{g}$, is generated by a Hamiltonian $H_X \in C^\infty(\mathcal{M})$, the mapping $X \mapsto H_X$ is linear and, moreover,

$$\{H_X, H_Y\} = H_{[X,Y]}. \tag{5.1}$$

Let us define the *moment map* $\mu: \mathcal{M} \to \mathfrak{g}$ by

$$\langle \mu(m), X \rangle = H_X(m), \quad m \in \mathcal{M}.$$

Formula (5.1) implies that μ is G-equivariant and is a Poisson mapping with respect to the Lie-Poisson bracket. Conversely, any Poisson mapping $\mu: \mathcal{M} \to \mathfrak{g}^*$ gives rise to a Hamiltonian action of $\mathfrak{g}$ on $\mathcal{M}$.

The properties of the maps $\pi: \mathcal{M} \to \mathcal{M}/G$ and $\mu: \mathcal{M} \to \mathfrak{g}^*$ deserve special study.

Definition. Let M be a symplectic manifold, $\mathcal{U}$ and $\mathcal{V}$ two Poisson manifolds. Two Poisson mappings $f: \mathcal{M} \to \mathcal{U}$, $g: M \to \mathcal{V}$ form a *dual pair* if the Lie algebras $f^*C^\infty(\mathcal{U})$ and $g^*C^\infty(\mathcal{V})$ are centralizers of each other in $C^\infty(M)$.

Remark. Without loss of generality we may assume that f and g are surjective, i.e. $f(g)$ maps M onto U (resp., onto V). A surjective Poisson mapping $f: M \to U$ is called a *symplectic realization* of the Poisson manifold $\mathcal{U}$. Symplectic realizations allow to treat dynamical systems on various symplectic leaves of a Poisson manifold in a uniform way at the expence of the introduction of some extra variables. This approach was probably used for the first time by Clebsch (for some special Poisson brackets in hydrodynamics) and for this reason the additional variables are sometimes referred to as Clebsch variables.

Symplectic realizations will play an important role in Sections 6, 7 where they are used to study the motion of multi-dimensional tops. Other examples related to the theory of Poisson Lie groups and of difference Lax equations are studied in Section 12.

The role of the notion of dual mappings is explained by the following general assertion.

Theorem 5.1. *Let $\mathscr{U} \xleftarrow{f} \mathscr{M} \xrightarrow{g} \mathscr{V}$ be a dual pair. Then the connected components of the sets $f(g^{-1}(v)) \subset \mathscr{U}$, $v \in \mathscr{V}$, and $g(f^{-1}(u)) \subset \mathscr{V}$, $u \in \mathscr{U}$, are symplectic leaves in $\mathscr{U}$ and $\mathscr{V}$, respectively.*

In our particular setting we have the following proposition.

Proposition 5.2. *Let $G \times M \to M$ be a Hamiltonian group action, $\pi: M \to M/G$ the canonical projection mapping, and $\mu: M \to \mathfrak{g}^*$ the corresponding moment mapping. Then $M/G \xleftarrow{\pi} M \xrightarrow{\mu} \mathfrak{g}^*$ is a dual pair.*

Corollary 5.3. *Symplectic leaves in M/G are the connected components of the sets $\pi(\mu^{-1}(f))$, $f \in \mathfrak{g}^*$.*

We shall say that the symplectic manifold $\overline{M}_f = \pi(\mu^{-1}(f))$ is obtained by reduction over $f \in \mathfrak{g}^*$. A more traditional definition of the reduced phase space is as follows. For $f \in \mathfrak{g}^*$ put $\mathscr{M}_f = \mu^{-1}(f)$ and let G_f be the identify component of the stationary subgroup of f. The action of G_f leaves M_f invariant. The restriction of the symplectic form to $\mathscr{M}_f$ is degenerate and its null foliation coincides with the orbits of G_f. The symplectic manifold $\mathscr{M}_f/G_f$ is called the reduced manifold obtained by reduction over f. Clearly, $\mathscr{M}_f/G_f$ coincides with $\overline{\mathscr{M}}_f = \pi(\mu^{-1}(f))$ and hence this definition is in agreement with Corollary 5.3.

A G-invariant Hamiltonian φ on $\mathscr{M}$ gives rise to the *reduced Hamiltonian* $\bar{\varphi}$ on $\mathscr{M}/G$ (which is the unique function on $\mathscr{M}/G$ such that $\varphi = \bar{\varphi} \circ \pi$). The projection π maps the trajectories of φ onto those of the reduced Hamiltonian. Frequently one of these dynamical systems is easier to study than the other.

As an example, let us sketch another proof of the Factorization Theorem 2.2. Under the hypotheses of Section 2 let us consider the action of the direct product $G_+ \times G_-$ on G:

$$(g_+, g_-) \circ g = g_+ g g_-^{-1}.$$

This action extends to a Hamiltonian action on T^*G. Consider the reduced space $\overline{\mathscr{M}} = T^*G/G_+ \times G_-$. Under some reasonable assumptions on the factorization in G (e.g. if the factorization $g = g_+ g_-$ is unique whenever it exists) the space $\mathfrak{g}_R^*$ is an open submanifold in $\overline{M}$, and reduction of bi-invariant Hamiltonians on T^*G yields Theorem 2.2. The reader will easily reconstruct the details. Observe an interesting fact: since the flows of bi-invariant Hamiltonians on T^*G are complete, so are the reduced flows on $\overline{\mathscr{M}}$. On the other hand, the flows of Lax equations in $\mathfrak{g}_R^*$ are quite often incomplete. For instance, the particles in a Toda lattice with the "wrong" signs of the exponential terms in the potential may go off to infinity in finite time. The reduction picture provides a universal completion of $\mathfrak{g}_R^*$ in which the trajectories of Lax equations are defined for all times.

5.2. Examples: Magnetic Cotangent Bundles. If a Lie group G acts on a manifold $\mathscr{V}$, there is a natural lift of this action to a Hamiltonian action on $T^*\mathscr{V}$. Moreover, this action preserves the canonical 1-form θ on $T^*\mathscr{V}$. For $X \in \mathfrak{g}$ let ξ_X be the associated vector field on $\mathscr{V}$ and $\hat{\xi}_X$ its canonical lift to $T^*\mathscr{V}$. Then the

Hamiltonian H_X of $\hat{\xi}_X$ is

$$H_X = \theta(\hat{\xi}_X). \tag{5.2}$$

In local coordinates (x, p) on $T^*\mathscr{V}$ we have $H_X(x, p) = \langle p, \xi_X(x)\rangle$. The Hamiltonians (5.2) are called *momenta*, which explains the term "moment map".

As an example, consider the action of G on itself by left or right translations. Let us fix the trivialization $T^*G \simeq G \times \mathfrak{g}^*$ identifying all tangent spaces by means of left translations. Then the "left" and "right" actions of G are given by

$$l_g\colon (h, \rho) \mapsto (gh, \rho),$$
$$r_g\colon (h, \rho) \mapsto (hg^{-1}, \operatorname{Ad}^* g \cdot \rho).$$

Let ω be the left-invariant Maurer-Cartan form on G. Then the canonical 1-form on T^*G is given by

$$\theta(x, F) = \langle F, \omega\rangle.$$

Hence the moment maps associated with the left and right translations are given by

$$\begin{aligned} \mu_l\colon (h, \rho) &\mapsto -\operatorname{Ad}^* h \cdot \rho, \\ \mu_r\colon (h, \rho) &\mapsto \rho. \end{aligned} \tag{5.4}$$

Clearly, μ_l and μ_r form a dual pair. This immediately implies that symplectic leaves in $\mathfrak{g}^*$ coincide with coadjoint orbits of G. In what follows we shall frequently deal with reduced spaces obtained by reduction of T^*G with respect to a subgroup $H \subset G$. We shall state these results in a slightly more general form.

Lemma 5.4. *Let $p\colon \mathscr{P} \to \mathscr{B}$ be a principal H-bundle. The reduced space obtained by reduction of $T^*\mathscr{P}$ with respect to the action of H over the zero momentum $0 \in \mathfrak{h}^*$ is isomorphic to $T^*\mathscr{B}$ as a symplectic manifold.*

Proof. Let $\mu\colon T^*\mathscr{P} \to \mathfrak{h}^*$ be the moment map. The set $\mu^{-1}(0)$ consists of the covectors which are annihilated on H-orbits in $\mathscr{P}$ and hence the reduced space $(T^*\mathscr{P})_0 = \mu^{-1}(0)/H$ may be identified with $T^*\mathscr{B}$. To compare the symplectic structures consider the canonical 1-form θ in $T^*\mathscr{P}$: $\theta_\xi(\eta) = \xi(\pi_* \cdot \eta)$, $\xi \in T^*\mathscr{P}$, $\eta \in T_\xi(T^*\mathscr{P})\pi\colon T^*\mathscr{P} \to \mathscr{P}$. The symplectic form is equal to $\Omega = d\theta$. It remains to observe that the restriction of θ to $\mu^{-1}(0)$ coincides with the pullback to $\mu^{-1}(0)$ of the canonical 1-form on $T^*\mathscr{B}$. □

Reduction over non-trivial one-point orbits gives rise to the so-called *twisted, or magnetic, cotangent bundles*. By definition, a twisted cotangent bundle is the ordinary cotangent bundle $\pi\colon T^*\mathscr{B} \to \mathscr{B}$ whose symplectic form differs from the canonical one by a perturbation $\pi^*\omega$ where ω is a closed 2-form on $\mathscr{B}$. (Physically, such perturbation may be associated with a magnetic field on $\mathscr{B}$.)

Proposition 5.5. *Let $p\colon \mathscr{P} \to \mathscr{B}$ be a principal H-bundle, $f \in \mathfrak{h}^*$ a character of $\mathfrak{h}$. Then the reduced phase space $(T^*\mathscr{P})_f$ obtained by reduction of $T^*\mathscr{P}$ over f is symplectically diffeomorphic to a magnetic cotangent bundle over $\mathscr{B}$.*

More precisely, fix an H-connection form α on $T^*\mathscr{P}$. There exists a diffeomorphism ψ_α: $T^*\mathscr{B} \to \overline{(T^*\mathscr{P})}_f$ such that $\psi_\alpha^*\omega_f - \omega = r^*\sigma$ where σ is a closed 2-form on $\mathscr{B}$ such that $p^*\sigma = d\langle f, \alpha\rangle$. In other words, σ is obtained by transgression from the 1-form $\langle f, \alpha\rangle$ on $\mathscr{P}$, $\sigma = \langle f, F(\alpha)\rangle$, where $F(\alpha)$ is the curvature of α. The cohomology class of σ has the meaning of magnetic charge. It is completely determined by f and the homotopy class of the fibration $\mathscr{P} \to \mathscr{B}$.

In particular, if $\mathscr{P} = K$ is a Lie group, and H is a subgroup of K, the reduced phase space $\overline{(T^*K)}_f$ is diffeomorphic to the cotangent bundle $T^*(K/H)$. Assume that H is reductive in K, so that there is an H-invariant complement $\mathfrak{p}$ to $\mathfrak{h}$ in $\mathfrak{k}$. Then there exists a left-invariant connection α in the principle bundle p: $K \to K/H$ whose horizontal space at the identity $e \in K$ is $\mathfrak{p}$. The magnetic form σ on K/H constructed by using this connection is K-invariant.

The reduced space of $T^*\mathscr{P}$ obtained by reduction over an arbitrary H-orbit $\mathcal{O}$ in $\mathfrak{h}^*$ is topologically a fibre bundle over $T^*\mathscr{B}$ with fibre $\mathcal{O}$. In physics such a space may be interpreted as the phase space of a particle in a Yang-Mills field with structure group H.

5.3. Orbits of Semi-direct Products. Let ρ be a linear representation of a Lie group K on a vector space V. Consider the semi-direct product $S = K \times_\rho V$. By definition, S is the product of K and V with multiplication law

$$(k_1, v_1)(k_2, v_2) = (k_1 k_2, v_1 + \rho(k_1)v_2).$$

The Lie algebra of S is the semi-direct sum $\mathfrak{s} = \mathfrak{k} +_\rho V$ with commutator

$$[(X_1, v_1), (X_2, v_2)] = ([X_1, X_2], \rho(X_1)v_2 - \rho(X_2)v_1),$$

where ρ denotes also the action of $\mathfrak{k}$ in V. We shall describe the Lie-Poisson structure and coadjoint orbits of the dual space $\mathfrak{s}^* = \mathfrak{k}^* + V^*$ in terms of Poisson mappings from various cotangent bundles to $\mathfrak{s}^*$.

Suppose that K acts on a manifold $\mathscr{M}$ and let μ: $T^*\mathscr{M} \to \mathfrak{k}^*$ be the corresponding moment map. Suppose we have a mapping j: $\mathscr{M} \to V^*$ which is equivariant with respect to the dual action ρ^* of K on V^*. Let π: $T^*\mathscr{M} \to \mathscr{M}$ be the natural projection.

Proposition 5.6. *The mapping* φ: $T^*\mathscr{M} \to \mathfrak{s}^*$ *defined by*

$$\varphi(p) = \mu(p) \oplus j(\pi(p)), \quad p \in T^*\mathscr{M}, \tag{5.5}$$

is a Poisson mapping.

Proof. Let $X, Y \in \mathfrak{s}$ be two linear functions on $\mathfrak{s}^*$. We must show that

$$\{X \circ \varphi, Y \circ \varphi\}_{T^*\mathscr{M}} = [X, Y] \circ \varphi. \tag{5.6}$$

The cases where X, Y belong to $\mathfrak{k}$ or to V are treated separately. If $X, Y \in V$, both sides of (5.6) vanish. If X, $Y \in \mathfrak{k}$, then (5.6) holds because μ is a moment map. Finally, if $X \in \mathfrak{k}$ and $Y \in V$, then (5.6) follows from the equivariance of j. □

In particular, take $\mathscr{M} = K$ with the action of K on itself by right multiplication. The corresponding moment map $T^*K \to \mathfrak{k}^*$ in terms of the left identificaton of T^*K with $K \times \mathfrak{k}^*$ is $\mu(k, l) = l$ (5.4). Fix an element $a \in V^*$ and let $j(k) = \rho(k^{-1})^* \cdot a$. Clearly, j is K-equivariant. Let K_a be the stationary subgroup of a, let $\mathfrak{k}_a$ be the Lie algebra of K_a, and let $\mathcal{M}_a$ be the K-orbit through a in V^*.

Theorem 5.7. (i) *The mapping* μ_a: $T^*K \to \mathfrak{s}^*$ *defined by*

$$\mu_a: (k, l) \mapsto l + \rho^*(k^{-1})a \tag{5.7}$$

is a Poisson mapping onto $\mathfrak{k}^* + \mathcal{M}_a$.

(ii) *Let* $\xi \in \mathfrak{k}^*$ *and let* $\bar{\xi} \in \mathfrak{k}_a^*$ *be the restriction of* ξ *to* $\mathfrak{k}_a$. *The S-orbit through* $\underline{\xi + a}$ *in* $\mathfrak{s}^*$ *is symplectically diffeomorphic (as a K-space) to the reduced manifold* $(T^*K)_{\bar{\xi}}$ *with respect to the action of* K_a *on* T^*K *by left translations at* $\bar{\xi} \in \mathfrak{k}_a^*$. *Hence the coadjoint orbits lying in* $\mathfrak{k}^* + \mathcal{M}_a$ *are in one to one correspondence with the orbits of* K_a *in* $\mathfrak{k}_a^*$.

Proof. Part (i) follows directly from Proposition 5.6. To show (ii) notice that the fibers of μ_a are precisely the left K_a-cosets in T^*K, so that $\mathfrak{k}^* + \mathcal{M}_a \simeq K_a \backslash T^*K$ and μ_a is the reduction map for the action of K_a on T^*K. Now part (ii) follows from Theorem 5.1 which describes the symplectic leaves of the reduced space $K_a \backslash T^*K$.

The mapping (5.7) plays an important role in the construction of Lax representations for multi-dimensional tops.

Corollary 5.8. *The orbit in* $\mathfrak{s}^*$ *passing through* $a \in V^*$ *is diffeomorphic as a symplectic K-space to the cotangent bundle* $T^*\mathcal{M}_a$. *The diffeomorphism is given by* (5.5): $\varphi(p) = \mu(p) \oplus \pi(p)$.

Corollary 5.9. *Let* $\xi \in \mathfrak{k}^*$ *be such that its restriction to* $\mathfrak{k}_a$ *is a character of* $\mathfrak{k}_a$. *Then the orbit in* $\mathfrak{s}^*$*-through* $\xi + a$ *is isomorphic to a magnetic cotangent bundle over* $\mathcal{M}_a$.

If $\mathfrak{k}_a$ is reductive we may assume with no loss of generality that ξ is a fixed point of K_a. In that case the diffeomorphism of the orbit with the cotangent bundle $T^*\mathcal{M}_a$ is given by

$$\varphi(p) = (\mu(p) + \mathrm{Ad}^*\, k_p \cdot \xi) \oplus \pi(p), \tag{5.8}$$

where k_p is any element of K such that $\rho^*(k_p)a = \pi(p)$. This mapping is K-equivariant, and the resulting magnetic form on $\mathcal{M}_a$ is K-invariant. In applications (see Sections 6, 7) the groups K, K_a are compact and this assumption is automatically fulfilled.

Corollary 5.10. *Under the hypotheses of Proposition 5.6 assume further that K acts transitively on* $\mathcal{M}$. *Let* $\alpha: \mathcal{M} \to \mathfrak{k}^*$ *be a mapping which commutes with the action of K. Then the mapping from* $T^*\mathcal{M}$ *to* $\mathfrak{s}^*$ *defined by*

$$(p, m) \mapsto (\mu(p) + \alpha(m)) \oplus j(m)$$

is a Poisson map with respect to a K-invariant magnetic structure on $T^*\mathcal{M}$.

The magnetic form ω on $\mathcal{M}$ can be described in terms of the projection $\pi: K \to \mathcal{M}$ given by $\pi(k) = k \cdot m_0$ for some point $m_0 \in \mathcal{M}$. We have $\pi^*\omega = d\alpha(m_0)$ where $\alpha(m_0) \in \mathfrak{k}^*$ is regarded as a left-invariant 1-form on K.

Finally, for a general $\xi \in \mathfrak{k}^*$, the orbit in $\mathfrak{s}^*$ passing through $\xi + a$ is a fiber bundle over $T^*\mathcal{M}_a$ with fiber $\mathcal{O}_{\bar{\xi}}$ where $\mathcal{O}_{\bar{\xi}}$ is the K_a-orbit of $\bar{\xi} = \xi|\mathfrak{k}_a$ in $\mathfrak{k}_a^*$.

Corollary 5.11. *If the stationary subgroup K_a is discrete, then all generic orbits in $\mathfrak{k}^* + M_a$ are symplectically isomorphic (up to a finite covering) to the space T^*K.*

5.4. Nonabelian Toda Lattices. Recall from Section 3.1 that the Poisson subspace $\mathfrak{g}_0^* + \mathfrak{g}_1^*$ of the R-bracket associated with a graded Lie algebra $\mathfrak{g} = \oplus \mathfrak{g}_i$ is canonically isomorphic to the dual of the semidirect product $\mathfrak{g}_0 \ltimes \mathfrak{g}_1$. Proposition 5.6. provides symplectic realizations for Poisson submanifolds in this subspace. This gives a mechanical interpretation for Lax systems with Lax matrices in $\mathfrak{g}_{-1}^* + \mathfrak{g}_0^* + \mathfrak{g}_1^*$.

The construction of Toda lattices described in Sections 3, 4 may be generalized in several ways, e.g. by considering different decompositions of split simple Lie algebras, or different real forms. Here we describe the GL(m)-lattices and quaternionic lattices which illustrate the range of possibilities available.

Let us represent the matrix algebra $\mathfrak{gl}(mn, \mathbb{R})$ as a tensor product $\mathfrak{gl}(mn) = \mathfrak{gl}(m) \otimes \mathfrak{gl}(n)$ and introduce the grading which is induced by the principal grading of the second factor:

$$\mathfrak{gl}(mn) = \bigoplus_i \mathfrak{g}_i, \quad \mathfrak{g}_i = \mathfrak{gl}(m) \otimes \mathfrak{g}_i', \quad \mathfrak{gl}(n) = \bigoplus_i \mathfrak{g}_i'.$$

Using the notations of Section 3 put $E = \mathbf{1}_m \otimes e$, $F = \mathbf{1}_m \otimes f$. The structure of the Poisson subspace $\mathfrak{g}_{-1} + \mathfrak{g}_0 + E$ is described by Theorem 5.7 in terms of reduction of the space T^*G_0, where $G_0 = \mathrm{GL}(m) \times \cdots \times \mathrm{GL}(m)$ (n times). Take the simplest Hamiltonian $\varphi(X) = 1/2 \operatorname{tr} X^2$ on $\mathfrak{gl}(mn)$ and pull it back to T^*G_0 by means of the mapping $\mu_F + E$ (5.7). This gives the Hamiltonian

$$H(g_1, \ldots, g_n, \xi_1, \ldots, \xi_n) = \frac{1}{2} \sum_{i=1}^{n} \operatorname{tr} \xi_i^2 + \sum_{i=1}^{n-1} \operatorname{tr} g_i g_{i+1}^{-1}. \tag{5.9}$$

Here $\xi_i = g_i^{-1} \dot{g}_i$ are left momenta on T^*G_0. This system may be regarded as a straightforward nonabelian analog of the ordinary Toda lattice. The Lax representation for it has the form

$$\frac{dL}{dt} = [L, L_{\pm}],$$

where $L(g, \xi) = \mu_F(g, \xi) + E$ is the block matrix

$$L = \begin{pmatrix} p_1 & 1 & \ldots & 0 \\ a_1 & p_2 & 1 & \\ & a_2 & \ddots & \\ & & \ddots & 1 \\ 0 & & a_{n-1} & p_n \end{pmatrix}, \quad p_i = g_i \xi_i g_i^{-1}, \quad a_i = g_i g_{i+1}^{-1}, \tag{5.10}$$

$$L_- = -\sum_i a_i \otimes e_{-\alpha_i}, \quad L_+ = L + L_-.$$

The quaternionic lattice is obtained in a similar way from the algebra of quaternionic matrices $\mathfrak{gl}(n, \mathbb{H})$ which is a real form of $\mathfrak{gl}(2n, \mathbb{C})$. The Lax matrix is given by (5.10) where p_i, a_i are quaternions, $a_i = g_i g_{i+1}^{-1}$.

There is a significant difference between the nonabelian and the ordinary Toda lattices: the trajectories of the nonabelian lattice may go off to infinity in finite time, i.e. its flow is incomplete. The reduction version of the factorization theorem mentioned in Section 5.1 provides a canonical completion of these flows. Also notice that the set of integrals of motion in involution provided by the spectral invariants in $\mathfrak{gl}(mn)$ is not sufficient for complete integrability (it seems, however, that complete integrability still holds, cf. Deift et al. [1986]).

Slightly more general lattices may be obtained by a different choice of E and F: if $E = \sum c_i \otimes e_{\alpha_i}$, $F = \sum f_i \otimes e_{-\alpha_i}$ the corresponding Hamiltonian has the form

$$H(g, \xi) = \frac{1}{2} \sum \operatorname{tr} \xi_i^2 + \sum \operatorname{tr} c_i g_i f_i g_{i+1}^{-1}. \tag{5.11}$$

In a similar way, the affine algebras $\mathfrak{L}(\mathfrak{gl}(mn))$, $\mathfrak{L}(\mathfrak{gl}(n, \mathbb{H}))$ give rise to periodic nonabelian and quaternionic lattices. For instance, the appropriate grading in $\mathfrak{L}(\mathfrak{gl}(mn)) \simeq \mathfrak{gl}(m) \otimes \mathfrak{L}(\mathfrak{gl}(n))$ is induced by the principal grading in $\mathfrak{L}(\mathfrak{gl}(n))$. The corresponding Hamiltonian on T^*G_0 is $\tilde{H} = H + \operatorname{tr} g_n g_1^{-1}$ where H is given by (5.9). The Lax matrix has a block form

$$L(\lambda) = \begin{pmatrix} p_1 & \lambda \cdot 1 & \dots & a_n \lambda^{-1} \\ a_1 \lambda^{-1} & p_2 & \lambda \cdot 1 \ \ddots & \vdots \\ \vdots & a_2 \lambda^{-1} \ \ddots & \ddots & \lambda \cdot 1 \\ \lambda \cdot 1 & \dots \ \ddots & a_{n-1} \lambda^{-1} & p_n \end{pmatrix}, \quad a_n = g_n g_1^{-1}.$$

In the Lax representation $\partial_t L = [L(\lambda), L_{\pm}(\lambda)]$ we have $L_-(\lambda) = -\sum_{b_i \in \hat{P}} a_i \otimes e_{-b_i} \lambda^{-1}$. Naturally, one can also consider more general lattices with Hamiltonians of the form (5.11). Unlike the non-periodic case one can show that the invariants of $L(\lambda)$ provide a complete set of integrals of motion on generic orbits in $\mathfrak{g}_{-1} + \mathfrak{g}_0 + \mathfrak{g}_1$ which correspond to periodic nonabelian Toda lattices.

5.5. Bibliographical Notes. The technique of Hamiltonian reduction goes back to the work of Euler and Jacobi on analytical mechanics. Its modern treatment was given by Marsden and Weinstein [1974], see also Abraham and Marsden [1978]. The Poisson version of reduction was used by Lie [1893]; for a modern exposition of his approach see Weinstein [1983]. (A similar approach was independently used by Karasev [1981].) In particular, Karasev and Weinstein showed that every Poisson manifold has a sympletic realization. Classically, the aim of reduction is to simplify the equations of motion. Kazhdan, Kostant and Sternberg [1978] reversed this point of view and used reduction to obtain complicated equations out of simple integrable ones. The R-matrix scheme can also be expressed in terms of Hamiltonian reduction (Reyman and Semenov-Tian-Shansky [1979]). The observation that magnetic field may be regarded as a perturbation of symplectic structure is due to Souriau [1970]. The general construction of the magnetic cotangent bundles and their non-abelian generalizations was given by Weinstein [1978]. Special examples of magnetic cotantent bundles were also given earlier by Kazhdan, Kostant and Sternberg [1978] who considered the symplectic counterparts of induced representations. Coadjoint orbits of semi-direct products were described by Rawnsley [1975]. A description based on Hamiltonian reduction was given by Reyman and Semenov-Tian-Shansky [1979], Ratiu [1982] and Marsden et al. [1984] (see also Guillemin and Sternberg [1980, 1984]). This description may be regarded as a symplectic version of the Wigner-Mackey representation theory for semi-direct products (cf. Mackey [1958]). The orbits of the group of Euclidean motions $E(3)$ were also described by Novikov and Schmeltzer [1981]. Non-abelian Toda lattices were introduced by Polyakov (unpublished). Their description based on the orbit method was given by Reyman [1980] who also introduced quaternionic Toda lattices.

§ 6. Lax Representations of Multi-dimensional Tops

In this section we describe a large family of "natural" integrable systems whose Hamiltonians are sums of kinetic and potential energies, in particular those whose configuration space is a compact Lie group. These systems can be therefore regarded as direct generalizations of the rotating rigid body. Their construction makes use of the twisted loop algebras $\mathfrak{L}(\mathfrak{g}, \sigma)$ where $\mathfrak{g}$ is a real semisimple Lie algebra and σ its Cartan automorphism. To simplify the comparison with realistic mechanical examples we first recall the description of the multi-dimensional rigid body.

6.1. Kinematics of the n-dimensional Rigid Body. The position of an n-dimensional rigid body fixed at a point is described by the orthogonal (moving) frame attached to the body or, which is the same, by the orthogonal matrix which relates this frame to the standard frame in $\mathbb{R}^n$. Hence the configuration space of the n-dimensional top is identified with the rotation group $\mathrm{SO}(n) = K$, and its phase space with the cotangent bundle $T^*\mathrm{SO}(n)$. We write $T^*\mathrm{SO}(n) = \mathrm{SO}(n) \times \mathfrak{so}(n)^*$ using left translations and identify $\mathfrak{so}(n)^*$ with $\mathfrak{so}(n)$ by means of the scalar product $(X, Y) = -\operatorname{tr} XY$. The rotation of the body is determined by prescribing its kinetic energy E, which is a left-invariant Riemannian metric on $\mathrm{SO}(n)$, and a potential function V on $\mathrm{SO}(n)$. It can be shown that the kinetic energy of a "realistic" body, i.e. the energy of a rotating rigid mass distribution in $\mathbb{R}^n$, has the following rather special form:

$$E(\omega) = -\operatorname{tr} J\omega^2, \tag{6.1}$$

where $\omega \in \mathfrak{so}(n)$ is the angular velocity and J is the *inertia tensor* (a symmetric positive definite $n \times n$-matrix). The angular momentum $\ell \in \mathfrak{so}(n)$ of the top is linearly related to the angular velocity by the Legendre transform:

$$\ell = J\omega + \omega J, \quad E(\ell) = -\frac{1}{2}\operatorname{tr}\omega\ell. \tag{6.2}$$

For instance, if J is diagonal, $J = \operatorname{diag}(a_1, \ldots, a_n)$, then

$$\omega_{ij} = \frac{1}{a_i + a_j}\ell_{ij}, \quad E(\ell) = \frac{1}{2}\sum_{i,j}\frac{1}{a_i + a_j}\ell_{ij}^2. \tag{6.3}$$

Clearly, for $n > 3$ the kinetic energies (6.1) constitute only a subspace in the space of all quadratic forms on $\mathfrak{so}(n)$.

We note for later use that the expression for ω in terms of ℓ can be written as

$$\omega = \operatorname{ad} J^2 \cdot (\operatorname{ad} J)^{-1} \cdot \ell. \tag{6.4}$$

The full Hamiltonian on $T^*\mathrm{SO}(n)$ is given by

$$H(k, \ell) = E(\ell) + V(k), \tag{6.5}$$

where $k \in \mathrm{SO}(n)$ is the frame variable. The associated equations of motion are

$$\begin{aligned} \dot{\ell} &= [\ell, \omega] - k^{-1}\, dV(k), \\ k^{-1}\dot{k} &= \omega. \end{aligned} \tag{6.6}$$

For the free top, i.e. when $V(k) = 0$, these equations reduce to the Euler equations

$$\dot{\ell} = [\ell, \omega]$$

on $\mathfrak{so}(n)$. If the top is placed in a gravitational field with potential $\varphi(x)$, $x \in \mathbb{R}^n$, which is a polynomial in x of degree m, then the potential energy $V(k)$ is easily computable and is in turn a polynomial of degree m in the matrix elements k_{ij}. In particular, for a linear potential $\varphi(x) = (x, g)$ we have

$$V(k) = (kr, g), \tag{6.7}$$

where $r = (r_1, \ldots, r_n)$ is the center of mass vector in the moving frame. For a quadratic potential $\varphi(x) = (Fx, x)$ we get (assuming that the top is fixed at $0 \in \mathbb{R}^n$)

$$V(k) = \operatorname{tr}(Jk^tFk) = \sum_{i,j,l} F_{ij}a_l k_{il}k_{jl}. \tag{6.8}$$

6.2. Integrable Top-Like Systems. It turns out that the n-dimensional top with kinetic energy (6.1) and potential energy (6.8) is completely integrable and, moreover, fits into the R-matrix scheme of Theorem 2.1. This results from a geometric construction which uses twisted loop algebras.

Let G be a Lie group, $\mathfrak{g}$ its Lie algebra, and σ an involution in G. We also denote by σ the induced involution in $\mathfrak{g}$. Let K be the subgroup of fixed points of σ and $\mathfrak{k}$ its Lie algebra. Let $\mathfrak{g} = \mathfrak{k} + \mathfrak{p}$ be the Cartan-type decomposition with respect to σ (i.e. $\sigma = \mathrm{id}$ on $\mathfrak{k}$ and $\sigma = -\mathrm{id}$ on $\mathfrak{p}$); the dual decomposition is $\mathfrak{g}^* = \mathfrak{k}^* + \mathfrak{p}^*$. Let $\mathfrak{L}(\mathfrak{g}, \sigma)$ be the twisted loop algebra associated with $\mathfrak{g}$ and σ.

Fix a point $a \in \mathfrak{p}^*$ and consider the Poisson subspace in $\mathfrak{L}(\mathfrak{g}, \sigma)^*$ (with respect to the standard R-bracket defined by the decomposition (4.4)) consisting of "Lax matrices" of the form

$$L(\lambda) = a\lambda + \ell + s\lambda^{-1}, \quad \ell \in \mathfrak{k}^*, \quad s \in \mathfrak{p}^*. \tag{6.9}$$

This subspace has the Lie-Poisson structure of the dual of the semidirect sum $\mathfrak{k} + \mathfrak{p}$ (see Section 3.1). By Theorem 2.1, the invariants of $L(\lambda)$ are in involution with respect to this Poisson structure, and the corresponding Hamiltonian equations have Lax representations. On the other hand, Proposition 5.6 and Theorem 5.7 provide a large family of Poisson mappings from cotangent bundles $T^*\mathcal{M}$ to $(\mathfrak{k} + \mathfrak{p})^*$ which, combined with (6.9), give rise to various integrable systems on $T^*\mathcal{M}$. More precisely, suppose that K acts on a manifold $\mathcal{M}$ and let $\mu\colon T^*\mathcal{M} \to \mathfrak{k}^*$ be the corresponding moment map. Let j be a K-equivariant mapping from $\mathcal{M}$ to $\mathfrak{p}^*$. To any cotangent vector p at $m \in \mathcal{M}$ we assign the Lax matrix

$$L(\lambda) = a\lambda + \mu(p) + j(m)\lambda^{-1}; \tag{6.10}$$

by Proposition 5.6 this is a Poisson mapping from $T^*\mathcal{M}$ to $\mathfrak{L}(\mathfrak{g}, \sigma)^*$.

In particular, let $\mathcal{M} = K$ with K acting by right multiplication. Fix a point $f \in \mathfrak{p}^*$ and define $j(k) = \mathrm{Ad}^*\, k^{-1} \cdot f$. Let $k \in K$, $\ell \in \mathfrak{k}^*$ be the configuration and momentum variables on T^*K identified with $K \times \mathfrak{k}^*$ by left translation. Then

$$(k, \ell) \mapsto L(\lambda) = a\lambda + \ell + \mathrm{Ad}^*\, k^{-1} \cdot f\lambda^{-1} \tag{6.11}$$

is a Poisson mapping from T^*K to $\mathfrak{L}(\mathfrak{g}, \sigma)^*$. (cf. Theorem 5.7). We have therefore the following proposition.

Proposition 6.1. *The coadjoint invariants of $L(\lambda)$ regarded as functions on $T^*\mathcal{M}$ (resp., T^*K) via the mapping* (6.10) *(resp.,* (6.11)*) Poisson commute under the canonical Poisson bracket on $T^*\mathcal{M}$ (resp., T^*K).*

The most interesting examples are obtained when G is a semisimple real Lie group and σ is a Cartan involution, so that K is compact. In particular, the n-dimensional top in the quadratic potential (6.8) is associated with the symmetric pair $(G, K) = (\mathrm{SL}(n, \mathbb{R}), \mathrm{SO}(n))$ (see Section 7 for more details).

Our next goal is to single out those invariants of $L(\lambda)$ which have the natural form $E(\ell) + V(s)$. This is easily achieved: for an invariant polynomial φ on $\mathfrak{g}^*$ put

$$H_\varphi(L) = \mathrm{Res}_{\lambda=0}\, \lambda\varphi(\lambda^{-1}L(\lambda))\, d\lambda. \tag{6.12}$$

From now on we shall assume that $\mathfrak{g}$ is semisimple and σ is a Cartan involution. We identify $\mathfrak{g}^*$ with $\mathfrak{g}$ by means of an invariant inner product $(\ ,\ )$ which is positive definite on $\mathfrak{k}$. To write the kinetic term in H_φ explicitly we need some notation. For $a \in \mathfrak{p}$ let K_a be the stationary subgroup of a in K and $\mathfrak{k}_a$ the corresponding stationary subalgebra. Consider the function $\varphi_a(\ell) = \varphi(a + \ell)$ on $\mathfrak{k}_a$ and let $\varphi_a''\colon \mathfrak{k}_a \to \mathfrak{k}_a$ be its second differential at $\ell = 0$. Denote $b = d\varphi(a)$ and put

$$\omega(\ell) = \begin{cases} \varphi_a'' \cdot \ell, & \ell \in \mathfrak{k}_a, \\ \operatorname{ad} b \cdot (\operatorname{ad} a)^{-1}\ell, & \ell \in \mathfrak{k}_a^{\perp}. \end{cases} \tag{6.13}$$

This is well-defined since Ker ad $a \subset$ Ker ad b. It is easily verified that $\ell \mapsto \omega(\ell)$ is a symmetric linear operator in $\mathfrak{k}$.

Proposition 6.2. *The Hamiltonian H_φ is*

$$H_\varphi(s, \ell) = \frac{1}{2}(\ell, \omega(\ell)) + (b, s), \tag{6.14}$$

where $b = d\varphi(a)$ and $\omega(\ell)$ is given by (6.13). *The corresponding Lax equation is $\dot{L} = [L, M_+]$ with*

$$\begin{aligned} L &= a\lambda + \ell + s\lambda^{-1}, \\ M_+ &= b\lambda + \omega(\ell). \end{aligned} \tag{6.15}$$

The Hamiltonian H_φ is invariant under the Ad**-action of K_a.*

Proof. Since H_φ is the coefficient of λ^{-2} in $\varphi(a + \ell\lambda^{-1} + s\lambda^{-2})$, we have $H_\varphi(L) = \frac{1}{2} d^2\varphi(a)(\ell) + (b, s)$. To evaluate the quadratic form $d^2\varphi(a)(\ell)$ we set $\ell = \ell_0 + \ell_1$ where $\ell_0 \in \mathfrak{k}_a$, and $\ell_1 \in \mathfrak{k}_a^{\perp}$. Noticing that $[a, \mathfrak{p}] = \mathfrak{k}_a^{\perp}$, let X be an element of $\mathfrak{p}$ such that $[a, X] = \ell_1$. Then for $g = \exp \lambda^{-1}X$ we have

$$\operatorname{Ad} g(a + \ell\lambda^{-1}) = a + \ell_0\lambda^{-1} + \frac{1}{2}[X, \ell + \ell_0]\lambda^{-2} \cdots$$

up to terms of order λ^{-3}. By invariance, $\varphi(a + \ell\lambda^{-1}) = \varphi(\operatorname{Ad} g(a + \ell\lambda^{-1}))$. Calculating the coefficient of λ^{-2} in the latter expression and using $(b, [X, \ell_0]) = 0$ we find

$$\begin{aligned} d^2\varphi(a)(\ell) &= (\varphi_a''\ell_0, \ell_0) + (d\varphi(a), [X, \ell]) \\ &= (\varphi_a''\ell_0, \ell_0) + ([d\varphi(a), X], \ell) = (\omega(\ell), \ell). \end{aligned}$$

To check (6.15) we apply Theorem 2.1. We must determine the positive part (with respect to the splitting (4.4) of $\mathfrak{L}(\mathfrak{g}, \sigma)$) of the full gradient M of H_φ regarded in the entire space $\mathfrak{L}(\mathfrak{g}, \sigma)^*$. Clearly, $M = \lambda\, d\varphi(\lambda^{-1}(L(\lambda))$ where $d\varphi$ is the gradient of φ in $\mathfrak{g}^*$. From the above expression for $d^2\varphi(a)$ it follows that $d\varphi(\lambda^{-1}L(\lambda)) = b + \omega(\ell)\lambda + \cdots$, hence $M_+ = b\lambda + \omega(\ell)$. Finally, the invariance property of H_φ is obvious. □

The Lax equation $\dot{L} = [L, M]$ with L, M given by (6.15) is equivalent to the system of equations

$$\begin{aligned} \dot{\ell} &= [\ell, \omega] + [s, b], \\ \dot{s} &= [s, \omega]. \end{aligned} \tag{6.16}$$

These are generalized *Euler-Poisson equations* for the semi-direct sum $\mathfrak{k} \oplus \mathfrak{p}$, the variable s plays the role of "*Poisson vector*". (Note that the potential $V(s) = (b, s)$ is linear in s.)

Substituting the Lax matrix (6.10) into H_φ gives a family of natural integrable Hamiltonians on $T^*\mathcal{M}$. More specifically, substituting the Lax matrix (6.11), i.e.

putting $s = \mathrm{Ad}^*\, k^{-1} \cdot f$, gives the following Hamiltonian on T^*K:

$$H_\varphi(k, \ell) = \frac{1}{2}(\ell, \omega(\ell)) + (b, \mathrm{Ad}^*\, k^{-1} \cdot f) \tag{6.17}$$

which is the sum of the left-invariant kinetic energy $E(l) = 1/2(\ell, \omega(\ell))$ and the potential $V(k) = (\mathrm{Ad}^*\, k \cdot b, f)$. If G is a matrix group, $V(k)$ is usually a linear or quadratic function of the matrix elements of k. Observe that H_φ is invariant under the action of K_a by right translations and under the action of K_f by left translations.

If $f = 0$, the potential term vanishes and the Hamiltonian (6.17) describes a free motion of a point on K, or a free rotating top. The equations of motion (6.16) reduce to the Euler equations in $\mathfrak{k}^* \simeq \mathfrak{k}$:

$$\dot{\ell} = [\ell, \omega].$$

The corresponding Lax pair $L = A\lambda + \ell$, $M = B\lambda + \omega$ with A and B related by $[A, \omega] = [B, \ell]$ was found by Manakov [1976] (for the case of $\mathfrak{g} = \mathfrak{gl}(n)$).

There is a criterion for the kinetic energy $E(\ell) = 1/2(\ell, \omega(\ell))$ to be positive definite. Let $\mathfrak{a}$ be a maximal Abelian subspace in $\mathfrak{p}$ containing a and let $\mathfrak{a}_+$ be a closed Weyl chamber in $\mathfrak{a}$ containing a.

Proposition 6.3. *If $b = d\varphi(a)$ lies in $\mathfrak{a}_+$ and the quadratic form φ_a'' is positive definite then E is nonnegative. If moreover the centralizer of b in $\mathfrak{k}$ coincides with $\mathfrak{k}_a$ then E is positive definite. In particular, this is the case if both a and b lie in the interior of $\mathfrak{a}_+$.*

The expression for the angular velocity (6.13) shows that the kinetic energy $E(\ell)$ decomposes as

$$E(\ell) = E_1(\ell_1) + E_0(\ell_0) = \frac{1}{2}(\ell_1, \mathrm{ad}\, b \cdot (\mathrm{ad}\, a)^{-1}\ell_1) + \frac{1}{2}(\varphi_a''\ell_0, \ell_0), \tag{6.18}$$

where $\ell_1 \in \mathfrak{k}_a^\perp$, $\ell_0 \in \mathfrak{k}_a$. Since the full Hamiltonian $H = E(\ell) + V(k)$ (6.17) is K_a-invariant on the right, it Poisson commutes with E_0. (Note that E_0 is a K_a-invariant quadratic form on $\mathfrak{k}_a$.) This can be used to express the trajectories of H in terms of the trajectories of the "simpler" Hamiltonian $H_1 = E_1 + V$. Indeed, the Hamiltonian flow E_0 is easily determined in view of the following lemma (Guillemin and Sternberg [1980]).

Lemma 6.4. *Suppose that we are given a Hamiltonian action of a Lie group H on a symplectic manifold $\mathcal{M}$. Let $\mu\colon \mathcal{M} \to \mathfrak{h}^*$ be the corresponding moment map, and let φ be an invariant function on $\mathfrak{h}^*$. Then the trajectory of the "collective" Hamiltonian $\varphi \circ \mu$ on $\mathcal{M}$ emanating from x is $x(t) = \exp[t\, d\varphi(\mu(x))] \cdot x$.*

The Hamiltonian flow of H is the superposition of the flow of H_1 and the flow of E_0.

There is another family of invariants of $L(\lambda)$ which have the form $\tilde{E}(\ell) + \tilde{V}(s)$:

$$\varphi_-(L) = \operatorname{Res} \lambda^3 \varphi(\lambda L(\lambda)) = \frac{1}{2}(\ell, \omega_s(\ell)) + (a, d\varphi(s)) \tag{6.19}$$

where $\omega_s(\ell)$ is defined as in (6.13) with a replaced by s. The "shorter" Lax representation for φ_- is $\dot{L} = [L, M_-]$ where

$$M_- = -d\varphi(s)\lambda^{-1}. \tag{6.20}$$

The kinetic form $\tilde{E}(\ell) = \frac{1}{2}(\ell, \omega_s(\ell))$ depends on s and is invariant under the Ad^*-action of K on $(\mathfrak{k} + \mathfrak{p})^*$. Substituting the Lax matrix (6.10) into φ_- we get a natural Hamiltonian on $T^*\mathcal{M}$ with a K-invariant kinetic term. In particular, substituting $L(\lambda) = a\lambda^{-1} + \ell + \mathrm{Ad}^*\, k^{-1} \cdot f\lambda$ gives a system with a right-invariant kinetic energy on T^*K. The inversion $k \to k^{-1}$ on K which sends the left momentum ℓ into the right momentum $\tilde{\ell} = -\mathrm{Ad}^*\, k \cdot \ell$, combined with the exchange of a and f, maps this Hamiltonian into the previous Hamiltonian H_φ (6.17). Hence (6.19) does not lead to new integrable systems on T^*K but provides another Lax pair $\tilde{L}$, $\tilde{M}$ for H_φ:

$$\begin{aligned} \tilde{L}(\lambda) &= f\lambda + \mathrm{Ad}^*\, k \cdot (-\ell + a\lambda^{-1}), \\ \tilde{M}(\lambda) &= -\mathrm{Ad}^*\, k \cdot b\lambda, \end{aligned} \tag{6.21}$$

where $b = d\varphi(a)$ as before ($d\varphi(s) = \mathrm{Ad}^*\, k \cdot b$ since $s = \mathrm{Ad}^*\, k \cdot a$). At first glance, the Lax pairs (6.21) and (6.15) differ considerably; they correspond to different descriptions of the top, one in the spatial frame and the other in the body frame. However, these two Lax pairs are related by the gauge transformation $k(t)$ which takes one frame to the other:

$$\begin{aligned} \tilde{L}(-\lambda^{-1}) &= \mathrm{Ad}\, k \cdot L(\lambda), \\ \tilde{M}(-\lambda^{-1}) &= \mathrm{Ad}\, k \cdot M_+(\lambda) - \frac{dk}{dt} k^{-1}. \end{aligned} \tag{6.22}$$

Remark. The Lax matrix (6.11) is invariant under the action of the isotropy subgroup K_f on T^*K by left translations and therefore actually corresponds to the reduced system with respect to K_f. In a similar way, the Lax pair (6.21) describes reduced systems on T^*K with respect to the action of K_a by right translations. The mapping $(k, \ell) \mapsto L(\lambda)$ embeds the reduced spaces as coadjoint orbits of the R-bracket in $\mathcal{L}(\mathfrak{g}, \sigma)^*$. By Lemma 5.4, one of these reduced spaces (at the zero momentum) is $T^*\mathcal{M}$, where $\mathcal{M} = K_f \backslash K$. In this case the Lax matrix (6.10) for $T^*\mathcal{M}$ is effectively the same as (6.11), where $j(k) = \mathrm{Ad}^*\, k^{-1} \cdot f$ and the angular momentum ℓ satisfies $\mathrm{Ad}^*\, k \cdot \ell \in \mathfrak{k}_f^\perp$.

6.3. Bibliographical Notes. For the kinematics of the n-dimensional rigid body see Arnol'd [1966, 1974]. A Lax pair for the Euler equations of the free n-dimensional top was found by Manakov [1976] and was studied in detail by Mishchenko and Fomenko [1978]. The general Lax representation (6.15) for multi-dimensional tops in potential fields is due to Reyman [1980] (see also Reyman and Semenov-Tian-Shansky [1979, 1986b]).

§ 7. Integrable Multi-dimensional Tops and Related Systems

In Section 6 we outlined a construction of top-like integrable systems which makes use of Riemannian symmetric pairs (G, K). Here we specialize it to the case where K is the rotation group $SO(n)$, or the product of two rotation groups. In all our examples G is a matrix group and we use the scalar product $(X, Y) = -\operatorname{tr} XY$ to identify $\mathfrak{g}^*$ with $\mathfrak{g}$. We recall for convenience the main formulae of Section 6.

Let $\mathfrak{g} = \mathfrak{k} + \mathfrak{p}$ be a Cartan decomposition, a and f two fixed elements of $\mathfrak{p}$, and φ an invariant function on $\mathfrak{g}$. The integrable Hamiltonian (6.17) on T^*K is

$$H_\varphi(k, \ell) = -\frac{1}{2}\operatorname{tr}(\ell\omega(\ell)) - \operatorname{tr}(bk^{-1}fk), \tag{7.1}$$

where $\omega(\ell)$ is given by (6.13) and $b = d\varphi(a)$. The corresponding Lax pair (6.15) becomes

$$\begin{aligned} L(\lambda) &= a\lambda + \ell + k^{-1}fk\cdot\lambda, \\ M(\lambda) &= b\lambda + \omega(\ell). \end{aligned} \tag{7.2}$$

In a slightly more general way, suppose that K acts on a manifold $\mathcal{M}$, let $\mu\colon T^*\mathcal{M} \to \mathfrak{k}^* \simeq \mathfrak{k}$ be the corresponding moment map, and let $j\colon \mathcal{M} \to \mathfrak{p}$ be a mapping which commutes with the action of K. Then the Lax matrix (6.10)

$$L(\lambda) = a\lambda + \mu(p) + j(m)\lambda^{-1}, \quad p \in T_m^*\mathcal{M}, \tag{7.3}$$

combined with (6.14) gives rise to integrable Hamiltonians on $T^*\mathcal{M}$:

$$H_\varphi = -\frac{1}{2}\operatorname{tr}(\mu\omega(\mu)) - \operatorname{tr}(bj(m)). \tag{7.4}$$

In particular, if $\mathcal{M}$ is a K-orbit through f in $\mathfrak{p}$ then Corollary 5.10 enables us to construct integrable systems on a magnetic cotangent bundle $T^*\mathcal{M}$ by using the Lax matrix (see (5.8))

$$L(\lambda) = a\lambda^{-1} + (\mu + k\xi k^{-1}) + m\lambda^{-1}, \tag{7.5}$$

where $\xi \in \mathfrak{k}$ is a fixed point for the stationary subgroup K_f of f. The magnetic form is K-invariant; it is nontrivial if ξ defines a nontrivial character of $\mathfrak{k}_f$.

We note finally that for a Lax matrix $L(\lambda) = a\lambda + \ell + s\lambda^{-1}$ the simplest Hamiltonian

$$H_0 = -\frac{1}{2}\operatorname{Res}\lambda^{-1}\operatorname{tr} L(\lambda)^2 = -\frac{1}{2}\operatorname{tr}\ell^2 - \operatorname{tr} as \tag{7.6}$$

is associated with the Lax equation $\dot{L} = [L, M_\pm]$ where

$$\begin{aligned} L &= a\lambda + \ell + s\lambda^{-1}, \\ M_+ &= a\lambda + \ell, \quad M_- = -s\lambda^{-1}. \end{aligned} \tag{7.7}$$

7.1. $(G, K) = (GL(n, \mathbb{R}), SO(n))$: Multidimensional Tops in Quadratic Potentials. In the Cartan decomposition $\mathfrak{g} = \mathfrak{k} + \mathfrak{p}$ we have $\mathfrak{k} = \mathfrak{so}(n)$ and $\mathfrak{p}$ consists of $n \times n$ symmetric matrices. Assume that a is a diagonal matrix, $a = \operatorname{diag}(a_1, \ldots, a_n)$, hence $b = \operatorname{diag}(b_1, \ldots, b_n)$. If, moreover, $a_i \neq a_j$ for $i \neq j$, then (6.13) gives $\omega(\ell)_{ij} = \frac{b_i - b_j}{a_i - a_j} \ell_{ij}$ and the Hamiltonian (7.1) becomes

$$H(k, \ell) = \sum \frac{b_i - b_j}{a_i - a_j} \ell_{ij}^2 - \sum b_i f_{js} k_{ij} k_{is}. \tag{7.8}$$

(If some of the a_j coincide one should either use the general formula (6.13) or take the limit of the regular case.) In particular, if $b^2 = a$, the kinetic term in (7.8) coincides with the realistic kinetic energy of the n-dimensional top (6.3). Thus the Hamiltonian (7.8) describes a general n-dimensional top in an arbitrary quadratic potential (cf. (6.8)). The corresponding Lax pair is given by (7.2). If $f = 0$, the potential term vanishes and we are left with a free top. The Lax representation in this case was found by Manakov [1976].

The Neumann System. Define a mapping from the $(n-1)$-sphere S^{n-1} to the space $\mathfrak{p}$ of symmetric $n \times n$-matrices by $j(x) = x \otimes x$ where $x \in \mathbb{R}^n$, $|x|^2 = 1$. Clearly, j commutes with the action of $SO(n)$. The corresponding moment map $\mu\colon T^*S^{n-1} \to \mathfrak{so}(n)$ is $\mu(x, p) = p \wedge x$, where $p \in \mathbb{R}^n$, $(p, x) = 0$. According to (7.3) we have the Lax matrix

$$L(\lambda) = a\lambda + p \wedge x + x \otimes x \lambda^{-1}. \tag{7.9}$$

(In fact, $L(\lambda)$ defines a symplectic mapping from T^*S^{n-1} onto a coadjoint orbit in the space of Lax matrices of the form $L(\lambda) = a\lambda + \ell + s\lambda^{-1}$.) From (7.6) we find the Hamiltonian of the Neumann system on S^{n-1}:

$$H = \frac{1}{2} \sum p_i^2 + \sum a_i x_i^2. \tag{7.10}$$

Other Hamiltonians (7.3) give a complete set of integrals of motion parametrized by $b_1, \ldots, b_n$ (Uhlenbeck's integrals):

$$F = \sum \frac{b_i - b_j}{a_i - a_j} (p_i x_j - p_j x_i) + \sum b_i x_i.$$

These integrals in turn describe motion on S^{n-1} in special Riemannian metrics. The Lax pair (7.7), (7.9) for the Neumann system was found by Moser [1975].

If $n = 3$, it is possible to define the *Neumann system in the field of a magnetic monopole*. The corresponding Lax matrix is

$$L'(\lambda) = a\lambda + (p \wedge x + \alpha\sigma(x)) + x \otimes x \lambda^{-1}, \tag{7.11}$$

where $\sigma(x)_{ij} = \sum \varepsilon_{ijk} x_k$ and α is some constant. The magnetic form is proportional to the area form on S^2. The Hamiltonian $H' = -\frac{1}{2} \operatorname{Res} \lambda^{-1} \operatorname{tr} L'(\lambda)^2 = H + \alpha^2$ remains the same as before (see (7.10)).

In a similar way one constructs the analogs of the Neumann system on other homogeneous SO(n)-spaces, in particular on the SO(n)-orbits in $\mathfrak{p}$. For instance, let St(k, n) be the Stiefel manifold consisting of k-tuples of orthogonal unit vectors $x^{(1)}, \ldots, x^{(k)}$ in $\mathbb{R}^n$. Define $j(x^{(1)}, \ldots, x^{(k)}) = \sum c_{ij} x^{(i)} \otimes x^{(j)}$ where (c_{ij}) is a symmetric matrix. This gives the Lax matrix

$$L(\lambda) = a\lambda + \sum p^{(i)} \wedge x^{(i)} + \sum c_{ij} x^{(i)} \otimes x^{(j)} \lambda^{-1} \tag{7.12}$$

where $(p^{(i)}, x^{(j)}) + (p^{(j)}, x^{(i)}) = 0$. (If the eigenvalues of (c_{ij}) are all distinct and nonzero, these Lax matrices range over a single coadjoint orbit.) The Hamiltonians (7.4) describe the motion on St(k, n) in quadratic potentials. All these systems can also be viewed as reductions of the n-dimensional top with Hamiltonian (7.8); for $k = n$ we recover the n-dimensional top itself.

For St($n-2$, n), it is possible to include a nontrivial magnetic form by using the Lax matrix $L' = L + \alpha x^{(n-1)} \wedge x^{(n)}$ where $x^{(n-1)} \wedge x^{(n)}$ is the unit bivector of the 2-plane orthogonal to $x^{(1)}, \ldots, x^{(n-2)}$.

We also mention analogous systems on the Grassmann manifold Gr(k, n) consisting of all k-planes in $\mathbb{R}^n$. If e is a k-plane and $x_1^{(1)}, \ldots, x_n^{(k)}$ an orthonormal frame in e, we set $j(e) = \sum x^{(i)} \otimes x^{(i)}$. Clearly, $j(e)$ is well defined and leads to the Lax matrix

$$L(\lambda) = a\lambda + \sum p_{(i)} \wedge x^{(i)} + \sum x^{(i)} \otimes x^{(i)} \lambda^{-1}, \tag{7.13}$$

where the momenta $p_{(i)}$ are orthogonal to e. The Hamiltonian (7.6) in this case becomes

$$H = \frac{1}{2} \sum_i p_{(i)}^2 + \sum_{i,j} a_j x_j^{(i)2}. \tag{7.14}$$

Other Hamiltonians of type (7.4) contain kinetic terms which (in general) are not SO(n)-invariant. All these systems can be regarded as reductions of similar systems on St(k, n) or on SO(n).

For Gr(2, n) it is possible to incorporate the magnetic form by setting

$$L'(\lambda) = L(\lambda) + \alpha x^{(1)} \wedge x^{(2)}.$$

The corresponding Hamiltonian (7.6) is the same as (7.14) up to a constant. It describes the motion on Gr(2, n) in the standard Riemannian metric under the influence of a quadratic potential and a magnetic monopole. The corresponding magnetic form is the unique (up to a factor) SO(n)-invariant symplectic form on Gr(2, n); it is also the imaginary part of an invariant Kähler metric.

Lax pairs for all the above systems are given by (7.7) or (7.2).

Multi-dimensional Rigid Body in Ideal Fluid: The Clebsch Case. Consider the motion of a free rigid body in an ideal incompressible fluid. The total angular momentum $\ell = (\ell_{ij})$ and linear momentum $p = (p_i)$ of the joint system obey the Poisson bracket relations described by the Lie algebra $\mathfrak{e}(n) = \mathfrak{so}(n) \oplus \mathbb{R}^n$ of all rigid motions of $\mathbb{R}^n$ (the semidirect product of $\mathfrak{so}(n)$ and $\mathbb{R}^n$). Under the assumption that the fluid is at rest at infinity and its velocity field is described by a potential function, the total energy of the system is a quadratic form of the variables (ℓ, p). In the Clebsch case this energy thas the special structure

$$H = \sum c_{ij}\ell_{ij}^2 + \sum b_i p_i^2 \tag{7.15}$$

with the additional relations

$$\frac{b_i - b_j}{c_{ij}} + \frac{b_j - b_k}{c_{jk}} + \frac{b_k - b_i}{c_{ki}} = 0$$

from which it follows that

$$c_{ij} = \frac{b_i - b_j}{a_i - a_j}.$$

To incorporate the Clebsch case into the previous discussion, consider the mapping from the dual of $\mathfrak{e}(n)$ into the twisted affine algebra $\mathfrak{L}(\mathfrak{g}, \sigma)$ given by

$$(\ell, p) \mapsto L(\lambda) = a\lambda + \ell + p \otimes p^t \lambda^{-1}, \tag{7.16}$$

where $a = \operatorname{diag}(a_1, \ldots, a_n)$. This is a Poisson map with respect to the R-bracket structure on the space of L-matrices of the form $L(\lambda) = a\lambda + \ell + s\lambda^{-1}$, as is shown by the following lemma.

Lemma 7.1. *Suppose we are given two linear representations of a Lie group K on vector spaces V and W. If c: $V^* \to W^*$ is a K-equivariant (in general, nonlinear) mapping then the mapping μ: $(\mathfrak{k} + V)^* \to (\mathfrak{k} + W)^*$ of the duals of the corresponding semi-direct sums given by $\mu(\ell, v) = (\ell, c(v))$ is a Poisson map.*

In particular, taking $c(p) = p \otimes p^t$ we conclude that (7.16) is a Poisson map. Substituting now (7.16) into the integrable Hamiltonian (6.14) we recover the Clebsch Hamiltonian (7.15). The corresponding Lax pair is given by (6.15).

7.2. $(G, K) = (SO(p, q), SO(p) \times SO(q))$: Interacting Tops and Kowalewski Tops. The Lie algebra $\mathfrak{so}(p, q)$ consists of $(p + q) \times (p + q)$-matrices X such that $X^t = -JXJ$, where $J = \operatorname{diag}(1, \ldots, 1, -1, \ldots, -1)$, $\operatorname{tr} J = p - q$. In the (p, q)-block notation, X has the form

$$X = \begin{pmatrix} \ell & s \\ s^t & m \end{pmatrix}, \quad \ell \in \mathfrak{so}(p), \quad m \in \mathfrak{so}(q). \tag{7.17}$$

The Cartan involution is $\sigma X = -X^t$, so that in the Cartan decomposition $\mathfrak{g} = \mathfrak{k} + \mathfrak{p}$ we have $\mathfrak{k} = \mathfrak{so}(p) \oplus \mathfrak{so}(q) = \left\{\begin{pmatrix} \ell & 0 \\ 0 & m \end{pmatrix}\right\}$ and $\mathfrak{p} = \left\{\begin{pmatrix} 0 & s \\ s^t & 0 \end{pmatrix}\right\}$. The compact subgroup K acts on $\mathfrak{p}$ by

$$\begin{pmatrix} 0 & s \\ s^t & 0 \end{pmatrix} \mapsto \begin{pmatrix} 0 & ksr^t \\ rs^tk^t & 0 \end{pmatrix}, \quad k \in SO(p), \quad r \in SO(q). \tag{7.18}$$

According to the general construction, the group $SO(p, q)$ gives rise to a system of two interacting tops of dimensions p and q. Let k and r denote the configuration variables on $SO(p)$ and $SO(q)$, and let ℓ, m be the corresponding angular momenta. Assuming that the leading coefficient in the Lax matrix (7.2) has the form $\begin{pmatrix} 0 & A \\ A^t & 0 \end{pmatrix}$, where $A_{ij} = \delta_{ij}a_j$, we have the following expression for the Hamiltonian (7.1) ($p \geqslant q$):

$$H(k, \ell; r, m) = \sum_{i<j\leqslant q} \frac{a_i b_i - a_j b_j}{a_i^2 - a_j^2}(\ell_{ij}^2 + m_{ij}^2) - 2\sum_{i<j\leqslant q} \frac{a_i b_j - a_j b_i}{a_i^2 - a_j^2}\ell_{ij}m_{ij}$$
$$+ \sum_{i=1}^{p}\sum_{j=q+1}^{p} \frac{b_i}{a_i}\ell_{ij}^2 + c\sum_{i,j=q+1}^{p} \ell_{ij}^2 - 2\sum_{i,j,s} b_i k_{ij} f_{js} r_{is}. \qquad (7.19)$$

The corresponding Lax pair is

$$L(\lambda) = \begin{pmatrix} 0 & A \\ A^t & 0 \end{pmatrix}\lambda + \begin{pmatrix} \ell & 0 \\ 0 & m \end{pmatrix} + \begin{pmatrix} 0 & k^t f r \\ r^t f^t k & 0 \end{pmatrix}\lambda^{-1},$$
$$M(\lambda) = \begin{pmatrix} 0 & B \\ B^t & 0 \end{pmatrix}\lambda + \begin{pmatrix} \omega(\ell) & 0 \\ 0 & \omega(m) \end{pmatrix}, \qquad (7.20)$$

where $A = a_i\delta_{ij}$, $B = b_i\delta_{ij}$ and the angular velocities $\omega(\ell)$ and $\omega(m)$ are determined from (7.19).

We shall consider separately three interesting series of integrable tops, where, respectively, $q = p$, $q = 2$, and $q = 1$.

Two interacting p-dimensional Tops: $\mathfrak{g} = \mathfrak{so}(p, p)$. From (7.19) we have

$$H = \sum_{i<j} \frac{a_i b_i - a_j b_j}{a_i^2 - a_j^2}(\ell_{ij}^2 + m_{ij}^2) - 2\sum_{i<j} \frac{a_i b_j - a_j b_i}{a_i^2 - a_j^2}\ell_{ij}m_{ij} - 2\sum b_i f_{js} k_{ij} r_{is}. \qquad (7.21)$$

This Hamiltonian describes two p-dimensional tops which interact through their angular momenta and a bilinear potential. In particular, when $A = B$ the interaction of the momenta disappears and we get a pair of spherical tops interacting via a bilinear potential. Notice that the top in a quadratic potential (7.8) is a subsystem of (7.21), where $k = r$ and $\ell = m$. (This corresponds to the canonical embedding of $\mathfrak{sl}(n, \mathbb{R})$ into $\mathfrak{so}(n, n)$.)

p-dimensional Top Interacting with a Rotator: $\mathfrak{g} = \mathfrak{so}(p, 2)$. Let θ be the angular variable of the rotator (i.e. a system with configuration space $SO(2)$) and γ its momentum: $r = \begin{pmatrix} \cos\theta & -\sin\theta \\ \sin\theta & \cos\theta \end{pmatrix}$, $m = \begin{pmatrix} 0 & \gamma \\ -\gamma & 0 \end{pmatrix}$. Then (7.19) becomes

$$H(k, \ell; \theta, \gamma) = \frac{a_1 b_1 - a_2 b_2}{a_1^2 - a_2^2}(\ell_{12}^2 + \gamma^2) + \sum_{i=3}^{p}\sum_{j=1}^{2} \frac{b_j}{a_j}\ell_{ij}^2$$
$$+ c\sum_{i,j=3}^{p} \ell_{ij}^2 + 2\frac{a_1 b_2 - a_2 b_1}{a_1^2 - a_2^2}\ell_{12}\gamma$$
$$- 2\sum_{i=1}^{p} \{(b_1 f_{i1} k_{i1} + b_2 f_{i2} k_{i2})\cos\theta + (b_1 f_{i2} k_{i1} - b_2 f_{i1} k_{i2})\sin\theta\} \qquad (7.22)$$

where c is an arbitrary constant. If $f = 0$ we can set $\theta = 0, \gamma = \text{const}$ which gives a free p-dimensional top whose kinetic energy contains a linear term $\gamma \ell_{12}$. In classical mechanics such systems are called *gyrostats.*

p-dimensional Lagrange's Heavy Top: $\mathfrak{g} = \mathfrak{so}(p, 1)$. Here the configuration space is $K = SO(p)$, and the symmetric subspace is identified with $\mathbb{R}^p$. From (7.19) we find the Hamiltonian

$$H(k, \ell) = \frac{1}{2} \sum_{i,j=1}^{p} \ell_{ij}^2 + \frac{\alpha}{2} \sum_{i,j=2}^{p} \ell_{ij}^2 - (k^t f, e_1), \tag{7.23}$$

where $f \in \mathbb{R}^p$. The kinetic term in (7.23) is a general quadratic form axially symmetric with respect to the axis e_1, and the potential term corresponds to a constant gravity field f, the centor of mass lying on the symmetry axis e_1. This is precisely the p-dimensional Lagrange case of axial symmetry.

Notice that the second term $H_1 = \frac{\alpha}{2} \sum_{i,j=2}^{p} \ell_{ij}^2$ Poisson commutes with H, and hence the motion of the Lagrange top can be expressed as the combination of two simpler flows: one for the Hamiltonian $H_0 = \frac{1}{2} \sum_{i,j=1}^{p} \ell_{ij}^2 - (k^t f, e_1)$ with a spherical kinetic energy, and the other for H_1. The latter flow is given explicitly by Lemma 6.4.

Two Interacting Points on a Stiefel Manifold. Let $\mathfrak{g} = \mathfrak{so}(n, n)$ and define a mapping from the product of two Stiefel manifolds to the symmetric subspace $\mathfrak{p}$: if $x^{(1)}, \ldots, x^{(k)}$ and $y^{(1)}, \ldots, y^{(k)}$ are two sets of unit orthogonal vectors in $\mathbb{R}^n$, we put $j(x^{(1)}, \ldots, x^{(k)}, y^{(1)}, \ldots, y^{(k)}) = \begin{pmatrix} 0 & s \\ s^t & \end{pmatrix}$, where $s = \sum c_{ij} x^{(i)} \otimes y^{(j)t}$, and c_{ij} are some constants. Clearly, j commutes with the action of $SO(n) \times SO(n)$. Using (7.9) we find the Lax matrix

$$L(\lambda) = \begin{pmatrix} 0 & A \\ A^t & 0 \end{pmatrix} \lambda + \begin{pmatrix} \ell & 0 \\ 0 & m \end{pmatrix} + \begin{pmatrix} 0 & s \\ s^t & 0 \end{pmatrix} \lambda^{-1}, \tag{7.24}$$

where s is as above and $\ell = \sum p_{(i)} \wedge x^{(i)}$, $m = \sum q_{(i)} \wedge y^{(i)}$ ($p_{(i)}$ and $q_{(i)}$ are the momenta conjugate to $x^{(i)}$ and $y^{(i)}$). Hamiltonians of the form (7.6), $H = -\frac{1}{2} \operatorname{Res} \operatorname{tr} \lambda^{-1} L(\lambda)^2$, describe a system of two points on $St(k, n)$ interacting via a bilinear potential. In particular, for $St(1, n) = S^{n-1}$ this gives a system of two interacting points on S^{n-1} with Hamiltonian

$$\begin{gathered} H = \frac{1}{2} \sum (p_i^2 + q_i^2) + \sum a_i x_i y_i, \\ \|x\|^2 = \|y\|^2 = 1, \quad (p, x) = (q, y) = 0. \end{gathered} \tag{7.25}$$

The Lax pair in this case is

$$L(\lambda) = \begin{pmatrix} 0 & A \\ A^t & 0 \end{pmatrix}\lambda + \begin{pmatrix} p \wedge x & 0 \\ 0 & q \wedge y \end{pmatrix} + \begin{pmatrix} 0 & x \otimes y^t \\ y \otimes x^t & 0 \end{pmatrix}\lambda^{-1},$$

$$M(\lambda) = \begin{pmatrix} 0 & A \\ A^t & 0 \end{pmatrix}\lambda + \begin{pmatrix} p \wedge x & 0 \\ 0 & q \wedge y \end{pmatrix}.$$

Spherical Pendulum. Let $\mathfrak{g} = \mathfrak{so}(n, 1)$, so that $\mathfrak{p} = \mathbb{R}^n$ with the usual action of $SO(n)$. Define $j\colon S^{n-1} \to \mathfrak{p}$ to be the standard embedding, $j(x) = x$. This leads to the Lax matrix

$$L(\lambda) = (a \otimes e_{n+1} + e_{n+1} \otimes a)\lambda + p \wedge x + (x \otimes e_{n+1} + e_{n+1} \otimes x)\lambda^{-1}.$$

The simplest Hamiltonian of type (7.6) is

$$H = \frac{1}{2}\sum p_i^2 + \sum a_i x_i \tag{7.26}$$

and describes a particle on the sphere moving under a constant force. This system is called the *spherical pendulum*. The corresponding Lax pair is given by (7.7).

If $n = 3$, one can modify $L(\lambda)$ according to (7.5), by adding $\alpha\sigma(x)$, where $\sigma(x)_{ij} = \varepsilon_{ijk}x_k$. This gives a Lax pair for the spherical pendulum on S^2 in a magnetic field.

Generalized Kowalewski Tops. It is well known that a 3-dimensional heavy rigid body with a fixed point rotating in the constant gravitational field is completely integrable only in the following three cases: (a) *the top is fixed at its center of mass* (Euler); (b) *the top is axially symmetric* (Lagrange); (c) *the moments of inertia have a fixed ratio* 2:2:1 *and the center of mass lies in the equatorial plane of the top* (Kowalewski). The peculiar geometry of the latter case is explained by the fact that the Kowalewski top is a reduction of a more symmetrical system with one extra degree of freedom consisting of a spherical top interacting with a rotator. This is one of the systems treated above, related to the group $SO(3, 2)$. The same construction applied to the Lie group $SO(p, q)$ gives a p-dimensional Kowalewski top in q homogeneous fields.

To describe these systems in more detail, consider again the system of two interacting tops with Hamiltonian (7.19), where we set $B = A$:

$$H = -\frac{1}{4}\operatorname{tr}\ell^2 - \frac{1}{4}\operatorname{tr}m^2 + c\sum_{i,j=q+1}^{p}\ell_{ij}^2 - \operatorname{tr}(k^tFrA^t). \tag{7.27}$$

(For instance, take $H = -\frac{1}{4}\operatorname{Res}\operatorname{tr}\lambda^{-1}L(\lambda)^2$ with L given by (7.20); in this case $c = 0$.) This Hamiltonian is invariant under the action of the stationary subgroup K_A by right translations on $SO(p) \times SO(q)$. The group K_A consists of orthogonal matrices $\begin{pmatrix} k & 0 \\ 0 & r \end{pmatrix}$, where $k \in SO(p)$, $r \in SO(q)$, such that $kAr^t = A$. We now choose A so as to make K_A large: set $A = E$ to be the truncated identity matrix, $E_{ij} = \delta_{ij}$. Then the equation $kEr^t = E$ implies that k has a $(q, p-q)$-block form: $k = \begin{pmatrix} r & 0 \\ 0 & v \end{pmatrix}$ where $v \in SO(p-q)$. Hence $K_A \simeq SO(q) \times SO(p-q)$. Note that the factor $SO(q)$ of K_A lies in $SO(p) \times SO(q)$ diagonally.

The generalized Kowalewski top is obtained by reducing the system (7.27) with respect to the action of the subgroup $SO(q)$ of K_A. The moment map $\mu: T^*(SO(p) \times SO(q)) \to \mathfrak{so}(q)$ for the diagonal action of $SO(q)$ by right translations is

$$\mu(k, \ell; r, m) = P\ell P + m,$$

where P is the orthgonal projection from $\mathbb{R}^p$ to $\mathbb{R}^q$ ($\mathbb{R}^q$ is spanned by $e_1, \ldots, e_q$). Reduction at the zero value of the moment map amounts to imposing the constraints

$$P\ell P + m = 0, \quad r = I \tag{7.28}$$

($r = I$ is a natural subsidiary condition). The reduced space is $T^*SO(p)$ with its canonical Poisson strucutre. Substituting these constraints into (7.27) gives the reduced Hamiltonian on $T^*SO(p)$:

$$H = \frac{1}{2}\sum_{i,j=1}^{p} \ell_{ij}^2 + \frac{1}{2}\sum_{i,j=1}^{q} \ell_{ij}^2 + c\sum_{i,j=q+1}^{p} \ell_{ij}^2 - \sum_{i=1}^{q} (ke_i, f_i), \tag{7.29}$$

where $e_1, \ldots, e_p$ is the orthogonal frame of the top and $f_1, \ldots, f_q$ are the column vectors of F. This Hamiltonian describes a p-dimensional top with a very special $SO(q) \times SO(p-q)$-symmetric inertia tensor, rotating under the influence of q constant forces $f_1, \ldots, f_q$, with the condition that the corresponding centers of charge lie on the orthogonal axes $e_1, \ldots, e_q$. The characteristic doubling of the terms ℓ_{ij}^2 for $i, j \leqslant q$ in (7.29) is due to reduction with respect to $SO(q)$. If $q = 2$, reduction can also be performed at a nonzero value γ of the $\mathfrak{so}(2)$-momentum with the same reduced space $T^*SO(p)$. The constraints (7.28) are then replaced by

$$P\ell P + m = \gamma e_1 \wedge e_2, \quad r = I, \tag{7.30}$$

which gives an extra linear "gyrostatic" term $\gamma\ell_{12}$ in the reduced Hamiltonian.

For $p = 3$, $q = 2$ we get a generalization of the classical Kowalewski case: the Kowalewski gyrostat in two fields. With the notation $\ell_i = \frac{1}{2}\varepsilon_{ijk}\ell_{jk}$, $g = k^t f_1$, $h = k^t f_2$, the Hamiltonian (7.29) becomes

$$H = \frac{1}{2}(\ell_1^2 + \ell_2^2 + 2\ell_3^2 + \gamma\ell_3) - g_1 - h_2. \tag{7.31}$$

To derive a Lax pair for the generalized Kowalewski top (7.29)we start with the Lax pair (6.21) for a system of two tops which in the present context takes the form

$$L(\lambda) = \begin{pmatrix} 0 & kAr^t \\ rA^tk^t & 0 \end{pmatrix}\lambda - \begin{pmatrix} k\ell k^t & 0 \\ 0 & r\mathfrak{m}r^t \end{pmatrix} + \begin{pmatrix} 0 & F \\ F^t & 0 \end{pmatrix}\lambda^{-1},$$
$$M(\lambda) = -\begin{pmatrix} 0 & kBr^t \\ rB^tk^t & 0 \end{pmatrix}\lambda. \tag{7.32}$$

It is invariant under the action of K_A by right translations. We can therefore impose the constraints (7.28) on this Lax pair which gives a Lax pair $\hat{L}$, $\hat{M}$ for the Kowalewski top. To write it in a more elegant way, we can go over to the rotating frame, i.e. perform a gauge transformation

$$L' = \hat{k}^t\hat{L}\hat{k}, \quad M' = \hat{k}^t\hat{M}\hat{k} + \hat{\omega},$$

where $\hat{k} = \begin{pmatrix} k & 0 \\ 0 & I \end{pmatrix}$, $\hat{\omega} = \begin{pmatrix} \omega & 0 \\ 0 & 0 \end{pmatrix}$, and ω is the angular velocity of the top given by $\omega_{ij} = 2\ell_{ij}$ for $i, j \leqslant q$, $\omega_{ij} = \ell_{ij}$ for $i > q, j \leqslant q$, $\omega_{ij} = (1 + c)\ell_{ij}$, for $i, j > q$. The resulting Lax pair is

$$L'(\lambda) = \begin{pmatrix} 0 & E \\ E^t & 0 \end{pmatrix}\lambda + \begin{pmatrix} -\ell & 0 \\ 0 & P\ell P \end{pmatrix} + \begin{pmatrix} 0 & k^tF \\ F^tk & 0 \end{pmatrix}\lambda^{-1},$$
$$M'(\lambda) = \begin{pmatrix} \omega & 0 \\ 0 & 0 \end{pmatrix} - \begin{pmatrix} 0 & E \\ E^t & 0 \end{pmatrix}\lambda. \tag{7.33}$$

The Hamiltonian (7.29), as well as the invariants of $L'(\lambda)$, have some residual symmetry: they are invariant under the group $SO(p - q)$ of rotations in the subspace spanned by $(e_{q+1}, \ldots, e_p)$ acting on $SO(p)$ by right translations, and under the stationary subgroup of the matrix F, acting by left translations. Therefore, the invariants of $L'(\lambda)$ do not form in general a complete set of integrals of motion in involution for the generalized Kowalewski top; still, this set is complete modulo the above two symmetry groups.

As explained in Section 6, the equations of the p-dimensional Kowalewski top in q fields can be written as the Euler-Poisson equation in the dual of the semi-direct sum $\mathfrak{k} \oplus \mathfrak{p} = \mathfrak{so}(p) \oplus \bigoplus_1^q \mathbb{R}^p$:

$$\dot{\ell} = [\ell, \omega] + \sum_{i=1}^{q} h_i \wedge e_i,$$
$$\dot{h}_i = -\omega h_i,$$

where $h_i = k^tf_i$, $i = 1, \ldots, q$, are the so-called Poisson vectors. The Lax pair (7.33) represents precisely these equations, and the invariants of $L'(\lambda)$ make up a complete involutive family in $\mathfrak{so}(p) \oplus \bigoplus^q \mathbb{R}^p$ modulo the symmetry subalgebra $\mathfrak{so}(p - q)$ spanned by $\{\ell_{ij}, i, j > q\}$.

To conclude, we give a more convenient 4×4 Lax pair for the 3-dimensional Kowalewski gyrostat in two fields (7.31). Since $\mathfrak{so}(3, 2)$ is isomorphic to $\mathfrak{sp}(4, \mathbb{R})$, the symplectic Lie algebra in $\mathbb{R}^4$, the Lax pair (7.33) can be represented in this case by symplectic 4×4 matrices. With the notation of (7.31) we have

$$L(\lambda) = \begin{bmatrix} g_1 - h_2 & g_2 + h_1 & g_3 & h_3 \\ g_2 + h_1 & -g_1 + h_2 & h_3 & -g_3 \\ g_3 & h_3 & -g_1 - h_2 & g_2 - h_1 \\ h_3 & -g_3 & g_2 - h_1 & g_1 + h_2 \end{bmatrix} \lambda^{-1}$$

$$+ \begin{bmatrix} 0 & -\gamma & -\ell_2 & -\ell_1 \\ \gamma & 0 & \ell_1 & -\ell_2 \\ \ell_2 & -\ell_1 & -2\lambda & -2\ell_3 - \gamma \\ \ell_1 & \ell_2 & 2\ell_3 + \gamma & 2\lambda \end{bmatrix}, \tag{7.34}$$

$$M(\lambda) = \tfrac{1}{2} \begin{bmatrix} 0 & -2\ell_3 - \gamma & \ell_2 & \ell_1 \\ 2\ell_3 + \gamma & 0 & -\ell_1 & \ell_2 \\ -\ell_2 & \ell_1 & 2\lambda & 2\ell_3 + \gamma \\ -\ell_1 & -\ell_2 & -2\ell_3 - \gamma & -2\lambda \end{bmatrix}.$$

The coefficients of $\operatorname{tr} L(\lambda)^4$ provide two additional integrals of the motion in involution which ensure complete integrability:

$$\begin{aligned} I_1 &= (\ell, g)^2 + (\ell, h)^2 + 2(\ell_3 + \gamma)(\ell, [g, h]) \\ &\quad + 2(g, h)(g_2 + h_1) - 2g_1|h|^2 - 2h_2|g|^2, \\ I_2 &= (\ell_1^2 - \ell_2^2 + 2g_1 - 2h_2)^2 + 4(\ell_1\ell_2 + g_2 + h_1)^2 \\ &\quad - 4\gamma((\ell_3 + \gamma)(\ell_1^2 + \ell_2^2) + 2\ell_1 g_3 + 2\ell_2 h_3). \end{aligned} \tag{7.35}$$

Observe that if $h = 0$ the integral I_1 reduces to $(\ell, g)^2$. The integral I_2 is an extension of the integral found by Kowalewski (1889).

Remark. The Lax pair (7.33) for the generalized Kowalewski top can also be derived by using the Lax matrix (7.3) for the mapping j: $\mathrm{SO}(p) \to \mathfrak{p}$ given by $j(k) = \begin{pmatrix} 0 & kA \\ A^t k^t & 0 \end{pmatrix}$, where $A_{ij} = \delta_{ij}$ as above. This mapping commutes with the action of $K = \mathrm{SO}(p) \times \mathrm{SO}(q)$, where the first factor acts on $\mathrm{SO}(p)$ by left multiplication and the second by right multiplication. The corresponding moment map μ: $T^*\mathrm{SO}(p) \to \mathfrak{so}(p) \oplus \mathfrak{so}(q)$ is $\mu(k, \ell) = -k\ell k^t \oplus P\ell P$, which by (7.3) gives precisely the Lax matrix (7.32) under the constraints (7.28).

Three-dimensional Tops Associated with the Lie Algebras $\mathfrak{so}(4, 4)$ and $\mathfrak{so}(4, 3)$. Since $\mathfrak{so}(4) = \mathfrak{so}(3) \oplus \mathfrak{so}(3)$, a pair of $\mathfrak{so}(4)$-tops associated with the algebra $\mathfrak{so}(4, 4)$ may be interpreted as 4 three-dimensional tops. We shall write only the corresponding kinetic energy.

In terms of the 3-dimensional angular momenta $u, v, w, z \in \mathfrak{so}(3) \simeq \mathbb{R}^3$, the 4-dimensional momenta ℓ_{ij}, m_{ij} are $\ell_{ij} = 1/2\varepsilon_{ijk}(u_k + v_u)$, $i, j \leqslant 3$, $\ell_{4k} = u_k - v_u$, and $m_{ij} = 1/2\varepsilon_{ijk}(w_u + z_k)$, $i, j \leqslant 3$, $m_{4k} = w_u - z_k$. Substituting this into (7.21) gives the kinetic energy

$$E(u, v, w, z) = \sum_k \{(c_{ij} + c_{4k})(u_k^2 + v_k^2 + w_k^2 + z_k^2)$$
$$+ 2(c_{ij} - c_{4k})(u_k v_u + w_k z_k) + 2(d_{ij} + d_{4k})(u_k w_k + v_k z_k)$$
$$+ 2(d_{ij} - d_{4k})(u_u z_k + v_k w_k)\}, \quad (7.36)$$

where $c_{lm} = (a_l b_l - a_m b_m)/(a_l^2 - a_m^2)$, $d_{lm} = (a_m b_l - a_l b_m)/(a_l^2 - a_m^2)$ and the subscripts i, j, k form a permutation of (1, 2, 3). The potential on SO(4) × SO(4) entering in (7.21) cannot be pushed forward to the product of 4 copies of SO(3). When lifted to the covering group, the product of 4 copies of SU(2), this potential becomes a 4-linear function of the matrix elements of the factors.

In a similar way, the Lie algebra $\mathfrak{so}(4, 3)$ gives rise to a system of 3 three-dimensional tops with kinetic energy

$$E(u, v, w) = \sum_{k=1}^{3} \left\{\left(c_{ij} + \frac{b_u}{a_k}\right)(u_k^2 + v_k^2) + c_{ij} w_k^2\right.$$
$$\left. + 2\left(c_{ij} - \frac{b_k}{a_k}\right) u_k v_k + 2d_{ij}(u_k w_k + v_k w_k)\right\}, \quad (7.37)$$

where c_{ij}, d_{ij} are the same as above and (i, j, k) is a permutation of (1, 2, 3).

7.3. The Lie Algebra G_2: an Exotic SO(4)-top. Let $\mathfrak{g}$ be the split real form of the simple Lie algebra of type G_2, dim $\mathfrak{g} = 14$. It can be represented as a subalgebra of $\mathfrak{so}(4, 3)$; the matrices in $\mathfrak{g}$ are conveniently parametrized by

$$X(u, w, a, y, z)$$
$$= \begin{pmatrix} 0 & -\frac{u_3 + w_3}{2} & \frac{u_2 + w_2}{2} & -\frac{u_1 - w_1}{2} & -y_2 & y_3 & a_1 \\ \frac{u_3 + w_3}{2} & 0 & -\frac{u_1 + w_1}{2} & -\frac{u_2 - w_2}{2} & y_1 & a_2 & z_3 \\ -\frac{u_2 + w_2}{2} & \frac{u_1 + w_1}{2} & 0 & -\frac{u_3 - w_3}{2} & a_3 & z_1 & -z_2 \\ \frac{u_1 - w_1}{2} & \frac{u_2 - w_2}{2} & \frac{u_3 - w_3}{2} & 0 & y_3 - z_3 & y_2 - z_2 & y_1 - z_1 \\ -y_2 & y_1 & a_3 & y_3 - z_3 & 0 & w_1 & -w_2 \\ y_3 & a_2 & z_1 & y_2 - z_2 & -w_1 & 0 & w_3 \\ a_1 & z_3 & -z_2 & y_1 - z_1 & w_2 & -w_3 & 0 \end{pmatrix}$$

where u, v, a, y, z are vectors in $\mathbb{R}^3$ and $\sum a_i = 0$. The Cartan involution on $\mathfrak{g}$ is given by $X \to -X^t$. Its fixed subalgebra $\mathfrak{k}$ is spanned by $X(u, w, 0, 0, 0)$ and is isomorphic to $\mathfrak{so}(4) = \mathfrak{so}(3) \oplus \mathfrak{so}(3)$. The symmetric matrices of the form $a = X(0, 0, a, 0, 0)$ make up a split Cartan subalgebra.

As before, we fix the inner product $(X, Y) = -\operatorname{tr} XY$ on $\mathfrak{g}$, which enables us to identify $\mathfrak{k}$ and $\mathfrak{k}^*$. Recall that the Poisson brackets of the linear coordinates x_ν on $\mathfrak{k}^*$ with respect to a given basis are $\{x_\mu, x_\nu\} = c_{\mu\nu}^\sigma x_\sigma$ where $c_{\mu\nu}^\sigma$ are the structure constants of $\mathfrak{k}$ in the dual basis. This gives $\{u_i, u_j\} = \varepsilon_{ijk}u_k$, $\{w_i, w_j\} = 1/3\varepsilon_{ijk}w_u$, $\{u_i, w_j\} = 0$, so that the standard $\mathfrak{so}(3)$-variables are u_i and $v_i = 3w_i$. Using (6.13) or (7.37) we find the kinetic energy of the $so(4)$-top related to the symmetric pair $(G_2, \mathrm{SO}(4))$:

$$E(u, v) = \frac{1}{2}\sum_{i=1}^{3}\left\{(3c_i + d_i)u_i^2 + 2(d_i - c_i)u_iv_i + \left(\frac{c_i}{3} + d_i\right)v_i^2\right\}, \qquad (7.38)$$

where $c_i = \dfrac{b_i}{a_i}$, $d_i = \dfrac{b_j - b_u}{a_j - a_u}$ and $\sum a_i = \sum b_i = 0$. The Euler top on SO(4) with kinetic energy (7.38) is different from the previously known cases of Manakov (related to $\mathfrak{sl}(4, \mathbb{R})$) and Steklov (see Section 11). In particular, it has no additional quadratic integrals of motion, but there is a quartic integral. Up to constant factors and Casimir terms, (7.38) is a one-parameter family. It can be parametrized by taking, for instance,

$$a = (1, a, 1 - a), \quad b = (1, 0, 1):$$

$$E(u, v) = \frac{3a + 2}{2a + 1}u_1^2 + \frac{1}{a + 2}u_2^2 + \frac{2 - a}{1 - a^2}u_3^2 - \frac{2a}{2a + 1}u_1v_1$$

$$+ \frac{2}{a + 2}u_2v_2 + \frac{2a}{1 - a^2}u_3v_3 + \frac{a + 2}{3(2a + 1)}v_1^2 + \frac{1}{a + 2}v_2^2 + \frac{a + 2}{3(1 - a^2)}v_3^2.$$

The corresponding potential energy on $K = \mathrm{SO}(4)$ is given by (7.1). Note that the 7×7-matrix $k \in K$ decomposes into two blocks of dimension 4 and 3. The first block is the ordinary vector representation of SO(4), while the second corresponds to the representation in the space of self-dual 2-forms on $\mathbb{R}^4$. Therefore the formula $V(k) = (f, \mathrm{ad}\, k \cdot b)$ becomes

$$V(k) = -2\sum_{i,j=1}^{3}\sum_{\mu=1}^{1} b_i f_{\mu, j+4} k_{\mu i} \pi_{j, 4-i}(k).$$

The matrix elements $\pi_{ij}(k)$, $i, j = 1, 2, 3$, are quadratic forms of the $k_{\mu\nu}$. Hence $V(k)$ is a cubic form of the matrix elements of $k \in \mathrm{SO}(4)$. Clearly, it cannot be reduced to the quotient $\mathrm{SO}(3) \times \mathrm{SO}(3) = \mathrm{SO}(4)/\{\pm I\}$.

7.4. The Anharmonic Oscillator, the Garnier System and Integrable Quartic Potentials. The *anharmonic oscillator* is defined by the Hamiltonian

$$H(p, x) = \frac{1}{2}\sum_i p_i^2 + \sum_i a_i x_i^2 + \left(\sum_i x_i^2\right)^2,$$

where p_i, x_i are canonical variables in $\mathbb{R}^{2n}$. Garnier (1919) studied an integrable system with $2n$ degrees of freedom defined by the Hamiltonian

$$H(p, x; q, y) = \frac{1}{2}\sum p_i q_i + \sum a_i x_i y_i + (\sum x_i y_i)^2,$$

where (p, x, q, y) are canonical coordinates in $\mathbb{R}^{4n}$. It contains the anharmonic oscillator as a subsystem for $x = y$. Both systems have Lax representations in affine Lie algebras. Unlike previous examples, these representations take place in a different Poisson subspace with respect to the R-bracket.

Let $\mathfrak{g} = \mathfrak{gl}(n+1, \mathbb{R})$. We are interested in the orbits of the R-bracket in the subspace $\mathfrak{L}(\mathfrak{g})^*_- = \left\{L(\lambda) = \sum_{i>0} u_i\lambda^i\right\}$, i.e. the orbits of the subalgebra $\mathfrak{L}(\mathfrak{g})_- = \left\{\sum_{i<0} u_i\lambda^i\right\}$. Let $e_1, \ldots, e_{n+1}$ be an orthogonal unit basis in $\mathbb{R}^{n+1}$; the subspace $\mathbb{R}^n \subset \mathbb{R}^{n+1}$ is spanned by $e_1, \ldots, e_n$.

Lemma 7.2. *The orbit in* $\mathfrak{L}(\mathfrak{g})^*_-$ *through the point* $A\lambda^3 + a\lambda$ *where* $A = \operatorname{diag}(0, \ldots, 0, 1)$ *is symplectically diffeomorphic to* $\mathbb{R}^{4n}$. *The diffeomorphism is given by*

$$\begin{aligned} L(\lambda) = {} & A\lambda^3 + (x \otimes e_{n+1} - e_{n+1} \otimes y)\lambda^2 \\ & + (x \otimes y + p \otimes e_{n+1} + e_{n+1} \otimes q + (x, y)e_{n+1} \otimes e_{n+1} + a)\lambda. \end{aligned} \tag{7.40}$$

Consider the simplest invariant $H = \frac{1}{2} \operatorname{Res} \operatorname{tr} \lambda^{-3} L(\lambda)^2$. Inserting (7.40) with $a = \operatorname{diag}(a_1, \ldots, a_n, 0)$ we recover, up to constants, the Hamiltonian of the Garnier system. The corresponding Lax equation $\dot{L} = [L, M_\pm]$ follows from Theorem 2.1: since $dH(L) = \lambda^{-2}L$, we have

$$M_+ = A\lambda + (x \otimes e_{n+1} - e_{n+1} \otimes y), \quad M_- = M_+ - \lambda^{-2}L.$$

The Lax pair for the anharmonic oscillator lives on a similar orbit for the twisted loop algebra $\mathfrak{L}(\mathfrak{gl}(n+1, \mathbb{R}), \sigma)$ where σ is the Cartan involution, $\sigma X = -X^t$. The orbit through $A\lambda^3 + a\lambda$ is symplectically diffeomorphic to $\mathbb{R}^{2n}$; its parametrization by canonical variables (p, x) is given by (7.40) with $p = q$ and $x = y$.

The Garnier system and the anharmonic oscillator can be generalized as follows. Let $\mathfrak{g}$ be a semisimple Lie algebra and A its semisimple element. Let $\mathfrak{g}_A$ be the centralizer of A and $\mathfrak{g}_A^\perp$ the orthogonal complement of $\mathfrak{g}_A$ with respect to the Killing form. We have

$$\mathfrak{g} = \mathfrak{g}_A + \mathfrak{g}_A^\perp, \quad [\mathfrak{g}_A, \mathfrak{g}_A] \subset \mathfrak{g}_A, \quad [\mathfrak{g}_A, \mathfrak{g}_A^\perp] \subset \mathfrak{g}_A^\perp.$$

Let $\{e_i\}$ be a basis of $\mathfrak{g}_A^\perp$, let $\{f_i\}$ be the dual basis, $(f_i, e_j) = \delta_{ij}$, and denote $g_i = [A, e_i]$.

Lemma 7.3. *Suppose that* A *is such that*

$$[\mathfrak{g}_A^\perp, \mathfrak{g}_A^\perp] \subset \mathfrak{g}_A,$$

i.e. $\mathfrak{g} = \mathfrak{g}_A + \mathfrak{g}_A^\perp$ *is a "symmetric" decomposition. For any* $a \in \mathfrak{g}$, *the orbit of* $\mathfrak{L}(\mathfrak{g})_-$ *in* $\mathfrak{L}(\mathfrak{g})^*_-$ *through* $A\lambda^3 + a\lambda$ *is symplectically diffeomorphic to* $T^*(\mathfrak{g}_A^\perp)$ *and has the following parametrization by the canonical variables* (p_i, q_i), $i = 1, \ldots, \dim \mathfrak{g}_A^\perp$:

$$L(\lambda) = A\lambda^3 + \sum_i q_i g_i \lambda^2 + \left(\sum p_i f_i + \frac{1}{2}\sum q_i q_j [g_i, e_j] + a\right)\lambda.$$

Assuming that $a \in \mathfrak{g}_A$ and calculating $\operatorname{Res} \lambda^{-3} \operatorname{tr} L(\lambda)^2$ we find the Hamiltonian

$$H = \sum (f_i, f_j) p_i p_j + \sum q_i q_j([e_i, g_j], a) + \frac{1}{4} \sum q_i q_j q_k q_l([e_i, g_j], [e_k, g_l])$$

which describes a system of particles on the line interacting via a quartic potential. Clearly, this is an extension of the Garnier system.

To recover the generalized anharmonic oscillator, consider an involution σ in $\mathfrak{g}$ (for instance, a Cartan involution) and assume that $\sigma A = -A$. Let $\mathfrak{g} = \mathfrak{k} + \mathfrak{p}$ be the eigenspace decomposition for σ; we have $A \in \mathfrak{p}$. Lemma 7.3 remains valid for the twisted algebra $\mathfrak{L}(\mathfrak{g}, \sigma)$ with $\mathfrak{g}_A^\perp$ replaced by $\mathfrak{p}_A^\perp = \mathfrak{g}_A^\perp \cap \mathfrak{p}$ (i.e. $\{e_i\}$ is a basis of $\mathfrak{p}_A^\perp$, etc.)

Notice that if $\mathfrak{g}$ is a compact semisimple Lie algebra, then $\mathfrak{g} = \mathfrak{g}_A \oplus \mathfrak{g}_A^\perp$ is a Hermitian symmetric decomposition and A spans the center of $\mathfrak{g}_A$. If $\mathfrak{g}$ is noncompact and σ is a Cartan involution, one can consider its compact form $\mathfrak{k} + i\mathfrak{p}$ and replace A by iA. Thus the integrable quartic potentials arising in this way are classified by compact Hermitian symmetric pairs $(\mathfrak{g}, \mathfrak{g}_A)$ together with an involution σ which leaves $\mathfrak{g}_A$ invariant. A list of such potentials can be found in the survey by Fomenko and Trofimov in this volume.

7.5. Bibiliographical Notes. The general Hamiltonian (7.1) and the corresponding Lax pair (7.2) were derived by Reyman [1980]. Examples treated in his paper include the n-dimensional top in a quadratic potential, two interacting tops, and motion on some homogeneous spaces in linear and quadratic potentials. Some top-like integrable systems were discussed by Adler and van Moerbeke [1980a] and Ratiu [1982] using affine Lie algebras. However, they missed the most interesting examples because they did not use twisted loop algebras $\mathfrak{L}(\mathfrak{g}, \sigma)$. Bogoyavlensky [1984] found independently a Lax pair for the top in a quadratic potential and elucidated its physical meaning. The Lax pair for the Neumann system is due to Moser [1980a, b] and for the Clebsch Hamiltonian (7.15) to Perelomov [1981a]. A system of two interacting tops was discovered by Reyman [1980] and by Perelomov, Ragnisco and Wojciechowsky [1986]. The derivation of the Kowalewski top and its generalization by reduction of a system of two tops and the associated Lax pair were given by Reyman and Semenov-Tian-Shansky [1987]; for a more detailed account see Bobenko, Reyman and Semenov-Tian-Shansky [1989]. Some partial results on the Kowalewski gyrostat in two fields were obtained by Bogoyavlensky [1984], Komarov [1987] and Yakhiya [1987]. Earlier Perelomov [1981b] found an artificial Lax pair without a spectral parameter for the Kowalewski top and for some of its multidimensional analogs. A different approach to the study of the 3-dimensional Kowalewski top has been developed by Adler and van Moerbeke [1988] and Haine and Horozov [1987] who used algebro-geometric methods. They established a birational isomorphism between the Kowalewski flow and the Manakov flow on SO(4), or the flow corresponding to the Clebsch case of the motion of rigid body in ideal fluid. As a by-product, this yields Lax pairs for the Kowalewski top (which are, however, rather complicated when expressed in terms of the dynamical variables of the top). This approach does not extend to higher dimensions, or to the motion of the top in two fields. The exotic SO(4) top related to the Lie algebra of type G_2 was originally discovered by Adler and van Moerbeke [1984] by completely different methods. Its relation to G_2 and the corresponding Lax pair were described by Reyman and Semenov-Tian-Shansky [1986b]. The Lax pair for the anharmonic oscillator was found by D.V. Chudnovsky and G.V. Chudnovski. More general systems with quartic potentials, based on Hermitian symmetric spaces, were studied by Fordy, Wojciechowsky and Marshall [1986]; their orbital interpretation was given by Reyman [1986]. Some interesting examples of multi-pole Lax equations are discussed by Adams, Harnad and Previato [1988].

§ 8. The Riemann Problem and Linearization of Lax Equations

Theorem 2.2 as applied to affine Lie algebras provides a link between the Hamiltonian scheme of Section 2 and the algebraic-geometric methods of finite-band integration theory (see Dubrovin et al. [1976]). Namely, the factorization problem (2.7) associated with the R-bracket coincides, for affine Lie algebras, with the *matrix Riemann problem*, i.e. the analytic factorization problem for matrix-valued functions. Moreover, formula (2.8) for the trajectories immediately implies that Lax equations linearize on the Jacobian of the spectral curve.

In this section we shall consider only complex Lie algebras.

8.1. The Riemann Problem. Let $\mathfrak{L}(\mathfrak{g}, \sigma)$ be the twisted loop algebra associated with an automorphism σ of order n of a complex Lie algebra $\mathfrak{g}$. Let us extend σ to an automorphism of the corresponding Lie group G which we denote by the same letter. The Lie group that corresponds to $\mathfrak{L}(\mathfrak{g}, \sigma)$ may be identified with the twisted loop group $\mathfrak{L}(G, \sigma)$ consisting of G-valued functions which are analytic in the punctured Riemann sphere $\mathbb{C}P_1\backslash\{0, \infty\}$ and satisfy the relation

$$g(\varepsilon\lambda) = \sigma g(\lambda), \quad \varepsilon = e^{2\pi i/n}.$$

The subalgebras $\mathfrak{L}(\mathfrak{g}, \sigma)_\pm$ (4.4) correspond to subgroups $\mathfrak{L}(G, \sigma)_\pm$ consisting of functions which are regular in $\mathbb{C}P_1\backslash\{\infty\}$ and $\mathbb{C}P_1\backslash\{0\}$, respectively. Functions in $\mathfrak{L}(G, \sigma)_-$ are normalized by the condition $g(\infty) = e$. In this setting Theorem 2.2 amounts to the following.

Let $\partial_t L = [L, M_\pm]$ be a Lax equation in the affine Lie algebra $\mathfrak{L}(\mathfrak{g}, \sigma)$ associated with the standard R-bracket (2.2) and a Hamiltonian φ. Put $M = d\varphi(L)$ and let

$$\exp tM(\lambda) = g_+(\lambda, t)^{-1} g_-(\lambda, t) \tag{8.1}$$

be the solution to the Riemann factorization problem, $g_\pm(\cdot, t) \in \mathfrak{L}(G, \sigma)_\pm$, $g_\pm(\lambda, 0) = e$. Then the solution of the Lax equation is given by

$$L(\lambda, t) = \mathrm{Ad}^* \, g_\pm(\lambda, t) \cdot L(\lambda). \tag{8.2}$$

The factorization problem (8.1) has the following geometric meaning. The projective line $\mathbb{C}P_1$ is covered by two domains $\mathbb{C}P_1 \setminus \{0\}$ and $\mathbb{C}P_1 \setminus \{\infty\}$. The function $\exp t\, d\varphi(L)$ is regular in their intersection and may be regarded as the transition function of a holomorphic G-bundle over $\mathbb{C}P_1$. The factorization problem (8.1) amounts to an analytic trivialization of this bundle. For instance, for $\mathfrak{g} = \mathfrak{gl}(n, \mathbb{C})$ we get an n-dimensional vector bundle over $\mathbb{C}P_1$. It is well known (see Pressley and Segal [1986]) that not all vector bundles over $\mathbb{C}P_1$ are analytically trivial: each n-dimensional bundle breaks up into a sum of line bundles, and their degrees $(k_1, \ldots, k_n)$ are the holomorphic invariants of the given bundle. In the language of transition functions this means that $\exp tM(\lambda)$ always admits a factorization of the form $\exp tM(\lambda) = g_+^{-1}(\lambda, t) d(\lambda) g_-(\lambda, t)$ where $d(\lambda) = \mathrm{diag}(\lambda^{k_1}, \ldots, \lambda^{k_n})$. Thus formula (8.2) requires that all indices $k_i = 0$. One can prove that this is true at least for t sufficiently small.

Proposition 8.1 (Gohberg and Fel'dman [1971]). *Put* $G = \mathrm{GL}\,(n, \mathbb{C})$ *and let* $\exp tM(\lambda) - I = \sum_{-\infty}^{\infty} c_i(t)\lambda^i$ *be the matrix Fourier expansions. If* $\sum_{-\infty}^{\infty} \|c_i(t)\| < 1$, *the matrix factorization problem* (8.1) *has a solution.*

Clearly, the solution of (8.1) is unique; indeed, if $g_+^{-1} g_- = h_+^{-1} h_-$, then the function $g_+ h_+^{-1} = g_- h_-^{-1}$ is holomorphic everywhere on $\mathbb{C}P_1$ and hence is a constant. The normalization condition implies that $g_- = h_-$.

Proposition 8.1 readily implies that factorization problems for other Lie groups are also solvable for small t.

Proposition 8.2. *Let* $G \subset \mathrm{GL}(n, \mathbb{C})$ *be a matrix group,* $M \in \mathfrak{L}(\mathfrak{g}, \sigma)$. *Let* $g_\pm(t)$ *be the solution of the factorization problem* (8.1) *in* $\mathfrak{L}(\mathrm{GL}(n, \mathbb{C}))$. *Then* $g_\pm(t) \in \mathfrak{L}(G, \sigma)_\pm$.

We shall see below that the exceptional values of $t \in \mathbb{C}$ for which the problem (8.1) does not admit a solution form a discrete set in $\mathbb{C}$. At these points the solution $L(t)$ has a pole: the trajectory of the Lax equation goes off to infinity.

8.2. Spectral Data and Dynamics. The inverse scattering method for nonlinear partial differential equations is based on the fact that evolution becomes linear when expressed in terms of the spectral data of the Lax operator. A similar statement is true for Lax equations with a spectral parameter.

Let $\mathfrak{g} = \mathfrak{gl}(n, \mathbb{C})$ and $L(\lambda) = \sum x_i \lambda^i$, $x_i \in \mathfrak{g}$. Consider the algebraic curve Γ_0 in $\mathbb{C} \setminus \{0\} \times \mathbb{C}$ defined by the characteristic equation

$$\det(L(\lambda) - v) = 0. \tag{8.3}$$

Assume that $L(\lambda)$ has simple spectrum for generic λ. For each nonsingular point $p \in \Gamma_0$ which is not a branching point of λ there is a one-dimensional eigenspace $E(p) \subset \mathbb{C}^n$ of $L(\lambda(p))$ with eigenvalue $v(p)$. This gives a holomorphic line bundle on Γ_0 defined everywhere except for singular points and branching points. Let

Γ be the nonsingular compact model of Γ_0. The eigenvector bundle extends to a holomorphic line bundle E on the whole smooth curve Γ. Indeed, the mapping $p \mapsto E(p)$ determines a meromorphic mapping $\Gamma \to \mathbb{C}P_{n-1}$. Since any such mapping is actually holomorphic, the eigenvector bundle extends to Γ.

The spectral curve Γ with two distinguished meromorphic functions λ, ν and the line bundle E on Γ constitute the set of spectral data for $L(\lambda)$.

The evolution determined by a Lax equation of motion leaves the spectral curve (8.3) invariant, and the dynamics of the line bundle E is easy to describe. Since $[L, M] = 0$, the eigenvectors for L are also eigenvectors for M:

$$M(\lambda(p))v = \mu(p)v, \quad p \in \Gamma, \quad v \in E(p), \tag{8.4}$$

where μ is a meromorphic function on Γ. Define the domains $U_\pm \subset \Gamma$ by $U_\pm = \{p \in \Gamma;\ \lambda^{\pm 1}(p) \neq \infty\}$. Clearly, $U_+ \cup U_- = \Gamma$ and μ is regular in the intersection $U_+ \cap U_-$. Let F_t be the line bundle on Γ determined by the transition function $\exp t\mu$ with respect to the covering $\{U_+, U_-\}$. The bundles F_t have degree zero and form a 1-parameter subgroup in the Picard group $\operatorname{Pic}_0 \Gamma$ of equivalence classes of holomorphic line bundles of degree zero on Γ which, by Abel's theorem, is canonically isomorphic to $\operatorname{Jac} \Gamma$, the Jacobian of Γ. Let d be the degree of E and let $\operatorname{Pic}_d \Gamma$ be the "shifted" Picard variety of line bundles of degree d. Clearly, $\operatorname{Pic}_d \Gamma = \operatorname{Jac} \Gamma$ as Abelian varieties.

Theorem 8.3. *The line bundle E regarded as a point of* $\operatorname{Pic}_d \Gamma$ *evolves linearly with time*: $E_t = E \otimes F_t$.

Proof. Let $g_\pm(t)$ be the solution of the Riemann problem (8.1). In view of (8.2), the eigenbundle E_t, regarded as a subbundle of $\Gamma \times \mathbb{C}^n$, is expressed as $E_t(p) = g_+(\lambda(p), t)E(p)$ in U_+ and $E_t(p) = g_-(\lambda(p), t)E$ in U_-. In other words, $g_\pm(\lambda, t)$ define isomorphisms of E and E_t over $U_\pm$. The transition function in $U_+ \cap U_-$ which matches these two isomorphisms is

$$g_+(\lambda(p), t)^{-1} g_-(\lambda(p), t)|_{E(p)} = \exp tM(\lambda(p))|_{E(p)} = \exp t\mu(p).$$

Hence $E_t = E \otimes F_t$. □

Let us give an expression for the velocity vector on $\operatorname{Pic}_d \Gamma$ which corresponds to the equation $\partial_t = [L, M_\pm]$. The tangent space to $\operatorname{Pic}_d \Gamma$ may be identified with the dual of the space of abelian differentials on Γ. Let ω be such a differential. Theorem 8.3 implies that the ω-component of the velocity vector V tangent to the 1-parameter subgroup F_t is given by

$$\omega(V) = \sum_{p:\lambda(p)=0} \operatorname{Res}_p \mu\omega. \tag{8.5}$$

A proof of this formula which uses the Baker-Akhiezer function is given in Section 10.

8.3. Reconstruction of $L(\lambda)$ from the Spectral Data. A Lax matrix can be reconstructed from its spectral data if it satisfies certain regularity conditions.

Definition. A Lax matrix $L(\lambda)$ is called *regular* if

(i) *Equation* (8.3) *is irreducible.*

(ii) *The curve* Γ_0 *is non-singular.*

(iii) *Let* $Q(p)$ *be the spectral projection map onto the eigenspace* $E(p)$. *Then the matrix differential* $Q(p)\,d\lambda$ *is regular at* $\lambda = 0$ *and* $Q(p)\lambda^{-2}\,d\lambda$ *is regular at* $\lambda = \infty$.

Condition (i) means, in particular, that the spectrum of $L(\lambda)$ at a generic point is simple. Condition (ii) is introduced for convenience, in order to simplify the algebro-geometric arguments. The analogue of Condition (iii) for $\lambda \neq 0, \infty$ is satisfied automatically. This condition is easy to verify by means of elementary perturbation theory. In particular, it is certainly true if the highest and the lowest coefficients of L are matrices with simple spectrum.

Since the eigenbundle E of a Lax matrix is a subbundle in $\Gamma \times \mathbb{C}^n$, linear coordinates on $\mathbb{C}^n$ give an n-dimensional space of sections of the dual bundle E^*. The bundle E itself has no holomorphic sections. For this reason it is more convenient to deal with the dual bundle E^*.

Let $\mathscr{L}(F)$ denote the space of holomorphic sections of the line bundle F over Γ. For a divisor D on Γ let $\mathscr{L}(F(D))$ be the space of meromorphic sections ψ of F such that $(\psi) \geqslant -D$. Let P_a be the divisor of zeros of the function $\lambda - a$, i.e. the set of points of Γ which lie over the point $\lambda = a$. Then $\mathscr{L}(F(-P_a))$ consists of those holomorphic sections ψ of F for which $(\lambda - a)^{-1}\psi$ is also holomorphic.

Definition. A line bundle F is called *λ-regular* if $\dim \mathscr{L}(F) = n$ and $\mathscr{L}(F(-P_a)) = \{0\}$ for all $a \in \mathbb{C}P_1$.

It is sufficient to require that $\mathscr{L}(F(-P_a)) = \{0\}$ for some point $a \in \mathbb{C}P_1$, since all divisors P_a are linearly equivalent.

Proposition 8.4. *If the matrix* $L(\lambda)$ *is regular, the line bundle* E^* *is also regular.*

Proof. Fix $\psi \in \mathscr{L}(E^*)$ and assume that a is not a branch point of $\lambda\colon \Gamma \to \mathbb{C}P_1$, and $a \neq 0, \infty$. Let $p_1, \ldots, p_n$ be the points of Γ which lie over a. Then the eigenspaces $E(p_1), \ldots, E(p_n)$ generate $\mathbb{C}^n$ and the linear functionals $\psi(p_1), \ldots, \psi(p_n)$ determine a linear functional $\psi(\lambda)$ on $\mathbb{C}^n$ which is a holomorphic function of λ. It is easy to check that $\psi(\lambda)$ remains bounded in the neighbourhood of branching points. The regularity condition (iii) implies that it remains bounded also for $\lambda = 0, \infty$. Hence $\psi(\lambda) = \text{const}$, i.e. the sections of E^* are exhausted by linear functions on $\mathbb{C}^n$. Now it is clear that if ψ vanishes at all points of the divisor P_a, then $\psi = 0$ and hence also $\psi = 0$. □

We shall now discuss the reconstruction of a regular matrix $L(\lambda)$ from its spectral data. Suppose we are given a curve Γ, an n-sheeted covering $\lambda\colon \Gamma \to \mathbb{C}P_1$, and a λ-regular line bundle E^* over Γ. Denote $\Gamma_0 = \Gamma \backslash \lambda^{-1}(0, \infty)$ and $R = \mathbb{C}[\lambda, \lambda^{-1}]$. Let $\mathscr{L}(\Gamma_0, E^*)$ be the space of those meromorphic sections of E^* which are regular in Γ_0. Multiplication of sections by functions defines the mapping $r\colon \mathscr{L}(E^*) \otimes R \to \mathscr{L}(\Gamma_0, E^*)$.

Proposition 8.5. *The mapping* r *is an isomorphism of* R*-modules.*

Proof. Let us first prove that r is injective. Assume that $\sum_{i=k}^{l} \psi_i \lambda^i = 0$, $\psi_i \in \mathscr{L}(E^*)$. Multiplying this equality by λ^{-k-1} we get $\psi_k \lambda^{-1} = -\sum_{k+1}^{l} \psi_i \lambda^{i-k-1}$. Hence $\psi_k \lambda^{-1} \in \mathscr{L}(E^*)$, i.e. $\psi_k \in \mathscr{L}(E^*(-P_0))$ and by the regularity condition $\psi_k = 0$. Repeating the argument we get $\psi_i = 0$ for all i. To prove that r is surjective observe that $\mathscr{L}(\Gamma_0, E^*)$ coincides with the union of an increasing sequence of its subspaces $\mathscr{L}_k = \mathscr{L}(E^*(kP_0 + kP_\infty))$. Hence it is enough to verify that $\mathscr{L}_k$ lies in the image of r. Let $\psi \in \mathscr{L}_k$. The regularity of E^* implies that there exist ψ_1, $\psi_2 \in \mathscr{L}(E^*)$ such that $\psi - \psi_1 \lambda^k - \psi_2 \lambda^{-k}$ lies in $\mathscr{L}_{k-1}$. Induction over k concludes the argument. □

Proposition 8.6. *The degree of a regular bundle is equal to $g + n - 1$ where g is the genus of Γ.*

Proof. The injectivity of r implies that $\dim \mathscr{L}(E^*(kP_0)) = (k+1)n$. Since $H^1(E^*(kP_0)) = 0$ for large k, the Riemann-Roch theorem yields $\dim \mathscr{L}(E^*(kP_0)) = 1 - g + \deg E^*(kP_0) = 1 - g + kn + \deg E^*$. Hence $\deg E^* = g + n - 1$. □

Proposition 8.5 allows us to reconstruct $L(\lambda)$ from the set of spectral data consisting of an n-sheeted covering $\lambda: \Gamma \to \mathbb{C}P_1$, a λ-regular line bundle E^*, and a meromorphic function v which is regular in Γ_0. Namely, multiplication by v defines an R-linear operator in $\mathscr{L}(\Gamma_0, E^*)$. Using the isomorphism r, we get an R-linear operator in $\mathscr{L}(E^*) \otimes R$ which, after choosing a basis in $\mathscr{L}(E^*)$ is represented by an $n \times n$-matrix with coefficients in $R = \mathbb{C}[\lambda, \lambda^{-1}]$. In other words, let $\psi^t = (\psi_1, \ldots, \psi_n)$ be a basis in $\mathscr{L}(E^*)$ (a vector Baker-Akhiezer function). Then, according to Proposition 8.5 we have $v\psi_i = \sum c_{ij}\psi_j$ where $c_{ij} \in R$ are the matrix coefficients of $L(\lambda)$, so that $L\psi = v\psi$. Thus we have proved the following assertion.

Proposition 8.7. *Let Γ be a smooth algebraic curve, $\lambda: \Gamma \to \mathbb{C}P_1$ an n-sheeted covering, $E^* \to \Gamma$ a λ-regular line bundle. Fix a basis $\psi = (\psi_1, \ldots, \psi_n)^t$ in the space $\mathscr{L}(E^*)$. For each function v on Γ which is regular on $\Gamma_0 = \Gamma \backslash \lambda^{-1}(0, \infty)$ there exists a matrix $L(\lambda) \in \mathrm{gl}(n, \mathbb{C}[\lambda, \lambda^{-1}])$ such that $L\psi = v\psi$. The mapping $v \mapsto L$ is an isomorphism of the algebra of meromorphic functions on Γ regular in Γ_0 onto a maximal commutative subalgebra in $\mathfrak{L}(\mathrm{gl}(n, \mathbb{C}))$.*

Remark. The algebro-geometric constructions of this section can be extended in an obvious way to the case of rational (multi-pole) matrices $L(\lambda)$. In particular, condition (iii) in the definition of a regular matrix must be modified to include all poles of $L(\lambda)$.

8.4. Bibliographical Notes. Algebro-geometric methods were first applied to the study of Lax equations with spectral parameter in a series of papers by Novikov, Durbrovin, Matveev, Its, Lax, McKean and van Moerbeke (see Dubrovin et al. [1976], Dubrovin [1981] for a review) in connection with the theory of finite-band solutions of nonlinear partial differential equations. Later the algebro-geometric scheme was developed by Krichever [1977a, 1977b].

Drinfeld [1977], Cherednik [1983], Mumford [1978] and van Moerbeke and Mumford [1979]. An intimate connection of algebro-geometric methods with the Hamiltonian approach and with factorization problem was pointed out by Reyman and Semenov-Tian-Shansky [1981]. The assumption that spectral curves are non-singular made in this section is in fact inessential. For singular curves the reconstruction of the Lax matrix from the spectral data may be performed in terms of coherent sheaves and generalized Jacobians (cf. van Moerbeke and Mumford [1979], Mumford [1984]).

§ 9. Completeness of the Integrals of Motion

The algebro-geometric results of Section 8 can be used to prove complete integrability of Lax equations with a spectral parameter. To give a precise formulation, one must take into account the ambiguity in reconstructing $L(\lambda)$ from its spectral data, due to the freedom in the choice of a basis in $\mathscr{L}(E^*)$. We do not aim at full generality and will only treat case by case several classical affine algebras. Other situations do not involve serious complications but can hardly be covered by a single statement.

9.1. The Case of $\mathscr{L}(\mathrm{gl}(n, \mathbb{C}))$. Let $L = \sum_{-p}^{q} x_i \lambda^i$ be a regular matrix and let T_L be the set of regular matrices isospectral with L. All matrices in T_L have the same characteristic equation (8.3) and the same spectral curve Γ. From Proposition 8.7 it follows that the eigenvector bundles E_L and $E_{L'}$ are isomorphic if and only if $L' = ALA^{-1}$ where A is a constant matrix. Therefore the mapping $L' \mapsto E_{L'}$ from T_L to Jac Γ is a fibration of T_L with fibre $PGL(n, \mathbb{C})$ over the regular part of Jac Γ.

Lemma 9.1. *Every linear flow on* Jac Γ *is induced by a Lax equation.*

Proof. It is easily shown that every line bundle of degree zero on Γ can be described by a transiton function of the form $\exp \mu$ with respect to the covering $\{U_+, U_-\}$, where μ is a meromorphic function on Γ which is regular in $U_+ \cap U_-$. Every such function is a polynomial in λ, λ^{-1} and v, hence $\mu = \sum \mu_{ij} \lambda^i v^j$ $(j \geqslant 0)$. Put $\varphi(L) = \sum \mu_{ij} \operatorname{Res} \operatorname{tr}\left(\frac{1}{j+1} L^{j+1} \lambda^{i-1}\, d\lambda\right)$. Then for $M = d\varphi(L)$ we have $M = \sum \mu_{ij} \lambda^i L^j$, so that the eigenvalue of $\mu(\lambda)$ at $p \in \Gamma$ is $\mu(p)$. We now refer to Theorem 8.3 to conclude the proof. □

Let $\mathcal{O}$ be the orbit of L with respect to the R-bracket defined by the standard decomposition (4.4) in $\mathfrak{L}(\mathrm{gl}(n, \mathbb{C}))$ and denote $T_{\mathcal{O}} = T_L \cap \mathcal{O}$. The Lax integrals of motion form a complete set at L if and only if the corresponding Hamiltonian vector fields span the whole tangent space to $T_{\mathcal{O}}$ at L. Lemma 9.1 shows that this is indeed true after projecting $T_{\mathcal{O}}$ to Jac Γ. To prove completeness if remains therefore to verify whether the fibers of the projection $T_{\mathcal{O}} \to$ Jac Γ are spanned by the flows of Lax equations in $\mathcal{O}$. Note that for $q \geqslant 1$ all matrices $L' = \sum_{-p}^{q} x_i \lambda^i$ in $\mathcal{O}$ have the same leading coefficient $x_q = f$ (see Lemma 3.1). Let G_f be the stationary subgroup of f in GL $(n, \mathbb{C})$. We shall consider two fairly general cases where the fibers of the mapping $T_{\mathcal{O}} \to \operatorname{Jac} \Gamma$ coincide with the orbits of G_f acting by conjugation.

Proposition 9.2. *Let $L(\lambda) = f\lambda^q + \sum_{-p}^{q-1} x_i \lambda^i$ be a regular matrix, $q \geqslant 1$.*

(i) *If f has simple spectrum then the invariants of $\mathfrak{L}(\mathrm{gl}(n, \mathbb{C}))$ form a complete system of functions in involution at L.*

(ii) *If $q = 1$, then for any f the invariants form a complete system on the quotient space $\mathcal{O}/G_f$.*

Proof. (i) It is enough to show that the action of G_f by conjugation is defined by Lax equations (and so, in particular, leaves $\mathcal{O}$ invariant). Let φ be an invariant of $\mathrm{gl}(n, \mathbb{C})$ and define $\varphi_q(L) = \operatorname{Res}(\lambda^{q-1}\varphi(\lambda^q L(\lambda))\, d\lambda)$. Then for $M = d\varphi_q(L)$ we have $M_+ = d\varphi(f)$ and the corresponding Lax equation $\dot{L} = [L, d\varphi(f)]$ gives rise to conjugation by $\exp(t\, d\varphi(f))$. Since f has simple spectrum, the matrices M_+, for various φ, span the subalgebra $\mathfrak{g}_f$. To show (ii) observe that $\mathcal{O} = f\lambda + \mathcal{O}_0$ where $\mathcal{O}_0$ is an orbit of $\mathfrak{L}_+(\mathrm{gl}(n, \mathbb{C}))$. Conjugation by $\exp t\xi$ on $\mathcal{O}_0$ is a Hamiltonian flow with Hamiltonian $H_\xi(L) = \operatorname{tr}(\xi x_0)$. Thus the action of G_f on $\mathcal{O}$ is Hamiltonian. Since its orbits are the fibers of $T_{\mathcal{O}} \to \operatorname{Jac} \Gamma$, the proof is complete. □

In the second case the set of integrals of motion on $\mathcal{O}$ may easily be completed. Namely, let $\mu: \mathcal{O} \to \mathfrak{g}_f^*$ be the moment map and $\{\varphi_\alpha\}$ a complete involutive set of functions on $\mathfrak{g}_f^*$. Then the functions $\{\varphi_\alpha \circ \mu\}$ together with the invariants form a complete set of integrals of motion in involution on $\mathcal{O}$.

9.2. Complete Integrability for Other Lie Algebras. Let $\mathfrak{L}$ be the fixed subalgebra of one of the following automorphisms of $\mathfrak{L}(\mathrm{gl}(n, \mathbb{C}))$:

(1) $\tau: L(\lambda) \mapsto -L(\lambda)^t$; $\mathfrak{L} = \mathfrak{L}(\mathfrak{so}(n, \mathbb{C}))$.

(2) $\tau: L(\lambda) \to -\Omega L(\lambda)^t \Omega$, $\Omega = \begin{pmatrix} 0 & I \\ -I & 0 \end{pmatrix} \in \mathrm{Mat}(2n)$; $\mathfrak{L} = \mathfrak{L}(\mathfrak{sp}(2n, \mathbb{C}))$.

(3) $\tau: L(\lambda) \mapsto -L(-\lambda)^t$; $\mathfrak{L} = \mathfrak{L}(\mathrm{gl}(n, \mathbb{C}), \sigma)$ where σ is a complexified Cartan automorphism.

(4) $\tau: L(\lambda) \mapsto QL(\varepsilon\lambda)Q^{-1}$, $\varepsilon = e^{2\pi i/n}$, $Q = \operatorname{diag}(1, \varepsilon, \ldots, \varepsilon^{n-1})$. Here $\mathfrak{L}$ is isomorphic to $\mathfrak{L}(\mathrm{gl}(n, \mathbb{C}))$ with principal grading.

Clearly, if $\tau L = L$ the spectral curve Γ_L admits the automorphism: $\tau(\lambda, \nu) = (\lambda, -\nu)$ in cases (1) and (2), $\tau(\lambda, \nu) = (-\lambda, -\nu)$ in case (3) and $\tau(\lambda, \nu) = (\lambda\varepsilon, \nu)$ in case (4).

Let φ be an invariant polynomial on $\mathfrak{L}$. Its gradient $M = d\varphi(L)$ lies in the center of the centralizer of L, hence $M(\lambda)$ is a polynomial of $L(\lambda)$ for each λ. It follows that the eigenvalue $\mu(p)$ of $M(\lambda)$ on the eigenspace $E(p)$, $p \in \Gamma$, satisfies $\mu(\tau p) = -\mu(p)$ in cases (1)–(3) and $\mu(\tau p) = \mu(p)$ in case (4). According to Theorem 8.3, the dynamics of the eigenvector bundle E is given by $E_t = E \otimes F_t$ where F_t is determined by the transition function $\exp t\mu$. Let τ^* denote the automorphism of Jac Γ induced by τ. The above properties of μ imply the following proposition.

Proposition 9.3. *$\tau^* F_t = F_t^{-1}$ in cases (1)–(3) and $\tau^* F_t = F_t$ in case (4).*
The next assertion specifies Lemma 9.1.

Lemma 9.4. *Every linear flow along the anti-invariant part of* Jac Γ *in cases (1)–(3), or along the invariant part of Jac Γ in case (4), is induced by a Lax equation in $\mathfrak{L}$.*

As before, Lemma 9.4 implies the complete integrability of Lax equations in $\mathfrak{L}$. One can also give a more direct proof using integrability in $\mathfrak{L}(\mathrm{gl}(n, \mathbb{C}))$. For simplicity let us consider the case where the highest coefficient of $L(\lambda)$ has simple spectrum (case (i) of Proposition 9.2). With the previous notations if suffices to show that Lax equations in $\mathfrak{L}$ span the tangent space of $T_L \cap \mathcal{O}_{\mathfrak{L}}$ at $L \in \mathfrak{L}$. As we already know, Lax equations in $\mathfrak{L}(\mathrm{gl}(n, \mathbb{C}))$ span the tangent space to $T_{\mathcal{C}}$ at L. Let $\dot{L} = [L, M_\pm]$, where $M = d\varphi(L)$, be a Lax equation in $\mathfrak{L}(\mathrm{gl}(n, \mathbb{C}))$ such that $\dot{L}$ is a given tangent vector to $T_L \cap \mathcal{O}_{\mathfrak{L}}$ at L. The gradient of φ restricted to $\mathfrak{L}$ is $\overline{M} = s^{-1} \sum_{k=1}^{s} \tau^k M$ where s is the order of τ. The corresponding Lax equation in $\mathfrak{L}$ is $\dot{L} = [L, \overline{M}_\pm]$ and coincides with the restriction to $\mathfrak{L}$ of the previous Lax equation in $\mathfrak{L}(\mathrm{gl}(n, \mathbb{C}))$.

Let us finally point out the simple geometric picture which was already mentioned in Section 2: generically the Liouville tori of Lax equations are intersections of the orbits of two Lie-Poisson structures in $\mathfrak{L}^*$ defined respectively by the original Lie bracket and by the R-bracket in $\mathfrak{L}$.

9.3. Bibliographical Notes. Already the first papers on the inverse scattering method have led to a general principle: existence of a Lax representation implies complete integrability.

A direct proof of the independence of a set of integrals leads to rather complicated calculations. This approach was developed by Mishchenko and Fomenko [1978] and their students. A different method which makes use of compatible Poisson brackets was proposed by A. Bolsinov (see the survey by Trofimov and Fomenko in this volume). The simple nontechnical proof given here was suggested by Reyman and Semenov-Tian-Shansky [1981].

§ 10. The Baker-Akhiezer Functions

10.1. Solution of the Matrix Riemann Problem. Let $\dot{L}(\lambda) = [L(\lambda), M_\pm(\lambda)]$ be a matrix Lax equation where $M_+ - M_- = M$ and $M = P(\lambda, \lambda^{-1}, L(\lambda))$ some polynomial of $L(\lambda)$. By Theorem 2.2, its solution $L(\lambda, t)$ is given by

$$L(\lambda, t) = g_\pm(\lambda, t) L(\lambda) g_\pm(\lambda, t)^{-1}, \tag{10.1}$$

where $g_\pm(\lambda, t)$ are matrix-valued functions which solve the Riemann factorization problem

$$e^{tM(\lambda)} = g_+(\lambda, t)^{-1} g_-(\lambda, t), \tag{10.2}$$

i.e. $g_\pm(\lambda, t)$ are analytic, respectively, in $\mathbb{C}P_1 \setminus \{\infty\}$, $\mathbb{C}P_1 \setminus \{0\}$ with the normalization condition $g_-(\infty, t) = I$.

Explicit expressions for $g_\pm(\lambda, t)$ and for the solution $L(\lambda, t)$ may be obtained in terms of the so called Baker-Akhiezer functions.

A *Baker-Akhiezer function for* $L(\lambda)$ is its eigenvector $\psi = (\psi_1, \ldots, \psi_n)^t$ parametrized by the spectral curve Γ; its components are regular sections of the dual eigenbundle E_L^*:

$$L(\lambda(p))\psi(p) = v(p)\psi(p). \tag{10.3}$$

From (10.1) it follows that the Baker-Akhiezer function $\psi(p, t)$ for $L(\lambda, t)$ in the domains $U_\pm = \{p \in \Gamma;\ \lambda^{\pm 1}(p) \neq \infty\}$ can be written in the form

$$\psi_\pm(p, t) = g_\pm(\lambda(p), t)\psi(p, 0), \tag{10.4}$$

so that

$$\psi_+(p, t) = e^{t\mu(p)}\psi_-(p, t), \tag{10.5}$$

where $\mu(p)$ is the eigenvalue of M on the subspace $E(p)$. (This agrees with the description of the evolution of the eigenbundle E in Theorem 8.3.) Since $\partial_t g_\pm \cdot g_\pm^{-1} = -M_\pm$ (see the proof of Theorem 2.2), we have

$$\frac{d}{dt}\psi_\pm(p, t) = -M_\pm(\lambda(p))\psi_\pm(p, t). \tag{10.6}$$

The point is that the Baker-Akhiezer function may be constructed (almost uniquely) from the algebraic data. This, in turn, allows us to obtain an explicit solution of the Riemann problem. Let us first explain how the matrices $g_\pm(\lambda, t)$ are constructed from the sections $\psi_\pm(p, t)$. Fix a basis $\{\psi^i\} = \{(\psi_\pm^i)\}$ of sections of the bundle $E_{L(t)}^* \to \Gamma$ which satisfy the normalization condition

$$\lambda\frac{d}{dt}\psi_-^i(p, t) \quad \text{is regular at } p\colon \lambda(p) = \infty. \tag{10.7}$$

(In other words, the sections $\psi_-^i(p, t) = \text{const}$ for any p which lies over $\lambda = \infty$). Such functions exist for all $t \in \mathbb{C}$ such that the bundle $E_{L(t)}^* = E_L^* \otimes F_{-t}$ is regular. These values of t are called regular and form an open dense set which contains some neighbourhood of $t = 0$.

Suppose that $\lambda \in \mathbb{C}^*$ is not a ramification point and let $p_1, \ldots, p_n$ be the points of Γ which lie over λ. Define the matrices $\Psi_\pm(\lambda, t)$ by

$$\Psi_\pm(\lambda, t)_{ij} = \psi_\pm^i(p_j, t) \tag{10.8}$$

and put

$$g_\pm(\lambda, t) = \Psi_\pm(\lambda, t)\Psi_\pm(\lambda, 0)^{-1} \tag{10.9}$$

Clearly, $g_\pm(\lambda, t)$ does not depend on the ordering of $p_1, \ldots, p_n$.

Proposition 10.1. *The functions $g_\pm(\lambda, t)$ are entire functions of $\lambda^{\pm 1}$ which are analytic in t for all regular values of t and solve the Riemann problem* (10.1).

Proof. Let $L(\lambda, t)$ and $M(\lambda, t)$ be the matrices reconstructed from the sections $\psi^i_\pm(p, t)$ and the functions v and μ, according to Proposition 8.7.

Let us show that

$$\frac{d}{dt}\psi_\pm(p, t) = -M_\pm(\lambda(p), t)\psi_\pm(p, t). \tag{10.9}$$

Indeed, (10.5) implies that $\partial_t\psi_- = e^{t\mu}(\partial_t\psi_+ + \mu\psi_+)$; hence $\partial_t\psi_\pm$ are meromorphic sections of $E^*_{L(t)}$ which are regular in $U_\pm$. Proposition 8.5 implies that there exist matrix polynomials $M'_\pm(\lambda, t) \in \mathrm{gl}(n) \otimes \mathbb{C}[\lambda^{\pm 1}]$ such that $\partial_t\psi_\pm = M'_\pm\psi_\pm$. The normalization condition (10.7) implies that $M'_\pm \in \mathfrak{L}(\mathrm{gl}(n))_\pm$. Moreover, we have $M'_-\psi_- = e^{t\mu}(M'_+ + \mu)\psi_+$, i.e. $(M'_+ - M'_-)\psi_- = -\mu\psi_-$. Hence $M'_+ - M'_- = -M$, i.e. $M'_\pm = -M_\pm$. Now, (10.9) implies that $g_\pm(\lambda, t)$ satisfies the differential equation $\partial_t g_\pm = -M_\pm g_\pm$ with the initial condition $g_\pm(\lambda, 0) = I$. Hence $g_\pm(\lambda^{\pm 1})$ are holomorphic non-degenerate matrix-valued function. Combined with the relation $M(\lambda, t) = g_\pm(\lambda, t)M(\lambda, 0)g_\pm(\lambda, t)^{-1}$, this shows that $g_\pm$ solve the Riemann problem (10.1). □

10.2. Explicit Formulae for the Baker-Akhiezer Functions. Let us now give explicit expressions for the functions $\psi^i_\pm(p, t)$. To simplify the calculations regarding the normalization condition (10.7) we assume that the coefficient of the highest power of λ in $L(\lambda)$ is a diagonal matrix with distinct eigenvalues. In that case Γ is unramified over $\lambda = \infty$. Let $P_\infty = \sum P_i$ be the divisor of poles of λ. We fix a point $p_0 \in \Gamma$ and define a divisor D of degree g by requiring that the line bundle E^* be associated with the divisor $P_\infty + D - p_0$. For this purpose it is enough to fix a section φ of E which has a simple pole at p_0 and zeros at $P_1, \ldots, P_n$. Then the functions $\varphi_i = \psi_i/\varphi$ have poles at $P_\infty + D$ and a zero at p_0. Moreover, by (10.7) the residues $\mathrm{Res}_{P_i}\, \varphi^i\, d\lambda \cdot \lambda^{-1} = d_i$ are time-independent; they may be regarded as normalization constants determined by the initial data. We must exhibit the functions $\varphi^i_-(p, t)$ which are meromorphic in U_-, have the same properties as φ_i above and for which $\varphi^i_+(p, t) = \exp(-t\mu)\varphi^i_-(p, t)$ are regular near $\lambda^{-1}(0)$.

Choose a canonical basis of cycles $(a_1, \ldots, a_g, b_1, \ldots, b_g)$ on Γ and a normalized basis $\omega_1, \ldots, \omega_g$ of Abelian differentials

$$\int_{a_i} \omega_j = 2\pi\sqrt{-1}\delta_{ij}, \quad \int_{b_i} \omega_j = B_{ij}.$$

The coefficients B_{ij} make up the period matrix B. Recall that the *Riemann theta function* on $\mathbb{C}^g$ is characterized by the quasi-periodicity conditions

$$\begin{gathered} \theta(z + 2\pi i N) = \theta(z), \quad i = \sqrt{-1}, \\ \theta(z + BN) = \exp\left\{-\frac{1}{2}(BN, N) - (z, N)\right\}\theta(z), \\ N \in \mathbb{Z}^g, \end{gathered} \tag{10.10}$$

and is explicitly given by

$$\theta(z) = \theta(z|B) = \sum_{m \in \mathbb{Z}^g} \exp\left\{\frac{1}{2}(Bm, m) + (z, m)\right\}. \tag{10.11}$$

Let σ be an Abelian differential of the second kind on Γ which is regular in U_- and such that $\sigma + d\mu$ is regular in U_+. We normalize σ by requiring that $\int_{a_i} \sigma = 0, i = 1, \ldots, g$. Let

$$A: p \mapsto \left\{\int_{p_0}^{p} \omega_i\right\} \in \mathbb{C}^g \tag{10.12}$$

be the *Abel transform* defined on Γ cut along the canonical cycles. Consider the function

$$\varphi_-^i(p, t) = \gamma_i(t) e^{t \int_{p_0}^{p} \sigma} \frac{\theta(A(p) - c_1 - Vt)\theta(A(p) - c_2)}{\theta(A(p) - c_3)\theta(A(p) - c_4)}. \tag{10.13}$$

We shall choose the vectors $c_1, c_2, c_3, c_4, V \in \mathbb{C}^g$ and the functions $\gamma_i(t)$ in such a way that

(i) φ_-^i has no discontinuity across the cuts;
(ii) φ_-^i is subordinate to the divisor $p_0 - P_i - D$;
(iii) $\operatorname{Res}_{P_i}(\varphi_-^i(p, t)\, d\lambda/\lambda) = d_i$.

If the integration contour in (10.12) is changed by adding a cycle

$$\gamma = \sum_i (m_i a_i + n_i b_i), \quad M = (m_i) \in \mathbb{Z}^g, \quad N = (n_i) \in \mathbb{Z}^g,$$

the arguments of all theta functions get an increment $2\pi i M + BN$. The functional equation (10.10) shows that φ_-^i has no discontinuities across the cuts if and only if $c_1 + c_2 - c_3 - c_4 = 0$, $V_i = \int_{b_i} \sigma$. Let D_i be the divisor of zeros of $\theta(A(p) - c_i)$. Choosing suitable constants c_2, c_3, c_4 we may ensure that $D_2 = p_0 + D'$, $D_3 = P_i + D'$ for some D', and $D_4 = D$ which implies (ii). The desired normalization at $p = P_i$ is achieved by the choice of $\gamma_i(t)$. Finally, by the standard properties of the Abelian differentials of the first and second kind we have

$$V_i = \int_{b_i} \sigma = -\sum_j \operatorname{Res}_{P_j} \mu\omega_i,$$

which coincides with the formula (8.5).

Remark. It is interesting to observe that the differential equation for the Baker-Akhiezer function

$$\partial_t \psi_\pm(p) = -M_\pm(\lambda(p))\psi_\pm(p)$$

contains more information than the Lax equation itself. Indeed, as we saw in Sections 6, 7, the Lax equation usually describes the reduced system modulo the isotropy subgroup of the highest or the lowest coefficient of $L(\lambda)$. In many cases (e.g. for the Euler-Poisson equations for heavy tops described in Section 7) the Baker-Akhiezer function describes the evolution of the full (non-reduced) system. (This is equivalent to saying that the normalization factors $\gamma_i(t)$ may be computed from the algebraic data wihtout extra quadratures.) In this way, for instance, one can obtain explicit formulae for the evolution of the moving frame attached to the heavy top. The details for the case of the Kowalewski top may be found in Bobenko, Reyman and Semenov-Tian-Shansky [1989].

10.3. Example: Algebraic Geometry of the Kowalewski Top. As an illustration let us describe the algebraic geometry of the generalized Kowalewski top using the Lax pair (7.33) (or (7.34) in the 3-dimensional case).

Recall that the spectral curve Γ is defined by the equation $\det(L(\lambda) - \mu) = 0$. The two symmetries of $L(\lambda)$,

$$X(-\lambda) = IL(\lambda)I,$$
$$L(\lambda)^t = -IL(\lambda)I,$$

give rise to two commuting involutions τ_1, τ_2 on Γ:

$$\tau_1(\lambda, \mu) = (-\lambda, \mu), \quad \tau_2(\lambda, \mu) = (\lambda, -\mu). \tag{10.14}$$

The involution τ_1 can be lifted to eigenbundle E_L:

$$\tau_1\psi(\lambda, \mu) = I\psi(-\lambda, \mu). \tag{10.15}$$

Let τ_1^*, τ_2^* be the induced involutions on Jac Γ. Then the velocity vector of the generalized Kowalewski flow on Jac Γ is invariant under τ_1^* and changes sign under τ_2^*:

$$\tau_1^* V = V, \quad \tau_2^* V = -V. \tag{10.16}$$

Consider the quotient curves $C = \Gamma/\tau_1$ and $E = C/\tau_2 = \Gamma/(\tau_1, \tau_2)$. Relations (10.15) and (10.16) imply that the Kowalewski flow on Jac Γ is confined to the subtorus Jac $C \subset Jac\ \Gamma$ and is parallel to the Prym variety of the covering $C \to E$.

For the 3-dimensional Kowalewski gyrostat we use the 4×4 Lax pair (7.34). The characteristic equation becomes

$$\mu^4 - 2d_1(\lambda^2)\mu^2 + d_2(\lambda^2) = 0, \tag{10.17}$$

where

$$\begin{aligned} d_1(z) &= (|g|^2 + |h|^2)z^{-1} - (2H + \gamma^2) + 2z, \\ d_2(z) &= ((|g|^2 - |h|^2)^2 + 4(g, h)^2)z^{-2} \\ &\quad + 4(I_1 - (H + \gamma^2/2)(|g|^2 + |h|^2))z^{-1} \\ &\quad + (I_2 + 4\gamma^2 H + \gamma^4) - 4\gamma^2 z \end{aligned}$$

and I_1, I_2 are given by (7.35). The curves C and E are defined by the equations

$$\mu^4 - 2d_1(z)\mu^2 + d_2(z) = 0 \tag{10.18}$$

and

$$y^2 - 2d_1(z)y + d_2(z) = 0, \tag{10.19}$$

respectively, and the coverings $\Gamma \to C$ and $C \to E$ are given by $z = \lambda^2$ and $y = \mu^2$. Note that the curves associated with the Lax matrix (7.34) are different from the classical Kowalewski curve (even in the special case when $h = 0, \gamma = 0$), or from algebraic curves introduced by Adler and Moerbeke [1988]. In a sense the curves associated with the Lax matrix (7.34) are the most 'natural', as confirmed by the relatively simple formulae for the solutions in terms of theta functions. The genera of typical curves depending on the values of the Casimir functions are given in the following table[4]:

	General case	$\gamma = 0$	$\lvert g\rvert = \lvert h\rvert, (g, h) = 0$		$[g, h] = 0, g \neq 0$			
			$\gamma \neq 0$	$\gamma = 0$	$\gamma \neq 0$		$\gamma = 0$	
					$(\ell, g) \neq 0$	$(\ell, g) = 0$	$(\ell, g) \neq 0$	$(\ell, g) = 0$
Γ	8	7	7	6	6	4	5	3
C	4	4	3	3	3	2	3	2
E	1	1	1	1	1	0	1	0

Note that E is in most cases elliptic. The genus of C falls down from 4 to 3 in two cases characterized by the existence of an additional SO(2)-symmetry of the system.

The Baker-Akhiezer function for the Kowalewski top satisfies (10.3) and (10.6) with L, M given by (7.34). We can also require ψ to be symmetric with respect to the involution τ_1

$$\psi(\tau_1 p) = I\psi(p),$$

where

$$I = \begin{pmatrix} \sigma_2 & 0 \\ 0 & \sigma_2 \end{pmatrix}, \quad \sigma_2 = \begin{pmatrix} 0 & -i \\ i & 0 \end{pmatrix}.$$

This enables us to regard ψ as a double-valued function on the curve $C = \Gamma/\tau_1$ and allows us to perform all calculations in terms of C rather than Γ.

The details of the construction of ψ for the classical 3-dimensional Kowalewski case (i.e., for $\gamma = 0, h = 0$) can be found in Bobenko, Reyman and Semenov-Tian-Shansky [1989]. We only give here the final expressions for the evolution of the angular momentum:

[4] Note that the curves in this table are nonsingular compactifications of the affine curves given by Eqs. (10.17), (10.18), and (10.19). For special values of the parameters the geometric genera of these curves drop down (while the arithmetic genera, of course, remain invariant)

$$\ell_1 + i\ell_2 = a\frac{\theta(Vt + P + \varepsilon - R)}{\theta(Vt + P + \varepsilon)},$$
$$\ell_1 - i\ell_2 = a\frac{\theta(Vt + P - R)}{\theta(Vt + P)}, \tag{10.20}$$
$$\ell_3 = -i\frac{\partial}{\partial t}\log\frac{\theta(Vt + P + \varepsilon)}{\theta(Vt + P)}.$$

Here θ is the 3-dimensional Riemann theta function associated with C, a is a constant, V is the velocity vector on Jac C, and P, R and ε are constant vectors. (P depends on the initial data, while R and ε depend only on the integrals of motion.) We may also completely characterize the geometry of the Liouville tori for the Kowalewski top. Let I_1, I_2 be the spectral invariants (7.35) of the Lax matrix (7.34). (In our case $I_1 = (l, g)^2$.)

Theorem 10.1. *If $I_1 \neq 0$, the common level surface of the spectral invariants H, I_1, I_2 consists of two components (Liouville tori) each of which is an affine part of an Abelian variety isomorphic to* Prym(C/E). *The singular loci on* Prym(C/E) *which correspond to the blow-up of solutions are the theta-divisors of $\theta(Vt + P)$ and $\theta(Vt + P + \varepsilon)$.*

10.4. Bibliographical Notes. First explicit formulae for solutions of Lax equations in terms of theta functions were obtained by Its and Matveev (see Dubrovin, Matveev and Novikov [1976] for a review).

Various modifications of these formulae are discussed by Dubrovin [1981], Krichever [1977a,b], van Moerbeke and Mumford [1979]. Chose relationship between the Baker-Akhiezer functions and the solution of the matrix Riemann problem was first observed by Reyman and Semenov-Tian-Shansky [1981]. The algebraic geometry of the Kowalewski top is described by Bobenko, Reyman and Semenov-Tian-Shansky [1989].

§ 11. Equations with an Elliptic Spectral Parameter

In this section we show how the general scheme of Section 2 incorporates Lax equations with a spectral parameter which ranges over an elliptic curve.

Ellitpic Lax matrices give rise to some interesting finite-dimensional systems including the Steklov integrable case of motion of a rigid body in an ideal fluid and various systems of interacting tops. We shall see that Lax equations with an elliptic parameter can be treated in terms of affine Lie algebras, and that the associated Poisson structures do not actually depend on the elliptic curve. The difference with the rational case is in the choice of the decomposition of the (completed) affine Lie algebra into a sum of two subalgebras. The new decomposition relies on the Mittag-Leffler problem of constructing a meromorphic function on an algebraic curve from a given a set of principal parts. Since in general such a function need not exist (Mittag-Leffler theorem) there are obstructions to defining a Lie algebra decomposition. In the elliptic case, and for the Lie algebra $\mathfrak{g} = \mathfrak{sl}(n)$, these obstructions can be eliminated by imposing suitable quasi-periodicity conditions.

11.1. Elliptic Decomposition. We shall first define the so-called *Heisenberg representation* of the group $\mathbb{Z}_n^2 = \mathbb{Z}/n\mathbb{Z} \times \mathbb{Z}/n\mathbb{Z}$ which is an irreducible projective representation in $\mathbb{C}^n$. Let

$$I_1 = \operatorname{diag}(1, \varepsilon, \ldots, \varepsilon^{n-1}), \quad I_2 = \begin{bmatrix} 0 & 1 & \ldots & 0 \\ \vdots & 0 & 1 \ddots & \vdots \\ \vdots & & \ddots \ \ddots & 1 \\ 1 & 0 & \ldots & 0 \end{bmatrix}, \quad \varepsilon = e^{2\pi i/n}. \tag{11.1}$$

For $a = (a_1, a_2) \in \mathbb{Z}_n^2$ we define $I_a = I_1^{a_1} I_2^{a_2}$. It is easily verified that

$$I_a I_b = I_b I_a e^{2\pi i/n \langle a, b \rangle}, \tag{11.2}$$

where $\langle a, b \rangle = a_2 b_1 - a_1 b_2$. Clearly, I is an irreducible representation. The associated representation of $\mathbb{Z}_n^2$ by automorphisms of $\mathfrak{gl}(n, \mathbb{C})$, $a \cdot X = I_a X I_a^{-1}$, is equivalent to the regular representation, i.e. contains all characters of $\mathbb{Z}_n^2$ with multiplicity one. In fact, the matrices I_a, $a \in \mathbb{Z}_n^2$, make up a basis of $\mathfrak{gl}(n, \mathbb{C})$, and (11.2) implies that $a \cdot I_b = \varepsilon^{\langle a, b \rangle} I_b$. The subalgebra $\mathfrak{sl}(n, \mathbb{C})$ is left invariant by this action.

Let C be an elliptic curve, $C = \mathbb{C}/(\mathbb{Z}\omega_1 + \mathbb{Z}\omega_2)$. We identify $\mathbb{Z}_n^2$ with the subgroup C_n of n-periodic points in C, so that $a = (a_1, a_2)$ is identified with $a = (a_1 \omega_1 + a_2 \omega_2)/n$. The group C_n acts on C by translations.

Let $\mathfrak{E}$ be the Lie algebra of meromorphic functions on C with values in $\mathfrak{sl}(n, \mathbb{C})$, which are regular away from C_n and satisfy the quasiperiodicity condition with respect to the action of $\mathbb{Z}_n^2 = C_n$:

$$X(\lambda + a) = I_a X(\lambda) I_a^{-1}. \tag{11.3}$$

Let $\mathfrak{L}$ be the Lie algebra of formal Laurent series with coefficients in $\mathfrak{sl}(n, \mathbb{C})$. The Laurent expansion of a meromorphic function at $\lambda = 0$ induces a natural embedding of $\mathfrak{E}$ into $\mathfrak{L}$. This allows us to identify the elliptic algebra $\mathfrak{E}$ with a subalgebra of $\mathfrak{L}$. Set

$$\mathfrak{L}_+ = \mathfrak{sl}(n, \mathbb{C}) \otimes \mathbb{C}[[\lambda]], \quad \mathfrak{L}_- = \mathfrak{sl}(n, \mathbb{C}) \otimes \lambda^{-1}\mathbb{C}[\lambda^{-1}]. \tag{11.4}$$

Obviously, $\mathfrak{L} = \mathfrak{L}_+ \dotplus \mathfrak{L}_-$. It turns out that $\mathfrak{E}$ is another subalgebra complementary to $\mathfrak{L}_+$.

Proposition 11.1. (i) *There is a decomposition*

$$\mathfrak{L} = \mathfrak{L}_+ \dotplus \mathfrak{E}. \tag{11.5}$$

(ii) *The subalgebras $\mathfrak{L}_\pm$, $\mathfrak{E}$ are isotropic with respect to the bilinear form on $\mathfrak{L}$*

$$\langle X, Y \rangle = \operatorname{Res}_{\lambda=0} \operatorname{tr} XY \, d\lambda. \tag{11.6}$$

Proof. For $X \in \mathfrak{L}$ let $X_- \in \mathfrak{L}_-$ be the projection of X parallel to $\mathfrak{L}_+$. A decomposition $X = X_+ + Y$ with $X_+ \in \mathfrak{L}_+$, $Y \in \mathfrak{E}$ exists if and only if there exists a function Y whose principal part at $\lambda = 0$ is equal to X_-. Denote $X_a = I_a X_- I_a^{-1}$ for $a \in C_n$. From (11.2), together with tr $X_- = 0$, if follows that $\sum_{a \in C_n} \operatorname{Res}_{\lambda=a} X_a \, d\lambda = 0$. Hence by the Mittag-Leffler theorem there exists a function Y' with principal parts X_a at $\lambda = a$. Then $Y(\lambda) = \sum_{a \in C_n} I_a Y'(\lambda - a) I_a^{-1}$ satisfies (11.3) and has the required principal part at $\lambda = 0$. Such a function is unique since the Heisenberg representation is irreducible and tr $Y = 0$.

(ii) Obviously, $\mathfrak{L}_+$ and $\mathfrak{L}_-$ are isotropic. The residue theorem shows that $\mathfrak{E}$ is also isotropic: for $X, Y \in \mathfrak{E}$ we have

$$\langle X, Y \rangle = \operatorname{Res}_0 \operatorname{tr} XY \, d\lambda = \frac{1}{n^2} \sum_{a \in C_n} \operatorname{Res}_a \operatorname{tr} XY \, d\lambda = 0. \; \square$$

It is useful to describe explicitly a basis in $\mathfrak{E}$. The above proof shows that for every $a \in C_n$, $a \neq 0$, there is a unique meromorphic quasiperiodic function φ_a with simple poles at the points of C_n, which satisfies the relation

$$\varphi_a(\lambda + b) = e^{2\pi i/n \langle a, b \rangle} \varphi_a(\lambda) \tag{11.7}$$

and is normalized by $\operatorname{Res}_0 \varphi(\lambda) \, d\lambda = 1$. Set $\varphi_a^{(0)} = \varphi_a$, $\varphi_a^{(k)} = \frac{(-1)^k}{k!} \frac{d^k \varphi_a}{d\lambda^k}$. Clearly, the principal part of $\varphi_a^{(k)}$ at $\lambda = 0$ is λ^{-k-1}. The functions $\varphi_a^{(k)} I_a$, where $k \geqslant 0$, $a \neq 0$, form a basis of $\mathfrak{E}$.

To write down Lax equations one must know the projection operator $P\colon \mathfrak{L} \to \mathfrak{E}$ parallel to $\mathfrak{L}_+$. It is given by

$$PX(\lambda) = \operatorname{Res}_0 r(\lambda, \mu) X(\mu) \, d\mu \tag{11.8}$$

where

$$r(\lambda, \mu) = \frac{1}{n} \sum_a \varepsilon^{a_1 a_2} \varphi_a(\lambda - \mu) I_{-a} \otimes I_a \tag{11.9}$$

and the operator $I_{-a} \otimes I_a$ acts in $\mathfrak{sl}(n, \mathbb{C})$ by $X \mapsto \operatorname{tr}(I_a X) \cdot I_{-a}$. The kernel (11.9) is the so called *elliptic r-matrix* discovered by Belavin [1980]. In the basis $\varphi_a^{(k)} I_a$ the projection P is written as follows: if $X = \sum_{k,a} c_{ak} \lambda^{-k-1} I_a$, then

$$PX(\lambda) = \sum_{k \geqslant 0} \sum_a c_{ak} \varphi_a^{(k)}(\lambda) I_a. \tag{11.10}$$

11.2. Lax Equations with an Elliptic Parameter. We shall now apply the general construction of Section 2 to the decomposition $\mathfrak{L} = \mathfrak{L}_+ \dotplus \mathfrak{E}$ (11.5). We shall be concerned only with equations in $\mathfrak{L}_+^*$; according to Proposition 11.1, $\mathfrak{L}_+^*$ can be identified with $\mathfrak{E}$. Comparing the decomposition $\mathfrak{L} = \mathfrak{L}_+ \dotplus \mathfrak{E}$ with the decomposition $\mathfrak{L} = \mathfrak{L}_+ \dotplus \mathfrak{L}_-$, where $\mathfrak{L}_+^* \simeq \mathfrak{L}_-$, we see that the mapping $P\colon \mathfrak{L}_- \to \mathfrak{E}$ which reconstructs a quasi-periodic elliptic function from its principal part identifies the Lie-Poisson brackets in $\mathfrak{L}_-$ and $\mathfrak{E}$. In particular, the functions with a pole at the origin of order not exceeding a given number make up a Poisson subspace in $\mathfrak{E}$. On the other hand, Lax equations in $\mathfrak{E}$ defined by the invariants of $\mathfrak{L}$ are essentially different from those in $\mathfrak{L}_-$. The invariants have the form

$$\varphi_f(L) = \operatorname{Res}_0 f(\lambda)\varphi(L(\lambda))\, d\lambda, \tag{11.11}$$

where $\varphi \in I(\mathfrak{sl}(n, \mathbb{C}))$ and $f \in \mathbb{C}((\lambda))$. The Lax equation in $\mathfrak{E}$ associated with the Hamiltonian (11.11) is

$$\frac{dL}{dt} = [L, M], \quad M = P(d\varphi_f[L]), \tag{11.12}$$

where the projection $P\colon \mathfrak{L} \to \mathfrak{E}$ is given by (11.10).

Remark. There are also invariant functions on $\mathfrak{E}$ defined by $L \mapsto \operatorname{Res}_\nu(\lambda - \nu)^{-k}\varphi(L(\lambda))\, d\lambda$ for $\nu \in C$. It is not hard to see that these can be expressed in terms of the invariants (11.11).

11.3. Multi-pole Lax Equations. The construction of multi-pole Lax equations outlined in Section 4.5 has an obvious analogue in the elliptic case.

Let $D = \{\nu_1, \ldots, \nu_N\}$ be a finite subset of C such that $\nu_i \neq \nu_j + a$ for $i \neq j$ and $a \in C_n$. Denote by $\mathfrak{E}_D$ the Lie algebra of meromorphic functions on C with values in $\mathfrak{sl}(n, \mathbb{C})$ which are regular away from $C_n \cdot D$ and satisfy the quasi-periodicity condition (11.3). Let $\mathfrak{L}_\nu$ be the Lie algebra of formal Laurent series of the variable $\lambda - \nu$ with coefficients in $\mathfrak{sl}(n)$, and set $\mathfrak{L}_D = \bigoplus_{\nu \in D} \mathfrak{L}_\nu$. Expanding functions in a Laurent series at $\nu \in D$ gives natural embedding $\mathfrak{E}_D \to \mathfrak{L}_D$. Put $\mathfrak{L}_D^\pm = \bigoplus_{\nu \in D} \mathfrak{L}_\nu^\pm$. Proposition 11.1 extends as follows:

Proposition 11.2 (i) *There is a decomposition*

$$\mathfrak{L}_D = \mathfrak{L}_D^+ \dotplus \mathfrak{E}_D. \tag{11.13}$$

(ii) *The subalgebras $\mathfrak{L}_D^\pm$, $\mathfrak{E}_D$ are isotropic with respect to the invariant bilinear form on $\mathfrak{L}_D$ defined by*

$$\langle X, Y \rangle = \sum_{\nu \in D} \operatorname{Res}_\nu \operatorname{tr} X_\nu Y_\nu \, d\lambda.$$

All constructions of the preceding subsection, including formulae (11.8) and (11.10), immediately extend to the multi-pole case. The dual of $\mathfrak{L}_D^+$ is identified with $\mathfrak{E}_D$. The functions $\{\varphi_a^{(k)}(\lambda - \nu)I_a\}_{\nu \in D, k \geqslant 0}$ form a basis of $\mathfrak{E}_D$. For a given set $n = (n_\nu)_{\nu \in D}$ of nonnegative integers the Lax matrices of the form

$$L(\lambda) = \sum_a \sum_{\nu \in D} \sum_{k=0}^{n_D - 1} c_{a\nu k}\varphi_a^{(k)}(\lambda - \nu)I_a, \tag{11.14}$$

i.e. such that the order of the pole at ν does not exceed n_ν, make up a Poisson subspace in $\mathfrak{E}_D = (\mathfrak{L}_D^+)^*$. The coefficients of $L(\lambda)$ at different poles Poisson commute: $\{c_{a\nu k}, c_{b\mu l}\} = 0$ if $\nu \neq \mu$. The Poisson structure at a given pole is the standard Lie-Poisson structure of $(\mathfrak{L}_\nu^+)^*$. The algebra of invariants of $\mathfrak{L}_D$ is generated by the functionals

$$\varphi_\nu[L] = \mathrm{Res}_{\lambda=\nu} f(\lambda_\nu)\varphi(L_\nu(\lambda_\nu))\, d\lambda, \tag{11.15}$$

where $\nu \in D$, φ is an invariant polynomial on $\mathfrak{sl}(n)$ and f is a Laurent series in $\lambda - \nu$. If $L \in \mathfrak{E}_D$ then φ_ν is defined for any point $\nu = \nu_0 \in C$. To deal with φ_ν in this case it is convenient to enlarge D by adding the point ν_0. The gradient of φ_{ν_0} is then $M_{\nu_0} = f(\lambda - \nu_0)\, d\varphi(L(\lambda - \nu_0))$ and the corresponding Lax equation is $\dot{L} = [L, P_{\nu_0} M_{\nu_0}]$ where P_{ν_0} is the projection operator (11.10) with λ replaced by $\lambda - \nu_0$.

In particular, this shows that it is often convenient not to specify D from the very beginning but rather take $D = C$ and use the adelic language (cf. Section 4.5). There is an equivalent construction of multi-pole Lax equations, where the basic Lie algebra consists of elliptic functions. Let D' be another subset of C such that $C_n D \cap C_n D' = \varnothing$. Then there is a direct vector space decomposition $\mathfrak{E}_{D \cup D'} = \mathfrak{E}_D + \mathfrak{E}_{D'}$ and the subalgebras $\mathfrak{E}_D$, $\mathfrak{E}_{D'}$, are isotropic with respect to the inner product $(X, Y) = \sum_{\nu \in D} \mathrm{Res}_\nu \operatorname{tr} XY\, d\lambda$. A convenient form of invariants is

$$\varphi_f(L) = \sum_{\nu \in D} \mathrm{Res}_\nu f(\lambda)\varphi(L(\lambda))\, d\lambda, \tag{11.16}$$

where $\varphi \in I(\mathfrak{sl}(n))$ and f is a C_n-invariant function on C regular away from $C_n \cdot (D \cup D')$. The gradient of φ_f is $M = f\, d\varphi(L)$, so that Lax equations with Hamiltonian φ_f are easy to write down. The equivalence between these two constructions is established by the natural embedding $\mathfrak{E}_{D \cup D'} \to \mathfrak{L}_D$ which assigns to a function the collection of its Laurent expansions at $\nu \in D$ (cf. Section 4.5).

11.4. Riemann Problem, Cousin Problem and Spectral Data. Associated with the Lie algebra $\mathfrak{E}_D$ is the group G_D of SL $(n, \mathbb{C})$-valued meromorphic functions on C which are regular away from $C_n \cdot D$ and are quasi-periodic: $g(\lambda + a) = I_a g(\lambda) I_a^{-1}$. Associated with the decomposition $\mathfrak{E}_{D \cup D'} = \mathfrak{E}_D + \mathfrak{E}_{D'}$ is the following factorization problem in the group $G_{D \cup D'}$ (a matrix Riemann problem on the elliptic curve). Let $M = d\varphi_f(L)$ (see (11.16)). Factorize $\exp tM$ as $\exp tM = g_+(t)^{-1} g_-(t)$ where $g_+(t)$ and $g_-(t)$ are smooth paths in the subgroups $G_{D'}$ and G_D, respectively. By Theorem 2.2, the trajectory of the Hamiltonian φ_f emanating from $L \in \mathfrak{E}_D$ is given by $L(t) = g_\pm(t) L g_\pm(t)^{-1}$.

The decomposition $\mathfrak{L}_D = \mathfrak{L}_D^+ + \mathfrak{E}_D$ also has an associated factorization problem. This case must be treated with caution since there is no Lie group which would correspond to the Lie algebra $\mathfrak{L}_D$. Let $M = (M_\nu)_{\nu \in D}$ be the gradient of an invariant Hamiltonian at $L \in \mathfrak{E}_D$. We may assume that the series M_ν are convergent (see Section 4.5).

Then, for $t \in \mathbb{C}$ small enough, there exists a unique matrix-valued function $g_0(t)$ on C which is holomorphic in $C \backslash C_n \cdot D$, satisfies the quasi-periodicity condition (11.3), and is such that the functions $e^{tM_\nu} g_0(t)^{-1}$ are regular in the vicinity of $\nu \in D$. (This is the matrix Cousin problem). The integral curve emanating from $L \in \mathfrak{E}_D$ has the form $L(t) = g_0(t) L g_0(t)^{-1}$. It is not hard to see that $g_0(t)$

coincides with the function $g_-(t)$ defined earlier, so that the Cousin problem agrees with the Riemann problem.

Let $L \in \mathfrak{E}_D$ be an elliptic Lax matrix. As in Section 8, the equation $\det(L(\lambda) - \nu) = 0$ defines an algebraic curve Γ which covers C. Assuming that $L(\lambda)$ has simple spectrum, we define a holomorphic line bundle E of eigenvectors of $L(\lambda)$. The symmetry group C_n acts on Γ in an obvious way, and this action commutes with the covering $\pi: \Gamma \to C$. The group C_n acts on the line bundle E as well, but this action is projective: it follows from (11.2) that the projective multiplier is $\varepsilon^{\langle a,b \rangle}$. The line bundle E on Γ with these properties constitutes the set of algebro-geometric data for the matrix $L(\lambda)$.

We now consider the evolution of E (notice that Γ is left invariant by Lax equations). Define a covering of Γ by two domains U and U': $U = \Gamma \backslash \pi^{-1}(C_n \cdot D)$ and $U' = \Gamma \backslash \pi^{-1}(C_n \cdot D')$ for the decomposition $\mathfrak{E}_{D \cup D'} = \mathfrak{E}_D + \mathfrak{E}_{D'}$, while U' is a small neighborhood of the set $\pi^{-1}(C_n \cdot D)$ for the decomposition $\mathfrak{L}_D = \mathfrak{L}_D^+ + \mathfrak{E}_D$. Let the Hamiltonian φ_f be given either by (11.16) or by (11.15). If $v \in E(p)$, where $p \in \Gamma$, is an eigenvector of $L(\lambda)$, then $M(\lambda)v = \mu(p)v$. Let the one-parameter group of line bundles F_t of degree 0 be defined by the transition function $\exp t\mu$ relative to the cover $\{U, U'\}$. Exactly as in Section 8 (Theorem 8.3), the factorization theorem implies the following proposition.

Proposition 11.3. *The evolution of the line bundle E is linear*: $E_t = E \otimes F_t$.

In view of some special properties of E, the corresponding linear flows on the Jacobian cannot be arbitrary. In fact, by construction, F_t is C_n-invariant and hence projects to a line bundle F_t' on the quotient curve $\Gamma' = \Gamma / C_n$. There is an obvious covering $\pi': \Gamma' \to C/C_n = C'$ which gives rise to the mapping of the Jacobians: $\pi'_*: J(\Gamma') \to J(C')$ (direct image). It turns out that $\pi'_* F_t'$ is a trivial bundle on C'. In fact, $\pi'_* F_t'$ is determined by the transition function $\psi(\lambda) = \prod_{P: \lambda(P) = \lambda} \exp t\mu(p)$, But $\sum_{p: \lambda(p) = \lambda} \mu(p) = \operatorname{tr} M(\lambda) = 0$; hence $\psi(\lambda) = 1$. Thus, the Lax flows sweep out a subtorus in $J(\Gamma)$, which is isomorphic to the subtorus $\operatorname{Ker} \pi'_*$ (of codimension 1) in $J(\Gamma')$.

Finally, the regularity conditions on E necessary for the reconstruction of $L(\lambda)$ are as follows (cf. Section 8.3):

$$\dim \mathscr{L}(E^*) = n, \quad \dim \mathscr{L}(E^*(-P_\nu)) = 0, \quad \dim \mathscr{L}(E^*(P_\nu)) = n$$

for $\nu \in C_n \cdot D$, where $P_\nu = \pi^{-1}(\nu)$. In particular, this implies $\deg E^* = g - 1$.

11.5. Real Forms and Reduction. The structure of the Heisenberg representation of $\mathbb{Z}_n^2$ does not allow to replace $\mathfrak{sl}(n, \mathbb{C})$ by some other simple Lie algebra in the construction of the elliptic decomposition. The only possibilities that remain are the elementary reductions to the real form $\mathfrak{sl}(n, \mathbb{R})$, or $\mathfrak{su}(2)$ for $n = 2$, and to the subalgebra of even functions of λ.

To discuss real forms we assume that the period lattice is rectangular, i.e. $i\omega_1$ and ω_2 are real, so that complex conjugation $\lambda \to \bar{\lambda}$ is well defined on C. Its action on the points of period n is $(a_1, a_2) \mapsto (-a_1, a_2)$. We have $I_{\bar{a}} = \overline{I_a}$ and

$\varphi_{\bar{a}}(\lambda) = \overline{\varphi_a(\bar{\lambda})}$. Let $\{\nu_1, \ldots, \nu_N\} = D$ be a set of real poles. Then the anti-involution $L(\lambda) \mapsto \overline{L(\bar{\lambda})}$ is well defined on $\mathfrak{L}_D$ and leaves $\mathfrak{E}_D$ invariant. The subalgebra of fixed points is identified with $\bigoplus_\nu \mathfrak{L}_\nu(\mathfrak{sl}(n, \mathbb{R}))$ and in this way we obtain its elliptic decomposition.

For $n = 2$ there is a more interesting anti-involution $L(\lambda) \mapsto -\overline{L(\bar{\lambda})}^t$ which leads to the elliptic decomposition of $\bigoplus_\nu \mathfrak{L}_\nu(\mathfrak{su}(2))$. The associated "unitary" elliptic subalgebra consists of matrices of the form

$$L(\lambda) = \sum_{i=1}^{3} \sum_{\alpha, k} c_{i\alpha k} \varphi_i^{(k)}(\lambda - \nu_\alpha)\sigma_i \tag{11.17}$$

where the coefficients $c_{i\alpha k}$ are real, $\sigma_1 = \begin{pmatrix} 0 & i \\ i & 0 \end{pmatrix}$, $\sigma_2 = \begin{pmatrix} 0 & 1 \\ -1 & 0 \end{pmatrix}$, $\sigma_3 = i\begin{pmatrix} 1 & 0 \\ 0 & -1 \end{pmatrix}$ are the Pauli matrices times i, and we have set $\varphi_1^{(k)} = \varphi_{11}^{(k)}$, $\varphi_2^{(k)} = \varphi_{01}^{(k)}$, $\varphi_3^{(k)} = \varphi_{10}^{(k)}$. The functions φ_i can be expressed in terms of the Jacobi elliptic functions (see Section 12.5). For $n > 2$ there seems to be no way to reduce to $\mathfrak{su}(n)$.

Another type of reduction is provided by the involution $L(\lambda) \to L(-\lambda)$. Let $D = \{0, \pm\nu_1, \ldots, \pm\nu_N\}$ so that the involution preserves the subalgebras $\mathfrak{L}_D$ and $\mathfrak{E}_D$. The fixed subalgebra $\mathfrak{L}_D^{\text{even}}$ clearly has a decomposition $\mathfrak{L}_D^{\text{even}} = (\mathfrak{L}_D^+)^{\text{even}} + \mathfrak{E}_D^{\text{even}}$. The dual space to $(\mathfrak{L}_D)^{\text{even}}$ can be identified with $\mathfrak{L}_D^{\text{odd}} = \{X \in \mathfrak{L}_D;\, X(\lambda) = -X(-\lambda)\}$. In particular, the dual to $(\mathfrak{L}_D^+)^{\text{even}}$ is identified with the space $\mathfrak{E}_D^{\text{odd}}$ of odd quasi-periodic functions. Since

$$(\mathfrak{L}_D^+)^{\text{even}} = \mathfrak{sl}(n, \mathbb{C}) \otimes \mathbb{C}[[\lambda^2]] + \bigoplus_i \mathfrak{L}_{\nu_i},$$

the space $\mathfrak{E}_D^{\text{odd}}$ inherits the already familiar Lie-Poisson structure of the direct sum of copies of $(\mathfrak{L}_\nu^+)^*$. Reduction to the subalgebra of even functions may be combined with reduction to a real form.

For special period lattices with extra symmetries (e.g. hexagonal) there exist further reductions.

11.6. Examples. We shall restrict ourselves to the case $n = 2$ which allows a reduction to the real form $\mathfrak{su}(2)$. Generalizations for $n > 2$ are straightforward but do not admit any obvious physical interpretation. With the above notation consider the 'compact' real form of $\mathfrak{E}_D$, $D = \{\nu_1, \ldots, \nu_N\}$. A Lax matrix with simple poles at the points of $C_2 \cdot D$ has the form

$$L(\lambda) = \sum_{i=1}^{3} \sum_{\alpha} \ell_{i\alpha} \varphi_i(\lambda - \nu_\alpha)\sigma_i. \tag{11.18}$$

The corresponding phase space is the direct sum of N copies of $\mathfrak{su}(2)^*$; the functions $\ell_{i\alpha}$ are linear coordinates on $\bigoplus^N \mathfrak{su}(2)^*$ with Lie-Poisson brackets $\{\ell_{i\alpha}, \ell_{j\beta}\} = \delta_{\alpha\beta}\varepsilon_{ijk}\ell_{k\alpha}$. The functions $H_\nu = -\frac{1}{2}\operatorname{tr} L(\nu)^2$ define a family of quadratic Hamiltonians in involution. Since $\operatorname{tr} \sigma_i^2 = -2$ and $\operatorname{tr} \sigma_i\sigma_j = 0$ for $i \neq j$, we have

$$H_\nu = \sum_i \sum_{\alpha,\beta} \ell_{i\alpha}\ell_{i\beta}\varphi_i(\nu - \nu_\alpha)\varphi_i(\nu - \nu_\beta). \tag{11.19}$$

These Hamiltonians describe systems of N interacting Euler tops. There are sufficiently many independent integrals to ensure complete integrability. The corresponding Lax equation is $\dot{L} = [L, M_\nu]$ with

$$M_\nu(\lambda) = \sum_i \sum_\alpha \ell_{i\alpha}\varphi_i(\nu - \nu_\alpha)\varphi_i(\lambda - \nu)\sigma_i \tag{11.20}$$

(see Section 11.3). For $N = 1$, the family H_ν contains two independent Hamiltonians. In fact, it is easily verified that $\varphi_i(\lambda)^2 = \varphi(\lambda) + J_i$ where $\varphi(\lambda)$ does not depend on i. Hence

$$H_\nu = \varphi(\nu)\sum_i \ell_i^2 + \sum_i J_i\ell_i^2. \tag{11.21}$$

The first term is a Casimir function while the second term describes an Euler top with moments of inertia J_i. For $N = 2$ we recover the Manakov top on $\mathfrak{so}(4) = \mathfrak{su}(2) \oplus \mathfrak{su}(2)$.

Interacting Euler-Poisson tops. Consider the L-matrix with second order poles:

$$L(\lambda) = \sum_i \sum_\alpha \{\ell_{i\alpha}\varphi_i(\lambda - \nu_\alpha) + p_{i\alpha}\varphi_i^{(1)}(\lambda - \nu_\alpha)\}\sigma_i. \tag{11.22}$$

The associated phase space is the direct sum of N copies of $\mathfrak{e}(3)^*$, where $\mathfrak{e}(3) = \mathfrak{so}(3) + \mathbb{R}^3$ is the Lie algebra of the group of Euclidean motions. The coefficients $\ell_{i\alpha}$ and $p_{i\alpha}$ are linear coordinates in $\mathfrak{so}(3)$ and $\mathbb{R}^3$, respectively, with the usual Lie-Poisson brackets. We have

$$\begin{aligned} -\frac{1}{2}\operatorname{tr} L(\nu)^2 = \sum_i \sum_{\alpha,\beta} \{&\ell_{i\alpha}\ell_{i\beta}\varphi_i(\nu - \nu_\alpha)\varphi_i(\nu - \nu_\beta) \\ &+ 2\,\ell_{i\alpha}p_{i\beta}\varphi_i(\nu - \nu_\alpha)\varphi_i^{(1)}(\nu - \nu_\beta) + p_{i\alpha}p_{i\beta}\varphi^{(1)}(\nu - \nu_\alpha)\varphi^{(1)}(\nu - \nu_\beta)\}. \end{aligned} \tag{11.23}$$

These Hamiltonians correspond to interacting Euler-Poisson tops. They may be interpreted as describing the motion of N rigid bodies in ideal fluid, where $\ell_{i\alpha}$ are the angular momenta of the bodies and $p_{i\alpha}$ their linear momenta. For $N = 1$ we recover the Clebsch integrable case whose multidimensional analogue was discussed in Section 7. Indeed, using that $\varphi_i(\lambda)^2 = \varphi(\lambda) + J_i$ we find from (11.23) four independent Hamiltonians. Two of them, $\sum p_i^2$ and $\sum p_i\ell_i$, are Casimir functions for $\mathfrak{e}(3)$. The other two are

$$\begin{aligned} H_1 &= \frac{1}{2}\sum_i (\ell_i^2 - J_i p_i^2), \\ H_2 &= \frac{1}{2}\sum_i \left(J_i\ell_i^2 + \frac{I}{J_i}p_i^2\right), \quad I = J_1J_2J_3. \end{aligned} \tag{11.24}$$

The corresponding matrices M_i in the Lax pair are

$$M_1(\lambda) = -\sum_i p_i\varphi_i(\lambda)\sigma_i,$$
$$M_2(\lambda) = -\sum_i \left\{\left(\frac{4}{3}J - J_i\right)p_i\varphi_i(\lambda) + \ell_i\varphi_i^{(1)}(\lambda) + p_i\varphi_i^{(2)}(\lambda)\right\}\sigma_i, \tag{11.25}$$
$$J = J_1 + J_2 + J_3.$$

Steklov case. Lax equations in the space $\mathfrak{E}_D^{\text{odd}}$ of odd elliptic functions of λ (see Section 11.5) lead to Steklov's integrable Hamiltonians on $\mathfrak{so}(4)$ and $\mathfrak{e}(3)^*$. Consider, for instance, Lax matrices with simple poles at $\pm\nu_1$, $\pm\nu_2$:

$$L(\lambda) = \sum_i \{\ell_i(\varphi_i(\lambda + \nu_1) + \varphi_i(\lambda - \nu_1)) + m_i(\varphi_i(\lambda + \nu_2) + \varphi_i(\lambda - v_2))\}\sigma_i. \tag{11.26}$$

The associated phase space is $\mathfrak{so}(4)^* = \mathfrak{su}(2)^* \oplus \mathfrak{su}(2)^*$ with the standard Poisson bracket. Calculating the residues of $\operatorname{tr} L(\lambda)^2$ at $\lambda = \nu_1$ and $\lambda = \nu_2$ gives

$$H_1 = \sum_i \{\ell_i^2\varphi_i(2\nu_1) + \ell_i m_i(\varphi_i(\nu_1 + \nu_2) + \varphi_i(\nu_1 - \nu_2))\},$$
$$H_2 = \sum_i \{m_i^2\varphi_i(2\nu_2) + \ell_i m_i(\varphi_i(\nu_1 + \nu_2) - \varphi_i(\nu_1 - \nu_2))\}. \tag{11.27}$$

By adding the Casimir functions $c_1 = \sum \ell_i^2$, $c_2 = \sum m_i^2$ these Hamiltonians can be written as

$$H_1 = \sum_i \left\{a_i^2\ell_i^2 - 2\frac{A}{a_i}\ell_i m_i\right\},$$
$$H_2 = c\sum_i \left\{a_i^{-2}m_i^2 - 2\frac{a_i}{A}\ell_i m_i\right\}, \tag{11.28}$$

where $A = a_1a_2a_3$ and a_i are independent constants. This is the Steklov integrable case for $\mathfrak{so}(4)$.

In a similar way, consider the Lax matrices with a third order pole at $\lambda = 0$:

$$L(\lambda) = \sum_i \{\ell_i\varphi_i(\lambda) + p_i\varphi_i^{(2)}(\lambda)\}\sigma_i. \tag{11.29}$$

The Poisson brackets of the variables ℓ_i, p_i are those of the Lie algebra $\mathfrak{e}(3)$. From $\operatorname{tr} L(\lambda)^2$ we derive the Hamiltonians

$$H_1 = \sum_i \{\ell_i^2 + J_i\ell_i p_i + (1/2J - 3/4J_i)J_ip_i^2\},$$
$$H_2 = \sum_i \{J_i\ell_i^2 + (J - J_i)\ell_i p_i + 1/4(J - J_i)^2J_ip_i^2\} \tag{11.30}$$

with the same notation as in (11.24)–(11.25). They describe the classical Steklov integrable case of motion of a rigid body in ideal fluid.

Remark. Lax matrices with simple poles at $\lambda = 0$, $\lambda = \pm\nu$ also give rise to Steklov's top on SO(4). In the limit $\nu \to 0$ the Lie algebra $\mathfrak{so}(4)$ contracts to $\mathfrak{e}(3)$, but the Lax matrix has no limit.

11.7. Hierarchies of Poisson Brackets. As in Sections 4.2, 4.5, one can define a hierarchy of compaticle Poisson structures on $\mathfrak{E}_D$ by multiplying the elliptic R-matrix by C_n-invariant meromorphic functions on C. It is convenient here not to fix the polar set D from the beginning but rather use the adelic point of view. Let $\mathfrak{E}_C$ be the Lie algebra of $\mathfrak{sl}(n)$-valued meromorphic functions on C satisfying (11.3). Let $Q = \sum n_i v_i$ be a C_n-invariant positive divisor on C and let $\mathfrak{E}(Q)$ be the subspaces of functions $\varphi \in \mathfrak{E}_C$ whose divisor of poles is subordinate to Q. Then $\mathfrak{E}(Q)$ is a Poisson subspace in $\mathfrak{E}_C$. If f is a C_n-invariant meromorphic function then $\mathfrak{E}(Q)$ is a Poisson subspace with respect to the $R \circ f$-bracket if and only if the divisor of poles of f is subordinate to Q. In this case $f^{-1}\mathfrak{E}(Q) = \mathfrak{E}(Q')$, where $Q' = Q - (f)$, is also a Poisson subspace for the standard R-bracket and the $R \circ f$-bracket on $\mathfrak{E}(Q)$ is the image of the standard bracket on $\mathfrak{E}(Q')$ under multiplication by f. Thus compatible $R \circ f$-brackets on $\mathfrak{E}(Q)$ are parametrized by C_n-invariant meromorphic function whose polar divisors are subordinate to Q.

For example, let $n = 2$ and consider the $\mathfrak{su}(2)$ real form of the elliptic algebra. Let $Q = C_2 \cdot \{v, -v\}$; there is a C_2-invariant function f with simple poles at $\lambda = \pm v$ and a double zero at $\lambda = 0$. The space $\mathfrak{E}(Q)$ has the Poisson structure $\mathfrak{su}(2)^* \oplus \mathfrak{su}(2)^*$ with respect to the standard bracket and the structure of $\mathfrak{e}(3)^*$ with respect to the $R \circ f$-bracket. Respectively, the Hamiltonian (11.21) gives rise to the Manakov top on SO(4) in the first case and to the Clebsch equations on $\mathfrak{e}(3)$ in the second case. There is a similar relationship between the Steklov integrable cases on $\mathfrak{so}(4)$ and $\mathfrak{e}(3)$.

11.8. Bibliographical Notes. The first Lax pair with a spectral parameter varying on an elliptic curve was found by Sklyanin [1979] and by Borovik and Robuk [1981] for the Landau-Lifschitz equation. General elliptic r-matrices related to finite Heisenberg groups were constructed by Belavin [1980]. Belavin and Drinfel'd [1982] proved the uniqueness theorem. The suggestion to use the Mittag-Leffler problem to generate Lax equation goes back to Cherednik [1983b] who also developed the algebro-geometric theory of Lax equations with an elliptic spectral parameter (Cherednik [1983a]). The decomposition $\mathfrak{E}_{D \cup D'} = \mathfrak{E}_D + \mathfrak{E}_{D'}$ was studied by Golod [1984] in a special case. Our exposition is based on Reyman and Semenov-Tian-Shansky [1986a]. Finite-dimensional integrable systems admitting a Lax representation with elliptic spectral parameter were also studied by Veselov [1984] and Bobenko [1986].

§ 12. Classical R-Matrices, Poisson Lie Groups and Difference Lax Equations

So far we considered finite-dimensional integrable systems with phase spaces that are coadjoint orbits of Lie groups. There is a large class of integrable systems with a different geometry which admit the so-called difference Lax representation

$$\frac{dL_m}{dt} = L_m M_{m+1} - M_m L_m, \quad m \in \mathbb{Z}. \tag{12.1}$$

In this case it is natural to suppose that the dynamics developes on submanifolds of Lie groups rather than Lie algebras or their duals. Typical dynamical systems of this kind are classical analogues of lattice models in quantum statistics, although some of the systems considered earlier (e.g. Toda lattices, certain tops, etc.) also admit difference Lax representations. An extension of the geometric scheme described in Section 2 to these systems is based on the theory of Poisson Lie groups introduced by Drinfel'd [1983] following the pioneering work of Sklyanin [1979]. (This latter paper was in turn motivated by the Quantum Inverse Scattering Method developed by Faddeev, Takhtajan and Sklyanin (cf. Faddeev [1984]) and by the works of Baxter on Quantum Statistical Mechanics (cf. Baxter [1982]).) The geometry of Poisson Lie groups has an interest of its own since it involves a non-trivial generalization of the Hamiltonian reduction technique. In this section we shall briefly discuss the theory of Poisson Lie groups, including such topics as the duality theory and dressing transformations, cf. Semenov-Tian-Shansky [1985], and its application to difference Lax equations. Applicatons to concrete systems will be considered in Section 12.6 (our main examples here are based on elliptic R-matrices). Finally, in Section 12.7 we present some additional comments on quadratic and cubic compatible Poisson brackets and Poisson structures on Lie groups associated with non-skew-symmetric R-matrices.

12.1. Poisson Lie Groups. Let $\mathcal{M}$, $\mathcal{N}$ be two Poisson manifolds. Their product is the manifold $\mathcal{M} \times \mathcal{N}$ equipped with the Poisson bracket

$$\{\varphi, \psi\}_{\mathcal{M}\times\mathcal{N}}(x, y) = \{\varphi(\cdot, y), \psi(\cdot, y)\}_{\mathcal{M}}(x) + \{\varphi(x, \cdot), \psi(x, \cdot)\}_{\mathcal{N}}(y); \quad \varphi, \psi \in C^\infty(\mathcal{M} \times \mathcal{N}). \tag{12.1}$$

In other words, (12.1) is the unique Poisson structure on $\mathcal{M} \times \mathcal{N}$ such that (i) natural projections $\pi_{\mathcal{M}}: \mathcal{M} \times \mathcal{N} \to \mathcal{M}$, $\pi_{\mathcal{N}}: \mathcal{M} \times \mathcal{N} \to \mathcal{N}$ are Poisson mappings and

(ii) $\{\pi^*_{\mathcal{M}} f, \pi^*_{\mathcal{N}} g\} = 0$ for any $f \in C^\infty(\mathcal{M})$, $g \in C^\infty(\mathcal{N})$.

Definition 12.1. A *Poisson Lie group* is a Lie group G equipped with a Poisson bracket such that multiplication in G defines a Poisson mapping $G \times G \to G$.

Poisson groups form a category where the morphisms are Lie group homomorphisms which are also Poisson mappings.

Example. (i) Any Lie group G equipped with the zero Poisson bracket satisfies the definition.

(ii) Let $\mathfrak{g}^*$ be the dual of a Lie algebra $\mathfrak{g}$ equipped with the Lie-Poisson bracket. We regard $\mathfrak{g}^*$ as an abelian group. Then the definiton is satisfied.

We shall see later that examples (i), (ii) are dual to each other. The duality theory for general Poisson Lie groups is the key part of their geometry which generalizes the theory of coadjoint orbits.

Let λ_x, ρ_x be the left and right translation operators on $C^\infty(G)$ by an element $x \in G$:

$$\lambda_x \varphi(y) = \varphi(xy), \quad \rho_x \varphi(y) = \varphi(yx).$$

Multiplication in G induces a Poisson mapping $G \times G \to G$ if

$$\{\varphi, \psi\}(xy) = \{\lambda_x \varphi, \lambda_x \psi\}(y) + \{\rho_y \varphi, \rho_y \psi\}(x). \tag{12.2}$$

Recall that any Poisson bracket is bilinear in derivatives of functions. It is convenient to write down Poisson brackets on a Lie group in the right- or left-invariant frame. Define the left and right differentials of a function $\varphi \in C^\infty(G)$ by the formulae

$$\begin{gathered}\langle D\varphi(x), X\rangle = \left(\frac{d}{dt}\right)_{t=0} \varphi(e^{tX}x), \quad \langle D'\varphi(x), X\rangle = \left(\frac{d}{dt}\right)_{t=0} \varphi(xe^{tX}),\\ X \in \mathfrak{g}, \quad D\varphi(x), \quad D'\varphi(x) \in \mathfrak{g}^*.\end{gathered} \tag{12.3}$$

Let us define the Hamiltonian operators $\eta, \eta'\colon G \to \operatorname{Hom}(\mathfrak{g}^*, \mathfrak{g})$ which correspond to our bracket by setting

$$\{\varphi, \psi\}(x) = \langle \eta(x) D\varphi, D\psi\rangle = \langle \eta'(x) D'\varphi, D'\psi\rangle. \tag{12.4}$$

Proposition 12.2. *Suppose G is a Poisson Lie group; then the functions η, η' satisfy the functional equations*

$$\begin{aligned}\eta(xy) &= \operatorname{Ad} x \circ \eta(y) \circ \operatorname{Ad}^* x^{-1} + \eta(x),\\ \eta'(xy) &= \operatorname{Ad} y^{-1} \circ \eta'(x) \circ \operatorname{Ad}^* y + \eta'(y).\end{aligned} \tag{12.5}$$

Proof. Obviously,

$$D(\lambda_x \varphi)(y) = \operatorname{Ad}^* x^{-1} \cdot D\varphi(xy), \quad D'(\lambda_x \varphi)(y) = D'\varphi(xy). \tag{12.6}$$

Clearly, (12.2) together with (12.4) and (12.6) imply (12.5). □

Functional equations (12.5) imply, in particular, that $\eta(e) = \eta'(e) = 0$, hence the Poisson structure on G is always degenerate at the unit element e. Linearizing the Poisson bracket at the point e gives a Lie-Poisson structure on $\mathfrak{g}$, i.e. a Lie algebra structure on $\mathfrak{g}^*$. To be more precise, let $\xi_1, \xi_2 \in \mathfrak{g}^*$ and choose $\varphi_1, \varphi_2 \in C^\infty(G)$ such that $d\varphi_i(e) = \xi_i$. Put

$$[\xi_1, \xi_2]_* = d\{\varphi_1, \varphi_2\}(e). \tag{12.7}$$

Proposition 12.3. *Formula* (12.7) *defines the structure of a Lie algebra on* $\mathfrak{g}^*$.

Proof. Formulae (12.4), (12.7) imply

$$[\xi_1, \xi_2]_* = \langle d\eta(e)\xi_1, \xi_2\rangle. \tag{12.8}$$

Hence the definition (12.7) is unambiguous. The Jacobi identity for (12.7) is obvious. □

Definition 12.4. Let $\mathfrak{g}$ be a Lie algebra, $\mathfrak{g}^*$ its dual. Suppose there is a Lie algebra structure on $\mathfrak{g}^*$, i.e. a mapping $\mathfrak{g}^* \wedge \mathfrak{g}^* \to \mathfrak{g}^*$ satisfying the Jacobi identity. The Lie brackets on $\mathfrak{g}$ and $\mathfrak{g}^*$ are said to be *consistent* if the dual mapping $\varphi: \mathfrak{g} \to \mathfrak{g} \wedge \mathfrak{g}$ is a 1-cocycle on $\mathfrak{g}$ (with respect to the adjoint action of $\mathfrak{g}$ on $\mathfrak{g} \wedge \mathfrak{g}$). A pair $(\mathfrak{g}, \mathfrak{g}^*)$ with consistent Lie brackets is called a *Lie bialgebra.*

Lie bialgebras form a category in which the morphisms are Lie algebra homomorphisms $p: \mathfrak{g} \to \mathfrak{h}$ such that the dual map $p^*: \mathfrak{h}^* \to \mathfrak{g}^*$ is also a Lie algebra homomorphism.

Proposition 12.5. *The Lie bracket* (12.7) *is consistent with the Lie bracket on* $\mathfrak{g}$. *Hence* $(\mathfrak{g}, \mathfrak{g}^*)$ *is a Lie bialgebra.* (We shall refer to it as the *tangent Lie bialgebra*).

Proof. The cobracket $\varphi: \mathfrak{g} \to \mathfrak{g} \wedge \mathfrak{g}$ which corresponds to (12.7) is given by

$$\varphi(X) = \left(\frac{d}{dt}\right)_{t=0} \eta(\exp tX). \tag{12.9}$$

(Note that since $\mathrm{Hom}(\mathfrak{g}^*, \mathfrak{g}) \simeq \mathfrak{g} \otimes \mathfrak{g}$, we may regard $\eta(x)$ as an element of $\mathfrak{g} \wedge \mathfrak{g}$.) The consistency condition now follows from the functional equation (12.5). □

The converse is also true.

Theorem 12.6. (i) *Let* $(\mathfrak{g}, \mathfrak{g}^*)$ *be a Lie bialgebra. Let G be a connected simply connected Lie group with Lie algebra* $\mathfrak{g}$. *There is a unique structure of a Poisson Lie group on G such that the tangent Lie bialgebra of G is* $(\mathfrak{g}, \mathfrak{g}^*)$. (ii) *The correspondence between Poisson Lie groups and Lie bialgebras is functorial.*

An important special case of Lie bialgebras are *coboundary bialgebras* for which the cocycle φ (the cobracket) is trivial, i.e.

$$\varphi(X) = \frac{1}{2}\operatorname{ad} X \cdot r = \frac{1}{2}[X \otimes 1 + 1 \otimes X, r], \tag{12.10}$$

where $r \in \mathfrak{g} \wedge \mathfrak{g}$ is a fixed element (*classical r-matrix*). In this case the linearized bracket on $\mathfrak{g}^*$ may be explicitly integrated to a Poisson structure on the corresponding Lie group. We may regard r as a skew linear map from $\mathfrak{g}^*$ into $\mathfrak{g}$. The bracket on $\mathfrak{g}^*$ which corresponds to φ is given by

$$[\xi_1, \xi_2]_* = \frac{1}{2}\operatorname{ad}^* r(\xi_1)\cdot \xi_2 - \frac{1}{2}\operatorname{ad}^* r(\xi_2)\cdot \xi_1. \tag{12.11}$$

Define a linear map $[r, r] \in \mathrm{Hom}(\mathfrak{g}^* \wedge \mathfrak{g}^*, \mathfrak{g})$ by

$$[r, r](\xi \wedge \eta) = [r\xi, r\eta] - r([\xi, \eta]_*). \tag{12.12}$$

Proposition 12.7. *The bracket* (12.11) *satisfies the Jacobi identity if and only if the mapping* (12.12) *is* $\mathfrak{g}$*-equivariant, i.e. for any* $X \in \mathfrak{g}$

$$[r, r](\operatorname{ad}^* X\cdot \xi \wedge \eta + \xi \wedge \operatorname{ad}^* X\cdot \eta) = [X, [r, r](\xi \wedge \eta)]. \tag{12.13}$$

Proposition 12.8. *Assume that condition* (12.13) *is satisfied. Then the formula*

$$2\{\varphi, \psi\} = \langle r(D\varphi), D\psi\rangle - \langle r(D'\varphi), D'\psi\rangle \tag{12.14}$$

defines the structure of a Poisson Lie group on G; its tangent Lie bialgebra is $(\mathfrak{g}, \mathfrak{g}^*)$ *with Lie bracket* (12.11).

Sketch of a proof. Consider first the left and right brackets

$$\{\varphi, \psi\}^l = \langle r(D\varphi), D\varphi\rangle, \quad \{\varphi, \psi\}^r = \langle r(D'\varphi), D'\psi\rangle. \tag{12.15}$$

The obstruction for the Jacobi identity is a trilinear form in differentials of functions which is given by

$$\begin{aligned} J_l(\xi_1, \xi_2, \xi_3) &= \langle [r, r](\xi_1 \wedge \xi_2), \xi_3\rangle, \quad \xi_i = D\varphi_i, \\ J_r(\xi_1', \xi_2', \xi_3') &= -\langle [r, r](\xi_1' \wedge \xi_2'), \xi_3'\rangle, \quad \xi_i' = D'\varphi_i. \end{aligned} \tag{12.16}$$

The difference of sign is due to the fact that Lie algebras of left- and right-invariant vector fields on G are anti-isomorphic. Condition (12.13) implies

$$[r, r](\mathrm{Ad}^* g \cdot \xi_1 \wedge \mathrm{Ad}^* g \cdot \xi_2) = \mathrm{Ad}\, g \cdot [r, r](\xi \wedge \eta). \tag{12.17}$$

Since $\xi_i'(g) = \mathrm{Ad}^* g \cdot \xi_i(g)$, both obstruction cancel each other. □

The Poisson bracket (12.14) is usually referred to as the *Sklyanin bracket.*

Let us now compare the notion of coboundary bialgebras and that of double Lie algebras (cf. the definition in Section 2.1). Formulae (2.1) and (12.11) show that they are close to each other but do not coincide, the difference being that in the latter case the second Lie bracket is defined on the same linear space $\mathfrak{g}$, while in the former case it is defined on the dual space $\mathfrak{g}^*$. It is natural to consider a class of double Lie algebras which satisfy both sets of axioms. Moreover, we shall assume that the R-matrix $R \in \mathrm{End}\, \mathfrak{g}$ satisfies the modified classical Yang-Baxter identity (2.12). Thus we are led to the following formal definition.

Definition 12.9. Let $(\mathfrak{g}, \mathfrak{g}_R)$ be a double Lie algebra such that (i) $\mathfrak{g}$ is equipped with a (fixed) invariant inner product and (ii) R is skew symmetric with respect to it and satisfies *mCYBE*. The pair $(\mathfrak{g}, \mathfrak{g}_R)$ is called a *Baxter Lie algebra*[5].

Proposition 12.10. *Identify* $\mathfrak{g}^*$ *and* $\mathfrak{g}$ *by means of the invariant inner product. Then* $(\mathfrak{g}, \mathfrak{g}^*)$ *is a Lie bialgebra.*

The definition above may seem too restrictive. However, as we shall see, any Lie bialgebra is canonically embedded into a larger one which is already a Baxter Lie algebra. This definition also proves to be appropriate for applications to lattice systems.

If $(\mathfrak{g}, \mathfrak{g}_R)$ is a Baxter Lie algebra, the definition of the Sklyanin bracket clearly makes sense. In this case the obstructions (12.16) are given by

$$\begin{aligned} J_l(\xi_1, \xi_2, \xi_3) &= ([\xi_1, \xi_2], \xi_3), \\ J_r(\xi_1', \xi_2', \xi_3') &= -([\xi_1', \xi_2'], \xi_3'). \end{aligned} \tag{12.18}$$

[5] Another name sometimes used in the literature is '*factorizable Lie bialgebra*'.

This follows immediately from the comparison of (12.12) and (12.17). More generally, we have

Proposition 12.11. *Let* $\mathfrak{g}$ *be a Lie algebra with an invariant inner product. Let* R, $R' \in \operatorname{End} \mathfrak{g}$ *be two skew linear operators satisfying the modified Yang-Baxter identity. Then the brackets*

$$\{\varphi, \psi\}_{R,R'} = \frac{1}{2}\langle R(D\varphi), D\psi\rangle + \frac{1}{2}\langle R'(D'\varphi), D'\psi\rangle \tag{12.19}$$

satisfy the Jacobi identity.

Indeed, it follows immediately from (12.19) that the obstructions J_l, J_r cancel each other.

12.2. Duality Theory for Poisson Lie Groups. Let $(\mathfrak{g}, \mathfrak{g}^*)$ be a Lie bialgebra. Put $\mathfrak{d} = \mathfrak{g} \dot{+} \mathfrak{g}^*$.

Proposition 12.12. *There is a unique structure of a Lie algebra on* $\mathfrak{d}$ *such that* (i) $\mathfrak{g}, \mathfrak{g}^* \subset \mathfrak{d}$ *are Lie subalgebras*; (ii) *The inner product on* $\mathfrak{d}$ *given by*

$$\langle\langle (X_1, f_1), (X_2, f_2)\rangle\rangle = f_1(X_2) + f_2(X_1) \tag{12.20}$$

is ad $\mathfrak{d}$*-invariant.*

Conversely, let $\mathfrak{d}$ *be a Lie algebra with an invariant inner product, and let* $\mathfrak{g}$, $\mathfrak{h}$ *be its Lie subalgebras such that* (i) $\mathfrak{d} = \mathfrak{g} \dot{+} \mathfrak{h}$; (ii) *both* $\mathfrak{g}$ *and* $\mathfrak{h}$ *are isotropic with respect to the inner product on* $\mathfrak{d}$. *Then* $\mathfrak{h} \simeq \mathfrak{g}^*$ *and both* $(\mathfrak{g}, \mathfrak{h})$ *and* $(\mathfrak{h}, \mathfrak{g})$ *are Lie bialgebras.*

Let $P_{\mathfrak{g}}$, $P_{\mathfrak{g}^*}$ be the projection operators onto $\mathfrak{g}, \mathfrak{g}^* \subset \mathfrak{d}$ parallel to the complementary subalgebra. Put

$$R_{\mathfrak{d}} = P_{\mathfrak{g}} - P_{\mathfrak{g}^*}. \tag{12.21}$$

Proposition 12.13. (i) *The operator* $R_{\mathfrak{d}}$ *equips* $\mathfrak{d}$ *with the structure of a Baxter Lie algebra.* (ii). *The canonical embeddings* $\mathfrak{g}, \mathfrak{g}^* \subset \mathfrak{d}$ *are Lie bialgebra morphisms.*

Corollary. *Let* $(\mathfrak{g}, \mathfrak{g}^*)$ *be a Lie bialgebra. Then* $(\mathfrak{g}^*, \mathfrak{g})$ *is also a Lie bialgebra.*

The Lie bialgebra $(\mathfrak{d}, \mathfrak{d}^*)$ is referred to as the *double* of $(\mathfrak{g}, \mathfrak{g}^*)$. Clearly, both $(\mathfrak{g}, \mathfrak{g}^*)$ and the dual bialgebra $(\mathfrak{g}^*, \mathfrak{g})$ have the same double.

Note. A triple $(\mathfrak{d}, \mathfrak{g}, \mathfrak{h})$ consisting of a Lie algebra $\mathfrak{d}$ with an invariant inner product and two complementary Lie subalgebras which are isotropic with respect to this inner product is called a *Manin triple.*

If $(\mathfrak{g}, \mathfrak{g}_R)$ is itself a Baxter Lie algebra, its double admits an alternative description. Put $\mathfrak{d} = \mathfrak{g} \oplus \mathfrak{g}$ (direct sum of two copies of $\mathfrak{g}$) and equip $\mathfrak{d}$ with the inner product

$$\langle\langle (X_1, Y_1), (X_2, Y_2)\rangle\rangle = \langle X_1, X_2\rangle - \langle Y_1, Y_2\rangle. \tag{12.22}$$

Recall from Proposition 2.4 that $\mathfrak{g}_R \xrightarrow{R_+ \oplus R_-} \mathfrak{g} \oplus \mathfrak{g}$ is a Lie algebra embedding. Clearly, if R is skew symmetric with respect to the inner product in $\mathfrak{g}$, the subalgebra $\mathfrak{g}_R \subset \mathfrak{d}$ is isotropic with respect to (12.22). Let $\mathfrak{g}^\delta = \{(X, X); X \in \mathfrak{g}\}$ be the diagonal subalgebra in $\mathfrak{d}$.

Proposition 12.14. *$(\mathfrak{d}, \mathfrak{g}^\delta, \mathfrak{g}_R)$ is a Manin triple.*

Corollary. *The double of a Baxter Lie algebra $(\mathfrak{g}, \mathfrak{g}_R)$ is isomorphic to the square of $\mathfrak{g}$ (direct sum of two copies).*

Let $(\mathfrak{g}, \mathfrak{g}^*)$ be a Lie bialgebra, $(\mathfrak{d}, \mathfrak{d}^*)$ its double. Let G, G^*, D be the local Lie groups which correspond to the Lie algebras $\mathfrak{g}, \mathfrak{g}^*, \mathfrak{d}$ respectively. We equip D with the Sklyanin bracket associated with the R-matrix $R_\mathfrak{d}$.

Proposition 12.15. (i) *The subgroups $G, G^* \subset D$ are Poisson submanifolds in D.* (ii) *The induced Poisson structure on G, G^* equips them with the structure of Poisson Lie groups and their tangent Lie bialgebras are $(\mathfrak{g}, \mathfrak{g}^*)$ and $(\mathfrak{g}^*, \mathfrak{g})$, respectively.* (iii) *The product map $G \times G^* \to D$ is an isomorphism of Poisson manifolds.*

If the tangent Lie bialgebra of G is a Baxter Lie algebra, the double of G admits a nice description which extends Proposition 12.14.

Proposition 12.16. *Let G be a local Poisson Lie group such that $(\mathfrak{g}, \mathfrak{g}^*)$ is a Baxter Lie algebra. Then* (i) $\mathfrak{D} = \mathfrak{G} \times \mathfrak{G}$. (ii) *The diagonal map $G \to D$: $x \mapsto (x, x)$ is a Poisson embedding.* (iii) *Let $R_\pm$: $G^* \to G$: $h \mapsto h_\pm$ be the homomorhisms induced by the Lie algebra homomorphisms $R_\pm$: $\mathfrak{g}^* \to \mathfrak{g}$. The mapping $G^* \to D$: $h \mapsto (h_+, h_-)$ is a Poisson embedding.* (iv) *Locally each element $(x, y) \in D$ admits a unique factorization $(x, y) = (\xi, \xi)(\eta_+, \eta_-)$ with $(\xi, \xi) \in G^\delta$, $(\eta_+, \eta_-) \in G^*$; this factorization establishes an isomorphism of Poisson manifolds $D \simeq G \times G^*$.*

12.3. Poisson Reduction, Dressing Transformations and Symplectic Leaves of Poisson Lie Groups. Let M be a Poisson manifold, G a Poisson Lie group. An action $G \times M \to M$ is called a *Poisson group action* if it is a Poisson mapping, the manifold $G \times M$ being equipped with the product Poisson structure.

We shall also need a (less restrictive) concept of admissible actions. Let $\mathcal{M}$ be a Poisson manifold. An action $H \times \mathcal{M} \to \mathcal{M}$ is called *admissible* if the subspace of H-invariants in $C^\infty(\mathcal{M})$ is a Lie subalgebra with respect to the Poisson bracket.

Proposition 12.17. (i) *Let $H \times \mathcal{M} \to \mathcal{M}$ be an admissible group action. Assume that the quotient space $\mathcal{M}/H$ is a smooth manifold. There is a unique Poisson structure on $\mathcal{M}/H$ such that the canonical projection π: $\mathcal{M} \to \mathcal{M}/H$ is a Poisson mapping.* (ii) *Assume that there is a Poisson group action $G \times \mathcal{M} \to \mathcal{M}$ which commutes with the action of H. Then the natural action of G on the quotient space $\mathcal{M}/H$ is also a Poisson group action.*

It is useful to have an infinitesimal criterion of admissibility. Let $G \times \mathcal{M} \to \mathcal{M}$ be a Lie group action. For $X \in \mathfrak{g}$ let $\hat{X} \in \operatorname{Vect} \mathcal{M}$ be the corresponding vector field on $\mathcal{M}$. For any $\varphi \in C^\infty(\mathcal{M})$ let $\xi_\varphi \in \mathfrak{g}^*$ be the covector defined by

$$\langle \xi_\varphi(x), X \rangle = \left(\frac{d}{dt}\right)_{t=0} \varphi(e^{tX} \cdot x). \tag{12.23}$$

Proposition 12.18. (i) *Assume that G is connected. An action $G \times \mathcal{M} \to \mathcal{M}$ is a Poisson group action if and only if for any $X \in \mathfrak{g}$ and $\varphi, \psi \in C^\infty(\mathcal{M})$*

$$\hat{X}\{\varphi, \psi\} - \{\hat{X}\varphi, \psi\} - \{\varphi, \hat{X}\psi\} = \langle [\xi_\varphi, \xi_\psi]_*, X \rangle. \tag{12.24}$$

(ii) *Let $H \subset G$ be a connected subgroup. Assume that $\mathfrak{h}^\perp \subset \mathfrak{g}^*$ is a Lie subalgebra. Then the restriction of the action of G to H is admissible.*

In particular, Poisson group actions are admissible.

Remark. If $\mathfrak{h}^\perp \subset \mathfrak{g}^*$ is an ideal, then $H \subset G$ is a Poisson Lie subgroup and the restriction of the action $G \times \mathcal{M} \to \mathcal{M}$ to H is again a Poisson group action.

An obvious example of a Poisson group action is given by the action of a Poisson Lie group on itself by left translations. The next example plays the key role in the description of symplectic leaves of Poisson Lie groups.

Let $(\mathfrak{g}, \mathfrak{g}^*)$ be a Lie bialgebra, $(\mathfrak{d}, \mathfrak{d}^*)$ its double, G, G^*, D the corresponding local Lie groups. We denote by $D_{(R_\mathfrak{d}, \pm R_\mathfrak{d})}$ the manifold D equipped with the Poisson bracket $\{\ ,\ \}_{R_\mathfrak{d}, \pm R_\mathfrak{d}}$ defined in (12.19).

Proposition 12.19. *The action of D on itself by left and right translations gives rise to left and right Poisson group actions*

$$D_{(R_\mathfrak{d}, -R_\mathfrak{d})} \times D_{(R_\mathfrak{d}, R_\mathfrak{d})} \to D_{(R_\mathfrak{d}, R_\mathfrak{d})},$$
$$D_{(R_\mathfrak{d}, R_\mathfrak{d})} \times D_{(R_\mathfrak{d}, -R_\mathfrak{d})} \to D_{(R_\mathfrak{d}, R_\mathfrak{d})}.$$

Restricting these actions to Poisson subgroups $G, G^* \subset D_{(R_\mathfrak{d}, -R_\mathfrak{d})}$ we get Poisson group actions

$$G \times D_{(R_\mathfrak{d}, R_\mathfrak{d})} \to D_{(R_\mathfrak{d}, R_\mathfrak{d})}, \quad D_{(R_\mathfrak{d}, R_\mathfrak{d})} \times G \to D_{(R_\mathfrak{d}, R_\mathfrak{d})},$$
$$G^* \times D_{(R_\mathfrak{d}, R_\mathfrak{d})} \to D_{(R_\mathfrak{d}, R_\mathfrak{d})}, \quad D_{(R_\mathfrak{d}, R_\mathfrak{d})} \times G^* \to D_{(R_\mathfrak{d}, R_\mathfrak{d})}$$

by left and right translations, respectively. By Proposition 12.18 all these these actions are admissible.

Proposition 12.20. (i) *The quotient spaces $D_{(R_\mathfrak{d}, R_\mathfrak{d})}/G^*$ and $G^* \backslash D_{(R_\mathfrak{d}, R_\mathfrak{d})}$ are anti-isomorphic as Poisson manifolds.* (ii) *Identify the quotient space $D_{(R_\mathfrak{d}, R_\mathfrak{d})}/G^*$ with G via the canonical map $G \times G^* \to D$. The induced Poisson structure on G coincides with the Sklyanin bracket.* (iii) *Canonical projections*

$$\begin{array}{ccc} & D_{(R_\mathfrak{d}, R_\mathfrak{d})} & \\ \swarrow & & \searrow \\ D/G^* & & G^* \backslash D \end{array}$$

form a dual pair.

Corollary. *Symplectic leaves in G are orbits of the action $G \times G^* \to G$ induced by right translations $D \times G^* \to D$.*

In the dual way, the quotient space D/G is identified with G^* and symplectic leaves in G^* are orbits of the action $G^* \times G \to G^*$ induced by right translations $D \times G \to D$.

The transformations $G \times G^* \to G$ and $G^* \times G \to G^*$ described in Proposition 12.20 are called *dressing transformations*, due to their relation to dressing transformations in soliton theory (see Semenov-Tian-Shansky [1985, 1987]). Proposition 12.17 implies that dressing transformations define Poisson group actions. Note that in general these actions do not preserve Poisson brackets and the corresponding vector fields are non-hamiltonian.

Example. Let $(\mathfrak{g}, \mathfrak{g}^*)$ be a trivial Lie bialgebra. In this case $\mathfrak{d}$ is the semidirect sum of $\mathfrak{g}$ and abelian ideal $\mathfrak{g}^*$, $D = G \times \mathfrak{g}^*$. The manifold $D_{(R_\mathfrak{d}, R_\mathfrak{d})}$ coincides with the cotangent bundle T^*G equipped with the canonical symplectic structure. Canonical projections $D \to D/G \simeq \mathfrak{g}^*$, $D \to G\backslash D \simeq \mathfrak{g}^*$ coincide with the standard left and right moment maps, and the dressing action $G \times \mathfrak{g}^* \to \mathfrak{g}^*$ coincides with the coadjoint representation. The dual action $G \times \mathfrak{g}^* \to G$ is trivial. Thus dressing transformations may be regarded as nonlinear analogues of the coadjoint representation. This analogy is further supported by the following result.

Proposition 12.21. *The linearization of the dressing action $G \times G^* \to G^*$ at the unit element coincides with the coadjoint action of the group G.*

In the dual way, linearization of the action $G^* \times G \to G$ coincides with the coadjoint action of G^*.

Now assume that the tangent bialgebra of G is a Baxter Lie algebra. In this case the description of symplectic leaves in G, G^* may be given in terms of the factorization problem associated with R described in Section 2.3.

Proposition 12.22. *The dressing action $G^* \times G \to G$ is given by*

$$h\colon x \mapsto (xh_+h_-^{-1}x^{-1})_\pm x h_\pm^{-1}.$$

Here $h \mapsto h_\pm$ are the canonical homomorphisms from G^* into G and for $g \in G$ we denote by $g_\pm$ the solutions to the factorization problem (2.15) *with the left-hand side equal to g, i.e. $g = g_+g_-^{-1}$.*

The symplectic leaves in the dual group G^* admit the following simple description.

Proposition 12.23. *Let $f\colon G^* \to G$ be the canonical factorization map described in Section* 2.1. *Symplectic leaves in G^* are mapped by f onto the conjugacy classes in G.*

Corollary. *The Casimir functions on G^* are the pullbacks of central functions on G.*

Proof. Recall that G^* is isomorphic, as a Poisson manifold, to the quotient space $D_{(R_\mathfrak{d}, R_\mathfrak{d})}/G$. Since in the present case $D \simeq G \times G$, the quotient space is also modelled on G and we have the following dual pair of Poisson mappings

$$\begin{array}{ccc} & D & \\ {}^{p'}\swarrow & & \searrow^{p''} \\ G & & G \end{array} \tag{12.25}$$

where $p'(x, y) = xy^{-1}$, $p''(x, y) = y^{-1}x$. The factorization map $f: G^* \to G$ establishes an isomorphism between these two models of the quotient space. Now, the symplectic leaves in G have the form $p'(p''^{-1}(x))$, $x \in G$, and clearly coincide with the conjugacy classes in G. □

Since G^* and G are locally homeomorphic, we may also speak of two Poisson structures on G. An explicit expression for the Poisson bracket on G^* pushed forward to G is given by

$$\{\varphi, \psi\} = \frac{1}{2}\langle R(D'\varphi), D\psi\rangle + \frac{1}{2}\langle R(D\varphi), D'\psi\rangle$$
$$+ \langle R_+(D\varphi), D'\psi\rangle - \langle R_-(D'\varphi), D\psi\rangle. \tag{12.26}$$

It is easy to check directly that central functions on G are the Casimir functions of (12.26). The situation is thus reminiscent of the involutivity theorem (Theorem 2.1). In the next Section we shall extend this theorem to the present setting and also prove its generalization suitable for the study of difference Lax equations.

12.4. Lax Equations on Lie Groups. In this Section we assume that G is a Poisson Lie group and that its tangent Lie bialgebra is a Baxter Lie algebra. To simplify the notation we shall assume, moreover, that G is a matrix group. Denote by $I(G)$ the space of central functions on G.

Theorem 12.24. (i) *Functions from $I(G)$ are in involution with respect to the Sklyanin bracket on G.*

(ii) *Let $\varphi \in I(G)$. The equation of motion defined by φ with respect to the Sklyanin bracket has the Lax form*

$$\frac{dL}{dt} = [L, M], \quad M = 1/2R(D\varphi(L)), \quad L \in G. \tag{12.27}$$

(iii) *Let $x_\pm(t)$ be the solutions to the factorization problem* (2.15) *associated with the R-matrix*:

$$\exp tD\varphi(L) = x_+(t)x_-(t)^{-1}. \tag{12.28}$$

Then the integral curve of the equation (12.27) *starting at $L \in G$ is given by*

$$L(t) = x_\pm(t)^{-1}Lx_\pm(t). \tag{12.29}$$

The proof is parallel to the proof of Theorem 2.2. Observe first of all that left and right gradients of a function $\varphi \in I(G)$ coincide. This makes (i), (ii) directly obvious from the definition of the Sklyanin bracket. Formula (12.29) may be proved by direct computation. It is interesting, however, to give a geometric proof since it gives an insight into the mutual relations of Poisson structures on G, G^* and D.

Lemma 1. *Let $p: D \to G: (x, y) \mapsto xy^{-1}$ be the standard projection, $\varphi \in I(G)$, $h_\varphi = \varphi \circ p$. The integral curves of the Hamiltonian h_φ on $D_{(R_D, R_D)}$ are given by*

$$(x_0 e^{tX'}, y_0 e^{tX'}), \quad X' = D'\varphi(x_0 y_0^{-1}). \tag{12.30}$$

Proof. Recall that p is included into a dual pair (12.25). Projections of the integral curve on D onto the quotient spaces D/G^δ, $G^\delta \backslash D$ reduce to points since the reduced Hamiltonians are Casimir functions. Since h_φ is both right- and left-G^δ-invariant we have $\nabla h_\varphi, \nabla'_{h_\varphi} \in \mathfrak{g}^\delta$ and for any $\psi \in C^\infty(D)$

$$\{h_\varphi, \psi\} = \langle\langle \nabla' h_\varphi, \nabla' \psi \rangle\rangle$$

Obviously, $\nabla' h_\varphi = (X', X')$, where $X' = \nabla' \varphi(xy^{-1})$, is time-independent. Now, (12.30) follows immediately. □

Consider the action $G_R \times D \to D$ defined by

$$h: (x, y) \mapsto (h_+ x h_-^{-1}, h_+ y h_-^{-1}), \tag{12.31}$$

where $h \mapsto h_\pm$ are the canonical homomorphisms from G_R into G. Notice that the subgroup $(G, e) \subset D$ is a cross-section of the action (12.31) on an open cell in D. Hence we get a projection

$$\pi: D \to G: (x, y) \mapsto y_+^{-1} x y_-, \quad y = y_+ y_-^{-1}, \tag{12.32}$$

whose fibers coincide with G_R-orbits in D.

Lemma 2. (i) *The action* (12.31) *is admissible.* (ii) *The quotient Poisson structure on $G \simeq D/G_R$ is given by*

$$\begin{aligned} 2\{\varphi, \psi\}_r &= (RX', Y') - (RX, Y) + 1/2((R^3 - R)(X' - X), Y' - Y), \\ X &= \nabla\varphi, \quad X' = \nabla'\varphi, \quad Y = \nabla\psi, \quad Y' = \nabla'\psi. \end{aligned} \tag{12.33}$$

In particular, if $R^3 = R$, it coincides with the Sklyanin bracket.

We shall prove a more general statement below (Proposition 12.31).

To finish the proof of Theorem 12.24 observe that for $\varphi \in I(G)$ the Hamiltonian h_φ is equal to $\varphi \circ \pi$. Hence (12.30) gives rise to the quotient flow on $G \simeq D/G_R$. Clearly, the quotient flow is given by

$$L \mapsto g_\pm(t)^{-1} L g_\pm(t), \quad g_+(t) g_-(t)^{-1} = e^{t\nabla_\varphi(L)}, \tag{12.34}$$

and hence satisfies the Lax equation. Now, if $R^3 = R$, this completes the proof, since the quotient Poisson structure on G coincides with the Sklyanin bracket. In the general case, the quotient Poisson structure on G differs from the Sklyanin bracket by a perturbation term

$$1/2((R^3 - R)(X' - X), Y' - Y). \tag{12.35}$$

Observe that it annihilates the gradient of any function $\varphi \in I(G)$, hence the equations of motion defined by φ with respect to the perturbed and the unperturbed bracket coincide.

Let us now discuss a generalization of Theorem 12.24 which is suited for the study of lattice systems. Recall from Proposition 12.11 that we may use more general Poisson brackets given by (12.19) with the left and right R-matrices not necessarily coinciding. We shall use this observation to twist the Poisson bracket on D.

Let τ be an automorphism of the Baxter Lie algebra $(\mathfrak{g}, \mathfrak{g}_R)$, i.e. an orthogonal operator $\tau \in \operatorname{Aut} \mathfrak{g}$ which commutes with R. It gives rise to an automorphism of G which we denote by $g \mapsto {}^\tau g$. Define the *twisted conjugation* $G \times G \to G$ by

$$g\colon h \mapsto gh^\tau g^{-1}. \tag{12.36}$$

Let ${}^\tau I(G)$ be the space of smooth functions on G invariant with respect to twisted conjugations.

Theorem 12.25. (i) *The functions* $\varphi \in {}^\tau I(G)$ *are in involution with respect to the Sklyanin bracket on* G. (ii) *The equations of motion defined by a Hamiltonian* $\varphi \in {}^\tau I(G)$ *have the generalized Lax form*

$$\frac{dL}{dt} = L^\tau A - AL, \quad L \in G, \quad A = 1/2R(V'\varphi(L)). \tag{12.37}$$

(iii) *Let* $x_+(t), x_-(t)$ *be the solutions to the factorization problem* (2.15) *associated with* R:

$$\exp tV\varphi(L) = x_+(t)x_-(t)^{-1}. \tag{12.38}$$

Then the integral curve of equation (12.37) *starting at* $L \in G$ *is given by*

$$L(t) = x_\pm(t)^{-1}L^\tau x_\pm(t). \tag{12.39}$$

The proof is based on the use of a twisted Poisson structure on D. Extend τ to $\mathfrak{d} = \mathfrak{g} \oplus \mathfrak{g}$ by setting

$$\hat{\tau}(X, Y) = (X, \tau Y) \tag{12.40}$$

and put

$${}^\tau R_{\mathfrak{d}} = \hat{\tau} \circ R_{\mathfrak{d}} \circ \hat{\tau}^{-1} \tag{12.41}$$

We also put

$${}^\tau G = {}^\tau(G^\delta) = \{(x, {}^\tau x);\, x \in G\} \subset D. \tag{12.42}$$

Equip D with the Poisson bracket (12.19) with $R = {}^\tau R_{\mathfrak{d}}$, $R' = R_{\mathfrak{d}}$.

Proposition 12.26. (i) *The natural action of* ${}^\tau G$ *on* $D_{({}^\tau R_{\mathfrak{d}}, R_{\mathfrak{d}})}$ *by left translations is a left Poisson action.* (ii) *The natural action of* G *on* $D_{({}^\tau R_{\mathfrak{d}}, R_{\mathfrak{d}})}$ *by right translations is a right Poisson action.*

Proposition 12.27. *The canonical projections*

$$\begin{array}{ccc} & D & \\ {}^{\rho}\swarrow & & \searrow^{\rho'} \\ D/G^{\delta} & & {}^{\tau}G\backslash D \end{array} \tag{12.43}$$

form a dual pair.

Both quotient spaces are modelled on G. The projections ρ, ρ' are given by

$$\rho\colon (x, y) \mapsto xy^{-1}, \quad \rho'\colon (x, y) \mapsto {}^{\tau^{-1}}y^{-1}x. \tag{12.44}$$

Proposition 12.28. (i) *The quotient Poisson structures on D/G^{δ} and ${}^{\tau}G\backslash D$ are anti-isomorphic.* (ii) *Symplectic leaves of the quotient Poisson structure on G are orbits of twisted conjugations.*

Proof. It suffices to compute $\rho(\rho'^{-1}(x))$. Clearly,

$$\rho'^{-1}(x) = \{({}^{\tau^{-1}}yx, y); y \in G\}, \quad \rho(\rho'^{-1}(x)) = \{{}^{\tau^{-1}}yxy^{-1}; y \in G\}. \ \square$$

Corollary. *The Casimir funcions of the quotient Poisson structure on G are invariants of twisted conjugations.*

An explicit formula for the quotient Poisson structure on G is given by

$$\begin{aligned} \{\varphi, \psi\}_{\text{red}} &= \langle R_+(\tau\nabla\varphi), \nabla'\psi\rangle + \langle R_-(\nabla'\varphi), \tau\nabla\psi\rangle \\ &\quad - 1/2\langle R(\nabla\varphi), \nabla\psi\rangle - 1/2\langle R(\nabla'\varphi), \nabla'\psi\rangle. \end{aligned} \tag{12.45}$$

Now everything is ready for the proof of Theorem 12.25.

Proposition 12.29. *Let $\varphi \in {}^{\tau}I(G)$, $h_\varphi = \varphi \circ \rho$. Integral curves of the Hamiltonian h_φ on $D_{({}^{\tau}R_D, R_D)}$ are given by* (12.38).

The proof is completely similar to that of Lemma 1.

Consider the action $G_R \times D \to D$ defined by

$$h\colon (x, y) \mapsto (h_+xh_-^{-1}, {}^{\tau}h_+yh_-^{-1}). \tag{12.46}$$

Proposition 12.30. *The action* (12.46) *is admissible.*

Proof. Observe that if there are two commuting Poisson actions $\Gamma \times \mathcal{M} \to \mathcal{M}$, $\mathcal{M} \times H \to \mathcal{M}$, their combination gives rise to a Poisson group action of $\Gamma \times H$ equipped with the product bracket. In the present setting we get the Poisson group action

$$D_{({}^{\tau}R_D, {}^{\tau}R_D)} \times D_{(-R_D, R_D)} \times D_{({}^{\tau}R_D, R_D)} \to D_{({}^{\tau}R_D, R_D)}$$

(we have changed the sign of the Poisson bracket on the second copy of D so as to consider left actions). Now, G_R is embedded into $D \times D$ via

$$h \mapsto (h_+, {}^{\tau}h_+, h_-, h_-)$$

To prove the admissibility of the action of G_R we use Proposition 12.18 (ii). Since the tangent Lie bialgebra of $D \times D$ is $\mathfrak{d} \oplus \mathfrak{d}$, $\mathfrak{d}_{-{}^\tau R_\mathfrak{d}} \oplus \mathfrak{d}_{R_\mathfrak{d}}$, our assertion follows from the next lemma.

Lemma 3. $\mathfrak{g}_R^\perp \subset (\mathfrak{d} \oplus \mathfrak{d})^*$ *is a Lie subalgebra in* $\mathfrak{d}_{-{}^\tau R_\mathfrak{d}} \oplus \mathfrak{d}_{R_\mathfrak{d}}$.

Proof. An element $(X_1, X_2, Y_1, Y_2) \in (\mathfrak{d} \oplus \mathfrak{d})^* \simeq \mathfrak{d} \oplus \mathfrak{d}$ annihilates $\mathfrak{g}_R$ if and only if

$$R_-(X_1 - \tau^{-1}X_2) + R_+(Y_1 - Y_2) = 0.$$

Equivalently,

$$\mathfrak{g}_R^\perp = \{((\xi, \tau\xi) + (\eta_+, \tau\eta_-), (\xi', \xi') + (\eta'_+, \eta'_-)); \xi \cdot \xi' \in \mathfrak{g}, \eta, \eta' \in \mathfrak{g}_R, \eta_- + \eta'_+ = 0\}.$$

Since $\mathfrak{g}_R, \mathfrak{g}^\delta \subset \mathfrak{d}_{R_\mathfrak{d}}$, $\mathfrak{g}, {}^\tau\mathfrak{g}_R \subset \mathfrak{d}_{-{}^\tau R_\mathfrak{d}}$ are Lie subalgebras, it suffices to check that

$$R_-\eta_1 = -R_+\eta'_1, \quad R_-\eta_2 = -R_+\eta'_2$$

implies

$$R_-([\eta_1, \eta_2]_R) = -R_+(-[\eta'_1, \eta'_2]_R).$$

The last assertion follows immediately from the modified Yang-Baxter identity. Indeed,

$$R_-([\eta_1, \eta_2]_R) = [R_{-\eta_1}, R_{-\eta_2}] = [R_+\eta'_1, R_+\eta'_2] = R_+([\eta'_1, \eta'_2]_R). \square$$

Observe that the subgroup $(G, e) \subset D$ is a cross-section of the action (12.46) on an open cell in D. Hence we may identify the quotient space D/G_R with G. The projection $\pi: D \to G$ is now given by:

$$\pi: (x, y) \mapsto {}^{\tau^{-1}}y_+^{-1}xy_-.$$

Proposition 12.31. *The quotient Poisson structure on G is given by*

$$2\{\varphi, \psi\}_{\text{red}} = (RX', Y') - (RX, Y) + 1/2((R^3 - R)(X' - \tau X), Y' - \tau Y),$$
$$X = \nabla\varphi, \quad X' = \nabla'\varphi, \quad Y = \nabla\psi, \quad Y' = \nabla'\psi. \tag{12.47}$$

In particular, if $R^3 = R$, *it coincides with the Sklyanin bracket.*

Proof. For $\varphi, \psi \in C^\infty(G)$ put $H_\varphi = \varphi \circ \pi$, $H_\psi = \psi \circ \pi$. Put $X = \nabla\varphi$, $Y = \nabla\psi$, $X' = \nabla'\varphi$, $Y' = \nabla'\psi$. It is easy to compute the gradients of H_φ. Their restrictions to the submanifold $(G, e) \subset D$ are given by

$$\nabla H_\varphi = (X, X'_+ - \tau X_-), \quad \nabla' H_\varphi = (X', X'_+ - \tau X_-).$$

Similar formulae hold for the gradients of H_ψ. Now,

$$R_\mathfrak{d}(\nabla' H_\varphi) = (X', X' - X'_- + \tau X_-),$$
$${}^\tau R_\mathfrak{d}(\nabla H_\varphi) = (2\tau^{-1}X'_+ - X_+ - X_-, X'_+ - \tau X_-).$$

After substituting these expressions into the definition of $\{H_\varphi, H_\psi\}_{({}^\tau R_\mathfrak{d}, R_\mathfrak{d})}$ we get, after some cancellations, the formula (12.47). $\square$

To finish the proof of Theorem 12.25, observe that for $\varphi \in {}^{\tau}I(G)$ we have $h_{\varphi} = \varphi \circ \pi$ and hence the flow (12.38) gives rise to the quotient flow on $D/G_R \simeq G$. It is clearly given by (12.39) and satisfies the generalized Lax equations (12.37). If $R^3 = R$, this completes the proof since the quotient Poisson bracket on G coincides with the Sklyanin bracket. In the general case, the perturbation term in (12.47) is equal to

$$\frac{1}{2}((R^3 - R)(X' - \tau X), Y' - \tau Y). \tag{12.48}$$

It annihilates the gradient of any function $\varphi \in {}^{\tau}I(G)$ and hence $\varphi \in {}^{\tau}I(G)$ defines the same equations of motion with respect to the perturbed and the unperturbed brackets. □

Let us now specialize Theorem 12.25 so as to apply it to lattice systems. Let $(\mathfrak{g}, \mathfrak{g}_R)$ be a Baxter Lie algebra, G, G_R the corresponding Lie groups. Put $\mathfrak{G} = \bigoplus^N \mathfrak{g}$, $\mathfrak{G}_R = \bigoplus^N \mathfrak{g}_R$, $\mathbb{G} = G^N$. We shall regard elements of $\mathbb{G}$ as functions on $\mathbb{Z}/N\mathbb{Z}$ with values in G. Equip G with the product Poisson structure. Clearly, $\mathbb{G}$ is a Poisson Lie group and its tangent Lie bialgebra is $(\mathfrak{G}, \mathfrak{G}_R)$. Equip $\mathfrak{G}$ with the natural inner product

$$\langle X, Y\rangle = \sum_n \langle X_n, Y_n\rangle \tag{12.49}$$

and extend $R \in \operatorname{End} \mathfrak{g}$ to $\mathfrak{G}$ by setting $(RX)_n = R(X_n)$. This makes $(\mathfrak{G}, \mathfrak{G}_R)$ a Baxter Lie algebra. The elements of $\mathbb{G}$ will be denoted by $L = (L_0, \ldots, L_{N-1})$. Define the mappings ψ_m, $T\colon \mathbb{G} \to G$ by

$$\psi_m(L) = \overset{\curvearrowleft}{\prod_{0 \leqslant k \leqslant m-1}} L_k, \quad T(L) = \overset{\curvearrowleft}{\prod_{0 \leqslant k \leqslant N-1}} L_k \tag{12.50}$$

The functions ψ_m satisfy the *linear difference system*

$$\psi_{m+1} = L_m \psi_m, \quad \psi_{-1} = 1, \tag{12.51}$$

and T is the *monodromy matrix* of this system.

Proposition 12.32. *The monodromy map $T\colon \mathbb{G} \to G$ is a Poisson mapping.*

This property of the Sklyanin bracket has served as a motivation of the whole theory. The quantum version of this statement goes back to R. Baxter (cf. Faddeev [1984]).

Let $\tau \in \operatorname{Aut} \mathfrak{G}$ be the cyclic permutation

$$\tau\colon (X_0, \ldots, X_{N-1}) \mapsto (X_1, \ldots, X_{N-1}, X_0). \tag{12.52}$$

Clearly, the twisted conjugations $L \mapsto gL^{\tau}g^{-1}$ coincide with the gauge transformations for (12.51) induced by left translations $\psi_m \mapsto g_m \psi_m$. The operator (12.52) preserves the inner product (12.49) and commutes with R. Hence Theorem 12.25 applies to the present situation. The space ${}^{\tau}I(G)$ is described by the following discrete version of the classical Floquet theorem.

Theorem 12.33. (i) *Two elements $L, L' \in \mathbb{G}$ lie on the same gauge orbit in $\mathbb{G}$ if and only if their monodromy matrices $T(L)$, $T(L')$ are conjugate in G.* (ii) *The algebra ${}^{\tau}I(\mathbb{G})$ is generated by the functions h_φ: $L \mapsto \varphi(T(L))$, $\varphi \in I(G)$.*

As a corollary of Theorem 12.25 we get

Theorem 12.34. (i) *The functions h_φ, $\varphi \in I(G)$, are in involution with respect to the Sklyanin bracket on $\mathbb{G}$.* (ii) *Hamilton's equations of motion with Hamiltonian h_φ are given by*

$$\frac{dL_m}{dt} = L_m M_{m+1} - M_m L_m,$$
$$M_m = \frac{1}{2} R(\psi_m^{-1} \nabla_\varphi(T(L))\psi_m). \tag{12.53}$$

(iii) *Let $g_m^{\pm}(t)$ be the solutions to the factorization problem* (2.15) *defined by R with the left-hand side given by*

$$g_m(t) = \psi_m^{0-1} \exp t\nabla_\varphi(T(L^0))\psi_m^0, \quad \psi_m^0 = \psi_m(L^0). \tag{12.54}$$

The integral curve of (12.53) *starting at $L = (L_0^0, \ldots, L_{N-1}^0)$ is given by*

$$L_m(t) = g_m(t)_{\pm}^{-1} L_m^0 g_{m+1}(t)_{\pm}. \tag{12.55}$$

Since $\mathbb{G}$ is equipped with the product Poisson structure, symplectic leaves in $\mathbb{G}$ are products of "local" symplectic leaves which were described in Proposition 12.22. Thus in principle we have a complete description of the geometry of Lax equations on $\mathbb{G}$. In typical applications we may take G to be the group of algebraic loops with values in a compact Lie group. In this way we get "difference Lax equations with a spectral parameter" which describe the interaction of spins on a one-dimensional lattice.

12.5. Poisson Structure on Loop Groups and Applications. R-matrices on loop algebras which are skew with respect to the inner product give rise to the structure of a Baxter Lie algebra and hence determine a Poisson structure on the corresponding loop groups. For concreteness we shall discuss in this section Poisson brackets associated with elliptic R-matrices. Other examples will be considered as limiting cases.

Fix an elliptic curve C and a set $D \subset C$ as in Section 11.3. For $\nu \in D$ put $L_\nu = \mathrm{SL}(n; \mathbb{C}((\lambda_\nu)))$, $L_D = \prod_{\nu \in D} L_\nu$. We may regard L_D as an algebraic Lie group with Lie algebra $\mathfrak{L}_D = \prod_{\nu \in D} \mathfrak{L}_\nu$. Let G_D be the group of rational matrix functions on C with values in $\mathrm{SL}(n, \mathbb{C})$ which are regular outside D and satisfy the quasi-periodicity condition

$$g(\lambda + a) = I_a g(\lambda) I_a^{-1}, \quad a \in C_n. \tag{12.56}$$

Clearly, G_D is canonically embedded into L_D. G_D is an affine algebraic group, and its affine ring $A(G_D)$ is generated by "tautological" functions $\varphi_{ij}(\lambda)$: $T \mapsto$

$T_{ij}(\lambda)$ which assign to an element $T \in G_D$ the values of its matrix coefficients at $\lambda \in \mathbb{C}$. By an abuse of language we shall write simply $\varphi_{ij}(\lambda)(T) = T_{ij}(\lambda)$ and suppress the matrix indices.

Let R be the classical R-matrix on $\mathfrak{L}_D$ associated with the decomposition (11.13).

Proposition 12.35. (i) *The R-matrix* (11.9) *equips L_D with the structure of a Poisson Lie group.* (ii) $G_D \subset L_D$ *is a Poisson Lie subgroup.* (iii) *Let $r(\lambda, \mu) \in$ Mat$(n) \otimes$ Mat(n) be the kernel of R defined by* (11.9). *The Poisson brackets of the generators of the affine ring $A(G_D)$ are given by*

$$\{\varphi(\lambda) \underset{,}{\otimes} \varphi(\mu)\}(T) = [r(\lambda, \mu), T(\lambda) \otimes T(\mu)]. \tag{12.57}$$

Here we use the obvious tensor notation to suppress the matrix indices. Usually, the above formula is further condensed to

$$\{T(\lambda) \underset{,}{\otimes} T(\mu)\} = [r(\lambda, \mu), T(\lambda) \otimes T(\mu)]. \tag{12.57'}$$

Let us now consider the special case of groups associated with 2×2-matrices. In this case the classical r-matrix may be expressed through the Jacobi elliptic functions of modulus k,

$$r(\lambda, \mu) = -\frac{1}{2} \sum_{a=1}^{3} w_a(\lambda - \mu) \sigma_a \otimes \sigma_a, \tag{12.58}$$

where σ_a are the standard Pauli matrices and $w_a(\lambda)$ are given by

$$w_1(\lambda) = \rho \frac{1}{\mathrm{sn}(\lambda, k)}, \quad w_2(\lambda) = \rho \frac{\mathrm{dn}(\lambda, k)}{\mathrm{sn}(\lambda, k)}, \quad w_3 = \rho \frac{\mathrm{cn}(\lambda, k)}{\mathrm{sn}(\lambda, k)}, \tag{12.59}$$

and $\rho(k)$ is the normalization constant such that $\mathrm{Res}_0\, w_i(\lambda)\, d\lambda = 1$. The associated period lattice is generated by $4K$, $4iK'$ where K, K' are complete elliptic integrals of moduli k, $k' = \sqrt{1 - k^2}$. It will be convenient to introduce another set of parameters J_1, J_2, J_3 such that

$$\rho = \sqrt{J_3 - J_1}, \quad k = \sqrt{\frac{J_2 - J_1}{J_3 - J_1}}. \tag{12.60}$$

Clearly, J_1, J_2, J_3 are defined uniquely up to an additive constant. In order to get appropriate real forms of the Poisson algebras let us assume that J_i are real and $J_3 > J_2 > J_1$. The functions $w_a(\lambda)$ satisfy the quadratic relations

$$w_a^2(\lambda) - w_b^2(\lambda) = J_b - J_a, \quad a, b = 1, 2, 3,$$

which determine the elliptic curve C.

Let D be the divisor of poles of $w_a(\lambda)$.

Proposition 12.36 (i) *The 4-dimensional affine submanifold in G_D consisting of matrices of the form*

$$T(\lambda) = t_0 I + \frac{1}{i} \sum_{a=1}^{3} w_a(\lambda) t^a \sigma_a, \quad t_0, t_a \in \mathbb{C}, \tag{12.61}$$

is a Poisson submanifold with respect to the Sklyanin bracket (12.57). (ii) *The Poisson brackets of the coordinate functions t_0, t_a are given by*

$$\begin{aligned}\{t_a, t_0\} &= J_{bc}t_bt_c,\\ \{t_a, t_b\} &= -t_0t_c.\end{aligned} \tag{12.62}$$

Here $J_{bc} = J_c - J_b$ and the indices (abc) form a cyclic permutation of (123). (iii) *If the structure constants J_1, J_2, J_3 are real and $J_3 > J_2 > J_1$, the Poisson algebra* (12.62) *admits a real form $t_0, t_a \in \mathbb{R}$.*

The proof of (12.62) is based on the addition formulae for the Jacobi elliptic functions which imply that

$$\begin{aligned}w_a(\lambda)w_b(\mu) &= w_a(\lambda-\mu)w_c(\mu) - w_b(\lambda-\mu)w_c(\lambda),\\ w_a(\lambda-\mu)w_b(\lambda)w_a(\mu) &- w_b(\lambda-\mu)w_a(\lambda)w_b(\mu) = J_{ab}w_c(\lambda).\end{aligned} \tag{12.63}$$

The quadratic Poisson algebra (12.62) is called the *Poisson-Sklyanin algebra* (it is the semiclassical limit of the quantum Sklyanin algebra, cf. Sklyanin [1982, 1983]).

Proposition 12.37. (i) *The Casimir functions of the Poisson bracket* (12.62) *are given by*

$$\mathscr{C}_0 = \sum_a t_a^2, \qquad \mathscr{C}_1 = t_0^2 - \sum_a J_a t_a^2. \tag{12.64}$$

(Note that the shift $J_a \to J_a + \text{const}$ does not affect the Poisson brackets (2.62) and hence the freedom in the choice of J_a (cf. (12.60)) does not lead to ambiguity.) (ii) *Assume that J_a satisfy $J_3 > J_2 > J_1 > 0$. The equations $\mathscr{C}_0 = c_0$, $\mathscr{C}_1 = c_1$ define a symplectic submanifolds $\mathscr{M}(J_a, c_0, c_1)$ in $\mathbb{R}^4$. For $c_1 > -J_1c_0$ or for $-J_3c_0 < c_1 < -J_2c_0$, $\mathscr{M}$ is homeomorphic to the disjoint union of two copies of S^2; for $-J_2c_0 < c_1 < -J_1c_0$, $\mathscr{M}$ is homeomorphic to the torus $\mathbb{T}^2$; for $c_1 < -J_3c_0$, the (real) level surface $\mathscr{M}(J_a, c_0, c_1)$ is empty.*

Let us consider the associated lattice system. For "physical" reasons we assume below that J_a are real, $J_3 > J_2 > J_1$, and fix the values c_1, c_0 of Casimir functions so that $c_1 > -J_1c_0$. The corresponding level surface consists of two copies of S^2. The "physical" component $\mathscr{M}_0$ is singled out by the condition $t_0 > 0$. The vector $t \in \mathscr{M}_0$ may be interpreted as a spin vector attached to a point on a lattice. The phase space of the corresponding lattice system is the product $\mathscr{M}_0 \times \cdots \times \mathscr{M}_0 \subset G_D \times \cdots \times G_D$ (N copies). It is natural to assume that the constants c_1, c_0 do not depend on the site of the lattice (the homogeneity condition).

Let $\varphi^{(n)}(\lambda) = \varphi_{ij}^{(n)}(\lambda)$ be the generators of the affine ring $A(G_D^N)$ defined by

$$\varphi_{ij}^{(n)}(\lambda)(L) = L_{ij}^{(n)}(\lambda), \quad L = (L^{(1)}, \ldots, L^{(N)}) \in G_D^N.$$

Proposition 12.38. (i) *The Poisson brackets of the generators $\varphi^{(n)}(\lambda)$ with respect to the product structure on G_D^N are given by*

$$\{\varphi^{(n)}(\lambda) \underset{,}{\otimes} \varphi^{(m)}(\mu)\}(L) = [r(\lambda - \mu), L^{(n)}(\lambda) \otimes L^{(m)}(\mu)]\delta_{mn}. \tag{12.65}$$

As before, we condense this formula to

$$\{L^{(n)}(\lambda) \underset{,}{\otimes} L^{(m)}(\mu)\} = [r(\lambda - \mu), L^{(n)}(\lambda) \otimes L^{(m)}(\mu)]\delta_{mn}. \tag{12.65'}$$

Consider the affine submanifold $\mathcal{M}_0 \times \cdots \times \mathcal{M}_0 \subset G_D^N$. Let

$$L^{(n)}(\lambda) = t_0^{(n)}I + \sum_a w_a(\lambda)t_a^{(n)}\sigma_a \tag{12.66}$$

be the corresponding "local" Lax matrices. The Poisson brackets of the coordinate functions $t_0^{(n)}$, $t_a^{(n)}$ are given by

$$\begin{aligned} \{t_0^{(n)}, t_a^{(m)}\} &= \delta_{mn}J_{bc}t_b^{(n)}t_c^{(n)}, \\ \{t_a^{(n)}, t_b^{(m)}\} &= -\delta_{mn}t_0^{(n)}t_0^{(n)}. \end{aligned} \tag{12.67}$$

Proposition 12.39. *Put*

$$T_N(\lambda) = \overset{\curvearrowleft}{\prod_n} L_n(\lambda), \quad H_N(\lambda) = \operatorname{tr} T_N(\lambda). \tag{12.68}$$

The functions $H_N(\lambda)$ are in involution on $\mathcal{M}_0^N$.

Physically meaningful Hamiltonians for lattice system describe interactions of the nearest neighbours. To single out such Hamiltonains from the family (12.68) observe that if $\det L_n(\lambda_0) = 0$ for some $\lambda_0 \in C$, we have

$$L^{(n)}(\lambda_0) = \alpha_n \otimes \beta_n^t, \tag{12.69}$$

where $\alpha_n, \beta_n \in \mathbb{C}^2$, and hence

$$H_N(\lambda_0) = \prod_{n=1}^N (\beta_{n+1}, \alpha_n).$$

In other words,

$$h_N(\lambda_0) = \log H_N(\lambda_0)$$

describes interaction of two nearest neighbours. In order to get a real Hamiltonian observe that if the constants J_a are positive, $J_3 > J_2 > J_1$, local Lax matrices obey the following symmetries

$$\overline{L_n(\lambda)} = \sigma_2 L_n(\bar{\lambda})\sigma_2, \tag{12.70}$$

$$L_n(-\lambda) = \sigma_2 L_n^t(\lambda)\sigma_2. \tag{12.71}$$

Hence $H_N(\bar{\lambda}) = \overline{H_N(\lambda)}$ and

$$H = \log \frac{1}{2}|\operatorname{tr} T_N(\lambda_0)|^2$$

belongs to the family (12.68). The equation $\det L_n(\lambda_0) = 0$ reduces to

$$c_1 + c_0(w_1^2(\lambda_0) + J_1) = 0. \tag{12.72}$$

Since we assumed that $c_1 > -J_1 c_0$ this means that

$$\bar{\lambda}_0 \equiv -\lambda_0 \,(\mathrm{mod}\, \Gamma)$$

and hence

$$H_N(\bar{\lambda}_0) = H_N(-\lambda_0) = \prod_{n=1}^{N} (\alpha_{n+1}, \beta_n). \tag{12.73}$$

This gives, after some calculations,

$$H = \sum_n \log h(t_a^{(n)}, t_a^{(n+1)}), \tag{12.74}$$

where

$$h(t_a^{(n)}, t_a^{(n+1)}) = t_0^{(n)} t_0^{(n+1)} + \sum_a \left(\frac{c_1}{c_0} + J_a \right) t_a^{(n)} t_a^{(n+1)}. \tag{12.75}$$

The Hamiltonian (12.74)–(12.75) describes the so called classical lattice Landau-Lifshitz model (Sklyanin [1982]). In the continuum limit when the lattice spacing $\Delta \to 0$ in such a way that $\Delta \cdot n = x$ and

$$t_0^{(n)} = 1 + 0(\Delta^2), \quad t_a^{(n)} = \Delta \cdot S_a(x) + O(\Delta^3), \tag{12.76}$$

the Poisson algebra (12.67) is replaced by

$$\{S_a(x), S_b(y)\} = \varepsilon_{abc} S_c(x) \delta(x - y) \tag{12.77}$$

and the Hamiltonian (12.74–75) has the following asymptotic behaviour:

$$-2H + 2N \log 2 = \frac{\Delta}{2} \int \left[\left(\frac{\partial S}{\partial x} \right)^2 + 4(JS, S) \right] dx + o(\Delta^2).$$

The Hamiltonian

$$\int \left[\left(\frac{\partial S}{\partial x} \right)^2 + 4(JS, S) \right] dx \tag{12.78}$$

with respect to the Poisson algebra (12.77) defines the continuous Landau-Lifshitz model (Sklyanin [1979], Borovik and Robuk [1981]). The Lax matrix

$$L(\lambda, x) = \sum_{a=1}^{3} w_a(\lambda) S_a(x) \sigma_a \tag{12.79}$$

enters into the zero curvature representation for the Landau-Lifshitz equation. (This zero curvature representation is the continuous counterpart of the difference Lax equations.)

The Poisson algebra (12.67) describes the most general anisotropic lattice system. Its degenerate cases correspond to systems with partial or complete isotropy. Letting $k \to 0$ in (12.59) we get $J_1 = J_2$. The corresponding Lax matrix is given by

$$L^{(n)}(\lambda) = t_0^{(n)} I + \frac{\rho}{i \sin \lambda} (t_1^{(n)} \sigma_1 + t_2^{(n)} \sigma_2 + \cos \lambda \cdot t_3^{(n)} \sigma_3). \tag{12.80}$$

In this case the variables $t_0^{(n)}$, $t_a^{(n)}$ may be expressed through ordinary spin variables $S_a^{(n)}$ satisfying

$$\{S_a^{(n)}, S_b^{(m)}\} = \delta_{mn}\varepsilon_{abc}S_c^{(n)}. \tag{12.81}$$

Namely, put

$$t_0 = \cosh \rho S_3, \quad t_3 = \frac{1}{\rho}\sinh \rho S_3,$$

$$t_1 = F(S_3)S_1, \quad t_2 = \frac{1}{\rho}F(S_3)S_2,$$

where

$$F(x) = \sqrt{\frac{\sinh^2 \rho R - \sinh^2 \rho x}{R^2 - x^2}},$$

$$R^2 = S_1^2 + S_2^2 + S_3^2.$$

It is easy to check that t_0, t_a satisfy the relations (12.62) with $k = 0$ (or, equivalently, with $J_1 = J_2$) and the values of the Casimir functions $\mathscr{C}_0, \mathscr{C}_1$ are given by

$$c_0 = \frac{\sinh^2 \rho R}{\rho^2}, \quad c_1 = 1 - J_1 c_0.$$

The r-matrix in the limit $k \to 0$ is given by

$$r(\lambda) = -\frac{\rho}{2\sin\lambda}(\sigma_1 \otimes \sigma_1 + \sigma_2 \otimes \sigma_2 + \cos\lambda\sigma_3 \otimes \sigma_3). \tag{12.82}$$

The corresponding lattice model is the partially anisotropic classical lattice Heisenberg model.

Putting $\lambda, \rho \to 0$ in such a way that $v = \frac{2\rho}{\lambda}$ is fixed we get $J_1 = J_2 = J_3$. In this limit the Poisson algebra (12.62) coincides with (12.81) and the r-matrix (12.82) reduces to

$$r(v) = -\frac{1}{v}(\sigma_1 \otimes \sigma_1 + \sigma_2 \otimes \sigma_2 + \sigma_3 \otimes \sigma_3). \tag{12.83}$$

Clearly, (12.83) is the kernel of the standard R-matrix on the loop algebra $\mathfrak{sl}(2; \mathbb{C}[\lambda, \lambda^{-1}])$. The lattice model we get in this case is the isotropic classical lattice Heisenberg ferromagnet. In the continuum limit we get the continuous Heisenberg ferromagnet with the Hamiltonian

$$H = \int \left(\frac{\partial S}{\partial x}\right)^2 dx$$

and the Lax matrix

$$L(x, v) = \frac{1}{v}\sum_{a=1}^{3} S_a(x)\sigma^a$$

Again, difference Lax representation for the lattice system gives rise to the zero curvature representation for the corresponding continuous system.

12.6. Additional Comments.

1. *Classical R-matrices and hierarchies of compatible Poisson brackets.* We have already seen in Section 4.2 that Lax equations are frequently Hamiltonian with respect to a linear family of compatible Poisson brackets. The linear families constructed in Section 4.2 were associated with intertwining operators for Lie algebras. Below we discuss another source of compatible Poisson brackets for Lax systems. We start with the definition of the so called Gel'fand-Dikij bracket. (The reader will see that this definition is quite close to that of Sklyanin bracket discussed in Section 12.1 above.)

Let $\mathscr{A}$ be an associative algebra. Assume that there is a nondegenerate bilinear form on $\mathscr{A}$ such that

$$(XY, Z) = (X, YZ) \quad \text{for any} \quad X, Y, Z \in \mathscr{A}.$$

Let $\mathfrak{g}_{\mathscr{A}}$ be the Lie algebra associated with $\mathscr{A}$, i.e. the space $\mathscr{A}$ equipped with the Lie bracket

$$[X, Y] = XY - YX.$$

Let $R \in \operatorname{End} \mathscr{A}$ be a skew symmetric linear operator satisfying the modified Yang-Baxter identity

$$[RX, RY] - R([RX, Y] + [X, RY]) = -[X, Y].$$

Define the gradient of a smooth function on $\mathscr{A}$ by

$$(\operatorname{grad} \varphi(L), \xi) = \left(\frac{d}{dt}\right)_{t=0} \varphi(L + t\xi), \quad L, \xi \in \mathscr{A}. \tag{12.84}$$

The (generalized) Gelfand-Dikij bracket on $C^\infty(\mathscr{A})$ is defined by

$$2\{\varphi, \psi\}_2(L) = (R(XL), YL) - (R(LX), LY), \quad X = \operatorname{grad} \varphi(L), \quad Y = \operatorname{grad} \psi(L). \tag{12.85}$$

Proposition 12.40. (i) *The bracket* (12.85) *is skew-symmetric and satisfies the Jacobi identity.*

(ii) *This bracket is compatible with the ordinary R-bracket on* $\mathscr{A}$

$$2\{\varphi, \psi\}_1(L) = ([R(X), Y] + [X, RY], L). \tag{12.86}$$

A comparison of the definition of left and right gradients with (12.84) shows that (12.85) is just a version of the Sklyanin bracket. The point is that when we deal with an associative algebra rather than with an arbitrary Lie group, the brackets $\{\ ,\ \}_1$ and $\{\ ,\ \}_2$ are defined on the same linear space and hence their linear combination makes sense.

Recently it was observed (see Li and Parmentier [1989]) that there exists another bracket compatible with $\{\ ,\ \}$ and $\{\ ,\ \}_2$ which is cubic in L. Namely, put

$$\{\varphi, \psi\}_3(L) = (R(LYL), (L, X)) - (R(LXL), (L, Y)) \tag{12.87}$$

Proposition 12.41. (i) *The bracket* (12.87) *is skew-symmetric and satisfies the Jacobi identity.* (ii) *It is compatible with* $\{\ ,\ \}_1$ *and* $\{\ ,\ \}_2$.

A direct proof of the Jacobi identity is, of course rather tedious. One may note, however, that (12.87) is obtained from (12.86) by the change of variables $L \mapsto L^{-1}$.

2. The brackets $\{\ ,\ \}_1$ and $\{\ ,\ \}_3$ make sense without the assumption that $R = -R^*$. By contrast, the bracket $\{\ ,\ \}_2$ is skew only when the R-matrix is skew. A milder assumption on R was recently proposed by Li and Parmentier [1989]. Put $2A = R - R^*$, $2S = R + R^*$ and suppose that both R and A satisfy the modified Yang-Baxter equation. Then the bracket

$$2\{\varphi, \varphi\}_2(L) = (A(LX), LY) - (A(XL), YL) + (S(XL), LY) - (S(LX), YL),$$
$$X = \text{grad}\,\varphi(L), \quad Y = \text{grad}\,\psi(L), \tag{12.88}$$

is skew, satisfies the Jacobi identity and is compatible with the ordinary R-bracket (12.86). A version of this definition is suited for the study of lattice systems. Let $\mathfrak{g}$ be a Lie algebra with an invariant inner product. Assume that $R \in \text{End}\,\mathfrak{g}$ is such that both R and $A = 1/2(R - R^*)$ satisfy *mCYBE*. Let $\tau \in \text{Aut}\,\mathfrak{g}$ be an orthogonal operator which commutes with R. Define the bracket on the Lie group G with Lie algebra $\mathfrak{g}$ by

$$\begin{gathered} 2\{\varphi, \psi\}_\tau = (AX', Y') - (AX, Y) + (S\tau X, Y') - (S\tau^{-1}X', Y), \\ X = \nabla\varphi, \quad Y = \nabla\psi, \quad X' = \nabla'_\varphi, \quad Y' = \nabla'_\psi. \end{gathered} \tag{12.89}$$

Theorem 12.42. (i) *The bracket* (12.89) *is skew-symmetric and satisfies the Jacobi identity.* (ii) *Invariants of twisted conjugations on G are in involution with respect to this bracket and generate generalized Lax equations on G.* (iii) *Solution of these equations is reduced to the factorization problem in G associated with R.* (*The exact statement is completely similar to that in Theorem* 12.3.)

Examples of R-matrices satisfying the hypotheses of the theorem may easily be constructed.

This theorem may be applied to lattice systems in a way similar to the one discussed in Section 12.5 (cf. Theorems 12.3–12.5) Note that the bracket (12.89) for lattice systems is non-ultralocal, i.e. the dynamical variables associated with different sites of the lattice do not commute (formula (12.89) implies that there is an interaction of the nearest neighbours).

12.7. Bibliographical Notes. The study of Poisson structures described in this section was started by Sklyanin [1979] who considered the semi-classical limit of the quantum commutation relations introduced earlier by Baxter (see Baxter [1982]). The general theory of classical R-matrices was developed by Belavin and Drinfel'd [1982] and Semenov-Tian-Shansky [1983]. The definition of

Poisson-Lie groups and Lie bialgebras is due to Drinfel'd [1983]. He also developed the duality theory and described the symplectic leaves in Poisson Lie groups. The notion of dressing transformations was introduced (in the context of the theory of Poisson Lie groups) by Semenov-Tian-Shansky [1985]. Further results on the geometry of Poisson Lie groups may be found in the papers of Weinstein [1988], Lu and Weinstein [1988], Kossmann – Schwarzbach and Magri [1988]. The theory of Lax equations described in Section 12.5 is due to Semenov-Tian-Shansky [1985]. The tensor formalism exposed in Section 12.6 goes back to Sklyanin [1979]. The Poisson algebra (12.62) is introduced by Sklyanin [1982]. The description of local Hamiltonians is based on the papers of Izergin and Korepin [1981, 1982]. The lattice models such as (12.74–75) were introduced as useful classical analogues of quantum lattice models which are one of the main topics of quantum statistical physics (cf. Baxter [1982]). Continuous models and zero curvature representations for them were actually studied prior to their lattice counterparts. The Lie algebraic approach to these models is simpler than in the lattice case (cf. Reyman and Semenov-Tian-Shansky [1980], Kulish and Reyman [1983], Faddeev and Takhtajan [1986]). However, in the present survey we only deal with finite-dimensional systems and so do not consider this theory.

Hierarchy of Poisson brackets $\{\ ,\ \}_1, \{\ ,\ \}_2$ was discussed by Semenov-Tian-Shansky [1983]. The definition given there is an abstract version of the one proposed by Gel'fand and Dikij [1978]. The bracket $\{\ \}_3$ was discovered by Li and Parmentier [1989] and Oevel. The definition of Poisson structures on groups associated with non-skew-symmetric R-matrices and Theorem 12.6 are due to Li and Parmentier [1989].

References*

Abraham, R., Marsden, J.E. [1978]: Foundations of Mechanics. (2nd ed.), Benjamin: New York, Zbl.393.70001

Adams, M.R., Harnad, J., Previato, E. [1988]: Isospectral Hamiltonian flows in finite and infinite dimensions. I. Commun. Math. Phys. *117*, 451–500, Zbl.659.58021

Adler, M. [1979]: On a trace functional for formal pseudodifferential operators and the symplectic structure for Korteweg-de Vries type equations. Invent. Math. *50*, 219–248, Zbl.393.35058

Adler, M., van Moerbeke, P. [1980a]: Completely integrable systems, Euclidean Lie algebras and curves. Adv. Math. *38*, 267–317, Zbl.455.58017

Adler, M., van Moerbeke, P. [1980b]: Linearizations of Hamiltonian systems, Jacobi varieties, and representation theory. Adv. Math. *38*, 318–379, Zbl.455.58010

Adler, M., van Moerbeke, P. [1982]: Kowalewski's asymptotic method, Kac-Moody Lie algebras, and regularization. Commun. Math. Phys. *83*, 83–106, Zbl.491.58017

Adler, M., van Moerbeke, P. [1984]: Geodesic flow on SO(4) and intersection of quadrics. Proc. Natl. Acad. Sci. USA *81*, 4613–4616, Zbl.545.58027

* For the convenience of the reader, references to reviews in Zentralblatt für Mathematik (Zbl.), compiled using the MATH database, have, as far as possible, been included in this bibliography.

Adler, M., van Moerbeke, P. [1988]: The Kowalewski and Henon-Heiles motions as Manakov geodesic flows on SO(4): a two-dimensional family of Lax pairs. Commun. Math. Phys. *113*, 659–700, Zbl.647.58022

Arnol'd, V.I. [1966]: Sur la géometrie différentielle des groupes de Lie de dimension infinie et ses applications à l'hydrodynamique des fluides parfaites. Ann. Inst. Fourier *16*, No. 1, 319–361, Zbl.148,453

Arnol'd, V.I. [1974]: Mathematical Methods of Classical Mechanics. Nauka: Moscow. English transl.: Graduate Texts in Math. *60*, Springer-Verlag: New York-Heidelberg-Berlin, 1978, 2nd ed. 1989, Zbl.386.70001

Baxter, R.J. [1982]: Exactly Solved Models in Statistical Mechanics. Academic Press: New York, Zbl.538.60093

Belavin, A. A. [1980]: Discrete groups and integrability of quantum systems. Funkts. Anal. Prilozh. *14*, No. 4, 18–26. English transl.: Funct. Anal. Appl. *14*, 260–267 (1980), Zbl.454.22012

Belavin, A.A., Drinfel'd V.G. [1982]: Solutions of the classical Yang-Baxter equation for simple Lie algebras. Funkts. Anal. Prilozh. *16*, No. 3, 1–29. English transl.: Funct. Anal. Appl. *16*, 159–180 (1982), Zbl.504.22016

Belavin, A.A., Drinfel'd, V.G. [1984]: Triangle equations and simple Lie algebras. Sov. Sci. Rev., Sect. C., Math. Phys. Rev. *4*, 93–165, Zbl.553.58040

Berezin, F.A. [1967]: Some remarks on the associative envelope of a Lie algebra. Funkts. Anal. Prilozh. *1*, No. 2, 1–14. English transl.: Funct. Anal. Appl. *1*, 91–102 (1968), Zbl.227.22020

Bobenko, A.I. [1986]: Euler equations in the algebras SO(4) and $e(3)$. Isomorphisms of integrable cases. Funkts. Anal. Prilozh. *20*, No. 1, 64–66. English transl.: Funct. Anal. Appl. *20*, 53–56 (1986), Zbl.622.58010

Bobenko, A.I., Reyman, A.G., Semenov-Tian-Shansky, M.A. [1989]: The Kowalewski top 99 years later: a Lax pair, generalizations and explicit solutions. Commun. Math. Phys. *122*, 321–354

Bogoyavlensky, O.I. (= Bogoyavlenskij, O.I.) [1976]: On perturbations of the periodic Toda lattice. Commun. Math. Phys. *51*, 201–209

Bogoyavlensky, O.I. [1984]: Integrable equations on Lie algebras arising in problems of mathematical physics. Izv. Akad. Nauk SSSR, Ser. Mat. *48*, No. 5, 883–938. English transl.: Math. USSR, Izv. *25*, 207–257 (1985), Zbl.583.58012

Borovik, A.E., Robuk, V.N. [1981]: Linear pseudopotentials and conservation laws for the Landau-Lifshitz equation describing the nonlinear dynamics of a ferromagnetic with axial anisothropy. Teor. Mat. Fiz. *46*, No. 3, 371–381. English transl.: Theor. Mat. Phys. *46*, 242–248 (1981)

Bourbaki, N. [1968]: Groupes et algèbres de Lie, Ch. IV–VI. Hermann: Paris, Zbl.186,330

Cherednik, I.V. [1983a]: Integrable differential equations and coverings of elliptic curves. Izv. Akad. Nauk SSSR, Ser. Mat. *47*, No. 2, 384–406. English transl.: Math. USSR, Izv. *22*, 357–377 (1984, Zbl.547.35109

Cherednik, I.V. [1983b]: Definition of τ-functions for generalized affine Lie algebras. Funkts. Anal. Prilozh. *17*, No. 3, 93–95. English transl.: Funct. Anal. Appl. *17*, 243–245 (1983), Zbl.528.17004

Deift, P.A., Li, L.C., Nanda, T., Tomei, C. [1986]: The Toda flow on a generic orbit is integrable. Commun. Pure Appl. Math. *39*, 183–232, Zbl.606.58020

Deift, P.A., Li, L.C. [1989]: Generalized affine Lie algebras and the solution of a class of flows associated with the QR eigenvalue algorithm. Commun. Pure Appl. Math. *42*, 963–991

Drinfel'd, V.G. [1977]: On commutative subrings of certain noncommutative rings. Funkts. Anal. Prilozh. *11*, No. 1, 11–14. English transl.: Funct. Anal. Appl. *11*, 9–12 (1977), Zbl.359.14011

Drinfel'd V.G. [1983]: Hamiltonian structures on Lie groups, Lie Bialgebras and the geometrical meaning of classical Yang-Baxter equations. Dokl. Akad. Nauk SSSR *268*, 285–287. English transl.: Sov. Math. Dokl. *27*, 68–71 (1983), Zbl.526.58017

Drinfel'd, V.G. [1987]: Quantum groups. In: Proc. Int. Congr. Math., Berkeley 1986, Vol. 1, 798–820, Zbl.667.16003

Dubrovin, B.A. [1981]: Theta functions and nonlinear equations. Usp. Mat. Nauk *36*, No. 2, 11–80. English transl.: Russ. Math. Surv. *36*, No. 2, 11–92 (1981), Zbl.478.58038

Dubrovin, B.A., Novikov, S.P., Matveev, V.B. [1976]: Nonlinear equations of the Korteweg-de Vries type, finite-band linear operators, and abelian varieties. Usp. Math. Nauk *31*, No. 1, 55–136. English transl.: Russ. Math. Surv. *31*, No. 1, 59–146 (1976), Zbl.326.35011

Faddeev, L.D. [1980]: Quantum completely integrable models in field theory. Sov. Sci. Rev. Sect. C., Math. Phys. Rev. *1*, 107–155, Zbl.569.35064

Faddeev, L.D. [1984]: Integrable models in 1 + 1-dimensional quantum field theory. In: Recent Advances in Field Theory and Statistical Mechanics. J.-B. Zuber and R. Stora (eds.) Les Houches, Elsevier Science Publishers: Amsterdam, 563–608

Faddeev, L.D., Takhtajan, L.A. [1986]: Hamiltonian methods in the theory of solitons. Nauka: Moscow, English transl.: Springer-Verlag: Berlin-Heidelberg-New York (1987), Zbl.632.58003

Flaschka, H. [1974a]: On the Toda lattice. I. Phys. Rev. *B9*, 1924–1925

Flaschka, H. [1974b]: On the Toda lattice. II. Progr. Theor. Phys. *51*, 703–716

Fordy, A.P., Wojciechowski, S., Marshall, I. [1986]: A family of integrable quartic potentials related to symmetric spaces. Phys. Lett. *A 113*, 395–400

Gel'fand, I.M., Dikij, L.A. [1978]: A family of Hamiltonian structures related to integrable nonlinear differential equations. Preprint IPM AN SSSR, No. 136. Institute of Applied Math.: Moscow (Russian)

Goddard, P., Olive, D. (Eds.) [1988]: Kac-Moody and Virasoro algebras. A reprint volume for physicists. Adv. Ser. Math. Phys. *3*. World Scientific: Singapore, Zbl.661.17001

Gohberg, I.Z., Fel'dman, I.A. [1971]: Convolution Equations and Projection Methods for Their Solution. Nauka: Moscow. English transl.: Transl. Math. Monogr. *41*, Providence (1974), Zbl.214,385

Golod, P.I. [1984]: Hamiltonian systems associated with anisotropic affine Lie algebras and higher Landau-Lifshitz equations. Dokl. Akad. Nauk Ukr. SSR, Ser. A, No. 5, 6–8 (Russian), Zbl.542,58010

Goodman, R., Wallach, N.R. [1982]: Classical and quantum mechanical systems of Toda lattice type. I. Commun. Math. Phys. *83*, 355–386, Zbl.503.22013

Goodman, R., Wallach, N.R. [1984]: Classical and quantum mechanical systems of Toda lattice type. II. Commun. Math. Phys. *94*, 177–217, Zbl.592.58028

Guillemin, V., Sternberg, S. [1980]: The moment map and collective motion. Ann. Phys. *127*, 220–253, Zbl.453.58015

Guillemin, V., Sternberg, S. [1984]: Symplectic Techniques in Physics. Cambridge University Press: Cambridge, Zbl.576.58012

Haine, L., Horozov, E. [1987]: A Lax pair for Kowalevski's top. Physica *D29*, 173–180, Zbl.627.58026

Helgason, S. [1978]: Differential Geometry, Lie Groups and Symmetric Spaces. Academic Press: New York, Zbl.451.53038

Izergin, A.G., Korepin, V.E. [1981]: The lattice quantum Sine-Gordon model. Lett. Math. Phys. *5*, 199–205

Izergin, A.G., Korepin, V.E. [1982]: Lattice versions of quantum field theory models in two dimensions. Nucl. Phys. *B 205*, 401–413

Kac, V.G. [1968]: Simple irreducible graded Lie algebras of finite growth. Ivz. Akad. Nauk SSSR, Ser. Mat. *32*, No. 6, 1323–1367. English transl.: Math. USSR, Izv. *2*, 1271–1311 (1968), Zbl.222.17007

Kac, V.G. [1969]: Automorphisms of finite order of semisimple Lie algebras. Funkts. Anal. Prilozh. *3*, No. 3, 94–96. English transl.: Funct. Anal. Appl. *3*, 252–254 (1970), Zbl.274.17002

Kac, V.G. [1984]: Infinite-Dimensional Lie Algebras. Progress in Math. *44*. Birkhäuser: Basel-Boston-Stuttgart, Zbl.537.17001

Karasev, M.V. [1981]: The Maslov quantization conditions in higher cohomology and the analogs of the objects of Lie theory for canonical fibre bundles of symplectic manifolds. MIEM: Moscow. VINITI, No. 1092–82, 1093–82. English transl.: Sel. Math. Sov. *8*, No. 3, 213–234, 235–258 (1989), Zbl.704.58019

Karasev, M.V. [1986]: Analogues of objects of the theory of Lie groups for nonlinear Poisson brackets. Izv. Akad. Nauk SSSR, Ser. Mat. *50*, No. 3, 508–538. English transl.: Math. USSR, Izv. *28*, 497–527 (1987), Zbl.608,58023

Kazhdan, D., Kostant, B., Sternberg, S. [1978]: Hamiltonian group actions and dynamical systems of Calogero type. Commun. Pure Appl. Math. *31*, No. 4, 491–508, Zbl.368.58008

Kirillov, A.A. [1972]: Elements of the Theory of Representations. Nauka: Moscow. English transl.: Grundlehren der Mathematischen Wissenschaften *220* Springer-Verlag: New York Berlin-Heidelberg. 1976. Zbl.264.22011

Kirillov, A.A. [1976]: Local Lie algebras. Usp. Mat. Nauk *31*, No. 4, 55–76. English transl.: Russ. Math. Surv. *31*, No. 4, 55–75 (1976), Zbl.352.58014

Komarov, I.V. [1987]: A generalization of the Kowalewski top. Phys. Lett. *A123*, 14–15

Kosmann-Schwarzbach, Y., Magri, F. [1988]: Poisson-Lie groups and complete integrability. I. Ann. Inst. Henri Poincaré, Phys. Théor. *49*, 433–460, Zbl.667.16005

Kostant, B. [1970]: Quantization and unitary representations, I: Prequantization. In: Lect. Notes Math. *170*, 87–208, Zbl.223.53028

Kostant, B. [1979a]: Quantization and representation theory. In: Representation theory of Lie groups, Proc. SRC/LMS Res. Symp., Oxford 1977, Lond Math. Soc. Lect. Note Ser., *34* (Atiyah, M.F. (Ed.)), 287–316, Zbl.474.58010

Kostant, B. [1979b]: The solution to a generalized Toda lattice and representation theory. Adv. Math. *34*, 195–338, Zbl.433.22008

Kostant, B. [1982]: Poisson commutativity and the generalized periodic Toda lattice. In: Differential Geometric Methods in Physics, Proc. Int. Conf., Clausthal 1980, In: Lect. Notes Math. *905* (Doebner, H.D., Andersson, S.I., Petry, H.R. (Eds.)), 12–28, Zbl.485.58010

Krichever, I.M. [1977a]: Integration of nonlinear equations by methods of algebraic geometry. Funkts. Anal. Prilozh. *11*, No. 1, 15–31. English transl.: Funct. Anal. Appl. *11*, 12–26 (1977), Zbl.346.35028

Krichever, I.M. [1977b]: Methods of algebraic geometry in the theory of nonlinear equations. Usp. Mat. Nauk *32*, No. 6, 183–208. English transl.: Russ. Math. Surv. *32*, No. 6, 185–213 (1977), Zbl.372.35002

Krichever, I.M. [1978]: Algebraic curves and nonlinear difference equations. Usp. Mat. Nauk *33*, No. 4, 215–216. English transl.: Russ. Math. Surv. *33*, No. 4, 255–256 (1978), Zbl.382.39003

Kulish, P.P., Reyman, A.G. [1983]: Hamiltonian structure of polynomial bundles. Zap. Nauchn. Semin. Leningr., Otd. Mat. Inst. Steklova *123*, 67–76. English transl.: J. Sov. Math. *28*, No. 4, 505–512 (1985), Zbl.511.47010

Lebedev, D.R., Manin, Yu.I. [1979]: The Gel'fand-Dikij Hamilton's operator and the coadjoint representation of the Volterra group. Funkts. Anal. Prilozh. *13*, No. 4, 40–46. English transl.: Funct. Anal. Appl. *13*, 268–273 (1980), Zbl.441.58007

Leznov, A.N., Savel'ev, M.V. [1979]: Representation of zero curvature for the system of non-linear partial differential equations $X_{\alpha,z\bar{z}} = \exp(KX)_\alpha$ and its integrability. Lett. Math. Phys. *3*, 489–494, Zbl.415.35017

Leznov, A.N., Savel'ev, M.V. [1985]: Group-theoretical methods for the solution of nonlinear dynamical systems. Nauka: Moscow (Russian), Zbl.667.58020

Li, L.-C., Parmentier, S. [1989]: Nonlinear Poisson structures and *r*-matrices. Commun. Math. Phys. *125*, 545–563, Zbl.695.58011

Lie, S. (unter Mitwirkung von F. Engel) [1893]: Theorie der Transformationsgruppen, Abschn. III. Teubner: Leipzig

Lu, J.-H. [1989]: Momentum mapping and reduction of Poisson actions. Preprint. Univ. of California: Berkeley

Lu, J.-H., Weinstein, A. [1988]: Poisson Lie groups, dressing transformations and Bruhat decompositions. Preprint PAM-414. Univ. of California: Berkeley. Appeared in: J. Differ. Geom. *31*, No. 2, 501–526 (1990), Zbl.673.58018

Mackey, G.W. [1958]: Unitary representations of group extensions. I. Acta Math. *99*, 265–311, Zbl.82,113

Magri, F. [1978]: A simple model of the integrable Hamiltonian equation. J. Math. Phys. *19*, No. 5, 1156–1162, Zbl.383.35065

Manakov, S.V. [1974]: Complete integrability and stochastization of discrete dynamical systems. Zh. Exp. Teor. Fiz. *40*, 269–274. English transl.: Sov. Phys. ZETP *40*, 269–274 (1975)

Manakov, S.V. [1976]: Note on the integration of Euler's equations of the dynamics of an n-dimensional rigid body. Funkts. Anal. Prilozh. *10*, No. 4, 93–94. English transl.: Funct. Anal. Appl. *10*, 328–329 (1977), Zbl.343.70003

Marsden, J., Ratiu, T., Weinstein, A. [1984]: Semidirect products and reduction in mechanics. Trans. Am. Math. Soc. *281*, No. 1, 147–177, Zbl.529.58011

Marsden, J., Weinstein, A. [1974]: Reduction of symplectic manifolds with symmetry. Rep. Math. Phys. *5*, 121–130, Zbl.327.58005

Mikhailov, A.V. [1981]: The reduction problem in the inverse scattering method. Physica *D 3*, 73–117

Mishchenko, A.S., Fomenko, A.T. [1978]: Euler equations on finite-dimensional Lie groups. Izv. Akad. Nauk SSSR, Ser. Mat. *42*, 396–415. English transl.: Math. USSR., Izv. *12*, 371–389. (1978), Zbl.383.58006

Moerbeke, P. van, Mumford, D. [1979]: The spectrum of difference operators and algebraic curves. Acta Math. *143*, 93–154, Zbl.502.58032

Moody, R.V. [1968]: A new class of Lie algebras. J. Algebra *10*, 221–230, Zbl.191,30

Moser, J. [1975]: Three integrable Hamiltonian systems, connected with isospectral deformations. Adv. Math. *16*, 197–220, Zbl.303.34019

Moser, J. [1980a]: Geometry of quadrics and spectral theory. In: Differential Geometry. Proc. Int. Chem. Symp., Berkeley 1979, 147–188, Zbl.455.58018

Moser, J. [1980b]: Various aspects of integrable Hamiltonian systems. In: Dynamical Systems, CIME Lect., Bressanone 1978, Prog. Math. *8*, 233–290, Zbl.468.58011

Mumford, D. [1978]: An algebro-geometric construction of commuting operators and of solutions to the Toda lattice equations, Korteweg-de Vries equation and related non-linear equations. In: Proc. Int. Symp. on Algebraic Geometry, Kyoto 1977, 115–153, Zbl.423.14007

Mumford, D. [1984]: Tata Lectures on Theta. II. Prog. Math. *43*, Zbl.549.14014

Novikov, S.P., Shmeltser, I. [1981]: Periodic solutions of the Kirchhoff equations for the free motion of a rigid body in an ideal fluid and the generalized Lyusternik-Shnirel'man-Morse theory. Funkts. Anal. Prilozh. *15*, No. 3, 54–66. English translation: Funct. Anal. Appl., *15*, No. 3, 197–207 (1982), Zbl.571,58009

Ol'shanetsky, M.A. (= Ol'shanetskij, M.A.), Perelomov, A.M. [1979]: Explicit solutions of the classical generalized Toda models. Invent. Math. *54*, 261–269, Zbl.419,58008

Ol'shanetsky, M.A., Perelomov, A.M. [1980]: The Toda lattice as a reduced system. Teor. Mat. Fiz. *45*, No. 1, 3–18. English transl.: Theor. Math. Phys. *45*, 843–854 (1981)

Perelomov, A.M. [1981a]: Several remarks on the integrability of the equations of motion of a rigid body in an ideal fluid. Funkts. Anal. Prilozh. *15*, No. 2, 83–85. English transl.: Funct. Anal. Appl. *15*, 144–146 (1981), Zbl.495.70016

Perelomov, A.M. [1981b]: Lax representation for systems of S. Kowalevskaya type. Commun. Math. Phys. *81*, 239–241, Zbl.478.70005

Perelomov, A.M., Ragnisco, P., Wojciechowski, S. [1986]: Integrability of two interacting n-dimensional rigid bodies. Commun. Math. Phys. *102*, 573–583, Zbl.596.58019

Pressley, A., and Segal, G. [1986]: Loop Groups. Clarendon Press: Oxford, Zbl.618.22011

Ratiu, T. [1982]: Euler-Poisson equations on Lie algebras and the n-dimensional heavy rigid body. Am. J. Math. *104*, 409–448, Zbl.509.58026

Rawnsley, J.H. [1975]: Representations of a semi-direct product by quantization. Math. Proc. Camb. Philos. Soc. *78*, No. 2, 345–350, Zbl.313.22014

Reyman, A.G. [1980]: Integrable Hamiltonian systems connected with graded Lie algebras. Zap. Nauchn. Semin. Leningr. Otd. Mat. Inst. Steklova *95*, 3–54. English transl.: J. Sov. Math. *19*, 1507–1545 (1982), Zbl.488.70013

Reyman, A.G. [1986]: Orbit interpretation of Hamiltonian systems of the type of an unharmonic oscillator. Zap. Nauchn. Semin. Leningr. Otd. Mat. Inst. Steklova *155*, 187–189. English transl.: J. Sov. Math. *41*, 999–1001 (1988), Zbl.619.58022

Reyman, A.G. [1988]: New results on the Kowalewski top. In: Nonlinear Evolutions, Proc. of the IV Workshop on Nonlinear Evolution Equations and Dynamical Systems, Balaruc-Les-Bains 1987. World Scientific: Singapore

Reyman, A.G., Semenov-Tian-Shansky, M.A. [1979]: Reduction of Hamiltonian systems, affine Lie algebras and Lax equations. I. Invent. Math. *54*, 81–100, Zbl.403.58004

Reyman, A.G., Semenov-Tian-Shansky, M.A. [1980]: Current algebras and nonlinear partial differential equations. Dokl. Akad. Nauk SSSR *251*, No. 6, 1310–1314. English transl.: Sov. Math., Dokl. *21*, 630–634 (1980), Zbl.501.58018

Reyman, A.G., Semenov-Tian-Shansky, M.A. [1981]: Reduction of Hamiltonian systems, Affine Lie algebras and Lax equations II. Invent. Math. *63*, 423–432, Zbl.442.58016

Reyman, A.G., Semenov-Tian-Shansky, M.A. [1986a]: Lie algebras and Lax equations with spectral parameter on an elliptic curve. Zap. Nauchn. Semin. Leningr. Otd. Mat. Inst. Steklova, *150*, 104–118. English transl.: J. Sov. Math. *46*, No. 1, 1631–1640 (1989), Zbl.603.35083

Reyman, A.G., Semenov-Tian-Shansky, M.A. [1986b]: A new integrable case of the motion of the four-dimensional rigid body. Commun. Math. Phys. *105*, 461–472, Zbl.606.58029

Reyman, A.G., Semenov-Tian-Shansky, M.A. [1987]: Lax representation with a spectral parameter for the Kowalewski top and its generalizations. Lett. Math. Phys. *14*, 55–61, Zbl.627.58027

Reyman, A.G., Semenov-Tian-Shansky, M.A. [1988]: Compatible Poisson structures for Lax equations: an *r*-matrix approach. Phys. Lett. *A 130*, 456–460

Reyman, A.G., Semenov-Tian-Shansky, M.A., Frenkel, I.B. [1979]: Graded Lie algebras and completely integrable systems. Dokl. Akad. Nauk SSSR *247*, 802–805. English transl.: Sov. Math., Dokl. *20*, 811–814 (1979), Zbl.437,58008

Semenov-Tian-Shansky, M.A. [1982]: Group-theoretical aspects of completely integrable systems. In: Twistor Geometry and Nonlinear Systems, 4th. Bulg. Summer Sch., Primorsko 1980, Lect. Notes Math. *970* (Doebner, H.D., Palev, T.D. (Eds.)), 173–185, Zbl.507.58028

Semenov-Tian-Shansky, M.A. [1983]: What is a classical *r*-matrix. Funkts. Anal. Prilozh. *17*, No. 4, 17–33. English transl.: Funct. Anal. Appl. *17*, 259–272 (1983), Zbl.535.58031

Semenov-Tian-Shansky, M.A. [1985]: Dressing transformations and Poisson group actions. Publ. Res. Inst. Math. Sci *21*, No. 6, 1237–1260, Zbl.673.58019

Semenov-Tian-Shansky, M.A. [1987]: Classical *r*-matrices, Lax equations, Poisson Lie groups and dressing transformations. In: Field Theory, Quantum Gravity and Strings. II. Proc. Semin., Mendon and Paris 1985/86, Lect. Notes Phys. *280*, (de Vega, H.J., Sanchez, N. (Eds.)), 174–214, Zbl.666.22010

Sklyanin, E.K. [1979]: On complete integrability of the Landau-Lifschitz equation. Preprint Leningr. Otd. Mat. Inst. E-3-79: Leningrad, Zbl.449.35089

Sklyanin, E.K. [1982]: Algebraic structures connected with the Yang-Baxter equation. Funkts. Anal Prilozh. *16*, No. 4, 27–34. English transl.: Funct. Anal. Appl. *16*, 263–270 (1982), Zbl.513.58028

Sklyanin, E.K. [1983]: Algebraic structures connected with the Yang-Baxter equation. II. Representations of the quantum algebra. Funkts. Anal. Prilozh. *17*, No. 4, 34–48. English transl.: Funct. Anal. Appl. *17*, 273–284 (1983), Zbl.536.58007

Sklyanin, E.K. [1987]: Boundary conditions for integrable equations. Funkts. Anal. Prilozh. *21*, No. 2, 86–87. English transl.: Funct. Anal. Appl. *21*, 164–166 (1987)., Zbl.643.35093

Souriau, J.-M. [1970]: Structure des systèmes dynamiques. Dunod: Paris, Zbl.186,580

Symes, W. [1980a]: Systems of Toda type, inverse spectral problems, and representation theory. Invent. Math. *59*, 13–51, Zbl.474.58009

Symes, W. [1980b]: Hamiltonian group actions and integrable systems. Physica *D1*, 339–374

Veselov, A.P. [1984]: Cnoidal solutions of the Landau-Lifshits equations. Dokl. Akad. Nauk SSSR *276*, No. 3, 590–593 (Russian)

Weinstein, A. [1978]: A universal phase space for particles in Yang-Mills fields. Lett. Math. Phys. *2*, 417–420, Zbl.388.58010

Weinstein, A. [1983]: Local structure of Poisson manifolds. J. Differ. Geom. *18*, No. 3, 523–557, Zbl.524.58011

Weinstein, A. [1988]: Some remarks on dressing transformations. J. Fac. Sci., Univ. Tokyo, Sect. 1A, *35*, No. 1, 163–167, Zbl.653.58012

Yakh'ya, H. [1987]: New integrable cases of the motion of a gyrostat. Vestn. Mosk. Univ., Ser. I. 1987, No. 4, 88–90. English transl.: Mosc. Univ. Mech. Bull. *42*, No. 4, 29–31 (1987), Zbl.661.70012

Chapter 3
Quantization of Open Toda Lattices

M.A. Semenov-Tian-Shansky

Translated by the author

Introduction

In the previous chapter we described a geometric realization of open Toda lattices. Already in his first paper on the subject Kostant [1979] indicated an extension of geometrical scheme to quantum Toda lattices. About the same time (1978–1981), a powerful and sophisticated technique of the Quantum Inverse Scattering Method was created for the study of quantum integrable systems (see Faddeev [1980, 1984]). As was soon realized, it goes beyond the ordinary theory of Lie groups and Lie algebras and introduces the new notion of quantum groups (Drinfel'd [1987]). From this general point of view the 'quantum Kostant-Adler scheme' is a certain limiting case which corresponds to linearization of the basic commutation relations. [It should be recalled that the basic commutation relations of the Quantum Inverse Scattering Method are quadratic; they may be regarded as the quantization of quadratic Poisson bracket relations considered in Section 12 of the previous chapter.]

If we confine ourselves to the linear case, we may compile the following vocabulary which establishes the correspondence between classical and quantum systems.

Poisson Lie algebras	Universal enveloping algebras
$C^\infty(\mathfrak{g}^*)$, $C^\infty(\mathfrak{g}_R^*)$	$U(\mathfrak{g})$, $U(\mathfrak{g}_R)$
Casimir functions	The center of $U(\mathfrak{g})$
Orbits of the R-bracket = phase spaces of integrable systems	Irreducible unitary representations of $U(\mathfrak{g}_R)$ = Hilbert spaces of quantum integrable systems

The main commutativity theorem asserts that the Casimir operators $z \in$ cent $U(\mathfrak{g})$ give rise to a commutative subalgebra in $U(\mathfrak{g}_R)$ and hence in any of its unitary representations. The key problem is to determine the spectrum of this subalgebra. Note that for affine Lie algebras even the definition of quantum integrals of motion is not obvious since the universal enveloping algebra of an affine Lie algebra does not have a bona fide center. The construction of quantum integrals of motion in this case was suggested by Reyman and Semenov-Tian-Shansky [1979], and their commutativity was recently proved by Malikov [1988]. Unfortunately, if the underlying Lie algebras are infinite-dimensional, their representation theory provides little information on the spectrum of quantum Hamiltonians (cf. Goodman and Wallach [1986]). In the Quantum Inverse

Scattering Method the spectrum is usually determined by means of the generalized Bethe Ansatz. This, however, imposes some restrictions on the systems in question (existence of a certain eigenstate, called vacuum or pseudovacuum). For systems without vacuum, e.g. for periodic Toda lattices, the problem becomes much more difficult. Over the past decade, the spectrum of the periodic Toda lattice was studied by different techniques (Gutzwiller [1980, 1981], Sklyanin [1985]). One of the important ingredients of these papers is the knowledge of joint eigenfunctions for the open Toda lattice. It is this latter problem that may be efficiently solved using the 'quantum Kostant-Adler scheme' and the representation theory of semisimple Lie groups. On the other hand, the technique of Quantum Inverse Scattering does not work for open Toda lattices, so that in a sense the two approaches are complementary.

In the present survey we confine ourselves to the study of open Toda lattices. Some of the methods we use are more general (e.g. reduction of line bundles has interesting applications in representation theory, see Guillemin and Sternberg [1982]). The analytic part of the problem amounts to the study of multiparticle scattering. The technique used here is an adaptation of Harish-Chandra's parabolic descent method (Harish-Chandra [1958, 1966, 1975, 1976a,b]).

§ 1. Reduction of Quantum Bundles

We shall use the Hamiltonian reduction technique exposed in Section 5.1 of the previous chapter.

1.1. Lagrangian Polarizations. Let $\mathcal{M}$ be a symplectic manifold. A Lagrangian polarization on $\mathcal{M}$ is a smooth fibering $\pi: \mathcal{M} \to B$ whose fibers are Lagrangian submanifolds. We recall the construction of G-invariant Lagrangian polarizations on coadjoint orbits of Lie groups.

Let G be a Lie group, $H \subset G$ its connected subgroup, $f \in \mathfrak{h}^*$ a character of its Lie algebra $\mathfrak{h}$ (i.e. f is a linear functional on $\mathfrak{h}$ which vanishes on $[\mathfrak{h}, \mathfrak{h}]$). Recall that the quotient symplectic manifold obtained by reduction of T^*G with respect to the action of H over the point $f \in \mathfrak{h}^*$ coincides with the magnetic cotangent bundle $T_f^*(G/H)$. Fix $F \in \mathfrak{g}^*$. A subalgebra $\mathfrak{l}_F \subset \mathfrak{g}$ is called a *Lagrangian subalgebra* subordinate to F if $F|[\mathfrak{l}_F, \mathfrak{l}_F] = 0$ and $\operatorname{codim} \mathfrak{l}_F = \frac{1}{2} \dim \mathcal{O}_F$ (the latter condition means that $\mathfrak{l}_F$ is maximal among all subalgebras which satisfy the former; the existence of $\mathfrak{l}_F$ is a condition imposed on F). Let $\mathfrak{l}_F$ be a Lagrangian subalgebra, $L_F \subset G$ the corresponding Lie subgroup. Then $G_F \subset L_F$ and the fibers of the canonical projection $\mathcal{O}_F \to G/L_F$ are Lagrangian submanifolds. Thus a choice of a Lagrangian subalgebra subordinate to F determines a *Lagrangian polarization* on the orbit $\mathcal{O}_F$. Obviously, this polarization is G-invariant.

Consider the reduction of $\mathcal{M} = T^*G$ with respect to the action of L_F over the point $f = F|_{\mathfrak{l}_F}$. The canonical projection

$$T_f^*(G/L_F) \to G/L_F$$

defines a Lagrangian polarization on the quotient space $\mathcal{M}_f \simeq T_f^*(G/L_f)$.

Proposition 1. *Assume that the Lagrangian subalgebra satisfies the Pukanszky condition*

$$F + \mathfrak{l}_F^{\perp} \subset \mathcal{O}_F.$$

Then $\mathcal{O}_F$ is canonically isomorphic to $T_f^*(G/L_F)$ as a polarized symplectic manifold.

1.2. Reduction of Quantum Bundles. The standard exposition of Geometric Quantization is given by Kostant [1970]. We refer the reader to these lectures for more detailed definitions. The reduction of quantum line bundles was first considered by Reyman and Semenov-Tian-Shansky [1979] and by Guillemin and Sternberg [1982].

Let $(\mathcal{M}, \omega)$ be a symplectic manifold. A *quantum bundle* over $\mathcal{M}$ is a principal $U(1)$-bundle with connection whose curvature form is $(2\pi)^{-1}\omega$. Such a bundle exists if the de Rham class of $(2\pi)^{-1}\omega$ is integral. Assume that there is a Hamiltonian group action $G \times \mathcal{M} \to \mathcal{M}$. Let $\mu: \mathcal{M} \to \mathfrak{g}^*$ be the corresponding moment map, $\pi: \mathcal{M} \to \mathcal{M}/G$ the canonical projection. Let $x \in \mathcal{M}/G$, $f \in \mathfrak{g}^*$ be regular values of π, μ, respectively, and put $\mathcal{M}_f = \mu^{-1}(f)$, $\mathcal{M}_{f,x} = \mu^{-1}(f) \cap \pi^{-1}(x)$.

Proposition 2. *The manifold $\mathcal{M}_{f,x}$ is isotropic with respect to ω.*

Let $E \to \mathcal{M}$ be a quantum bundle, $E_f = E|_{\mathcal{M}_f}$. We shall say that E_f satisfies the *Bohr-Sommerfeld condition* if the monodromy of flat bundles $E|_{\mathcal{M}_{f,x}}$, $x \in \pi(\mathcal{M}_f)$, is trivial.

Theorem 3. *Assume that E_f satisfies the Bohr-Sommerfeld condition and the manifolds $\mathcal{M}_{f,x}$ are connected. In this case there exists a unique quantum bundle $\bar{E}_f \to \bar{\mathcal{M}}_f$ over the quotient symplectic space $\bar{\mathcal{M}}_f$ such that $E_f = \pi^* \bar{E}_f$.*

Sketch of a proof (cf. Guillemin and Sternberg [1982]). The bundle $\bar{E}_f$ is uniquely determined by the sheaf of germs of its sections. We define this sheaf as the sheaf of germs of sections of E_f which are covariantly constant on the fibers of $\pi|_{\mathcal{M}_f}$. Due to the Bohr-Sommerfeld condition the bundles $E|_{\mathcal{M}_{f,x}}$ admit global covariantly constant sections. One can check that the connection form on E_f projects down to a uniquely defined connection form on $\bar{E}_f$ and that its curvature coincides with the quotient symplectic form on $\bar{\mathcal{M}}_f$.

The bundle $\bar{E}_f$ is called the *reduced quantum bundle.*

Remark. If the Bohr-Sommerfeld condition is satisfied, the deRham class of the quotient symplectic form $(2\pi)^{-1}\bar{\omega}_f$ is a fortiori integral. Indeed, there is a simple relation between the cohomology class of $\bar{\omega}_f$ and the one-dimensional cohomology class of $\mathcal{M}_{f,x}$ determined by the connection form (transgression in the fiber bundle $\mathcal{M}_f \to \bar{\mathcal{M}}_f$).

Let G_f be the stabilizer of $f \in \mathfrak{g}^*$ in the group G. Assume that the action of G_f on $\mathcal{M}_f$ is free. For simplicity we assume that G_f is connected.

Proposition 4. *The quantum bundle $\bar{E}_f$ satisfies the Bohr-Sommerfeld condition if and only if the function $\chi_f(\exp X) = \exp if(X)$ extends to a character of G_f.*

Let $p: E \to \mathcal{M}$ be a quantum bundle, α its connection form. For $h \in C^\infty(\mathcal{M})$ denote by $\lambda(h)$ the vector field on E such that $p_*\lambda(h) = \xi_h$ is the Hamiltonian vector field on $\mathcal{M}$ generated by h and $\alpha(\lambda(h)) = h \circ p$.

Proposition 5. (i) *The Lie derivative of the connection form along $\lambda(h)$ is zero.*
(ii) *The mapping*

$$\lambda: C^\infty(\mathcal{M}) \to \text{Vect}\, E: h \mapsto \lambda(h)$$

is a Lie algebra homomorphism.

The homomorphism λ is called the *prequantization map.*

Let G be a connected Lie group, $G \times \mathcal{M} \to \mathcal{M}$ a Hamiltonian group action, (E, α) a quantum line bundle over $\mathcal{M}$. For $X \in \mathfrak{g}$ denote by h_X the Hamiltonian of the vector field on $\mathcal{M}$ generated by X.

Proposition 6. (i) *The mapping*

$$\lambda: \mathfrak{g} \to \text{Vect}\, E: X \mapsto \lambda(h_X)$$

defines an action of $\mathfrak{g}$ by infinitezimal automorphisms of (E, α).
(ii) *There exists a unique action $G \times E \to E$ whose differential at the identity coincides with λ. This action preserves the connection α.*

Assume that there are commuting Hamiltonian actions of two groups G and H on $\mathcal{M}$. Let $\mu_H: \mathcal{M} \to \mathfrak{h}^*$ be the moment map, $f \in \mathfrak{h}^*$, $\mathcal{M}_f = \mu_H^{-1}(f)$. Let $\bar{\mathcal{M}}_f$ be the corresponding quotient space. Assume that $E_f = E|_{\mathcal{M}_f}$ satisfies the Bohr-Sommerfeld condition, and let $\bar{E}_f \to \bar{\mathcal{M}}_f$ be the quotient quantum bundle.

Proposition 7. *The group G acts by automorphisms on E_f and $\bar{E}_f$. These actions commute with the reduction with respect to H.*

Recall now the simplest version of the Kirillov-Kostant quantization (cf. Kostant [1970, 1979], Kirillov [1972, 1987]). This construction assigns to a group of automorphisms of a quantum bundle its unitary representation. The key condition (that often restricts the range of applications) is the demand that these automorphisms should preserve a Lagrangian polarization.

Let $(\mathcal{M}, \pi)$ be a polarized symplectic manifold, $E \to \mathcal{M}$ a quantum bundle. Denote by $S_\pi(E)$ the space of sections of the associated Hermitian line bundle which are covariantly constant on the fibers of π[1]. Let Λ be the space of half-densities on $B = \pi(\mathcal{M})$, i.e. the space of sections of the line bundle $|\Lambda^n|^{1/2} \to B$ associated with the frame bundle over B with respect to the representation $g \mapsto |\det g|^{1/2}$ of the general linear group. The space $S_\pi^\Lambda(E) = S_\pi(E) \otimes \Lambda$ carries

[1] If the fibers of π are not simply connected, there may be no such sections. In examples we have in mind the fibers of π are affine spaces, and so there are no nontrivial conditions of the Bohr-Sommerfeld type.

the natural inner product

$$(s_1 \otimes \lambda_1, s_2 \otimes \lambda_2) = \int_B \langle s_1, s_2 \rangle \lambda_1 \lambda_2$$

and may be completed to a Hilbert space.

Let G be a connected Lie group, $G \times \mathcal{M} \to \mathcal{M}$ a Hamiltonian action. Assume that G preserves the Lagrangian polarization on $\mathcal{M}$. In this case the space $S_\pi^A(E)$ is G-invariant.

Proposition 8. *The natural action of G in $S_\pi^A(E)$ extends to a unitary representation.*

Examples. We shall consider representations associated with reduced quantum bundles described in Section 1.2.

Let G be a connected simply connected unimodular Lie group. Let θ be the canonical 1-form on T^*G. The quantum bundle over T^*G is the trivial (and trivialized) bundle $E = T^*G \times U(1)$ with connection form $\alpha = \theta + d\varphi$, where $d\varphi$ is the standard 1-form on $U(1)$. Let π: $T^*G \to G$ be the canonical projection. The space $S_\pi^A(E)$ may be canonically identified with $C^\infty(G)$ as a left and right G-module.

Let $H \subset G$ be a closed connected subgroup, $\mathfrak{h} \subset \mathfrak{g}$ its Lie algebra. Let $f \in \mathfrak{h}^*$ be a character of $\mathfrak{h}$. Put $X = G/H$ and let T_f^*X be the magnetic contangent bundle. The quantum bundle E may be reduced to T_f^*X if and only if f defines a character χ_f of H. Assume that this condition is fulfilled and denote by $\bar{E}_f$ the reduced quantum bundle.

Let π: $T_f^*X \to X$ be the canonical projection. Its fibers are Lagrangian and simply connected. Thus the space $S_\pi(\bar{E}_f)$ is defined and its elements may be canonically identified with sections from $S_\pi(E_f)$ which are covariantly constant on left coset classes of H. Since we have already identified $S_\pi(E)$ with $C^\infty(G)$ we obtain

Proposition 9. *The space $S_\pi(\bar{E}_f)$ is canonically isomorphic to the space of smooth functions on G satisfying the functional equation*

$$\varphi(hg) = \chi_f(h)\varphi(g).$$

Define the functional $\delta \in \mathfrak{h}^*$ by

$$\delta(X) = \tfrac{1}{2}\operatorname{tr}(\operatorname{ad} X|\mathfrak{g}/\mathfrak{h}), \quad X \in \mathfrak{h}.$$

Let Δ be the corresponding (non-unitary) character of H: $\Delta(\exp X) = \exp \delta(X)$. We shall call Δ the *modular character*.

Proposition 10. *The space $S_\pi^A(\bar{E}_f)$ is canonically isomorphic to the space of functions on G satisfying the functional equation*

$$\varphi(hg) = \chi_f(h)\Delta(h)\varphi(g), \quad g \in G, \quad h \in H.$$

The action of G on itself by right translations commutes with left translations from H and extends to the Hamiltonian action on T^*G. According to Propositions 9, 10, this action defines a natural representation of G in $S_\pi^A(\bar{E}_f)$.

Corollary. *The representation of G in $S_\pi^\Delta(\bar{E}_f)$ is canonically isomorphic to* $\operatorname{ind}_H^G(\chi_f \otimes \Delta)$.

Fix $f \in \mathfrak{g}^*$; assume that the orbit $\mathcal{O}_f$ admits an invariant polarization satisfying the Pukanszky condition. Let L_f be the Lagrangian subgroup subordinate to f, $\mathfrak{l}_f$ its Lie algebra. We assume for simplicity that the stabilizer G_f is connected. In this case $\mathcal{O}_f$ is simply connected and hence the quantum bundle over $\mathcal{O}_f$ is unique if it exists.

Proposition 11. (i) *Assume that the character $\bar{f} = f|_{\mathfrak{l}_f}$ extends to a 1-dimensional representation χ_f of L_f. In that case the orbit $\mathcal{O}_f$ is integral.*

(ii) *Let Δ be the modular character of L_f. The representation T_f of G associated with the orbit $\mathcal{O}_f$ is canonically isomorphic to* $\operatorname{ind}_{L_f}^G(\chi_f \otimes \Delta)$.

In applications to quantum integrable systems we shall not deal with the representation T_f itself, but rather with the corresponding representation of the universal enveloping algebra $U(\mathfrak{g})$; we shall denote this representation by the same letter.

1.3. The Quantum Version of Kostant's Commutativity Theorem. In this section we shall use the notation of Section 2 of the previous survey.

Let $\mathfrak{g}$ be a Lie algebra, $\mathfrak{g}_+, \mathfrak{g}_- \subset \mathfrak{g}$ its Lie subalgebras such that $\mathfrak{g} = \mathfrak{g}_+ \dot{+} \mathfrak{g}_-$. Let $P_\pm$ be the projection operator onto $\mathfrak{g}_\pm$ parallel to the complementary subalgebra, $R = P_+ - P_-$.

It is natural to regard the universal enveloping algebra $U(\mathfrak{g})$ as the quantum mechanical analogue of the space $C^\infty(\mathfrak{g}^*)$ equipped with the Lie-Poisson bracket. Let us describe the relation between multiplications in $U(\mathfrak{g})$ and in $U(\mathfrak{g}_R)$. Observe that $R_\pm = \pm P_\pm\colon \mathfrak{g}_R \to \mathfrak{g}$ are Lie algebra homomorphisms. Extend them to the homomorphisms $R_\pm\colon U(\mathfrak{g}_R) \to U(\mathfrak{g})$. Let $\Delta\colon U(\mathfrak{g}_R) \to U(\mathfrak{g}_R) \otimes U(\mathfrak{g}_R)$ be the standard comultiplication which is defined on the generators $X \in \mathfrak{g}_R$ by

$$\Delta(X) = X \otimes 1 + 1 \otimes X.$$

Put $\mu = (R_+ \otimes R_-) \circ \Delta$. Let $x \mapsto {}'x$ be the principal antiautomorphism of $U(\mathfrak{g})$ (i.e. the unique antiautomorphism of $U(\mathfrak{g})$ which is equal to $-\mathrm{id}$ on $\mathfrak{g}$). Define the mapping $p\colon U(\mathfrak{g}) \otimes U(\mathfrak{g}) \to U(\mathfrak{g})$ by $p(x \otimes y) = x \cdot {}'y$. The composition $p \circ \mu$: $U(\mathfrak{g}_R) \to U(\mathfrak{g})$ is a linear isomorphism. Put $\delta = (p \circ \mu)^{-1}$ and

$$x * y = \delta^{-1}(\delta(x)\delta(y)).$$

The operation $*$ is the multiplication in $U(\mathfrak{g}_R)$ pulled back to $U(\mathfrak{g})$ by means of the isomorphism δ.

Theorem 12 (i). *For any $x, y \in U(\mathfrak{g})$ we have*

$$x * y = \sum_k x_k^+ \cdot y \cdot {}'x_k^-, \tag{1}$$

where

$$x = \sum_k x_k^+ \cdot {}'x_k^-, \quad x_k^\pm \in U(\mathfrak{g}_\pm).$$

(ii) *Let $\mathscr{Z}$ be the center of $U(\mathfrak{g})$. The restriction of δ to $\mathscr{Z}$ is an algebra homomorphism.*

Remark. Formula (1) is the multiplicative version of the formula

$$[X, Y]_R = \tfrac{1}{2}([RX, Y] + [X, RY])$$

which relates the Lie brackets in $\mathfrak{g}$ and in $\mathfrak{g}_R$.

It is natural to demand that symmetric elements in $\mathscr{Z}$ (i.e. such that $z = {}'z$) should correspond to self-adjoint operators in unitary representations of $U(\mathfrak{g}_R)$. To this end the mapping δ should be slightly modified. Define the linear form $\rho \in \mathfrak{g}_R^*$ by

$$\rho(X) = \tfrac{1}{2} \operatorname{tr} \operatorname{ad}_{\mathfrak{g}_R} X, \quad X \in \mathfrak{g}_R.$$

Let α be the automorphism of $U(\mathfrak{g}_R)$ which is given by

$$\alpha(X) = X - \rho(X),$$

for $X \in \mathfrak{g}_R$. Put $\gamma = \alpha \circ \delta$.

Proposition 13. *Assume that* $\mathfrak{g}$ *is unimodular, i.e. that* $\operatorname{tr} \operatorname{ad}_{\mathfrak{g}} X = 0$ *for all* $X \in \mathfrak{g}$; *then the homomorphism* $\gamma\colon \mathscr{Z} \to U(\mathfrak{g}_R)$ *is symmetric, i.e.* $\gamma('z) = {}'\gamma(z)$.

1.4. The Generalized Kostant Theorem and Reduction of Quantum Bundles. The geometric meaning of Theorem 12 is clarified by its comparison with the results of Section 1.2. We keep to the notations introduced above.

Define the action of G_R on G by

$$h\colon g \longmapsto h_+ g h_-^{-1} \tag{2}$$

and extend it to the Hamiltonian action $G_R \times T^*G \to T^*G$. Let $\mathcal{O}_f \subset \mathfrak{g}_R^*$ be an $\operatorname{Ad}^* G_R$-orbit of the point $f \in \mathfrak{g}_R^*$. Assume that $\mathcal{O}_f$ possesses an invariant Lagrangian polarization satisfying the Pukanszky condition. Let L_f be the Lagrangian subgroup which is subordinate to f, let $\mathfrak{l}_f$ be its Lie algebra and $L_f \times T^*G \to T^*G$ the restriction of the action of G to L_f.

Proposition 14. *The quotient manifold obtained by reduction of T^*G with respect to the action of L_f over the point $\bar{f} = f|_{\mathfrak{l}_f}$ contains $\mathcal{O}_f$ as an open cell. If the canonical mapping $G_R \to G$ is surjective, the quotient manifold is canonically isomorphic to $\mathcal{O}_f$ as a polarized symplectic manifold.*

Let $E_f \to \mathcal{O}_f$ be the quantum bundle, let $S_\pi^\Delta(E_f)$ be the associated pre-Hilbert space of half-densities. Let χ_f be the character of L_f defined by $\bar{f} = f|_{\mathfrak{l}_f}$. Let Δ be the modular character of L_f.

Proposition 15. (i) *The bundle E_f is canonically isomorphic to the quotient bundle obtained by reduction of the standard quantum bundle $E \to T^*G$ with respect to the action of L_f.*

(ii) *The space $S_\pi^\Delta(E_f)$ is canonically isomorphic to the space $\mathscr{H}_f$ of functions on G satisfying the functional equation*

$$\varphi(h_+ x h_-^{-1}) = \chi_f(h)\Delta(h)\varphi(x),$$
$$x \in G, \quad h = (h_+, h_-) \in L_f \subset G_R. \tag{3}$$

Proposition 11 implies that the space $S_\pi^\Delta(E_f)$ has the structure of a G_R-module. The representation T_f of G_R in $S_\pi^\Delta(E_f)$ is isomorphic to $\operatorname{ind}_{L_f}^{G_R}(\chi_j \otimes \Delta)$.

Theorem 16. *Let i_f: $S_\pi^\Delta(E_f) \to \mathscr{H}_f$ be the canonical isomorphism. The space $\mathscr{H}_f$ is invariant with respect to the natural action of Casimir operators on G and*

$$z \cdot i_f \varphi = i_f(T_f \circ \delta(z) \cdot \varphi), \quad z \in \mathscr{Z}, \quad \varphi \in S_\pi^\Delta(E_f). \tag{4}$$

More briefly we can say that the operator $\delta_f(z) = T_f \circ \delta(z)$ coincides with the *radial part* of $z \in \mathscr{Z}$ in $\mathscr{H}_f$.

Theorem 16 is the main result of the quantization theory for integrable systems that we are exposing here. It reduces the study of spectra of quantum Hamiltonians to the study of spectra of *Casimir operators* on G in the space $\mathscr{H}_f$ of functions satisfying the functional equation (3).

1.5. Quantization. The Duflo Homomorphism. Let us examine more thoroughly the correspondence between the quantum operators $\gamma(z)$, $z \in \mathscr{Z}$, and their symbols.

Let gr be the canonical functor from the category of filtered algebras into the category of graded algebras. Recall that $\operatorname{gr} U(\mathfrak{g}) = S(\mathfrak{g})$ is the symmetric algebra of $\mathfrak{g}$. Let $U = (U_n)_{n \geqslant 0}$ be the canonical filtration, $S = \bigoplus_n S_n$ the corresponding canonical grading. An element $\operatorname{gr}_k u \in S_k$ is called the principal symbol of $u \in U_k$. It is easy to see that the mapping γ preserves canonical filtration and that the principal symbols of z and $\gamma(z)$ coincide. In particular, the symbol of $\gamma(z)$, $z \in \mathscr{Z}$, is a Casimir element of $\mathfrak{g}$, i.e. it belongs to the subspace $I(\mathfrak{g}) = S(\mathfrak{g})^G$.

The symmetric algebra $S(\mathfrak{g}_R)$ is equipped with the natural Lie bracket which coincides with the Lie-Poisson bracket under the canonical isomorphism $S(\mathfrak{g}_R) \simeq P(\mathfrak{g}_R^*)$. For any $u \in U(\mathfrak{g}_R)_k$, $v \in U(\mathfrak{g}_R)_l$ we have

$$\operatorname{gr}_{k+l-1}(u \cdot v - v \cdot u) = \{\operatorname{gr}_k u, \operatorname{gr}_l u\}_R \tag{5}$$

("the principal part of the commutator is the Poisson bracket", cf. Berezin [1967]). Using (5) it is easy to deduce from Theorem 12 that elements of $P(\mathfrak{g}_R^*)$ are in involution with respect to R-bracket (cf. Theorem 2.1 from the previous survey).

A quantization of the symmetric algebra $S(\mathfrak{g}_R)$ is a mapping q: $S(\mathfrak{g}_R) \to U(\mathfrak{g}_R)$ which is compatible with the filtration and preserves the principal symbol. In other words, if $x \in S_k$, then $q(x) \in U_k$ and $\operatorname{gr}_k q(x) = x$. An example of quantization is given by the symmetrization map β. Let us call a quantization q symmetric if it commutes with the principal antiautomorphism $x \mapsto {}'x$ i.e. $q({}'x) = {}'q(x)$. Assume that the quantization Q is symmetric, maps $I(\mathfrak{g})$ onto the commutative algebra $\gamma(\mathscr{Z})$ and, moreover, its restriction to $I(\mathfrak{g})$ is an algebra isomorphism. These properties determine Q uniquely.

In order to construct Q let us first recall the definition of the *Duflo isomorphism.*

Let $P(\mathfrak{g})$ be the algebra of polynomials on $\mathfrak{g}$, $\hat{P}(\mathfrak{g})$ its completion consisting of formal power series. Define the formal series $j \in \hat{P}(\mathfrak{g})$ by

$$j(X) = \det(\operatorname{sh} \operatorname{ad} X/\operatorname{ad} X).$$

The series j is invertible in the ring of formal power series. Let j^{-1} be its inverse. The canonical pairing $\hat{P}(\mathfrak{g}) \times S(\mathfrak{g}) \to \mathbb{C}$ allows to consider formal series from $\hat{P}(\mathfrak{g})$ as (infinite order) differential operators on $S(\mathfrak{g})$ (this definition is unambiguous, since these operators are locally nilpotent). Denote by $D(j^{-1})$ the differential operator which corresponds to j^{-1}. Let $\beta: S(\mathfrak{g}) \to U(\mathfrak{g})$ be the symmetrization map.

Theorem 17 (Duflo [1977]). *The mapping*

$$\Delta: S(\mathfrak{g})^G \to \mathscr{Z}: u \mapsto \beta(D(j^{-1})u)$$

is an algebra isomorphism.

For semisimple Lie algebras this result (in a slightly different form) was obtained by Harish-Chandra. We shall give an alternative description of the Harish-Chandra homomorphism in the next section.

It is easy to see that if $u \in S_k$ then $\operatorname{gr}_k \Delta u = u$. Hence the mapping Δ is a quantization of $S(\mathfrak{g})$. Let $Q = \gamma \circ \Delta$ be the composition map

$$Q: S(\mathfrak{g}) \to U(\mathfrak{g}_R).$$

Proposition 18. (i) *Q is a quantization of $S(\mathfrak{g})$.*
(ii) *The restriction of Q to $I = S(\mathfrak{g})^G$ is an algebra homomorphism.*

Let $\mathcal{O}_f \subset \mathfrak{g}_R^*$ be a polarized orbit of G_R, let T_f be the corresponding representation of $U(\mathfrak{g}_R)$, $i: \mathcal{O}_f \hookrightarrow \mathfrak{g}_R^*$ the canonical embedding. Assume that the dual map $i^*: I \to C^\infty(\mathcal{O}_f)$ is monomorphic. In this case the correspondence between classical and quantum Hamiltonians on $\mathcal{O}_f$ is also one-to-one. Thus the quantization may be defined on $I_f = i^*(I) \subset C^\infty(\mathcal{O}_f)$ and is given by $Q_f = T_f \circ Q \circ (i^*)^{-1}$.

§ 2. Quantum Toda Lattices

Generalized quantum Toda lattices yield an application of the techniques developed in Section 1.

2.1. Semisimple Lie Groups and Lie Algebras. Notation. The notation we introduce below will be used up to the end of this review. Let $\mathfrak{g}$ be a real split semisimple Lie algebra, τ a Cartan involution in $\mathfrak{g}$, $\mathfrak{g} = \mathfrak{k} + \mathfrak{p}$ the corresponding Cartan decomposition, $\mathfrak{a} \subset \mathfrak{p}$ a split Cartan subalgebra. For $\alpha \in \mathfrak{a}^*$ put

$$\mathfrak{g}_\alpha = \{X \in \mathfrak{g}: \operatorname{ad} H \cdot X = \alpha(H) \cdot X, H \in \mathfrak{a}\}.$$

The nonzero vectors α such that $\mathfrak{g}_\alpha \neq \{0\}$ form the root system of $(\mathfrak{g}, \mathfrak{a})$. Fix an order in the root system and let $\Delta^\pm$ be the set of positive (negative) roots. Let

$P \subset \Delta^+$ be the set of simple roots. Put $\mathfrak{n} = \bigoplus_{\alpha \in \Delta^+} \mathfrak{g}^\alpha$, $\mathfrak{v} = \tau(\mathfrak{n})$. The subalgebra $\mathfrak{b} = \mathfrak{n} + \mathfrak{a}$ is a maximal solvable subalgebra in $\mathfrak{g}$ (the Borel subalgebra). We have

$$\mathfrak{g} = \mathfrak{n} + \mathfrak{a} + \mathfrak{k}$$

(the Iwasawa decomposition).

Let us identify the spaces $\mathfrak{g}^*$ and $\mathfrak{g}$ by means of the Killing form. This enables us to realize the duals of various subalgebras in $\mathfrak{g}$ as linear subspaces in $\mathfrak{g}$. For instance, $\mathfrak{a}^* \simeq \mathfrak{a}$, $\mathfrak{v}^* \simeq \mathfrak{n}$. When we regard an element $f \in \mathfrak{n}$ as a linear functional on $\mathfrak{v}$, we always have in mind this canonical duality.

Let us denote by (,) the inner product on $\mathfrak{a}$ induced by the Killing form.

Let G be a real semisimple Lie group with finite center which corresponds to $\mathfrak{g}$. The Cartan involution on G which corresponds to $\tau \in \operatorname{Aut} \mathfrak{g}$ will be denoted by the same letter. Let K be the fixed point subgroup of this involution, $A = \exp \mathfrak{a}$, $N = \exp \mathfrak{n}$. We have $G = NAK$ (the Iwasawa decomposition). If $x = nak$, $n \in N$, $a \in A$, $k \in K$, we set $H(x) = \log a$, $\kappa(x) = k$. Let M, M' be the centralizer and the normalizer of $\mathfrak{a}$ in K and $W = M'/M$ the Weyl group of $(\mathfrak{g}, \mathfrak{a})$. Let $B = MAN$ be the Borel subgroup (the normalizer of $\mathfrak{b}$ in G). Let $\rho = 1/2 \sum_{\alpha \in \Delta^+} \alpha$ be half the sum of positive roots. Let C_+ be the positive Weyl chamber in $\mathfrak{a}$ defined by Δ_+, and $\bar{C}_+$ its closure.

2.2. Toda Lattices: a Geometric Description. A generalized Toda lattice may be associated with any real split semisimple Lie algebra. Its phase space is $\mathfrak{a} \times \mathfrak{a}^*$ with the canonical symplectic structure. Its Hamiltonian in the canonical coordinates (x, p) is given by

$$H = \frac{1}{2}(p, p) + \sum_{\alpha \in P} e^{-2\alpha(x)}. \tag{6}$$

The geometric treatment of classical and quantum Toda lattices is based on the Hamiltonian reduction technique. Some versions of such treatment were already discussed in the preceding chapter. The one exposed below is best suited to the study of the quantum case.

Recall that the space T^*G is equipped with the canonical symplectic structure. Let us fix its trivialization by means of left translations. This allows to identify the algebra $P(\mathfrak{g}^*)^G$ of G-invariant polynomials with the subalgebra I of functions on T^*G which are polynomial on each fiber and invariant with respect to left and right translations (actually, this isomorphism does not depend on the choice of the trivialization and is canonical).

Let $\mathfrak{d} = \bigoplus_{\alpha \in P} \mathfrak{g}_\alpha$. Clearly, $\mathfrak{n} = \mathfrak{d} + [\mathfrak{n}, \mathfrak{n}]$. An element $f \in \mathfrak{d}$ is called a *principal nilpotent element* if it determines a nondegenerate character of $\mathfrak{v}$, i.e. if $f_\alpha = f|_{\mathfrak{g}_{-\alpha}} \neq 0$ for all $\alpha \in P$.

Let us define the action of $V \times K$ on G by

$$(v, k)\colon x \mapsto vxk^{-1}.$$

This action extends to the Hamiltonian action of $V \times K$ on T^*G. Let μ: $T^*G \to \mathfrak{w}^* \oplus \mathfrak{k}^*$ be the corresponding moment map. Consider the dual pair

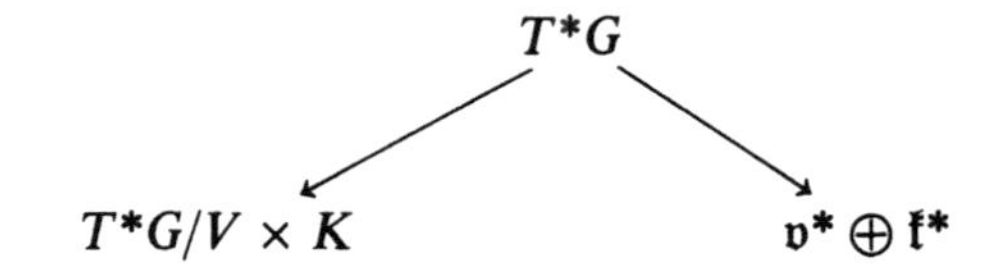

Fix $f \in \mathfrak{b}$ and let $\bar{f} = f|_{\mathfrak{v}}$. Let $\bar{\mathcal{M}}_f$ be the symplectic leaf in the quotient space $T^*G/V \times K$ which corresponds to $(\bar{f}, 0) \in \mathfrak{v}^* \oplus \mathfrak{k}^*$.

Proposition 19. (i) *The quotient phase space $\bar{\mathcal{M}}_{\bar{f}}$ is isomorphic to $\mathfrak{a} \times \mathfrak{a}^*$ as a polarized symplectic manifold.*

(ii) *Let $2h$ be the invariant Hamiltonian on T^*G which corresponds to the Killing quadratic form on $\mathfrak{g}^* \simeq \mathfrak{g}$. The reduced Hamiltonian $H = \bar{h}$ on $\bar{\mathcal{M}}_{\bar{f}}$ in canonical coordinates $(x, p) \in \mathfrak{a} \times \mathfrak{a}^*$ is given by*

$$H = \frac{1}{2}(p, p) + \sum_{\{\alpha \in P;\, f_\alpha \neq 0\}} c_\alpha^2 e^{-2\alpha(x)}. \tag{7}$$

The constants c_α depend on f.

(iii) *Let f be a principal nilpotent element. For a suitable choice of $c \in \mathfrak{a}$, the canonical mapping $(x, p) \mapsto (x + c, p)$ transforms (7) into (6).*

Functions from $I \subset C^\infty(T^*G)$ are constant on the fibers of the projection π: $T^*G \to T^*G/V \times K$ and hence give rise to reduced Hamiltonians on $\bar{\mathcal{M}}_{\bar{f}}$. (More precisely, for any $\varphi \in I$ there exists a unique $\bar{\varphi} \in C^\infty(\bar{\mathcal{M}}_{\bar{f}})$ such that $\pi^*\bar{\varphi} = \varphi|_{\mu^{-1}((f,0))}$.) Let $I_f \subset C^\infty(\bar{\mathcal{M}}_{\bar{f}})$ be the algebra generated by reduced Hamiltonians. We obtain an algebra homomorphism γ_f: $I \to I_f$. For $f = 0$, the reduced Hamiltonians depend only on the momenta. This defines a homomorphism γ_0: $I \to P(\mathfrak{a}^*) \simeq S(\mathfrak{a})$.

Theorem 20. *The homomorphism γ_0 is injective and its image coincides with the subalgebra $S(\mathfrak{a})^W$ of the Weyl group invariants.*

This result is a reformulation of the classical Chevalley theorem (Bourbaki [1968, Ch. 5]).

Proposition 21. *There exists a canonical isomorphism h^f: $S(\mathfrak{a})^W \to I_f$ such that the diagram*

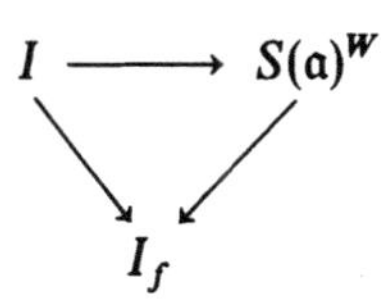

is commutative.

The polynomial $\sigma \in S(\mathfrak{a})^W$ is called the *symbol* of the Hamiltonian $h_\sigma^f \in I_f$.

Clearly, all functions $h_\sigma^f \in I_f$ are in involution with each other and with the Hamiltonian (7).

Remark. For degenerate Toda lattices (i.e. when $f \in \mathfrak{d}$ is not a principal nilpotent element) the full algebra of integrals of the motion is slightly larger than I_f, namely it is a finite rank module over I_f. We shall return to this question is Section 3.3 below.

2.3. Toda Lattices: Geometric Quantization. The main result of this section is a specialization of Theorem 16. We keep to the notation introduced above.

Let $E \to T^*G$ be the standard quantum line bundle.

Proposition 22. (i) *For each $f \in \mathfrak{d}$ the line bundle E may be reduced to the quotient space $\bar{\mathcal{M}}_{\bar{f}}$.*

(ii) *The completion of $S_\pi^A(\bar{E}_{\bar{f}})$ with respect to the natural inner product is isomorphic to $L_2(\mathfrak{a})$.*

(iii) *Let ψ_f be the character of V given by*

$$\psi_f(\exp X) = \exp i f(X), \quad X \in \mathfrak{v}. \tag{9}$$

The natural representation T_f of the group $S = VA$ in $S_\pi^A(\bar{E}_{\bar{f}})$ is isomorphic to $\operatorname{ind}_V^S \psi_f$. This representation is irreducible if and only if f is a principal nilpotent element.

Let $\mathcal{H}_f$ be the space of smooth functions on G which satisfy the functional equation

$$\varphi(vxk) = \psi_f(v)\varphi(x), \quad v \in V, \quad k \in K. \tag{10}$$

Define an embedding $i_f\colon C^\infty(\mathfrak{a}) \to \mathcal{H}_f$ by

$$i_f\varphi(x) = \psi_f(v)\varphi(\log a), \quad x = vak, \quad v \in V, \quad a \in A, \quad k \in K.$$

Let $\mathcal{Z}$ be the center of the universal enveloping algebra $U(\mathfrak{g})$. Define the radial part, $\delta_f(z)$, of $z \in \mathcal{Z}$ by

$$i_f(\delta_f(z)\cdot\varphi) = z \cdot i_f\varphi, \quad \varphi \in C^\infty(\mathfrak{a}). \tag{11}$$

Put $d^2(X) = \det(\operatorname{Ad}\exp X|_{\mathfrak{n}}) = \exp 2\rho(X)$, $X \in \mathfrak{a}$. Let

$$\hat{\gamma}_f(z) = d \circ \delta_f(z) \circ d^{-1} \tag{12}$$

The right hand side is understood as the composition of differential operators in $C^\infty(\mathfrak{a})$.

Exercise. Check that the definition of $\hat{\gamma}_f$ agrees with the one in Section 1.4.

Let $\mathcal{Z}^f$ be the image of $\mathcal{Z}$ under the mapping $\hat{\gamma}_f$. For $f = 0$, the operators $\hat{\gamma}_0(z)$, $\delta_0(z)$ $(z \in \mathcal{Z})$ are differential operators with constant coefficients on $\mathfrak{a}$. Thus $\hat{\gamma}_0$, δ_0 may be regarded as homomorphisms from $\mathcal{Z}$ into $U(\mathfrak{a}) \simeq S(\mathfrak{a})$.

Theorem 23. (i) *The homomorphism $\delta_0\colon \mathcal{Z} \to U(\mathfrak{a})$ is the restriction to $\mathcal{Z}$ of the projection onto $U(\mathfrak{a})$ in the decomposition*

$$U(\mathfrak{g}) = U(\mathfrak{a}) \oplus (\mathfrak{n}U(\mathfrak{g}) + U(\mathfrak{g})\mathfrak{k}). \tag{13}$$

(ii) *The image of $\hat{\gamma}_0\colon \mathcal{Z} \to U(\mathfrak{a})$ coincides with $U(\mathfrak{a})^W$.*

The mapping $\hat{\gamma}_0$ is called the *Harish-Chandra homomorphism.* One can show that the mapping

$$\gamma_0^{-1} \circ \hat{\gamma}_0 \colon \mathscr{Z} \to S(\mathfrak{g})^G$$

is the inverse of the Duflo isomorphism described in Theorem 17.

Proposition 24. *There exists a canonical isomorphism of algebras $\hat{h}^f \colon S(\mathfrak{a})^W \to \mathscr{Z}^f$ such that the diagram*

$$\begin{array}{ccc} \mathscr{Z} & \xrightarrow{\hat{\gamma}_0} & S(\mathfrak{a})^W \\ & {\scriptstyle \hat{\gamma}_f} \searrow \quad \swarrow {\scriptstyle \hat{h}^f} & \\ & \mathscr{Z}^f & \end{array}$$

is commutative.

We shall say that $\sigma \in S(\mathfrak{a})^W$ is the *symbol* of the operator $\hat{h}^f_\sigma$.

Proposition 25. *Let $\sigma_2 \in S(\mathfrak{a})^W$ be the Casimir element associated with the Killing form in $\mathfrak{g}$. The Hamiltonian $H = -\frac{1}{2}\hat{h}^f_{\sigma_2}$ is given by*

$$\hat{H} = -\frac{1}{2}\Delta + \sum_{\{\alpha \in P;\, f_\alpha \neq 0\}} c_\alpha^2 e^{-2\alpha(x)}. \tag{14}$$

Here Δ is the Laplacian on $\mathfrak{a}$ associated with the Killing form.[2]

Hence the geometric quantization of generalized Toda lattices agrees with the standard Schroedinger quantization of the Hamiltonian (7).

§ 3. Spectral Theory of the Quantum Toda Lattice

The results of Section 2 reduce the description of the eigenfunctions of the quantum Toda lattice Hamiltonian to the construction of the eigenfunctions of Casimir operators on a semi-simple Lie group satisfying the functional equation (10). An obvious way to perform this latter task is to realize these eigenfunctions as matrix coefficients of irreducible representations of G.

3.1. Representations of the Principal Series and Whittaker Functions. The subject of this section is classical (see Gel'fand et al. [1962, 1966], Knapp [1971], Warner [1972] for more details on the principal series representations and the role of the intertwining operators). The Whittaker functions were first studied by Jacquet [1967], Schiffmann [1971].

---

[2] Under our conventions the operator on $\mathfrak{a}$ with the symbol σ_2 is negative, hence the sign flip in (14).

Fix $\lambda \in \mathfrak{a}_{\mathbf{C}}^*$. Define a 1-dimensional representation of the Borel subgroup $B = MAN$ by

$$\chi_\lambda(man) = e^{(\lambda+\rho)\log a}, \quad m \in M, \quad a \in A, \quad n \in N. \tag{15}$$

The induced representation $\pi_\lambda = \operatorname{ind}_B^G \chi_\lambda$ is called the spherical principal series representation (or simply *principal series representation*, for short). The representation π_λ is realized in the space $\mathfrak{H}_\lambda$ of functions on G satisfying the functional equation

$$\varphi(bx) = \chi_\lambda(b)\varphi(x), \quad b \in B, \tag{16}$$

and is defined by the formula

$$\pi_\lambda(g)\varphi(x) = \varphi(xg). \tag{17}$$

Define an Hermitian bilinear form on $\mathfrak{H}_\lambda \times \mathfrak{H}_{-\bar{\lambda}}$

$$(\varphi, \psi) = \int_{B\backslash G} \varphi \cdot \bar{\psi}, \quad \varphi \in \mathfrak{H}_\lambda, \quad \psi \in \mathfrak{H}_{-\bar{\lambda}}. \tag{18}$$

This definition is unambiguous, since there is a natural correspondence between the functions on G satisfying the functional equation

$$\mu(manx) = e^{2\rho(\log a)}\mu(x), \quad m \in M, \quad a \in A, \quad n \in N,$$

and the measures on the quotient space $B\backslash G$. For λ purely imaginary the pairing (18) defines an inner product in $\mathfrak{H}_\lambda$ and the representation π_λ is unitary.

The function φ_λ given by $\varphi_\lambda(x) = e^{(\lambda+\rho, H(x))}$ belongs to $\mathfrak{H}_\lambda$. Clearly,

$$\pi_\lambda(k)\varphi_\lambda = \varphi_\lambda, \quad k \in K.$$

The Iwasawa and Bruhat decompositions yield realizations of π_λ in the space of functions on $M\backslash K$ and on V, respectively. These realizations are completely determined by normalization of measures on these spaces. Recall that a representation T of a Lie group is called infinitesimally irreducible if the operators $T(z)$ representing the center of the universal enveloping algebra are scalars.

Let $\hat{\gamma}_0: \mathscr{Z} \to S(\mathfrak{a})^W$ be the Harish-Chandra homomorphism. Denote by $\gamma_0(z: \lambda)$ the value of $\hat{\gamma}_0(z) \in S(\mathfrak{a})^W$ at $\lambda \in \mathfrak{a}_{\mathbf{C}}^*$.

Proposition 26. *The representations π_λ are infinitesimally irreducible for all $\lambda \in \mathfrak{a}_{\mathbf{C}}^*$; $\pi_\lambda(z) = \gamma_0(z: \lambda) \cdot \mathrm{id}$.*

Definition 27. Let π be a representation of G in a linear space $\mathfrak{H}$. A vector $w_\psi \in \mathfrak{H}$ is called the *Whittaker vector* corresponding to a character ψ of V if

$$\pi(v)w_\psi = \psi(v)w_\psi, \quad v \in V. \tag{19}$$

Proposition 28. *The Gårding space of the representation π_λ contains precisely one Whittaker vector $w_{\psi,\lambda}$. The function $w_{\psi,\lambda} \in \mathfrak{H}_\lambda$ is given by*

$$w_{\psi,\lambda}(b \cdot v) = \psi(v)\chi_\lambda(b), \quad b \in B, \quad v \in V. \tag{20}$$

By contrast with K-invariant vectors φ_λ, the Whittaker vectors are not square summable, which leads to some analytical difficulties in the theory of the Whittaker functions (as compared to the theory of spherical functions).

Definition 29. *The class* 1 *Whittaker function* (or simply *Whittaker function*, for short) corresponding to the character ψ of V is the matrix coefficient

$$W(g: \lambda: \psi) = (\pi_\lambda(g)\varphi_\lambda, w_{\psi, -\bar{\lambda}}). \tag{21}$$

Formula (21) implies the following integral representation:

$$W(g: \lambda: \psi) = \int_V e^{(\lambda+\rho)H(vg)}\overline{\psi(v)}\, dv. \tag{22}$$

Theorem 30. (i) *The integral* (22) *converges in the tubular domain* $\operatorname{Re} \lambda \in C_+$.
(ii) *The Whittaker function satisfies the functional equation*

$$W(vgk: \lambda: \psi) = \psi(v) W(g: \lambda: \psi) \tag{23}$$

and is the eigenfunction of Casimir operators on G,

$$zW(\lambda: \psi) = \gamma_0(z: \lambda) W(\lambda: \psi). \tag{24}$$

The convergence of (22) was first studied by Jacquet on the basis of a theorem due to Gindikin and Karpelevich (see below).

Fix $f \in \mathfrak{b}$ and let ψ_f be the character of V given by $\psi_f(\exp X) = \exp if(X)$, $X \in \mathfrak{v}$. Put $d(x) = e^{\rho(x)}$, $x \in \mathfrak{a}$.

Proposition 31. *The function*

$$w(x: \lambda: f) = d(x) W(e^x: \lambda: \psi_f), \quad x \in \mathfrak{a}, \tag{25}$$

is an eigenfunction of the Hamiltonians of the quantum Toda lattice,

$$\hat{h}_\sigma^f w(\lambda: f) = \sigma(\lambda) w(\lambda: f), \quad \sigma \in S(\mathfrak{a})^W. \tag{26}$$

Thus the Whittaker functions differ from the eigenfunctions of the quantum Toda lattice only by a simple factor.

3.2. Analytic Properties of the Whittaker Functions. The analytic properties of the Whittaker functions were studied by Jacquet [1967] and Schiffmann [1971]. See also Hashizume [1979, 1981].

Theorem 32. *Let* $f \in \mathfrak{b}$ *be a principal nilpotent element.*
(i) *The function* $w(x: \lambda: f)$ *extends to an entire function in* $\mathfrak{a}_{\mathbb{C}}^*$.
(ii) *There exist functions* $M(s: \lambda: f)$ *which are meromorphic in* $\mathfrak{a}_{\mathbb{C}}^*$ *and such that, for any* $s \in W$,

$$w(x: s\lambda: f) = M(s: \lambda: f) w(x: s\lambda: f). \tag{27}$$

(iii) *The functions* $M(s: \lambda: f)$ *satisfy the functional equation*

$$M(s_1 s_2: \lambda: f) = M(s_2: \lambda: f) M(s_1: s_2 \cdot \lambda: f). \tag{28}$$

Fix $\alpha \in P$ and let $s_\alpha \in W$ be the corresponding simple reflection. Then

$$M(s_\alpha\colon \lambda\colon f) = e_\alpha(\lambda)e_\alpha(-\lambda)^{-1}\left(\frac{|f_\alpha|}{2\sqrt{2(\alpha, \alpha)}}\right)^{2\lambda_\alpha}, \tag{29}$$

where $\lambda_\alpha = \dfrac{(\lambda, \alpha)}{(\alpha, \alpha)}$ *and*

$$e_\alpha(\lambda)^{-1} = \Gamma\left(\frac{\lambda_\alpha}{2} + \frac{1}{4}\right)\Gamma\left(\frac{\lambda_\alpha}{2} + \frac{3}{4}\right). \tag{30}$$

The functional equation (28) for the Whittaker functions is closely related to the properties of intertwining operators for representations of principal series (Schiffmann [1971], Knapp [1971]).

The Whittaker functions which correspond to degenerate characters of V may be reduced to Whittaker functions on a lower rank group (a precise statement is given in Section 3.4 below).

3.3. The Spectral Decomposition Theorem. Let

$$c(\lambda) = \int_V e^{(\lambda+\rho)H(v)}\, dv \tag{31}$$

be the *Harish-Chandra function.*

Theorem 33 (Gindikin and Karpelevich [1962]). (i) *The integral* (31) *is absolutely convergent for* Re $\lambda \in C_+$.

(ii) *Normalize the Haar measure on V by the condition* $\int \exp 2\rho(v)\, dv = 1$. *Put*

$$c_\alpha(\lambda) = B(\lambda_\alpha, \tfrac{1}{2}), \quad \lambda_\alpha = (\lambda, \alpha)/(\alpha, \alpha). \tag{32}$$

Then

$$c(\lambda) = \prod_{\alpha \in \Delta^+} c_\alpha(\lambda)/c_\alpha(\rho). \tag{33}$$

For $s \in W$ define a unitary operator M_s in $L_2(i\mathfrak{a}^*)$, $i = \sqrt{-1}$, by the formula

$$M_s a(\lambda) = M(s^{-1}\colon \lambda\colon f)a(s\cdot\lambda), \quad \lambda \in i\mathfrak{a}^*. \tag{34}$$

Put $M = \sum_{s \in W} M_s$. (Operators of this type were first introduced by Gel'fand [1963].)

The main spectral decomposition theorem for the quantum Toda lattice is as follows.

Theorem 34. (i) *For any $f \in \mathfrak{v}$, the functions $w(x\colon \lambda\colon f)$ form an orthogonal system in $L_2(\mathfrak{a})$.*

(ii) *Let $f \in \mathfrak{v}$ be a principal nilpotent element. Let*

$$w(x\colon a\colon f) = \frac{1}{(2\pi i)^l}\int_{i\mathfrak{a}^*} d\lambda\, a(\lambda)w(x\colon \lambda\colon f). \tag{35}$$

Then

$$\|w(a: f)\|^2_{L_2(\mathfrak{a})} = \frac{1}{(2\pi i)^l} \int_{i\mathfrak{a}^*} d\lambda\, |Ma(\lambda)|^2 |c(\lambda)|^{-2}. \tag{36}$$

As usual, the proof of the spectral decomposition theorem for a multi-particle Schroedinger operator is rather complicated. Besides the analytic properties of the Whittaker functions, we also need a precise information on their coordinate asymptotics. The statement and the proof of the theorem may serve a good introduction into the so called Harish-Chandra philosophy (Harish-Chandra [1958, 1966, 1975, 1976a, 1976b]). Below we give a sketch of a proof which is based on the non-stationary technique of Semenov-Tian-Shansky [1976][3].

To prove the theorem we need an exact asymptotics of the wave packets (35) for large x. It is technically convenient to introduce into the integral (35) one more parameter $t \in \mathfrak{a}$ and to consider the asymptotics of the integral

$$w(x: t: a: f) = \frac{1}{(2\pi i)^l} \int_{i\mathfrak{a}^*} d\lambda\, a(\lambda) e^{-\lambda(t)} w(x: \lambda: f) \tag{37}$$

for large t and x. For $G = \mathrm{SL}(2, \mathbb{R})$ the function (37) satisfies the wave equation $w_{tt} = -\hat{H}w$, where $\hat{H}$ is the Toda lattice Hamiltonian. In general, the function (37) satisfies an overdetermined system of equations in t, x.

As first observed by Semenov-Tian-Shansky [1976], its properties are completely analogous to those of the massless wave equation.

Replacing the L_2-space by the energy space for this system of equations simplifies the spectral decomposition theorem and leads to its simple proof. In the course of the proof we shall also study the scattering problem for the quantum Toda lattice.

3.4. Degenerate Toda Lattices. To define the wave operators for the quantum Toda lattice we shall need a thorough study of degenerate Toda lattices. We assume that the reader is familiar with the theory of parabolic subgroups (Harish-Chandra [1958, 1966, 1975, 1976a, 1976b], Warner [1972]).

Recall that there is a natural correspondence between the parabolic subgroups in G containing B and the subsets of the set of simple roots. Denote by Q_F the parabolic subgroup associated with a subset $F \subset P$. Let $Q_F = M_F A_F N_F$ be its Langlands decomposition, $L_F = M_F A_F$, ${}^*B_F = M_F \cap B$. Then ${}^*B_F$ is a Borel subgroup in M_F. Let ${}^*B_F = {}^*M_F {}^*A_F {}^*N_F$ be its Langlands decomposition. Then $A = {}^*A_F A_F$, $N = {}^*N_F N_F$. The group L_F is reductive, and, A is its split Cartan subgroup. Let W_F be the Weyl group of the pair $({}^*M_F, {}^*A_F)$ or, equivalently, of (L_F, A). The group W_F is canonically isomorphic to the subgroup of W generated by reflections s_α, $\alpha \in F$. Put $V_F = \tau(N_F)$, $V_F^* = \tau({}^*N_F)$. Denote by $\mathfrak{l}_F$, $\mathfrak{v}_F$, ${}^*\mathfrak{v}_F$, $\mathfrak{n}_F$, ${}^*\mathfrak{n}_F$, $\mathfrak{a}_F$, ${}^*\mathfrak{a}_F$ the Lie algebras of the groups L_F, V_F, ${}^*V_F$, N_F, ${}^*N_F$, A_F,

[3] At the present moment a complete proof is available only for $G = \mathrm{SL}(n, \mathbb{R})$ (Semenov-Tian-Shansky [1984]).

$^*A_F$, respectively. We have

$$\mathfrak{a}_F = \{x \in \mathfrak{a} \colon \alpha(x) = 0 \ \forall \alpha \in F\}, \quad {}^*\mathfrak{a}_F = \bigoplus_{\alpha \in F} \mathbb{R} \cdot \alpha.$$

Let Δ_F^+ be the set of positive roots which are identically zero on $\mathfrak{a}_F$, let $^*\Delta_F^+$ be the set of positive roots lying in the linear span of F. Then

$$\mathfrak{n}_F = \bigoplus_{\alpha \in \Delta^+ \setminus \Delta_F^+} \mathfrak{g}_\alpha, \quad {}^*\mathfrak{n}_F = \bigoplus_{\alpha \in {}^*\Delta_F^+} \mathfrak{g}_\alpha, \quad \mathfrak{v}_F = \bigoplus_{\alpha \in \Delta^+ \setminus \Delta_F^+} \mathfrak{g}_{-\alpha}, \quad {}^*\mathfrak{v}_F = \bigoplus_{\alpha \in {}^*\Delta_F} \mathfrak{g}_{-\alpha}.$$

Put $K_F = L_F \cap K$. One can show that K_F intersects each connected component of L_F.

Put $\mathfrak{d}_F = \mathfrak{d} \cap \mathfrak{n}_F$. Fix $f \in \mathfrak{d}_F$ and $\bar{f} \in \mathfrak{v}^*$, $^*f \in {}^*\mathfrak{v}_F$ be the linear functionals determined by f. Define the action of the group $^*V_F \times K_F$ on L_F by

$$(v, k) \colon x \mapsto vxk^{-1}.$$

This action is canonically extended to the Hamiltonian action of $^*V_F \times K_F$ on T^*L_F. Let $\mu_F \colon T^*L_F \to {}^*\mathfrak{v}_F^* \oplus \mathfrak{k}_F^*$ be the corresponding moment map. Consider the dual maps

$$\begin{array}{ccc} & T^*L_F & \\ \swarrow & & \searrow \\ T^*L_F/{}^*V_F \times K_F & & {}^*\mathfrak{v}_F^* \oplus \mathfrak{k}_F^* \end{array}$$

Let $\bar{\mathcal{M}}_{*f}$ be the symplectic leaf in the quotient space $T^*L_F/{}^*V_F \times K_F$ which corresponds to $({}^*f, 0) \in {}^*\mathfrak{v}_F^* \oplus \mathfrak{k}_F^*$. Let $\bar{\mathcal{M}}_{\bar{f}}$ be the quotient space defined in Section 2.2.

Proposition 35. *The quotient spaces $\bar{\mathcal{M}}_{\bar{f}}$, $\bar{\mathcal{M}}_{*f}$ are canonically isomorphic as polarized symplectic manifolds.*

As in Section 2.2, we identify the algebra $P(\mathfrak{l}_F)^{L_F}$ with the algebra I_F of invariant functions on T^*L_F which are polynomial on each fiber. The Chevalley theorem (Bourbaki [1968, Ch. 5]) implies that there is a canonical embedding $\nu \colon I \to I_F$ which makes I_F a free I-module of rank $[W \colon W_F]$.

The reduction map π induces a canonical homomorphism γ_{*f} of I_F onto the algebra $I_F^f \subset C^\infty(\mathcal{M}_{*f})$ which consists of reduced Hamiltonians (cf. Section 2.2).

Proposition 36. (i) *The following diagram is commutative.*

$$\begin{array}{ccc} I & \xrightarrow{\ \nu\ } & I_F \\ {\scriptstyle \gamma_f}\downarrow & & \downarrow{\scriptstyle \gamma_{*f}} \\ I^f & \xrightarrow{\ i\ } & I_F^f \end{array}$$

Here the map i is induced by the isomorphism described in Proposition 35.

(ii) *Let $\gamma_0 \colon I_F \to P(\mathfrak{a}^*)^{W_F} \simeq S(\mathfrak{a})^{W_F}$ be the Chevalley isomorphism. There exists a canonical isomorphism $h^f \colon S(\mathfrak{a})^W \to I_F^f$ such that the diagram*

$$\begin{array}{ccc} I_F & \longrightarrow & S(\mathfrak{a})^{W_F} \\ & \searrow \quad \swarrow & \\ & I_F^f & \end{array}$$

is commutative.

Let $\bar{E}_{*f} \to \bar{\mathcal{M}}_{*f}$ be the quotient quantum line bundle obtained by reduction from the standard quantum line bundle $E \to T^*L_F$. Let $\bar{E}_{\bar{f}}$ be the quantum line bundle defined in Proposition 22.

Proposition 37. *The line bundles $\bar{E}_{*f}$ and $\bar{E}_{\bar{f}}$ are canonically isomorphic.*

Denote by $\mathcal{H}_F^{*f}$ the space of smooth functions on L_F satisfying the functional equations

$$\varphi(vxk) = \psi_{*f}(v)\varphi(x), \quad v \in {}^*V_F, \quad k \in K_F; \tag{38}$$

Here the character ψ_{*f} is defined by $\psi_{*f}(\exp X) = \exp i\, f(X)$, $X \in {}^*\mathfrak{v}_F$. Define the embedding $i_f\colon C^\infty(\mathfrak{a}) \to \mathcal{H}_F^{*f}$ by

$$i_f\varphi(x) = \psi_{*f}(v)\varphi(\log a),$$

where $x = vak$, $v \in {}^*V_F$, $a \in A$, $k \in K$, is the Iwasawa decomposition in L_F. Let $\mathcal{Z}_F$ be the center of the universal enveloping algebra $U(\mathfrak{l}_F)$. Define the radial part of $z \in \mathcal{Z}_F$ by

$$i_f(\delta_f(z)\cdot\varphi) = z \cdot i_f(\varphi).$$

Put $d_F^2(x) = \det(\operatorname{Ad} \exp x|_{*\mathfrak{n}_F})$, $x \in \mathfrak{a}$. Put $\hat{\gamma}_f(z) = d_F \circ \delta_f(z) \circ d_F^{-1}$, $z \in \mathcal{Z}_F$.

For $f = 0$, the operators $\hat{\gamma}_0(z)$, $z \in \mathcal{Z}_F$, are differential operators on $\mathfrak{a}$ with constant coefficients.

Proposition 38. *The image of $\hat{\gamma}_0\colon \mathcal{Z}_F \to U(\mathfrak{a}) \simeq S(\mathfrak{a})$ coincides with the subalgebra $S(\mathfrak{a})^{W_F}$.*

Let $\nu\colon \mathcal{Z} \to \mathcal{Z}_F$ be the homomorphism defined by the commutative diagram

$$\begin{array}{ccc} \mathcal{Z} & \xrightarrow{\ \nu\ } & \mathcal{Z}_F \\ {\scriptstyle \hat{\gamma}_0}\downarrow & & \downarrow{\scriptstyle \hat{\gamma}_0} \\ S(\mathfrak{a})^W & \longrightarrow & S(\mathfrak{a})^{W_F} \end{array}$$

Denote by $\mathcal{Z}_F^f$ the image of $\mathcal{Z}_F$ under the mapping $\hat{\gamma}_f$.

Proposition 39. (i) *The algebra $\mathcal{Z}^f$ lies in $\mathcal{Z}_F^f$, and the diagram*

$$\begin{array}{ccc} \mathcal{Z} & \xrightarrow{\ \nu\ } & \mathcal{Z}_F \\ {\scriptstyle \hat{\gamma}_f}\downarrow & & \downarrow{\scriptstyle \hat{\gamma}_f} \\ \mathcal{Z}^f & \longrightarrow & \mathcal{Z}_F^f \end{array}$$

is commutative.

(ii) *There exists a canonical homomorphism* $\hbar^f\colon S(\mathfrak{a})^{W_F} \to \mathscr{Z}_F^f$ *such that the diagram*

$$\begin{array}{ccc} \mathscr{Z}_F & \xrightarrow{\;\gamma_0\;} & S(\mathfrak{a})^{W_F} \\ & {\scriptstyle \gamma_f}\searrow \quad \swarrow{\scriptstyle \hbar^f} & \\ & \mathscr{Z}_F^f & \end{array}$$

is commutative.

Proposition 39 (i) implies that the eigenfunctions of the quantum Toda lattice associated with a degenerate character f may be reduced to Whittaker functions on a group of lower rank. An exact statement is as follows.

Denote by x_F, $^*x_F$ the components of $x \in \mathfrak{a}$ in the decomposition $\mathfrak{a} = \mathfrak{a}_F \oplus {}^*\mathfrak{a}_F$.

Proposition 40 (Hashizume [1982]). *Fix* $f \in \mathfrak{b}_F$. *Then*

$$w(x\colon \lambda\colon f) = e^{\lambda(x_F)} c_F(\lambda) w({}^*x_F\colon {}^*\lambda\colon {}^*f), \tag{39}$$

where ${}^*\lambda = \lambda|_{{}^*\mathfrak{a}_F}$,

$$c_F(\lambda) = \prod_{\alpha \in \Delta^+ \setminus \Delta_F^+} c_\alpha(\lambda). \tag{40}$$

In particular, for $f = 0$ *we have*

$$w(x\colon \lambda\colon 0) = c(\lambda) e^{\lambda(x)}. \tag{41}$$

3.5. The Wave Packets and their Properties.

For $\lambda \in \mathfrak{a}_{\mathbb{C}}^*$, put

$$\pi(\lambda) = \prod_{\alpha \in \Delta^+} (\lambda, \alpha). \tag{42}$$

Let $\mathscr{A}$ be the space of functions of $\lambda \in \mathfrak{a}_{\mathbb{C}}^*$ which are holomorphic in the strip $|\mathrm{Re}\,\lambda| < \delta$, rapidly decrease for $|\mathrm{Im}\,\lambda| \to \infty$ uniformly in this strip and such that the function $\lambda \mapsto \pi(\lambda)^{-1} a(\lambda)$ is regular. The function

$$w(x\colon t\colon a\colon f) = \frac{1}{(2\pi i)^l} \int_{i\mathfrak{a}^*} d\lambda\, e^{-\lambda(t)} a(\lambda) w(x\colon \lambda\colon f) \tag{43}$$

will be called the wave packet with the amplitude a. It is easily checked that for $a \in \mathscr{A}$ the wave packet is a smooth function in t, x.

Fix a parabolic subgroup $Q_F \supset B$ and assume that $f \in \mathfrak{b}_F$ is an F-principal nilpotent element (i.e. $f|_{\mathfrak{g}_{-\alpha}} \neq 0$ for all $\alpha \in F$).

Proposition 41. *The wave packet* (43) *satisfies the overdetermined system of equations*

$$'\sigma\left(\frac{\partial}{\partial t}\right) w(x\colon t\colon a\colon f) = \hbar_\sigma^f w(x\colon t\colon a\colon f), \quad \sigma \in S(\mathfrak{a})^{W_F}. \tag{44}$$

(Recall that $\sigma \mapsto {}'\sigma$ is the automorphism of $S(\mathfrak{a})$ induced by the map $X \mapsto -X$).

The properties of the system (44) are closely connected with the properties of the algebraic extension $S(\mathfrak{a})^{W_F} \subset S(W)$.

Theorem 42 (Bourbaki [1968, Ch. 5]). (i) *S^{W_F} is a graded free algebra of polynomials.*

(ii) *S is a graded free S^{W_F}-module.*

(iii) *Put $S_+^{W_F} = S^{W_F} \cap \mathfrak{a}S(\mathfrak{a})$, $J_F = S_+^{W_F}S$. There exists a graded subspace of S complementary to the ideal J_F which is stable with respect to W_F.*

(iv) *Let U_F be such a complement. Then the canonical homorphism $U_F \otimes_{\mathbb{C}} S^{W_F} \to S$ is an isomorphism. The representation of W_F in S (in U_F) is isomorphic to the regular representation of W_F over S^{W_F} (respectively, over $\mathbb{C}$).*

(v) *Let R be the quotient field of S. Then R^{W_F} is the quotient field of S^{W_F}; $R^{W_F} \subset R$ is the Galois extension with the Galois group W_F.*

The mapping $\hat{h}^f\colon S^{W_F} \to \operatorname{End} C^\infty(\mathfrak{a})\colon \sigma \mapsto \hat{h}^f_\sigma$ equips $C^\infty(\mathfrak{a})$ with the structure of an S^{W_F}-module. Put

$$\mathscr{H}_F^f = \operatorname{Hom}_{S^{W_F}}(S, C^\infty(\mathfrak{a})).$$

Let $\mathscr{L}_F^f$ be the space of solutions of the system (44) which have the form (43) with amplitudes $a \in \mathscr{A}$. Define the mapping $i_t\colon \mathscr{L}_F^f \to \mathscr{H}_F^f$ by

$$\langle i_t(\varphi), p\rangle = p\left(\frac{\partial}{\partial t}\right)\varphi(t, \cdot), \quad p \in S. \tag{45}$$

Definition 43. The space $\mathscr{H}_F^f$ is called the space of Cauchy data, or the *energy space*, for the system (44).

The mapping i_t assigns to the solution φ the set of all its derivatives, all the relations between them implied by the system (44) being automatically taken into account.

We shall now define an inner product in $\mathscr{H}_F^f$, the so called energy form. Recall that the relative trace $\operatorname{tr}_{R/R^{W_F}}$ is the R^{W_F}-linear map

$$\operatorname{tr}_{R/R^{W_F}}\colon R \to R^{W_F}\colon p \mapsto \sum_{s \in W_F} s \cdot p.$$

The bilinear form $\operatorname{tr}_{R/R^{W_F}} x \cdot {}'y$ on R is nondegenerate and allows to identify R with its dual. Put

$$D_F = \{x \in R\colon \operatorname{tr} x \cdot {}'y \in S^{W_F}\ \forall y \in S\}.$$

It is clear that $\mathscr{H}_F^f = C^\infty \otimes_{S^{W_F}} D_F$.

Let π_F be the product of positive roots of $(\mathfrak{l}_F, \mathfrak{a})$.

Proposition 44 (Harish-Chandra [1958]). *The set $\pi_F \cdot D_F$ lies in S.*

Hence the bilinear form on D_F

$$(x, y) \mapsto \pi_F^2 \operatorname{tr}_{R/R^{W_F}}(x \cdot {}'y)$$

takes its values in S^{W_F}.

For an arbitrary S-module V denote by $'V$ the same module twisted by the automorphism $x \mapsto 'x$: $'V = V$ as a linear space, but $x \cdot v = 'xv$ for $v \in V$, $x \in S$. Define the pairing

$$e\colon \mathscr{H}_F \otimes_{S^{W_F}} {}'\mathscr{H}_F \to C^\infty \otimes_{S^{W_F}} {}'C^\infty$$

by the formula

$$e((\varphi \otimes x) \otimes (\psi \otimes y)) = \pi_F^2 \operatorname{tr}(x \cdot 'y) \cdot (\varphi \otimes \psi),$$

for any $\varphi, \psi \in C^\infty(\mathfrak{a})$, $x, y \in D_F$. Let $\delta\colon \mathfrak{a} \to \mathfrak{a} \times \mathfrak{a}$ be the diagonal map. It induces the linear map $\delta^*\colon C^\infty(\mathfrak{a} \times \mathfrak{a}) \to C^\infty(\mathfrak{a})$ (restriction of smooth functions to the diagonal).

Definition 45. The *energy form on $\mathscr{H}_F^f$* is the hermitian form given by

$$(\varphi, \psi)_{\mathscr{H}_F^f} = \frac{1}{|W_F|} \int_{\mathfrak{a}} dx\, \delta^* \circ e(\varphi \otimes '\overline{\psi}). \tag{46}$$

The definition (46) is slightly inaccurate, since $e(\varphi \otimes '\overline{\psi})$ is an element not in $C^\infty(\mathfrak{a}) \otimes_{\mathbf{C}} C^\infty(\mathfrak{a})$, but rather in $C^\infty(\mathfrak{a}) \otimes_{S^{W_F}} C^\infty(\mathfrak{a})$. The latter space is the quotient of $C^\infty(\mathfrak{a}) \otimes_{\mathbf{C}} C^\infty(\mathfrak{a})$ over the submodule $\mathscr{N}$ generated by the elements of the form

$$h_\sigma^f \varphi \otimes '\overline{\psi} - \varphi \otimes h_{\bar\sigma}^{f\prime}\overline{\psi}, \quad \varphi, \quad \psi \in C^\infty(\mathfrak{a}), \quad \sigma \in S^{W_F}.$$

However, integration by parts shows that $\int dx\, \delta^* \varphi = 0$ for any $\varphi \in \mathscr{N}$. This proves that (46) is unambiguous.

We shall call

$$\|u\|_F^2 = (i_t(u), i_t(u))_{\mathscr{H}_F^f}. \tag{47}$$

the energy of a solution $u \in \mathscr{L}_F^f$.

As in the classical case, the energy space is introduced in order to reduce the system (44) to a system of evolution equations.

For $\chi \in S$ let L_χ be the linear operator in $\mathscr{H}_F^f$ defined by

$$\langle L_\chi \cdot \varphi \cdot p \rangle = \langle \varphi, \chi \cdot p \rangle, \quad p \in S. \tag{48}$$

Proposition 46. *Let $e_1, \ldots, e_l$ be a basis in $\mathfrak{a}$. The system* (44) *is equivalent to the system of evolution equations for the Cauchy data*

$$\frac{\partial}{\partial t_k} \varphi = L_{e_k} \varphi, \quad \frac{\partial}{\partial t_k} = e_k \left(\frac{\partial}{\partial t} \right), \quad k = 1, \ldots, l. \tag{49}$$

The generators L_{e_k} are skew Hermitian with respect to the energy form.

In complete analogy with the classical case (i.e. with the case of the wave equation), the skew symmetry of the generators L_{e_k} provides for the conservation of energy of the wave packets.

The space $\mathscr{H}_F^f$ is equipped with the natural action of the Weyl group W_F. Let ${}^a\mathscr{H}_F^f \subset \mathscr{H}_F^f$ be the subspace of W_F-antiinvariant elements (cf. Theorem 42 (iv)).

Proposition 47. *The restriction of the energy form to ${}^a\mathscr{H}_F^f$ coincides (up to a factor $|W_F|^{-1}$) with the L_2-inner product.*

Thus, as in the classical case, the L_2-spectral decomposition theorem may be easily derived from the corresponding theorem for the energy space.

In order to get a more explicit description of the energy space and the form (46) one has to choose a splitting subspace U_F as in Theorem 42 (iii) and a basis in U_F. Coordinate formulae for the energy of a wave packet are given in Semenov-Tian-Shansky [1976].

Let $Q \supset B$ be a parabolic subgroup. We shall write $t \underset{Q}{\to} \infty$ if $t \in \mathfrak{a}_{F(Q)}$ and

$$\beta(t) = \min_{\alpha \in \Delta^+ \setminus \Delta^+_{F(Q)}} \alpha(t) \to +\infty.$$

Let $f \in \mathfrak{d}$ be arbitrary. Define the element f_F by

$$f_F|_{\mathfrak{g}_{-\alpha}} = \begin{cases} f_\alpha, & \alpha \in F, \\ 0, & \alpha \in P \setminus F. \end{cases}$$

Theorem 48. *For any $a \in \mathscr{A}$ there exists the limit*

$$\lim_{\substack{t \to \infty \\ Q}} w(x + t: t + \tau: a: f) = w(x: \tau: a: f_{F(Q)}). \tag{50}$$

We are able now to state the main decomposition theorems for the wave packets.

Fix a parabolic subgroup $Q_{F_0} \supseteq B$ and a character $f \in \mathfrak{d}_{F_0}$.

Theorem 49. *For any $a \in \mathscr{A}$, the wave packet $w(a: f)$ has finite energy in each of the spaces $\mathscr{H}^f_{F'}$, $F' \supseteq F_0$.*

Let Q be a parabolic subgroup, $F = F(Q)$ the associated set of simple roots, we define the wave operator $U_{F|F'}$ acting from $\mathscr{H}^f_{F'}$ into $\mathscr{H}^{f_F}_F$ by the formula

$$(U_{F|F'} w)(x: t: a: f) = \lim_{\substack{t \to \infty \\ Q_F}} \pi_{F|F'}\left(\frac{\partial}{\partial t}\right) w(x + t, \tau + t: a: f),$$
$$\pi_{F|F'} = \pi_{F'} \cdot \pi_F^{-1} \in S(\mathfrak{a}). \tag{51}$$

Using the language of many-body problems we may call $U_{F|F'}$ the *cluster wave operator*. It describes the transition of the system into a state where several groups of particles are "widely separated" and do not interact with each other (cf. formulae (50), (39), (14)).

Theorem 50. (i) *The wave packet $U_{F|F'} w$ satisfies the system* (44) *associated with the character f_F and has a finite energy in the space $\mathscr{H}^{f_F}_F$.*

(ii) *The following Plancherel equality is valid*

$$\|w(a: f)\|^2_{\mathscr{H}^f_{F'}} = \|U_{F|F'} w(a: f)\|^2_{\mathscr{H}^{f_F}_F}. \tag{52}$$

(iii) *For any two parabolics $Q_{F_1} \subseteq Q_{F_2} \subseteq Q_{F'}$,*

$$U_{F_1|F'} = U_{F_1|F_2} \circ U_{F_2|F'}. \tag{53}$$

Theorem 50 is closely connected with the factorized scattering for the Toda lattice.

The proof of Theorems 49, 50 goes by induction over the rank of groups. The difficult point is to get sharp enough estimates which are valid uniformly in t up to the walls of the Weyl chambers.

For $G = SL(n, \mathbb{R})$ one can get an elementary proof which uses rather rough estimates of the integral (22) in its convergence domaine.

3.6. Scattering Theory. In this section f is a principal nilpotent element. Define the Fourier transform in $\mathscr{H}_P^f \equiv \mathscr{H}^f$ by

$$F_+ u(\lambda) = (u, w(x\colon \lambda\colon f)e^{-\lambda t})_{\mathscr{H}^f}. \tag{54}$$

Theorem 51. *The wave packets* (27) *satisfy the Plancherel equality*

$$\|w\|^2_{\mathscr{H}^f} = \int_{i\mathfrak{a}^*} d\mu(\lambda)|F_+ w(\lambda)|^2 \tag{55}$$

with the Plancherel measure

$$d\mu(\lambda) = \frac{d\lambda}{(2\pi i)^l}\cdot\frac{1}{|b(\lambda)|^2}, \quad b(\lambda) = \pi(\lambda)c(\lambda). \tag{56}$$

For any $s \in W$ define the wave operator U_s by

$$U_s w(x\colon \tau\colon a\colon f) = \lim_{\substack{t\to\infty \\ B}} \pi\left(\frac{\partial}{\partial t}\right) w(x + t, \tau + s\cdot t\colon a\colon f). \tag{57}$$

The existence of the limit for any $a \in \mathscr{A}$ is an easy consequence of Theorems 48, 32.

The *scattering operators* are defined by

$$\hat{S}_s = U_+ U_s^{-1}, \quad U_+ \equiv U_1. \tag{58}$$

It is more convenient to deal with scattering operators in the spectral representation, i.e. to put

$$S_s = \Phi \hat{S}_s \Phi^{-1}, \tag{59}$$

where Φ is the Laplace transform. Let R_s be the reflection operator,

$$R_s\varphi(\lambda) = \varphi(s^{-1}\cdot\lambda).$$

Proposition 52. *The operator $S_s \circ R_s$ is the multiplication operator by the function*

$$S_s(\lambda) = \frac{b(\lambda)}{b(s^{-1}\lambda)} M(s^{-1}\colon \lambda), \tag{60}$$

where $b(\lambda) = \pi(\lambda)c(\lambda)$, and the functions $c(\lambda)$, $M(s\colon \lambda)$ are given by (32), (33), (28), (30).

Formulae (60), (28) easily imply the functional equation for the scattering operator

$$S_{s_1 s_2} = S_{s_2} R_{s_2}^{-1} S_{s_1} R_{s_2}. \tag{61}$$

Formulae (61), (28), (34) are typical for the so called factorized scattering (Olshanetsky and Perelomov [1983]).

Operators (57), (60) are scattering operators for the system (44) in the Lax-Phillips sense (Lax and Phillips [1967], Semenov-Tian-Shansky [1976]). However, the Birman-Kato invariance principle implies that the ordinary scattering operator for the stationary Schroedinger equation

$$-\frac{1}{2}\Delta\psi + \sum_{\alpha \in P} e^{-2\alpha(x)}\psi = \lambda^2\psi \tag{62}$$

coincides with S_{w_0} where w_0 is the longest element of the Weyl group (i.e. the unique element which permutes the positive and negative Weyl chambers). This agrees with the elementary scattering theory for classical Toda lattices, since as we saw in Section 3.3 of the previous chapter the scattering transformation that relates incoming and outgoing momenta is precisely w_0. One can easily generalize the classical scattering theory to multi-dimensional time setting, and in that case there will also be scattering transformations which correspond to arbitrary elements of the Weyl group.

Let us finally comment on the proof of L_2-Plancherel theorem. To replace the energy space $\mathscr{H}^f$ with L_2 we use Proposition 47. The knowledge of scattering operators allows to describe the range of the L_2-Fourier transform, and this yields Theorem 34.

3.7. Sketch of a Proof of the Main Theorem. The Case $G = \mathrm{SL}(2, \mathbb{R})$. As we have seen, the decomposition theorems follow from two "difficult" theorems on wave packets (Theorems 49, 50). In this section we shall explain the idea of the "non-stationary" proof of these theorems on the elementary example $G = \mathrm{SL}(2, \mathbb{R})$, and in the next section we shall briefly discuss the induction over the rank. Rough estimates that we shall indicate suffice to prove the theorems for $G = \mathrm{SL}(n)$.

The Toda Hamiltonian for $G = \mathrm{SL}(2, \mathbb{R})$ is equal to

$$\hat{H} = -\frac{d^2}{dx^2} + e^{-2x}. \tag{63}$$

Its eigenfunctions are the classical Whittaker functions. The integral representation (22) in the present case reduces to

$$w(x: \lambda) = e^{2\lambda x} \int_{-\infty}^{\infty} dv(1 + v^2)^{-1/2-\lambda} \exp(-ie^{-2x}v). \tag{64}$$

This integral may be evaluated in terms of the McDonald functions

$$w(x: \lambda) = 2^{-\lambda+1} \frac{\sqrt{\pi}}{\Gamma(\lambda + 1/2)} e^{-x} K_\lambda(e^{-2x}). \tag{65}$$

However, for the qualitative method we have in mind we do not need this formula.

Consider the wave packet

$$w(x\colon t\colon a) = \frac{1}{2\pi i}\int_{-i\infty}^{i\infty} d\lambda\, e^{-2\lambda t} a(\lambda) w(x\colon \lambda). \tag{66}$$

The amplitude $a(\lambda)$ is assumed to be regular in the strip $|\mathrm{Re}\,\lambda| < \delta$ and to satisfy $a(0) = 0$. The wave packet (66) satisfies the wave equation

$$w_{tt} = w_{xx} - e^{-2x} w. \tag{67}$$

The energy space $\mathscr{H}$ associated with (67) is the space of two-component functions $\varphi = (\varphi_0, \varphi_1)$ with the norm

$$\|\varphi\|^2_{\mathscr{H}} = \frac{1}{2}\int_{-\infty}^{\infty} dx(\hat{H}\varphi_0\bar{\varphi}_0 + \varphi_1\bar{\varphi}_1). \tag{68}$$

Equation (67) is equivalent to the evolution equation for φ:

$$\varphi_t = L\varphi, \quad L = \begin{pmatrix} 0 & 1 \\ -\hat{H} & 0 \end{pmatrix}. \tag{69}$$

We leave it to the reader to check that formulae (68), (69) are in agreement with the general definitions of Section 3.5.

Proposition 53. *For the wave packet* (66) *there exist the limits*

$$w_{\pm}(x\colon \tau\colon a) = \lim_{t\to+\infty} w(x + t\colon \tau \pm t\colon a). \tag{70}$$

Proof. Let us shift the integration contour in (66) into the right half-plane. Formulae (64), (66) imply that

$$\begin{aligned} &w(x + t\colon \tau + t\colon a) \\ &= \frac{1}{2\pi i}\int_{\mathrm{Re}\,\lambda=\varepsilon} d\lambda\, e^{-2(\tau-x)} a(\lambda) \int_{-\infty}^{\infty} dv(1 + v^2)^{-1/2-\lambda} \exp(-ie^{-2(x+t)}v). \end{aligned}$$

The integral $\int dv(1 + v^2)^{-1/2-\varepsilon}$ is convergent for $\varepsilon > 0$ and bounds the integrand. Using the Fubini theorem and the Lebesgue theorem on the bounded convergence we get

$$\lim_{t\to+\infty} w(x + t\colon t + \tau\colon a) = \frac{1}{2\pi i}\int d\lambda'\, e^{-2\lambda(\tau-x)} a(\lambda) c(\lambda), \tag{71}$$

$$c(\lambda) = \int_{-\infty}^{\infty} dv(1 + v^2)^{-1/2-\lambda} = B(1/2, \lambda). \tag{72}$$

To compute the second limit in (70) we must use the functional equation for the Whittaker function which in the present case takes the form

$$w(x\colon -\lambda) = M(\lambda) w(x\colon \lambda), \quad M(\lambda) = 2^{2\lambda}\frac{\Gamma(\lambda + 1/2)}{\Gamma(-\lambda + 1/2)}. \tag{73}$$

The estimates of the wave packets that are required for the proof of the finite energy theorem are performed in a different way on the left and on the right half-lines. The estimates that are good on the left half-line immediately follow from (64), (66).

Proposition 54. *We have*

$$\left|\left(\frac{\partial}{\partial\tau}\right)^n w(x+t:\tau+t:a)\right| \leqslant C_{\varepsilon,n}e^{2\varepsilon x}, \quad \varepsilon>0, \tag{74}$$

uniformly in $t\to+\infty$.

The proof is immediately obtained by shifting the integration contour in (66) into the right half-plane.

The estimates that work on the opposite half-line are more cumbersome.

Proposition 55. *We have*

$$\left|\left(\frac{\partial}{\partial\tau}\right)^n w(x+t:\tau+t:a)\right| \leqslant C_{\delta,n}e^{-\delta x}, \quad \delta>0, \tag{75}$$

uniformly in $t\to+\infty$.

Observe first that $w_+(x:t:a)$ is a Schwartz function in x (this makes use of the condition $a(0)=0$ which cancels the pole of $c(\lambda)$ at $\lambda=0$). Obviously,

$$|w(x+t:t+\tau:a)| \leqslant |w_+(x:\tau:a)| + |w(x+t:t+\tau:a) - w_+(x:\tau:a)|. \tag{76}$$

We have

$$\begin{aligned} &w(x+t:t+\tau:a) - w_+(x:\tau:a)\\ &= \frac{1}{2\pi i}\int d\lambda\, e^{2\lambda(x-\tau)}a(\lambda)\int_{-\infty}^{\infty} dv(1+v^2)^{-1/2-\lambda}[\exp(-ie^{-2(x+t)}v)-1]. \end{aligned}$$

Shift the integration contour in the outer integral into the right half-plane and divide the integration domain in the inner one into two parts $|v|<e^T$, $|v|>e^T$. We get an estimate which is uniform in $t\to+\infty$:

$$\begin{aligned} &|w(x+t:t+\tau:a) - w_+(x:\tau:a)|\\ &\leqslant e^{2\varepsilon(x-\tau)}\cdot\frac{1}{2\pi}\int ds\,|a(is+\varepsilon)|\\ &\quad\times\left\{\int_{|v|<e^T} dv(1+v^2)^{-1/2-\varepsilon}2\sin(e^{-2(x+t)}v/2) + \int_{|v|>e^T} dv(1+v^2)^{-1/2-\varepsilon}\right\}\\ &\leqslant C_1e^{2\varepsilon(x-\tau)}e^{-2x+T} + C_2e^{2\varepsilon(x-\tau)-2\varepsilon T}. \end{aligned}$$

Choose T so that $x<T<2x$. Then for $\varepsilon>0$ sufficiently small the above inequality and (76) imply (75).

The estimates (74), (75) are sufficient to prove the finiteness of energy.

Proposition 56. *The energy of the wave packet* (66) *is equal to*

$$2\|w\|^2_{\mathscr{H}} = (w_t, w_t) - (w_{tt}, w) \tag{77}$$

(the inner product in the r.h.s. is in $L_2(\mathbb{R})$). Using the conservation of energy we get

$$2\|w\|^2_{\mathscr{H}} = \lim_{t\to\infty} \int dx[|w_t(x\colon t\colon a)|^2 - w_{tt}(x\colon t\colon a)\overline{w(x\colon t\colon a)}]. \tag{78}$$

Let us make in (78) the change of variables $x = y + t$.

Lemma. *The integrand in* (78) *has a summable bound uniformly in* $t \to +\infty$.

This assertion immediately follows from (74), (75). Evaluating the limit in (3.78) under the integral sign and using (71) we get, after integration by parts,

$$\|w\|^2_{\mathscr{H}} = (\partial_t w_+, \partial_t w_+)_{L_2} = \|\partial_x w_+\|^2_{L_2}. \tag{79}$$

Formula (79) is the Plancherel equality that we need (cf. (52)).

3.8. Induction Over Rank. We begin with some formulae for the wave packets that serve to evaluate their asymptotics and to prove the estimates which ensure the finiteness of energy. The calculation of $\lim\limits_{\substack{t\to\infty\\ Q}}$ in (50) for a wave packet is based on the following formula for the Whittaker function.

Lemma. *For* Re $\lambda \in C_+$ *we have*

$$w(x + t\colon \lambda\colon f) = e^{\lambda(x)} \int_{{}^*V_Q} dv_0\, e^{({}^*\lambda + {}^*\rho, H(v_0))}\overline{\psi}_{f_Q}(e^{{}^*x} v e^{-{}^*x}) \cdot \int_{V_Q} dv_1\, e^{(\lambda+\rho, H(v_1))}\overline{\psi}_f(e^{x+t} v_1^{k_0} e^{-x-t}), \tag{80}$$

where $v_0 = a_0 n_0 k_0$ *is the Iwasawa decomposition.*

For $t \underset{Q}{\to} \infty$ the inner integral converges to $\int dv_1\, e^{(\lambda+\rho, H(v_1))} = c_Q(\lambda)$ and this proves Theorem 48.

In order to get the estimates of wave packets for large x we shall use another integral representation. Let us fix a parabolic subgroup $Q \supset B$ and put $N_1 = {}^*N_Q$, $N_2 = N_Q$, $A_1 = {}^*A_Q$, $A_2 = A_Q$. Fix $f \in \mathfrak{d}$ and put

$$W_1(x\colon \lambda\colon f) = \int_{V_1} dv_1\, e^{(\lambda+\rho, H(v_1 x))}\overline{\psi}_f(v_1). \tag{81}$$

Proposition 57. *For* Re $\lambda \in C_+$ *we have*

$$W(x\colon \lambda\colon f) = \int_{V_2} dv_2\, \overline{\psi}_f(v_2) W_1(v_2 x\colon \lambda\colon f). \tag{82}$$

Proposition 58. (i) *The function* (81) *satisfies the functional equation*

$$W_1(n_2 a_2 yk\colon \lambda\colon f) = e^{(\lambda+\rho, \log a_2)} W_1(y\colon \lambda\colon f),$$
$$y \in G, \quad n_2 \in N_2, \quad a_2 \in A_2, \quad k \in K. \tag{83}$$

(ii) *The restriction of* (81) *to the subgroup* M_Q *coincides with the Whittaker function* $W(\lambda_1\colon f_1)$, *where* $\lambda_1 = \lambda|_{\mathfrak{a}_1}$, $f_1 = f|_{\mathfrak{v}_1}$.

Proof. (i) Since V_1 normalizes N_2 and A_2 centralizes N_1, we have

$$H(v_1 n_2 a_2 yk) = H(v_1 n_2 v_1^{-1} v_1 a_2 y) = H(v_1 a_2 y) = \log(a_2) + H(v_1 y),$$

which gives (i). To prove (ii) observe that the functional $\rho|_{\mathfrak{a}_1}$ coincides with the half-sum of positive roots of $\mathfrak{m}_Q$. □

Corollary. *The function* (81) *is determined by its restriction to* A *in the Iwasawa decomposition* $G = AN_2 V_1 K$.

If $x = an_2 v_1 k$, $a \in A$, $n_2 \in N_2$, $v_1 \in V_1$, $k \in K$, we put $\tilde{H}(x) = \log a$. Let w_1 be the maximal element of the Weyl group $W_Q = W_1$. We denote its representative in $K \cap M_Q$ by the same letter.

Lemma. *For* $v_2 \in V_2$ *we have*

$$\tilde{H}(v_2) = H(w_1 v_2 w_1^{-1})^{w_1}. \tag{84}$$

Put

$$\rho_1 = \frac{1}{2} \sum_{\alpha \in \Delta_Q^+} \alpha, \quad \rho_2 = \frac{1}{2} \sum_{\alpha \in \Delta^+ \setminus \Delta_Q^+} \alpha. \tag{85}$$

Consider the function

$$w_1(x\colon \lambda\colon f) = e^{-(\rho_2 - \rho_1, x)} W_1(e^x\colon \lambda\colon f). \tag{86}$$

We immediately get

$$w_1(x\colon \lambda\colon f) = e^{\lambda(x)} \int_{V_1} dv_1\, \overline{\psi}_f(e^x v_1 e^{-x}) e^{(\lambda+\rho, H(v_1))};$$

thus $w_1(x\colon \lambda\colon t)$ behaves for $x \in C_+$, $\operatorname{Re}\lambda \in C_+$ like a plane wave.

Proposition 59. *For* Re $\lambda \in C_+$ *we have*

$$w(x\colon \lambda\colon f) = \int_{V_2} dv_2\, \overline{\psi}_f(e^x v_2 e^{-x}) w_1(x + \tilde{H}(v_2)\colon \lambda\colon f). \tag{87}$$

The proof is straightforward; it is sufficient to notice that $\rho^{w_1} = \rho_2 - \rho_1$. Hence we get the following formula for wave packets:

$$w(x\colon a\colon f) = \int_{V^2} dv_2\, e^{(\rho, H(v_2))} \overline{\psi}_f(e^x v_2^{w_1} e^{-x}) w_1(x + H(v_2)^{w_1}\colon a\colon f). \tag{88}$$

Let us discuss the special case of this formula for $Q = B$. Let

$$\hat{a}(x) = \frac{1}{(2\pi i)^l} \int_{i\mathfrak{a}^*} e^{\lambda(x)} a(\lambda)\, d\lambda$$

be the Laplace transform of a. Then

$$w(x\colon a\colon f) = \int_V dv\, e^{(\rho, H(v))} \overline{\psi}_f(e^x v e^{-x}) \hat{a}(x + H(v)). \tag{89}$$

Putting in (89) $f = 0$ we get, by (41), the following formula for the Laplace transform of the Harish-Chandra function (cf. Harish-Chandra [1958])

$$(c \cdot a)\hat{\ }(x) = \int_V dv\, \hat{a}(x + H(v)) e^{(\rho, H(v))}. \tag{90}$$

Put

$$w_+(x\colon \tau\colon a\colon f) = \lim_{\substack{t \to \infty \\ Q}} w(x + t\colon \tau + t\colon a\colon f). \tag{91}$$

Formula (88) and Theorem 48 imply the following integral formula for the difference of w and w_+:

$$w(x\colon \tau\colon a\colon f) - w_+(x\colon \tau\colon a\colon f)$$
$$= \int_{V_2} dv_2\, e^{(\rho, H(v_2))} [\overline{\psi}_f(e^x v_2^{w_1} e^{-x}) - 1] w_1(x + H(v_2)^{w_1}\colon \tau\colon a\colon f). \tag{92}$$

We shall use (88), (92) to prove the following main estimate.

Theorem 60. *Let $S \subset \mathfrak{a}^*$ be the unit sphere. For any $\varepsilon \in S$ there exist positive constants A, c (depending on a) such that*

$$|w(x + t\colon t\colon a)| \leqslant A e^{c\varepsilon(x)}$$

uniformly in $t \in \overline{C}_+$.

The proof goes by induction over the rank of group. Observe first of all that for $f = 0$ the theorem is obvious. We may prove it for non-degenerate f assuming that it is already proved for all groups of lower rank and for all degenerate characters. We begin with the estimates that generalize Proposition 54 for $G = \mathrm{SL}(2, \mathbb{R})$. The second step will be to generalize Proposition 55.

Let $Q_1, \ldots, Q_l$ be the maximal rank parabolics, $Q_i = Q_{P \setminus \{\alpha_i\}}$. Let W_i be their Weyl groups, $w_i \in W_i$ the corresponding maximal elements,

$$\Delta_i = \Delta(w_i w_0) = \{\alpha \in \Delta^+\colon w_0 w_i \alpha < 0\}.$$

Let C_i be the corresponding Weyl chambers,

$$C_i = \{x\colon \alpha(x) > 0 \text{ for } \alpha \in \Delta_i\}.$$

Proposition 61. *For any $\varepsilon \in S \cap \bigoplus_i C_i$, we have*

$$|w(x + t\colon t\colon a\colon f)| \leqslant A_\varepsilon e^{c\varepsilon(x)} \tag{93}$$

uniformly in $t \to \overline{C}_+$.

Proof. Fix a parabolic $Q = Q_k$, $k = 1, \ldots, l$. By the induction hypothesis,

$$|w_1(e^{x+t}: t: a: f)| \leqslant \tilde{A}_\varepsilon e^{c\varepsilon(x)}.$$

Choose $\varepsilon \in C_k$. Then the integral

$$I_\varepsilon = \int_{V_2} dv_2 \, e^{(\varepsilon+\rho, H(v_2))}$$

is convergent, and we get the estimate

$$|w(x + t: t: a: f)| \leqslant A_\varepsilon e^{\varepsilon(w_k(x))}$$

with the constant $A_\varepsilon = \tilde{A}_\varepsilon I_\varepsilon$. It is clear that w_k maps C_k into itself.

The estimate (93) implies that the wave packet exponentially decays outside the dual cone $C_k^* = \{x \in \mathfrak{a}: (\varepsilon, x) \geqslant 0 \ \forall \varepsilon \in C_k\}$ and hence outside $\bigcap_{k=1}^{l} C_k^*$.

Example. For $\mathfrak{g} = A_n$ the intersection $\bigcap_{k=1}^{n} C_k^*$ is the half-line $\mathbb{R}_+ \cdot \tilde{\alpha}$ spanned by the maximal root.

For other series the cone $\bigcap_k C_k^*$ is non-degenerate.

Let us now turn to the second step of induction. Our argument here will be close to the standard proof of the Gindikin-Karpelevich formula (Gindikin and Karpelevich [1962], Schiffmann [1971]).

For $s \in W$ put $\Delta(s) = \{\alpha \in \Delta^+; s^{-1}\alpha \in \Delta^-\}$. Recall that the length $l(s)$ of s is the number of elements $\alpha_i \in P$ in the shortest decomposition $s = s_{\alpha_1} \ldots s_{\alpha_l}$. Such a decomposition is called a reduced decomposition.

Lemma (Bourbaki [1968, Ch. 4]). (i) *If* $l(s_1 s_2) = l(s_1) + l(s_2)$ *we have* $\Delta(s_1 s_2) = \Delta(s_1) \cup s_1 \Delta(s_2)$.

(ii) *Put* $\mathfrak{v}_s = \bigoplus_{\alpha \in \Delta(s)} \mathfrak{g}_{-\alpha}$, $V_s = \exp \mathfrak{v}_s$. *Then* $V_s = V_{s_1} \cdot s_1 V_{s_2} s_1^{-1}$.

Fix a maximal parabolic $Q_F \supset B$, $F = P \backslash \{\alpha\}$ and let W_1 be its Weyl group, $w_1 \in W_1$ the maximal element, $s = w_1 w_0$. Put $\beta = -w_0\alpha$. Clearly, $\beta \in P$.

Lemma. $l(sw_\beta) = l(s) - 1$.

Put $s_1 = sw_\beta$.

Corollary. $V_s = V_{s_1} \cdot s_1 V_{w_\beta} s_1^{-1}$.

We shall apply these geometric observations in order to estimate the integral (92). Put $C(s) = \{H \in \mathfrak{a}: \alpha(H) > 0 \text{ for } \alpha \in \Delta(s)\}$.

Proposition 62. *For any* $\varepsilon \in C(s)$ *we have*

$$|w(x + t: t: a: f) - w_+(x + t: t: a: f)|$$
$$\leqslant A_\varepsilon e^{\varepsilon w_1(x)} \int_{V_{w_\beta}} dv \, e^{(\varepsilon w_1 s_1^{-1} + \rho_\beta, H(v))} \times |\exp\{-if_\alpha(\log v^{s_1 w_1}) e^{-\alpha(x+t)}\} - 1|. \quad (94)$$

Proof. We use the induction hypothesis to estimate $w_1(x\colon a\colon f)$. The rest of the argument is parallel to the proof of the Gindikin-Karpelevich formula. Let us make in (92) the change of variables $v_2 = v's_1 v s_1^{-1}$ and observe that $\psi_f(v_2)$ actually depends only on v. The integral over V_{s_1} is the standard integral of Gindikin-Karpelevich which converges due to the condition $\varepsilon \in C(s)$. □

A rough estimate of the integral over V_{w_β} similar to the one in Proposition 55 now yields the following result.

Proposition 63. *For any $\varepsilon \in C(s)$ we have*

$$|w(x+t\colon t\colon a\colon f) - w_+(x+t\colon t\colon a\colon f)| \leqslant A_1 e^{\varepsilon^{w_1}(x)} e^{-\alpha(x)+T} + A_2 e^{\varepsilon^{w_1}(x)} e^{-(2(\varepsilon^{w_1},\alpha)/(\alpha,\alpha))T} \tag{95}$$

uniformly in t.

Let us now combine the two estimates (93), (95). It is at this final stage that we should restrict ourselves to the case $\mathfrak{g} = A_n$. Choose $\varepsilon = \delta\cdot\tilde{\alpha}$ where $\delta > 0$ and $\tilde{\alpha}$ is the maximal root. Put $T = d(x_n - x_{n+1})$, $0 < d < 1$. Let us apply (95) choosing as the parabolic $Q_F = Q_{P\setminus\{\alpha_n\}}$. The r.h.s. of (95) takes the form

$$A_1 e^{-(1-d-\delta)(x_n - x_{n+1})} + A_2 e^{(1-2d)\delta(x_n - x_{n+1})}.$$

We may now choose d, δ in such a way that $1 - 2d > 0$, $0 < \delta < 1 - d$. Thus we get the following estimate which is uniform in t

$$|w(x+t\colon t\colon a\colon f)| \leqslant A e^{-c(x_n - x_{n+1})}. \tag{96}$$

By combining this result with Proposition 61 we get the following

Corollary. *For $\mathfrak{g} = A_n$ the wave packet exponentially decays in each half-space in a uniformly in $t \in \overline{C}_+$.*

This assertion is equivalent to Theorem 60 and is sufficient to prove the finiteness of energy for the wave packets. The Plancherel theorem now follows in the routine way as in Section 3.7.

Bibliographical Notes

The reduction of quantum bundles was introduced by Reyman and Semenov-Tian-Shansky [1980], Guillemin and Sternberg [1982]. The commutativity theorem for quantum integrals of motion for the Toda lattice is proved by Kostant [1978]. The role of Whittaker functions in the theory of quantum Toda lattices was made clear already by Kostant [1979]. The algebraic theory of Whittaker modules was elaborated in Kostant [1979]. The scattering problem for Toda lattices was solved by the author in 1978; the complete proofs were published in Semenov-Tian-Shansky [1984]. The quantization of open Toda lattices and related systems was also studied by Goodman and Wallach [1982, 1984]. The analytic properties of Whittaker functions were studied already in the 60's by Jacquet [1967] and Schiffmann [1971]. However, the spectral decomposition theorems were not established at that time. A related problem for spherical functions on a semisimple Lie group was solved in the fundamental paper of Harish-Chandra [1958], with a substantial contribution made by Gindikin and Karpelevich [1962]. The Plancherel theorems exposed here are due to the author. They are based on the non-stationary approach which was proposed in Semenov-Tian-Shansky [1976]. One key ingredient is the analogy between the intertwining operators for representations of principal series and the scattering operators which was pointed out by Gel'fand [1963] in his Stockholm talk. This talk served also as one of the principal sources for Jacquet [1967] and Schiffmann [1971] where the functional equations for Whittaker functions were established. More recently, important contributions to the theory of Whittaker functions were made by Hashizume [1980, 1982] who also studied the Plancherel theory for Toda lattices (Hashizume [1984]).

References*

Berezin, F.A. [1967]: Some remarks on the associative envelope of a Lie algebra. Funkts. Anal. Prilozh. *1*, No. 2, 1–14. English transl.: Funct. Anal. Appl. *1*, 91–102 (1968), Zbl.227.22020

Bourbaki, N. [1968]: Groupes et algèbres de Lie, Ch. IV–VI. Hermann: Paris, Zbl.186,330

Drinfel'd, V.G. [1987]: Quantum Groups. In: Proc. Int. Congr. Math., Berkeley 1986, Vol. 1, 798–820, Zbl.667.16003

Duflo, M. [1977]: Opérateurs différentiels bi-invariants sur un groupe de Lie. Ann. Sci. Ec. Norm. Super., IV, Sér. *10*, 265–288, Zbl.353.22009

Faddeev, L.D. [1980]: Quantum completely integrable systems in quantum field theory. Sov. Sci. Rev., Sect. C., Math. Phys. Rev. *1*, 107–155, Zbl.569.35064

Faddeev, L.D. [1984]: Integrable models in 1 + 1 – dimensional quantum field theory. In: Recent Advances in Field Theory and Statistical Mechanics. Les Houches, Amsterdam, Sess. 39, 563–608

Gel'fand, I.M. [1963]: Automorphic functions and the theory of representations. In: Proc. Intern. Congr. Math. 1962, 74–85, Zbl.138,71

Gel'fand, I.M., Graev, M.I., Vilenkin, N.Ya. [1962]: Integral geometry and the related questions of representation theory (Generalized Functions, Vol. 5) Nauka: Moscow. English transl.: Academic Press: New York, London 1966, Zbl.115,167

Gel'fand, I.M., Graev, M.I., Piatetskij-Shapiro, I.I. [1966]: Representation theory and automorphic functions. (Generalized Functions, Vol. 6). Nauka: Moscow. English transl.: B. Saunders Co.: Philadelphia 1969, Zbl.138,72

Gindikin, S.G., Karpelevich, F.I. [1962]: The Plancherel measure for Riemannian symmetric spaces of non-positive curvature. Dokl. Akad. Nauk SSSR *145*, 252–255. English. transl.: Sov. Math. Dokl. *3*, 962–965 (1962), Zbl.156,32

Goodman, R., Wallach, N.R. [1982]: Classical and quantum mechanical systems of Toda lattice type. I. Commun. Math. Phys. *83*, 355–386, Zbl.503.22013

Goodman, R., Wallach, N.R. [1984]: Classical and quantum mechanical systems of Toda lattice type. II. Commun. Math. Phys. *94*, 177–217, Zbl.592.58028

*For the convenience of the reader, references to reviews in Zentralblatt für Mathematik (Zbl.), compiled using the MATH database, have, as far as possible, been included in this bibliography.

Goodman, R., Wallach, N.R. [1986]: Classical and quantum mechanical systems of Toda lattice type. III. Joint eigenfunctions of the quantized systems. Commun. Math. Phys. *105*, No. 3, 473–509, Zbl.616.22010

Guillemin, V., Sternberg, S. [1982]: Homogeneous quantization and multiplicities of group representations. J. Funct. Anal. *47*, No. 3, 344–380

Gutzwiller, M.C. [1980]: The quantum mechanical Toda lattice. I. Ann. Phys. *124*, 347–381

Gutzwiller, M.C. [1981]: The quantum mechanical Toda lattice. II. Ann. Phys. *133*, 304–331

Harish-Chandra [1958]: Spherical functions on a semisimple Lie group. I, II. Am. J. Math. *80*, 241–310, 553–613, Zbl.93,128

Harish-Chandra [1966]: Discrete series for semisimple Lie groups. II. Acta Math. *116*, 1–111, Zbl.199,201

Harish-Chandra [1975]: Harmonic analysis on real reductive groups. I. J. Funct. Anal. *19*, 104–204, Zbl.315.43002

Harish-Chandra [1976a]: Harmonic analysis on real reductive groups. II. Invent. Math. *36*, 1–55, Zbl.341.43010

Harish-Chandra [1976b]: Harmonic analysis on real reductive groups. III. Ann. Math., II. Ser. *104*, 117–201, Zbl.331.22007

Hashizume, M. [1979]: Whittaker models for real reductive groups. Jap. J. Math., New Ser. *5*, 349–401, Zbl.506.22016

Hashizume, M. [1982]: Whittaker functions on semisimple Lie groups. Hiroshima Math. J. *12*, 259–293, Zbl.524.43005

Hashizume, M. [1984]: The spectral decomposition of the generalized quantum Toda lattice. Preprint, Hiroshima University

Jacquet, H. [1967]: Fonctions de Whittaker associées aux groupes de Chevalley. Bull. Soc. Math. Fr. *95*, 243–309, Zbl.155,59

Knapp, A.W., Stein, E.M. [1971]: Intertwining operators for semisimple Lie groups. Ann. Math., II. Ser. *93*, 489–578, Zbl. 257.22015

Kostant, B. [1970]: Quantization and unitary representation. I. Prequantization. In: Lect. Notes Math. *170*, 87–208, Zbl.223.53028

Kostant, B. [1979]: Quantization and representation theory. In: Representation Theory of Lie Groups, Proc. SRC/LMS Res. Symp., Oxford 1977. Lond. Math. Soc. Lect. Notes Ser. *34*, 287–316, Zbl.474.58010

Kostant, B. [1978]: Whittaker vectors and representation theory. Invent. Math. *48*, 101–184, Zbl.405.22013

Lax, P., Phillips, R.S. [1967]: Scattering Theory. Academic Press: New York-London, Zbl.186,163

Malikov, F.G. [1988]: On quantum integrals of motion. Usp. Mat. Nauk *43*, No. 4, 209–210. English transl.: Russ. Math. Surv. *43*, No. 4, 217–218 (1988)

Ol'shanetsky, M.A. (= Ol'shanetskij, M.A.), Perelomov, A.M. [1983]: Quantum integrable systems related to Lie algebras. Phys. Rep. *94C*, 313–404

Reyman, A.G., Semenov-Tian-Shansky, M.A. [1979]: Reduction of Hamiltonian systems, affine Lie algebras and Lax equations. I. Invent. Math. *54*, 81–100, Zbl.403.58004

Schiffmann, G. [1971]: Intégrales d'entrelacement et fonctions de Whittaker. Bull. Soc. Math. Fr. *99*, 3–72, Zbl.223.22017

Semenov-Tian-Shansky, M.A. [1976]: Harmonic analysis on Riemannian symmetric spaces of negative curvature and scattering theory. Izv. Akad. Nauk SSSR, Ser. Math. *40*, No. 3, 562–592. English transl.: Math. USSR, Izv. *10*, 535–563 (1976), Zbl.354.35071

Semenov-Tian-Shansky, M.A. [1984]: Quantum Toda lattices. Spectral theory and scattering. Preprint R-3-84. LOMI: Leningrad (Russian)

Sklyanin, E.K. [1985]: The quantum Toda chain. In: Nonlinear equations in classical and quantum field theory, Proc. Seurin., Paris, 1983–1984, Lect. Notes Phys. *226*, 196–233, Zbl.601.58039

Warner, G. [1972]: Harmonic Analysis on Semi-Simple Lie Groups. I, II. Die Grundlehren der Math. Wissenschaften *188*, *189*. Springer-Verlag: New York-Berlin-Heidelberg Zbl.265.22020, Zbl.265.22021

III. Geometric and Algebraic Mechanisms of the Integrability of Hamiltonian Systems on Homogeneous Spaces and Lie Algebras

V.V. Trofimov, A.T. Fomenko

Translated from the Russian
by A.G. Reyman

Contents

Chapter 1
Geometry and Topology of Hamiltonian Systems

§ 1. Symplectic Geometry

1.1. Symplectic Manifolds. A pair (M, ω) consisting of a $2n$-dimensional manifold M together with a closed 2-form ω is called a *symplectic manifold* if the form ω is nondegenerate, i.e. if $\omega^n = \omega \wedge \cdots \wedge \omega \not\equiv 0$.

Let f be a smooth function on a symplectic manifold (M, ω). The *skew gradient* sgrad f of f is a smooth vector field on M uniquely determined by the equation $\omega(v, \operatorname{sgrad} f) = v(f)$ where v ranges over the smooth vector fields on M and $v(f)$ is the value of the differential operator (vector field) v on the function f.

A smooth vector field v on a symplectic manifold (M, ω) is said to be *Hamiltonian* if it has the form $v = \operatorname{sgrad} F$ for some smooth function F on M which is called the *Hamiltonian*. The *Poisson bracket* of two smooth functions f and g on the symplectic manifold (M, ω) is a function $\{f, g\}$ defined by $\{f, g\} = \omega(\operatorname{sgrad} f, \operatorname{sgrad} g)$. The operation of Poisson bracket is bilinear, skew-symmetric, and satisfies the Jacobi identity, so that the space of smooth functions on a symplectic manifold in a natural way acquires the structure of an infinite-dimensional Lie algebra with respect to the Poisson bracket.

A smooth function f on a manifold M is an *integral* of a vector field v if it is constant along all integral curves of v. If $v = \operatorname{sgrad} F$ is a Hamiltonian field and f Poisson commutes with the Hamiltonian F, then f is an integral of v. In particular, the Hamiltonian F is always an integral of the field $v = \operatorname{sgrad} F$.

Two functions on a symplectic manifold are said to be *in involution* if their Poisson bracket vanishes.

An exposition of the topological concepts and theorems used in this survey can be found in the review papers by S.P. Novikov, Topology, and D.B. Fuks, Classical Manifolds, in Volume 12 of this Encyclopaedia.

Let (M^{2n}, ω) be a symplectic manifold. The form ω determines the following topological data: the cohomology class $a = [\omega] \in H^2(M, \mathbb{R})$ and a homotopy class of reductions of the structural group of the tangent bundle of M to the group $\operatorname{Sp}(2n, \mathbb{R})$, hence a homotopy class $[J]$ of almost complex structures on M. Gromov has shown that, if M is open, every pair $(a, [J])$ can be realized by some symplectic form (see Heafliger [1971]). If M is closed, a^n must be a generator of the group $H^{2n}(M, \mathbb{R})$ which is positive with respect to the orientation defined by $[J]$. To give some examples, let us mention that any two-dimensional surface is a symplectic manifold with respect to the area form, any Kähler manifold has the structure of a symplectic manifold, any cotangent bundle T^*M possesses a canonical symplectic structure. The details of these constructions can be found, for instance, in Arnol'd [1974], Arnol'd and Givental' [1985], Fomenko [1983a], Guillemin and Sternberg [1984a].

Thurston [1976] constructed a closed symplectic manifold which has no Kähler structure; his example is the quotient $\mathbb{R}^n/\Gamma$ of $\mathbb{R}^n$ by a discrete group Γ and so is not simply connected. Later McDuff [1984] constructed a simply connected non-Kähler manifold. Let us briefly describe her method. For that purpose we need the operation of "blowing up" a manifold X along a submanifold M. If a symplectic manifold (M, ω) is embedded in a symplectic manifold (X, σ), one may blow up X along M, to obtain a new symplectic manifold $(\tilde{X}, \tilde{\omega})$. Let M be a compact submanifold of codimension $2k$ in X such that the structural group of its normal bundle $E \to M$ is reduced to the unitary group $U(k)$. We assume that $k \geqslant 2$ since otherwise blowing up does not change X. Let U be a tubular neighborhood of M in X, and V a subdisc bundle of $\tilde{E}$ which is homeomorphic to U. Let PE be the projectivization of E, and $\widetilde{PE}$ the associated complex line bundle over PE. We have the natural mapping φ: $\widetilde{PE} \to E$ since the fiber over $x \in M$ can be thought of as a subspace in $\mathbb{C}^k$.

By definition, the *blow-up* $\tilde{X}$ of X along M is the smooth manifold

$$\tilde{X} = \overline{X \backslash U} \bigcup_{\partial\varphi^{-1}(V)} \varphi^{-1}(V),$$

where $\partial\varphi^{-1}(V)$ is identified with ∂V in the obvious way. For more details, see, for instance, McDuff [1984].

Let us now outline the construction of a non-Kähler simply connected symplectic manifold. To do so we need the example found by Thurston [1976]. *Thurston's manifold* is the quotient space $M = \mathbb{R}^4/\Gamma$ where Γ is the discrete group generated by the unit translations along the x_1, x_2, x_3 axes together with the transformation $(x_1, x_2, x_3, x_4) \mapsto (x_1 + x_2, x_2, x_3, x_4 + 1)$. Its symplectic form σ lifts to $dx_1 \wedge dx_2 + dx_3 \wedge dx_4$ on $\mathbb{R}^4$. In this way, M is fibered over the two-dimensional torus T^2 with fiber T^2. Recall that, for a topological space W, $\beta_k(W)$ denotes its k-th Betti number. In Thurston's example we have $\beta_1(M) = \beta_3(M) = 3$. Since these are odd, M has no Kähler structure.

Thurston's manifold can be symplectically embedded in $\mathbb{C}P^5$ with the standard symplectic structure (see Subsection 1.2). Let $\tilde{X}$ be the manifold obtained by blowing up $\mathbb{C}P^5$ along M as described above. We then have the following theorem.

Theorem 1.1. *$\tilde{X}$ is a simply connected symplectic manifold with $\beta_1(M) = \beta_3(\tilde{X}) = 3$. Hence, $\tilde{X}$ is not Kähler.*

The proof of this theorem, as well as the construction of a symplectic structure on the blow-up $\tilde{X}$ referred to above, can be found in McDuff [1984].

1.2. Embeddings of Symplectic Manifolds. As we saw in the preceding section, to be able to produce concrete examples of symplectic manifolds it is important to learn to embed one symplectic manifold into another: we then can define the blow-up $\tilde{X}$.

Consider the canonical homomorphism ε: $H^k(M, \mathbb{Z}) \to H^k(M, \mathbb{R})$ which maps the integer cohomology groups of the manifold M into its real cohomology groups. This homomorphism is induced by embedding $\mathbb{Z}$ into $\mathbb{R}$ and regarding integral cochains as real ones. A closed exterior k-form on M is *integral* if its cohomology class is the image of an integral class under this homomorphism.

Let (M, ω) and (N, σ) be symplectic manifolds. A mapping $f: M \to N$ is *symplectic* if $\omega = f^*\sigma$. Tischler [1977], Narasimhan, and Ramanan [1961], and Gromov [1971] showed that the complex projective space $\mathbb{C}P^n$ with its standard Kähler form Ω is a universal integral symplectic manifold. To put it another way, any manifold with integral symplectic form ω may be embedded in $\mathbb{C}P^n$ for suitably large n by a map f such that $f^*\Omega = \omega$. In fact, Gromov showed that if $\dim M = 2m$, one can take $n = 2m + 1$. In particular, if M is Thurston's manifold, then its symplectic structure is integral and so M embeds in $\mathbb{C}P^5$.

Tischler [1977] proved the following theorem.

Theorem 1.2. *Let M be a closed manifold and ω a closed integral 2-form on M. Then for N large enough there is a mapping $g: M \to \mathbb{C}P^N$ such that $g^*\Omega = \omega$ where Ω is the standard symplectic structure on the projective space $\mathbb{C}P^N$.*

By elaborating Tischler's proof, V.A. Popov was able to dispense with the hypothesis that M is closed and proved the following proposition.

Theorem 1.3. *Let M be any smooth manifold of dimension $m = 2n$ (or $m = 2n + 1$) and ω a closed integral 2-form. Then for $N = n(m + 2)$ (or $N = n(m + 2) + 1$ if $m = 2n + 1$) there is a mapping $f: M \to \mathbb{C}P^N$ such that $f^*\Omega = \omega$.*

The last two theorems do not assume that ω is a symplectic form, so that they only claim the existence of a smooth mapping, not necessarity an embedding, which induces the form ω from the standard symplectic structure Ω on the projective space $\mathbb{C}P^N$.

We now briefly discuss symplectic embeddings of noncompact symplectic manifolds. Let $\mathbb{R}^{2N}$ be endowed with symplectic form $\Omega = \sum_{i=1}^{N} dp_i \wedge dq_i$. A necessary condition for the existence of a symplectic embedding of (M, ω) into $(\mathbb{R}^{2N}, \Omega)$ is, of course, the exactness of ω. The results of Gromov imply that this condition is also sufficient.

Theorem 1.4. *Let (M^{2n}, ω) be a real-analytic noncompact symplectic manifold such that its symplectic structure ω is exact: $\omega = d\sigma$. Then there is always a symplectic embedding of (M^{2n}, ω) in $(\mathbb{R}^{2N}, \Omega)$ for some $N < \infty$. Moreover, one can take $N = n(2n + 1)$.*

1.3. Symplectic Geometry of the Coadjoint Representation. Let $\mathfrak{g}$ be a finite-dimensional Lie algebra, $\mathfrak{g}^*$ its dual space, and $\mathrm{Ad}^*: G \to \mathrm{GL}(\mathfrak{g}^*)$ the coadjoint representation of the Lie group G on $\mathfrak{g}^*$. Then $\mathfrak{g}^*$ becomes foliated into orbits, $\mathcal{O}$. Each orbit is an embedded smooth submanifold in $\mathfrak{g}^*$.

Theorem 1.5 (see Kirillov [1978], Arnol'd [1974]). *Every orbit of the coadjoint representation of a Lie group is a symplectic manifold.*

We recall the definition of the symplectic structure on coadjoint orbits (this structure will be called *Kirillov's form*). Let $x \in \mathcal{O}$ be a point of the orbit, and ξ_1, $\xi_2 \in T_x\mathcal{O}$ any two tangent vectors to the orbit. We shall determine the value $\omega_x(\xi_1, \xi_2)$ of ω. Every vector ξ tangent to the orbit can be expressed as $\xi = \mathrm{ad}_g^* x$. Hence there are elements $g_1, g_2 \in \mathfrak{g}$ such that $\xi_i = \mathrm{ad}_{g_i}^* x$, $i = 1, 2$. This representation is not in general unique but this does not affect the construction below. We define the value $\omega_x(\xi_1, \xi_2)$ of the form ω at $x \in \mathcal{O}$ on the tangent vectors ξ_1, ξ_2 to the orbit $\mathcal{O}$ as follows: $\omega_x(\xi_1, \xi_2) = \langle x, [g_1, g_2] \rangle$ where $x \in \mathfrak{g}^*$, $g_1, g_2 \in \mathfrak{g}$.

This is a bilinear form and its value does not depend on the arbitrariness in the choice of g_1, g_2. Also, this form is skew-symmetric and nondegenerate on the orbit, and, moreover, is closed. The proofs of these facts can be found, for instance, in Kirillov [1978], Arnol'd [1974].

1.4. Poisson Structures on Lie Algebras. Let G be a Lie group, $\mathfrak{g}$ its Lie algebra, $\mathfrak{g}^*$ the dual space of $\mathfrak{g}$. The natural pairing between $\mathfrak{g}$ and $\mathfrak{g}^*$ is denoted by $\langle\ ,\ \rangle$. The Poisson structure (for the definition see, for instance, Arnol'd and Givental' [1985]) on $\mathfrak{g}^*$ can be defined in three equivalent ways. To define a Poisson structure on $\mathfrak{g}^*$ we have to define, for any two functions F, H on $\mathfrak{g}^*$, their Poisson bracket $\{F, H\}$ which is bilinear, skew symmetric, and satisfies the Jacobi identity.

First Definition. Given a function $F\colon \mathfrak{g}^* \to \mathbb{R}$ we define $dF(x) \in \mathfrak{g}$ by $\langle dF(x), f \rangle = dF(x)(f)$ for any $f \in \mathfrak{g}^*$, i.e., we identify $\mathfrak{g}^{**}$ with $\mathfrak{g}$, so that $dF(x) \in \mathfrak{g}^{**}$ belongs to the Lie algebra $\mathfrak{g}$. We then define the Poisson bracket by

$$\{F, H\}(x) = \langle x, [dF(x), dH(x)] \rangle, \quad x \in \mathfrak{g}^*,$$

where $[X, Y]$ is the commutator in the Lie algebra $\mathfrak{g}$. This definition goes back to Berezin [1967]; the same bracket was also used by Kirillov and Arnol'd, see Arnol'd [1966], [1974].

Proposition 1.1. *The operation $\{F, H\}$ defined above is bilinear and antisymmetric, satisfies the Jacobi identity, and is a derivation with respect to each argument. Hence if defines a Poisson structure.*

This proposition can be proved by a straightforward computation.

Second Definition. As was noted in Theorem 1.5, every coadjoint orbit is a symplectic manifold. Let $\{F, H\}_{\mathcal{O}(x)}$ be the Poisson bracket on the symplectic manifold $\mathcal{O}(x)$. The whole space $\mathfrak{g}^*$ is the union of disjoint symplectic submanifolds: $\mathfrak{g}^* = \bigcup_{f \in \mathfrak{g}^*} \mathcal{O}(f)$ where $\mathcal{O}(f)$ is the coadjoint orbit passing through f. We define the Poisson bracket of the functions $F, H\colon \mathfrak{g}^* \to \mathbb{R}$ by $\{F, H\}(x) = \{F|_{\mathcal{O}(x)}, H|_{\mathcal{O}(x)}\}_{\mathcal{O}(x)}(x)$ where $x \in \mathfrak{g}^*$, $\mathcal{O}(x)$ is the orbit passing through $x \in \mathfrak{g}^*$, and $F|_{\mathcal{O}(x)}$ is the restriction of F to $\mathcal{O}(x)$.

Third Definition. Given two functions $f, g\colon \mathfrak{g}^* \to \mathbb{R}$, we extend them to left-invariant functions $\hat{f}, \hat{g}\colon T^*G \to \mathbb{R}$. Since the cotangent bundle T^*M of any manifold M has a canonical symplectic structure, we may consider the canonical Poisson bracket of functions on T^*M (see Arnol'd [1974], Fomenko [1983a], Sternberg [1964]). We then set, by definition,

$$\{f, g\} = \{\hat{f}, \hat{g}\}|_{\mathfrak{g}^*}.$$

Proposition 1.2. *The three definitions of the Poisson bracket on the dual space* $\mathfrak{g}^*$ *of the Lie algebra* $\mathfrak{g}$ *are equivalent.*

1.5. Euler Equations. Let F be a smooth function on the orbit $\mathcal{O}(f)$ of the coadjoint representation Ad_G^* of the Lie group G. The Hamiltonian vector field $v = \operatorname{sgrad} F$ on $\mathcal{O}(f)$ relative to the Kirillov form on $\mathcal{O}(f)$ is given by $v = \mathrm{ad}^*_{dF(x)}(x)$, $x \in \mathfrak{g}^*$. If F is a smooth function on $\mathfrak{g}^*$, then on each orbit there is a Hamiltonian vector field $v = \mathrm{ad}^*_{dF(x)}(x)$. The system of differential equations $\dot{x} = \mathrm{ad}^*_{dF(x)}(x)$ on $\mathfrak{g}^*$ associated to this vector field will be called the *Euler equations* on $\mathfrak{g}^*$. The Euler equations on $\mathfrak{g}^*$ have the remarkable property that the associated vector field is tangent to coadjoint orbits $\mathcal{O}(f)$, and on each orbit the equations are Hamiltonian.

We point out the following simple method for finding first integrals of Euler's equations. Let $e_1, \ldots, e_n$ be a basis of the Lie algebra $\mathfrak{g}$, $e^1, \ldots, e^n$ the dual basis of the dual space $\mathfrak{g}^*$, $x = x_i e^i$, and c_{ij}^k the structure tensor of $\mathfrak{g}$ relative to the basis $e_1, \ldots, e_n$, i.e. $[e_i, e_j] = c_{ij}^k e_k$.

Proposition 1.3. *Suppose that a function* $H \in C^\infty(\mathfrak{g}^*)$ *is constant on the coadjoint orbits of the Lie group* G *associated with the Lie algebra* $\mathfrak{g}$. *Then* H *is a first integral of Euler's equations* $\dot{x} = \mathrm{ad}^*_{dF(x)}(x)$. *In coordinate form this can be stated as follows: if* H *satisfies a system of partial differential equations*

$$c_{ij}^k x_k \frac{\partial H}{\partial x_j} = 0 \quad (i = 1, \ldots, \dim \mathfrak{g}),$$

then H *is a first integral of the system of Euler's equations.*

It is of considerable interest to study Euler's equations where the Hamiltonian F is a quadratic function on $\mathfrak{g}^*$. Let us discuss this case in more detail. Given a linear operator $C\colon \mathfrak{g}^* \to \mathfrak{g}$ we can define a quadratic function $F(x) = \frac{1}{2}\langle x, C(x)\rangle$. We shall assume that C is self-adjoint, i.e., $\langle y, C(x)\rangle = \langle C(y), x\rangle$. Then $dF(x) = C(x)$. In this way, any self-adjoint linear operator $C\colon \mathfrak{g}^* \to \mathfrak{g}$ gives rise to a system of nonlinear differential equations $\dot{x} = \mathrm{ad}^*_{C(x)}(x)$ on $\mathfrak{g}^*$. This is a Hamiltonian system on each coadjoint orbit of G; one can take for the Hamiltonian the function $F(x) = \frac{1}{2}\langle x, C(x)\rangle$ restricted to the coadjoint orbit of G.

In coordinate form, the operator C is given by a matrix a^{ij} where $C(e^i) = a^{ij} e_j$. Euler's equations are then written as

$$\dot{x}_s = a^{ij} c_{js}^k x_k x_i, \quad s = 1, \ldots, \dim \mathfrak{g},$$

where $[e_i, e_j] = c_{ij}^k e_k$.

1.6. Euler Equations Arising in Problems of Mathematical Physics. We shall give here a list of Euler equations which arise in various problems of mechanics and mathematical physics. The Hamiltonian F in these examples need not be a quadratic form on $\mathfrak{g}^*$. For instance, in order to include a potential, we consider Hamiltonians which are the sum of a quadratic form and a linear functional.

1. The equations of motion of the three-dimensional rigid body about a fixed point are Euler's equations for the Lie algebra SO(3) of real skew-symmetric 3×3 matrices. They can be written in commutator form $\varphi(\dot{X}) = [\varphi X, X]$, $X \in \mathrm{SO}(3)$, where $\varphi(X) = IX + XI$, $I = \mathrm{diag}(\lambda_1, \lambda_2, \lambda_3)$, $\lambda_1 = \frac{1}{2}(-A + B + C)$, $\lambda_2 = \frac{1}{2}(A - B + C)$, $\lambda_3 = \frac{1}{2}(A + B - C)$, and A, B, C are the moments of inertia of the rigid body relative to the axes Ox, Oy, Oz, respectively; for more details see Arnol'd [1974].

Definition 1.1. *The equations of motion of the n-dimensional rigid body* with a fixed point are $\varphi(\dot{X}) = [\varphi X, X]$, $X \in \mathrm{SO}(n)$, where $\varphi(X) = IX + XI$, $I = \mathrm{diag}(\lambda_1, \ldots, \lambda_n)$ ($\lambda_i + \lambda_j \neq 0$ for all i, j).

An invariant description of the operator φ in terms of semisimple Lie algebras was given by Mishchenko and Fomenko [1976], [1978b], [1979a] (see Section 1.4 of Chapter 2).

2. The equations of motion of a free rigid body are Euler's equations for the Lie algebra $E(3)$ of the group of motions of three dimensional Euclidean space $\mathbb{R}^3$ (for more details see Novikov and Schmeltser [1981]).

3. The equations describing the motion of a rigid body about a fixed point in the Newtonian gravitational field are Euler's equations for the Lie algebra L_{12} defined as follows: L_{12} is the semidirect sum $\mathrm{SO}(3) + \mathbb{R}^3 + \mathbb{R}^3 + \mathbb{R}^3$ with basis X_i, Y_j^α, $i, j, \alpha = 1, 2, 3$, in which the commutation relations have the form $[X_i, X_j] = \varepsilon_{ijk} X_k$, $[X_i, Y_j^\alpha] = \varepsilon_{ijk} Y_k^\alpha$, $[Y_i^\alpha, Y_j^\beta] = 0$ (for more details see Bogoyavlensky [1984]).

4. The dynamics of a rigid body with distributed electric charge in an ideal incompressible fluid in gravitational and electric fields under the assumption that the hydrostatic lift is equal to the force of gravity and the total electric charge is zero, is described by Euler's equations on the dual space L_{12}^* of the Lie algebra L_{12} defined above (for more details see Bogoyavlensky [1984]).

5. The rotation of a magnetized rigid body about a fixed point in homogeneous gravitational and magnetic fields is described by Euler's equations for the Lie algebra $L = \mathrm{SO}(3) + \mathbb{R}^3 + \mathbb{R}^3$ with the commutation relations analogous to those written in Subsection 3 (for details see Bogoyavlensky [1984]).

6. The dynamics of a rigid body with an ellipsoidal cavity filled with a magnetic fluid which is in homogeneous vortex motion is described by Euler's equations for the Lie algebra $\mathrm{SO}(3) + \mathbb{R}^3$ (see Bogoyavlensky [1984]).

7. Let $A_{k,m}$ denote the direct sum $\mathrm{SO}(3) \oplus \cdots \oplus \mathrm{SO}(3) \oplus E(3) \oplus \cdots \oplus E(3)$ of k copies of the Lie algebra SO(3) and m copies of the Lie algebra $E(3)$. The dynamics of a rigid body with n ellipsoidal cavities filled with a magnetic fluid

which is in homogeneous vortex motion is described by Euler's equations for the Lie algebra $L = A_{1,n}$. If in k of these cavities the magnetic field is absent, the Lie algebra L reduces to $A_{k+1,n-k}$, see Bogoyavlensky [1984].

8. Rotation of a rigid body with n ellipsoidal cavities filled with magnetic fluid about a fixed point in a Newtonian gravitational field is described by Euler's equations for the Lie algebra $L = L_{12} \oplus A_{0,n}$. If in k of these cavities the magnetic field is absent the Lie algebra L reduces to $L_{12} \oplus A_{k,n-k}$, see Bogoyavlensky [1984].

9. Kirchhoff's equations of motion of a heavy rigid body in fluid are Euler's equations for the Lie algebra $E(3)$ of the group of motions of three-dimensional Euclidean space $\mathbb{R}^3$. We recall that $E(3)$ is the semidirect sum of the Lie algebra SO(3) and the commutative Lie algebra $\mathbb{R}^3$ with the standard action of SO(3) on $\mathbb{R}^3$. For more details see Novikov and Schmeltser [1981], Arnol'd [1969], Trofimov and Fomenko [1983a].

10. Finite-dimensional analogues of the equations of magnetohydrodynamics for an incompressible perfectly conducting fluid were obtained by Vishik and Dolzhansky [1978]. These equations were expressed as Euler's equations by Trofimov [1982a], [1983a].

Let $\mathfrak{g}$ be a Lie algebra. By $\Omega(\mathfrak{g})$ we denote the Lie algebra which as a linear space is $\mathfrak{g} \oplus \mathfrak{g}$ with the operation of commutation given by $[x_1 \oplus y_1, x_2 \oplus y_2] = [x_1, x_2] \oplus ([y_1, x_2] + [x_1, y_2])$. Finite-dimensional analogues of the equations of magnetohydrodynamics for an incompressible perfectly conducting fluid are Euler's equation for the Lie algebra $\Omega(\mathrm{SO}(3))$ (for more details see Trofimov [1983a]).

Some interesting Euler equations result from infinite-dimensional Lie algebras. We shall give two examples of this kind.

11. Let $L = L^2(\mathbb{R}^1, dx)$ where dx is the usual Lebesgue measure and let $H = L \oplus L$. We define a skew-symmetric bilinear form on H by $((f_1, g_1), (f_2, g_2))_s = (f_1, g_2) - (f_2, g_1)$ where (f, g) is the standard scalar product in L. Let $\mathrm{Sp}^2(H)$ denote the Lie algebra of real operators on H which have a common domain of definition, preserve the form $(f, g)_s$, and are Hilbert-Schmidt operators. Euler's equations for the Lie algebra $\mathrm{Sp}^2(H)$ include the classic *Korteweg-de Vries equation* $v_t = v_{xxx} - 6vv_x$ (see Berezin and Perelomov [1980], Guillemin and Sternberg [1984a]).

12. Let $D(N)$ denote the Lie algebra of vector fields on a manifold N, let $\mathbb{R}^p$ be the Euclidean space with coordinates $y_1, \ldots, y_p$, and let $D^p(N)$ be the subalgebra of $D(N \times \mathbb{R}^p)$ consisting of the vector fields of the form $X = X' + \sum_{s=1}^{p} f_s \frac{\partial}{\partial y_s}$, $X' \in D(N)$, $f_s \in C^\infty(N)$. By $\Lambda^{i,p}(N)$ we denote the space of i-forms on N pulled back to $N \times \mathbb{R}^p$ by means of the projection $N \times \mathbb{R}^p \to N$. Let $\mathfrak{g} = D^p(N) + \Lambda^{i,p}(N)$ be the semidirect sum with commutator $[(X, \omega), (Y, \nu)] = ([X, Y], X(\nu) - Y(\omega))$ where $X(\nu)$ is the Lie derivative of the form ν with respect to the vector field X.

Euler's equations for $\mathfrak{g}$ in the case $p = 2$, $i = n - 1$, are the equations of magnetohydrodynamics (see Holm and Kupershmidt [1983]).

§ 2. Some Classical Mechanisms of Integrability

2.1. The Hamilton-Jacobi Equation. Let $H(t, a_1, \dots, a_n, b_1, \dots, b_n)$ be a smooth function of the variables $t, a_1, \dots, a_n, b_1, \dots, b_n$. The equation of the form $\frac{\partial S}{\partial t} + H\left(t, \frac{\partial S}{\partial q_1}, \dots, \frac{\partial S}{\partial q_n}, q_1, \dots, q_n\right) = 0$ will be called the *Hamilton-Jacobi equation.*

Definition 1.2. A solution $S(t, q, \alpha)$ of the Hamilton-Jacobi partial differential equation containing n arbitrary constants $\alpha = (\alpha_1, \dots, \alpha_n)$ is said to be a *complete integral* of this equation if $\det(\partial^2 S/\partial q \partial \alpha) \neq 0$.

Theorem 1.6 (Jacobi). *If $S(t, q, \alpha)$ is a complete integral of the Hamilton-Jacobi equation $\frac{\partial S}{\partial t} + H\left(t, \frac{\partial S}{\partial q}, q\right) = 0$, then the equations of motion of the Hamiltonian system*

$$\begin{cases} \dot{q}_i = \dfrac{\partial H(t, p, q)}{\partial p_i}, \\ \dot{p}_i = -\dfrac{\partial H(t, p, q)}{\partial q_i} \end{cases}$$

with Hamiltonian H can be written in the form

$$\frac{\partial S}{\partial q_i} = p_i, \quad \frac{\partial S}{\partial \alpha_i} = \beta_i \quad (i = 1, \dots, n),$$

where α_j and β_j are arbitrary constants $(j = 1, \dots, n)$.

In this way, the problem of integrating a Hamiltonian system in replaced by the equivalent problem of determining the complete integral of the Hamilton-Jacobi equation.

The general solution of a partial differential equation depends on several arbitrary functions. Therefore, a complete integral of the Hamilton-Jacobi equation is by no means its general solution. Compared to the general solution, a complete integral embraces only a "handful" of solutions of the Hamilton-Jacobi equation.

Let us now outline the coordinate-free Hamilton-Jacobi theory.

Definition 1.3. Let Y be a smooth manifold and (M^{2n}, ω) a symplectic manifold. A mapping $s\colon Y \to M^{2n}$ is said to be isotropic if $s^*\omega = 0$; if, moreover, s is an immersion, i.e. if the differential ds is everywhere injective, then Y, together with s, is called an immersed *Lagrangian submanifold* of M^{2n} provided that $\dim Y = \frac{1}{2} \dim M^{2n} = n$.

To give an example of a Lagrangian submanifold we note that for $\varphi \in C^\infty(M)$, the graph $\Lambda = d\varphi(M) \subset T^*M$ of the differential $d\varphi$ is a Lagrangian submanifold in the cotangent bundle T^*M of M.

Definition 1.4. Let (M, ω) be a symplectic manifold. A *Hamilton-Jacobi equation* on M is an arbitrary hypersurface V in M. Any Lagrangian submanifold L lying in V is called a *solution* of the equation V.

The problem of finding Lagrangian submanifolds contained in the hypersurface V is called the *Hamilton-Jacobi problem.*

The relationship between the solutions of the Hamilton-Jacobi equation in the classical sense and in the sense of Definition 1.4 is discussed by Vinogradov and Kupershmidt [1977].

To establish the relationship between the solutions of the Hamilton-Jacobi equation and the solutions of Hamiltonian equations we consider the following construction. Suppose we are given a one-parameter family $V_c = \{H = c\}$, $c \in \mathbb{R}$, of level surfaces of the Hamiltonian H and the associated family of Hamilton-Jacobi problems in the symplectic manifold K^{2n}. Let U be a domain in K^{2n}, let V be a domain in the parameter space $\mathbb{R}^n$, and let $\varphi: U \to V$ be a mapping of U onto V such that a) φ is regular, i.e. $d\varphi$ has maximal rank; b) for every $v \in V$, $\varphi^{-1}(v) = L_v$ is a Lagrangian submanifold contained in some level surface of H. Let $Q_i: V \to \mathbb{R}$ $(i = 1, \ldots, n)$ be the coordinate functions on V. If $V \ni v = (c_1, \ldots, c_n)$ then $L_v = \bigcap_{i=1}^{n} \{\bar{Q}_i = c_i\}$, where $\bar{Q}_i = \varphi^*(Q_i) = Q_i \circ \varphi$, are parameters of the Lagrangian manifolds L_v regarded as functions on U.

Theorem 1.7. *The functions $\bar{Q}_i$ constitute a set of n Poisson commuting functions in U, i.e. $\{\bar{Q}_i, \bar{Q}_j\} = 0$, and their differentials are linearly independent.*

The situation described above, in which there exist n functionally independent Poisson commuting integrals, will be studied in more detail in the following sections. Motivated by the importance of this situation we formulate it as a special definition.

Definition 1.5. A system sgrad H on a symplectic manifold (M^{2n}, ω) is said to be *completely integrable* if there exists a set of functions $f_1, \ldots, f_n$ such that: a) f_i, $i = 1, \ldots, n$, are integrals of the flow sgrad H, i.e. $\{f_i, H\} \equiv 0$; b) f_i are functionally independent on M^{2n}; c) $\{f_i, f_j\} \equiv 0$ $(i, j = 1, \ldots, n)$; d) $n = \frac{1}{2} \dim M^{2n}$.

2.2. Integration of the Equations of Motion According to Liouville and Stäckel. If the kinetic energy T and the potential U of a mechanical system do not involve time and are defined by

$$2T = B \sum_{j=1}^{k} A_j(q_j)\dot{q}_j^2, \quad U = \frac{1}{B} \sum_{j=1}^{k} U_j(q_j),$$

where $B = \sum_{j=1}^{k} B_j(q_j)$ and each of the functions A_j, U_j and B_j depends only on the single variable q_j, $j = 1, \ldots, k$, then Hamilton's equations with Hamiltonian H can be integrated by quadratures.

The above integrability case was indicated by Liouville. Here one can find explicitly the complete integral of the Hamilton-Jacobi equation:

$$S = \sum_{j=1}^{k} \int \sqrt{2A_j(q_j)[U_j(q_j) + hB_j(q_j) + \alpha_j]}\, dq_j.$$

The next integrable case of Hamilton's equations was described by Stäckel. Suppose we are given $k(k+1)$ functions $\psi_{ji}(q_j)$, $U_j(q_j)$, $(i, j = 1, \ldots, k)$ each of which depends only on one of the variables q_j, and assume that $\Delta = \det(\varphi_{ji}(q_j)) \not\equiv 0$. If T and U do not involve time and are defined by

$$T = 2^{-1} \sum_{j=1}^{k} A_j p_j^2, \quad U = \sum_{j=1}^{k} A_j U_j(q_j),$$

where $A_j = \Delta^{-1} \dfrac{\partial \Delta}{\partial \varphi_{j1}}$ $(j = 1, \ldots, k)$, then Hamilton's equations can be integrated by quadratures.

In this case the Hamilton-Jacobi equation has a complete integral

$$S = \sum_{j=1}^{k} \int \sqrt{2U_j(q_j) + 2h\varphi_{j1}(q_j) + \sum_{i=2}^{k} \alpha_i \varphi_{ji}(q_j)}\, dq_j.$$

2.3. Lie's Theorem. Let us consider a canonical system $\dot{p} = -\dfrac{\partial H}{\partial q}$, $\dot{q} = \dfrac{\partial H}{\partial p}$, $H = H(p, q)$, where $p = (p_1, \ldots, p_n)$, $q = (q_1, \ldots, q_n)$. Suppose that we are given a system of n functionally independent functions $F_1(p, q) = H, \ldots, F_n(p, q)$ such that $\{F_i, F_j\} = c_{ij}^k F_k$ where $c_{ij}^k = \text{const}$. The linear space A spanned by $F_1, \ldots, F_n$ is then a finite-dimensional Lie algebra with c_{ij}^k being its structure constants in the basis $F_1, \ldots, F_n$.

A subspace C of A is called an ideal if for $f \in C$, $g \in A$ the Poisson bracket $\{f, g\}$ lies in C. The Lie algebra A is solvable if there is a sequence $A = A_0 \supset A_1 \supset \cdots \supset A_k = \{0\}$ of subalgebras such that A_{i+1} is an ideal of codimension 1 in A_i $(i = 0, 1, \ldots, n-1)$. In particular, commutative Lie algebras $\{f, g\} \equiv 0$ are solvable (see the Appendix).

Theorem 1.8. *Suppose that*

1) *the functions $F_1, \ldots, F_n$ are functionally independent on the set $M_a = \{p, q \mid F_1 = a_1, \ldots, F_n = a_n\}$;*

2) $\sum_{k=1}^{n} c_{ij}^k a_k = 0$ *for all $i, j = 1, \ldots, n$;*

3) *the Lie algebra A is solvable and $\{F_1, F_i\} = c_{1i}^1 F_1$. Then the solutions of the system $\dot{p} = -\dfrac{\partial H}{\partial q}$, $\dot{q} = \dfrac{\partial H}{\partial p}$ that lie on the surface M_a can be found by quadratures.*

2.4. Liouville's Theorem. We shall describe here the structure of completely integrable Hamiltonian systems (see Definition 1.5).

Theorem 1.9. *Let (M^{2n}, ω) be a symplectic manifold and let $\rho\colon M^{2n} \to B^n$ be a fibration whose fibers are compact connected Lagrangian submanifolds. Then for every point $b \in B^n$ there is a transitive action of the vector space $T_b^*B^n$ on the fiber*

$\rho^{-1}(b)$, and all the fibers are tori. Each linear differential form α on B^n defines an automorphism k_α of the fibration ρ such that $k_\alpha^*\omega = \omega + d(\rho^*\alpha)$.

The proof of this theorem can be found in Guillemin and Sternberg [1984a].

The action mentioned in the theorem can be described as follows. We have a surjective mapping of linear spaces: $d\rho_x$: $T_xM^{2n} \to T_bB^n$. The dual mapping $d\rho_x^*$ $(b = \rho(x))$ sends $X = T_b^*B^n$ into those covectors in $T_x^*M^{2n}$ that vanish on T_xF, the tangent space to the fiber F of ρ. Since F is Lagrangian, the symplectic form ω gives an identification of $d\rho^*(X)$ with T_xF. In this way an element $v \in T_b^*B^n$ gives rise to a tangent vector $\hat{v} \in T_xF$. Since F is compact and connected, we get an action of the commutative group $T_b^*B^n$ on F: an element $\alpha \in T_b^*B^n$ acts as the shift k_α along the trajectories of the vector field $\hat{\alpha}$.

In the setting of Theorem 1.9 the symplectic form ω can be given a canonical description. Let us assume that the fibration ρ: $M^{2n} \to B^n$ admits a sections: $B \to M$ such that the de Rham cohomology class of the form $s^*\omega$ vanishes. Then by modifying s we can assume further that $s^*\omega = 0$. Define a mapping χ: $T^*B \to M$ by $\chi(v) = -vs(x)$, $v \in T_b^*B$, where va is the action described above.

Theorem 1.10. *Under the above assumptions, $\chi^*\omega$ is the standard symplectic form on the cotangent bundle T^*B.*

The proof of this theorem can be found in Guillemin and Sternberg [1984a].

Theorem 1.11. *Under the hypotheses of Theorem 1.9, in an open neighbourhood of the fiber there exists a system of coordinates $s_1, \ldots, s_n, \varphi_1, \ldots, \varphi_n$ with $0 \leqslant \varphi_i < 2\pi$ (known as action-angle variables) such that*

a) *the symplectic structure ω is expressed in the simplest way $\omega = \sum_{i=1}^{n} ds_i \wedge d\varphi_i$, which is equivalent to the relations $\{s_i, s_j\} = \{\varphi_i, \varphi_j\} = 0$, $\{s_i, \varphi_j\} = \delta_{ij}$;*

b) *the functions $s_1, \ldots, s_n$ are a set of coordinates in the directions transversal to the torus.*

c) *the functions $\varphi_1, \ldots, \varphi_n$ are a set of coordinates along the torus $T^n = S^1 \times \cdots \times S^1$ where φ_i is the angular coordinate on the i-th circle.*

If σ is a linear differential form on M such that $\omega = d\sigma$ and $\gamma_i(c)$ are smoothly varying curves in the fiber above c whose homotopy classes $[\gamma_i(c)]$, $i = 1, \ldots, n$, form a basis of the fundamental group of the fiber above c for each c, then the functions

$$I_i(c) = \frac{1}{2\pi} \int_{\gamma_i(c)} \sigma, \quad i = 1, \ldots, n,$$

give action variables, whose conjugate variables give angle variables.

For the proof see, for instance, Guillemin and Sternberg [1984a] or Arnol'd [1974]. For completely integrable systems we have the following theorem.

Theorem 1.12 (see Arnol'd [1974], Dubrovin et al. [1985], Abraham and Marsden [1978]). *Let M^{2n} be a symplectic manifold and let $f_1, \ldots, f_n$ be a set of smooth functions that are in involution on M^{2n}, i.e. $\{f_i, f_j\} \equiv 0$ for $1 \leqslant i, j \leqslant n$. Let*

M_ξ denote the common level set of $f_1, \ldots, f_n$, i.e. $M_\xi = \{x \in M | f_i(x) = \xi_i, 1 \leqslant i \leqslant n\}$. Assume that the functions $f_1, \ldots, f_n$ are functionally independent on M_ξ. If M_ξ is compact and connected, then

1) M_ξ is a smooth manifold diffeomorphic to the n-torus T^n;

2) In an open neighbourhood of M_ξ one can introduce regular coordinates $s_1, \ldots, s_n, \varphi_1, \ldots, \varphi_n$ where $0 \leqslant \varphi_i < 2\pi$ (known as action-angle coordinates) such that a) *the symplectic structure ω takes the canonical form $\omega = \sum_{i=1}^{n} ds_i \wedge d\varphi_i$;* b) *the functions $s_1, \ldots, s_n$ are a set of coordinates in the direction transverse to the torus and can be expressed as functions of the integrals $f_1, \ldots, f_n$;* c) *the functions $\varphi_1, \ldots, \varphi_n$ are a set of coordinates on the torus $T^n = S^1 \times \cdots \times S^1$, where φ_i is the angular coordinate on the i-th circle S^1;* d) *in terms of the coordinates $\varphi_1, \ldots, \varphi_n$ on T^n every vector field $v =$ sgrad F, where F is one of the functions $f_1, \ldots, f_n$, takes the form $\dot{\varphi}_i = q_i(\xi_1, \ldots, \xi_n)$, i.e. its components are constant along the torus and the integral curves give rise to conditionally periodic motion described by a "rectilinear winding" on the torus. The functions $q_1, \ldots, q_n$ are defined in a neighbourhood of T^n, and on nearby level sets one has $\dot{\varphi}_i = q_i(s_1, \ldots, s_n)$. Thus in the neighbourhood of the torus the system $v =$ sgrad F can be written in the form $\dot{s}_i = 0$, $\dot{\varphi}_i = q_i(s_1, \ldots, s_n)$.*

§3. Non-commutative Integration According to Liouville

3.1. Non-commutative Lie Algebras of Integrals. We consider the situation described in Theorem 1.12. Let V be the linear space spanned by the functions $f_1, \ldots, f_n$, i.e. $V = \bigoplus_{i=1}^{n} \mathbb{R} f_i$. Then V is a commutative Lie algebra with respect to Poisson bracket. In applications one often encounters the situation where the first integrals of a Hamiltonian system do not Poisson commute but nevertheless form a finite-dimensional Lie algebra with respect to Poisson bracket. Liouville's Theorem 1.12 can be generalized to include this case. We shall call the classical Liouville theorem the commutative Liouville theorem to emphasize that V is an Abelian Lie algebra.

Definition 1.6. Let $\mathfrak{g}$ be a finite-dimensional Lie algebra. The *annihilator of a covector* $f \in \mathfrak{g}^*$ is the space $\mathrm{Ann}(f) = \{\xi \in \mathfrak{g} | \mathrm{ad}_\xi^* f = 0\}$. The *rank of a covector* $f \in \mathfrak{g}^*$ is the number $r(f) = \dim \mathrm{Ann}(f)$. The number $r = \inf_g \dim \mathrm{Ann}(g)$ is called the *index* of $\mathfrak{g}$ and is denoted by ind $\mathfrak{g}$. A covector is called *regular* if $r(f) = r$.

We have the following important theorem whose proof can be found, for instance, in Dixmier [1974].

Theorem 1.13. *If $f \in \mathfrak{g}^*$ is a regular covector then* $\mathrm{Ann}(f)$ *is an Abelian subalgebra of* $\mathfrak{g}$.

3.2. Non-commutative Liouville Theorem. The classical Liouville theorem can be generalized to the case of non-commutative algebras of integrals of Hamiltonian systems. The following theorem was proved by Mishchenko and Fomenko [1978a] (see also Fomenko [1983a]).

Theorem 1.14. *Let* $f_1, \ldots, f_k$ *be a collection of* k *smooth functions on a symplectic manifold* M^{2n} *whose linear hull is a* k*-dimensional Lie algebra relative to the Poisson bracket, i.e.* $\{f_i, f_j\} = c_{ij}^q f_q$ *where the* c_{ij}^q *are constants. Let* T^r *be a non-singular common level surface of the* f_i*, i.e.* $f_i(x) \equiv \xi_i$ *on* T^r*, where the* ξ_i *are constants. We suppose that* $\dim \mathfrak{g} + \operatorname{ind} \mathfrak{g} = \dim M$*, i.e.* $k + r = 2n$*, where* $r = \operatorname{ind} \mathfrak{g}$. *If the* f_i *are functionally independent on* T^r*, then* T^r *is a smooth* r*-dimensional submanifold and is invariant relative to each vector field sgrad* h*, where* h *belongs to the annihilator* $\mathfrak{H}_\xi$ *of the covector* $\xi = (\xi_1, \ldots, \xi_k)$*. In other words, the Hamiltonian* h *is an element of* $\mathfrak{g}$ *and lies in the annihilator of the covector in general position that defines the non-singular level surface* T^r*. Then each compact connected level surface* T^r *is diffeomorphic to an* r*-dimensional torus, and on this torus and all tori sufficiently close to it one can introduce regular coordinates* $\varphi_1, \ldots, \varphi_r$ *such that the vector field* v *takes the form* $\dot{\varphi}_i = q_i(\xi_1, \ldots, \xi_k)$*, i.e., the components of the field are constant on the torus (in these coordinates) and the integral curves of the field define a conditionally periodic motion; they give a "rectilinear winding" of the torus.*

By now several other analogues of Theorem 1.14 have been obtained. For a detailed survey of non-commutative integration methods see Trofimov and Fomenko [1984]. We shall formulate the situation described in Theorem 1.14 as a separate definition.

Definition 1.7. Let $v = \operatorname{sgrad} h$ be a Hamiltonian system on a symplectic manifold M^{2n} satisfying the hypotheses of Theorem 1.14, i.e. its commutative Lie algebra of integrals admits an embedding as an annihilator in a larger, generally speaking non-commutative, Lie algebra $\mathfrak{g}$ such that $\dim \mathfrak{g} + \operatorname{ind} \mathfrak{g} = \dim M$. Here $\mathfrak{g}$ does not have to be an algebra of integrals for v. Then we say that the system v is *completely integrable in the non-commutative sense.*

3.3. Interrelationships of Systems with Commutative and Non-commutative Symmetries

Definition 1.8. Let $\mathfrak{g}$ be a finite-dimensional Lie algebra. A finite-dimensional space F of functions defined on the dual space $\mathfrak{g}^*$ of $\mathfrak{g}$ is called a *complete involutive family of functions* for the Lie algebra $\mathfrak{g}$ if the following three conditions are fulfilled: 1) Any two functions $f, g \in F$ are in involution on all coadjoint orbits in $\mathfrak{g}^*$. 2) There is a linear basis in F which consists of functionally independent functions. 3) The relation $\dim \mathfrak{g} + \operatorname{ind} \mathfrak{g} = 2 \dim F$ holds.

The following two conjectures were formulated by Mishchenko and Fomenko [1981].

Conjecture 1 (the equivalence conjecture). *Suppose that a Hamiltonian system* $v = \operatorname{sgrad} f$ *on a symplectic manifold is completely integrable in the non-commutative sense. Then this system is also completely integrable in the usual commutative (Liouville) sense, i.e. there is another commutative Lie algebra* $\mathfrak{g}_0$ *of functionally independent integrals such that* $2 \dim \mathfrak{g}_0 = \dim M$ *and* $f \in \mathfrak{g}_0$. *Here the integrals in* $\mathfrak{g}_0$ *can be expressed as functions of the integrals in* $\mathfrak{g}$.

It turns out that this conjecture is closely connected with another conjecture on the existence of complete involutive collections of functions on generic coadjoint orbits (see Mishchenko and Fomenko [1981]).

Conjecture 2 (the conjecture on the fibering of orbits into the Liouville tori). *For any finite-dimensional Lie algebra* $\mathfrak{g}$ *there exists a complete involutive family of functions on* $\mathfrak{g}^*$.

There is a close connection between Conjectures 1 and 2. Namely, the following theorem holds.

Theorem 1.15 (see Mishchenko and Fomenko [1978a], [1981]). *Suppose that on a symplectic manifold* M^{2n} *there is given a non-commutative collection of functions which form a Lie algebra* $\mathfrak{g}$ *with the condition* $\dim \mathfrak{g} + \operatorname{ind} \mathfrak{g} = 2n$. *If Conjecture 2 holds for* $\mathfrak{g}$, *then there is a complete involutive collection* $\mathfrak{g}_0$ *of functions on* M *(i.e.* $\dim \mathfrak{g}_0 = n$*) and the commutative Lie algebra* $\mathfrak{g}_0$ *consists of functions that can be expressed functionally in terms of functions in the original non-commutative Lie algebra* $\mathfrak{g}$. *In other words, if Conjecture 2 holds, then from the non-commutative integrability of a Hamiltonian system of the form* $\operatorname{sgrad} h$, *where* $h \in \mathfrak{H}_\xi$, *it follows that it is integrable in the commutative (Liouville) sense.*

Thus, a proof of Conjecture 2 would enable one, first, to produce a substantial supply of completely integrable systems on coadjoint orbits in finite-dimensional Lie algebras, and second, to reduce the non-commutative integration of Hamiltonian systems on various symplectic manifolds to classical commutative integration.

Conjecture 2 was proved by Mishchenko and Fomenko [1978b] for all semisimple Lie algebras. Since every finite-dimensional Lie algebra of functions on a compact symplectic manifold is reductive (see, for example, Mishchenko and Fomenko [1981]) this implies the following theorem.

Theorem 1.16 (Mishchenko and Fomenko [1981]). *Let* $\mathfrak{g}$ *be a complete non-commutative collection of functions on a closed compact symplectic manifold* M^{2n}. *Then there is a complete involutive collection* $\mathfrak{g}_0$ *of functions on* M *that can be expressed functionally in terms of the functions in the original collection* $\mathfrak{g}$. *In particular, if* $v = \operatorname{sgrad} h$ *is a Hamiltonian system on a compact symplectic manifold that admits complete non-commutative integration, then it is always completely integrable in the classical commutative sense. Consequently, Conjecture 1 is always valid for compact manifolds.*

Remark 1.1. At present Conjecture 2 on the fibering of the orbits by the Liouville tori has been proved for all semisimple and reductive Lie algebras

(Mishchenko and Fomenko [1976], [1978b], [1979a]), for many nipotent Lie algebras (Vergne [1972]), for many infinite series of solvable Lie algebras (Arkhangel'sky [1979]), and for some non-solvable Lie algebras with a non-trivial radical (Belyaev [1981], Brailov [1985], Pevtsova [1982], Trofimov [1982a], Trofimov and Fomenko [1983a]).

3.4. Local Equivalence of Commutative and Non-commutative Integration. The validity of Conjecture 1 on the equivalence of commutative and non-commutative integration can be deduced locally (i.e. in a neighbourhood of a given level surface) from the results of E. Cartan (see Cartan [1927]). We shall discuss this problem below.

Definition 1.9. We say that a family B of smooth functions on a manifold M or in some open domain in M is of *class* l and is generated by the collection of functions $G = (G_1, \ldots, G_l)$ if for a function Γ to belong to B it is necessary and sufficient that $d\Gamma \wedge dG_1 \wedge \cdots \wedge dG_l \equiv 0$ and $G_1, \ldots, G_l$ are independent at every point.

Thus, a set B of a given class l contains all the functions that can be functionally expressed in terms of $G_1, \ldots, G_l$. For what follows it is especially important that B is an infinite-dimensional linear space which consists of all the functions that can be functionally expressed in terms of $G_1, \ldots, G_l$. It is natural to call such families functionally complete. Generally speaking, this space does not have to be a Lie algebra under Poisson bracket.

Definition 1.10. We say that a set of functions B of class l has *symplectic rank* $2q$ if rank $\{G, G\} = 2q$.

We consider those families of functions of a given class l and constant symplectic rank $2q$ which are Lie algebras under Poisson bracket. Locally, a family B has the form $\{\Gamma = \Psi(G_1, \ldots, G_l)\}$. Here we understand by the term "locally" some open neighbourhood of a common level surface of $G_1, \ldots, G_l$.

Theorem 1.17. *For a functionally complete set of functions B of class l and constant symplectic rank $2q$ the following conditions are locally equivalent*:

a) *B is an infinite-dimensional Lie algebra under Poisson bracket.*

b) *The Poisson brackets $\{G_\alpha, G_\beta\}$ can be functionally expressed in terms of $G_1, \ldots, G_l$, that is, $\{G_\alpha, G_\beta\} = \varphi_{\alpha\beta}(G_1, \ldots, G_l)$.*

c) *The distribution of planes $\Pi_z(G)$ is integrable ($\Pi_z(G)$ is the linear hull of the skew gradients* sgrad $G_1, \ldots,$ sgrad G_l).

d) *Consider the mapping $g: M^{2n} \to \mathbb{R}^l$, $y = g(z) = (G_1(z), \ldots, G_l(z))$. Then the distribution of planes $\Theta_y = dg(\Pi_z) \subset T_y\mathbb{R}^l$, $y = g(z)$, is well-defined and integrable.*

e) *There are functions $F_1, \ldots, F_{2q}, F_{2q+1}, \ldots, F_l \in B$ such that $\{F_\alpha, F_{q+\alpha}\} = -\{F_{q+\alpha}, F_\alpha\} = -1$ ($\alpha = 1, \ldots, q$) and the remaining Poisson brackets vanish. In other words, the matrix of Poisson brackets of the functions $F = (F_1, \ldots, F_l)$ has the form*

$$\begin{array}{r} 2q\left\{\begin{array}{l} \\ \\ \end{array}\right. \\ l-2q\{ \end{array} \left[\begin{array}{cc|c} 0 & -E & 0 \\ E & 0 & 0 \\ \hline 0 & 0 & 0 \end{array}\right]$$

We now return to Conjecture 1 on the equivalence of noncommutative and commutative integration. Its validity locally, i.e., in a neighbourhood of a single level surface, follows easily from Theorem 1.17, i.e., in fact, from the observations of Cartan [1927].

Theorem 1.15. *Suppose that a completely integrable Hamiltonian system in the non-commutative sense is given on a symplectic manifold. Then locally, in a neighbourhood of a non-singular common level surface, this system is integrable in the Liouville sense. Here the new integrals, which are in involution, can be functionally expressed in a neighbourhood of the level surface in terms of the original functions that generate a non-commutative Lie algebra.*

Thus, locally non-commutative integrability always implies commutative integrability.

§4. The Geometry of the Moment Map

4.1. The Moment Map. A diffeomorphism $f\colon M \to M$ of a symplectic manifold (M, ω) is called a *symplectomorphism* if $f^*\omega = \omega$. A Lie group G acts *symplectically* on the symplectic manifold M if all transformations $\hat{g}\colon M \to M$, $g \in G$, that determine the action are symplectomorphisms. Let $D(M)$ be the Lie algebra of vector fields on M. An action of G on M gives rise to a homomorphism σ of the Lie algebra $\mathfrak{g}$ of G into the Lie algebra $D(M)$, defined by

$$\hat{X}(m) = \sigma(X)(m) = \frac{d}{dt}\bigg|_{t=0} \widehat{\exp tX}(m), \quad X \in G, \quad m \in M.$$

The space of vector fields of the form sgrad f on the symplectic manifold (M, ω) is a Lie algebra with respect to the commutator. We shall denote this Lie algebra by $\mathrm{Ham}(M)$.

Definition 1.11. A symplectic action of a Lie group G on a symplectic manifold (M, ω) is said to be *strictly symplectic* if $\sigma(\mathfrak{g}) \subset \mathrm{Ham}(M)$, and is said to be Hamiltonian if there is a Lie algebra homomorphism $\lambda\colon \mathfrak{g} \to C^\infty(M)$ (which we call the *lift* of σ, or the covering homomorphism for σ) such that the diagram

$$\begin{array}{ccccccccc} 0 & \longrightarrow & \mathbb{R} & \longrightarrow & C^\infty(M) & \xrightarrow{\ p\ } & \mathrm{Ham}(M) & \longrightarrow & 0 \\ & & & & & \nwarrow^{\lambda} & \uparrow{\scriptstyle\sigma} & & \\ & & & & & & \mathfrak{g} & & \end{array}$$

is commutative, where the mapping $p: C^\infty(M) \to \mathrm{Ham}(M)$ is defined by $p(f) = \operatorname{sgrad} f$.

Definition 1.12. Suppose that we are given a strictly symplectic group action of G on a symplectic manifold (M, ω) with the lift $\lambda: \mathfrak{g} \to C^\infty(M)$. The map $\Phi: M \to \mathfrak{g}^*$ defined by $\Phi(m)(X) = \lambda(X)(m)$ is called the *moment map.*

The following theorem gives a criterion for the existence of the lift λ.

Theorem 1.19. *Given a central extension $\mathfrak{F}$ of a Lie algebra $\mathfrak{H}$ by $\mathbb{R}$, i.e. an exact sequence of Lie algebras $0 \to \mathbb{R} \to \mathfrak{F} \xrightarrow{\rho} \mathfrak{H} \to 0$, $[\mathbb{R}, \mathfrak{F}] = 0$, and a homomorphism $k: \mathfrak{g} \to \mathfrak{H}$, there is a well-defined cohomology class $[c]$ that measures the obstruction to the existence of a homomorphism $\lambda: \mathfrak{g} \to \mathfrak{F}$ covering k (i.e. $k = \rho \circ \lambda$). Such a λ exists if and only if $[c] = 0$. If the second cohomology group $H^2(\mathfrak{g})$ is zero, then λ always exists; if $H^1(\mathfrak{g}) = 0$, then λ is unique.*

If $\mathfrak{g}$ is a semisimple Lie algebra, then $H^1(\mathfrak{g}) = H^2(\mathfrak{g}) = 0$ (see, for instance, Jacobson [1962]). The proof of Theorem 1.19 can be found, for instance, in Guillemin and Sternberg [1984a]. For the cohomology theory of Lie algebras the reader may consult, for instance, Jacobson [1962], or Fuks [1984].

We shall give two cohomological criteria for the existence of the moment map (see Guillemin and Sternberg [1984a]).

Theorem 1.20. *Let G be a connected Lie group and $\mathfrak{g}$ its Lie algebra. Suppose that $H^1(\mathfrak{g}) = H^2(\mathfrak{g}) = 0$. If we are given a symplectic action of G on a symplectic manifold (M, ω), then there is a unique homomorphism $\lambda: \mathfrak{g} \to C^\infty(M)$, where $C^\infty(M)$ denotes the Lie algebra of smooth functions on M with respect to Poisson bracket, which satisfies $\omega(\hat{\xi}, X) = d\lambda(\xi)(X)$, where $\hat{\xi}$ is the vector field on M corresponding to $\xi \in \mathfrak{g}$. This allows us to define the moment map $\Phi: M \to \mathfrak{g}^*$ by $\langle \Phi(m), \xi \rangle = \lambda(\xi)(m)$. The moment map is equivariant with respect to the given action of G on M and the coadjoint action of G on $\mathfrak{g}^*$.*

This theorem gives criteria for a symplectic group action to be Hamiltonian in terms of Lie algebra cohomology. One can also give similar criteria involving the topology of M. The proof of the following theorem can be found in Guillemin and Sternberg [1984a].

Theorem 1.21. *Suppose that we are given a symplectic action of a Lie group G on a compact symplectic manifold M. If $H^1(M, \mathbb{R}) = 0$, then $\lambda(\xi)$ can be uniquely determined by the condition $\int \lambda(\xi)\omega^n = 0$.*

Various examples of moment maps can be found in Kazhdan et al. [1978], and Guillemin and Sternberg [1984a].

4.2. Convexity Properties of the Moment Map. Our exposition here follows Atiyah [1982], Guillemin and Sternberg [1982]; for more details the reader is referred to Guillemin and Sternberg [1984a].

We introduce some notation. Let V be a vector space and $a_1, \ldots, a_n \in V$. Then by $S(a_1, \ldots, a_n)$ we denote the convex set $\left\{ \sum_{i=1}^{n} s_i a_i | s_1, \ldots, s_n \geqslant 0 \right\}$.

Definition 1.13. Suppose that a Lie group G acts on a manifold M and that $x_0 \in M$ is a fixed point of this action. The homomorphism from G to the group of linear transformations of $T_{x_0}M$ which to an element $g \in G$ assigns its differential dg at x_0 is called the *isotropy representation.*

Theorem 1.22 (local convexity theorem). *Let X be a symplectic manifold, T^r an r-dimensional torus, $T^r \times X \to X$ a Hamiltonian action of T^r on X, and Φ: $X \to \mathfrak{t}^*$ the associated moment map. Let x be a fixed point of T^r and let $p = \Phi(x)$. Then there exists a neighbourhood U of x in X and a neighbourhood U' of p in $\mathfrak{t}^*$ such that $\Phi(U) = U' \cap \{p + S(\alpha_1, \ldots, \alpha_n)\}$ where $\alpha_1, \ldots, \alpha_n$ are the weights associated with the linear isotropy representation of T^r on the tangent space T_xX.*

Atiyah [1982] and Guillemin and Sternberg [1982] proved the following theorem.

Theorem 1.23. *Suppose that we are given a Hamiltonian action of the r-dimensional torus T on a symplectic manifold M with the associated moment map Φ: $M \to \mathfrak{t}^*$. Then the image of the moment map is a convex polytope.*

The above theorems can be used to study the structure of coadjoint orbits of compact connected Lie groups. The proofs of the following theorems can be found in Guillemin and Sternberg [1984a].

Theorem 1.24. *Coadjoint orbits of a compact connected Lie group are simply connected.*

Theorem 2.5. *In the coadjoint representation of a compact connected Lie group G the stability subgroup G_x of every element $x \in \mathfrak{g}^*$ is connected.*

Remark 1.2. An example of the moment map with a non-convex image is given in Section 5.3.

4.3. Multiplicity-free Representations. The algebra of operators which commute with a representation of a Lie group G is commutative if and only if every irreducible representation of G occurs in the given representation with multiplicity at most one. By analogy with this "quantum" property we give the following definition.

Definition 1.14. A Hamiltonian action of a Lie group G on a symplectic manifold M is called *multiplicity-free* if the algebra of all G-invariant functions on M is commutative under Poisson bracket.

This notion is related to completely integrable systems. Consider the moment map Φ: $M^{2n} \to \mathfrak{g}^*$ associated to a Hamiltonian action of a Lie group G on a symplectic manifold M^{2n}. Functions on M^{2n} of the form $f \circ \Phi$, where f: $\mathfrak{g}^* \to \mathbb{R}$, will be called *collective* (we follow the terminology of Guillemin and Sternberg

[1980]). A completely integrable system whose Hamiltonian and all first integrals are collective will be called *collective completely integrable* (see Mishchenko and Fomenko [1978a], [1979b], Guillemin and Sternberg [1980]).

Theorem 1.26. *Suppose that we are given a Hamiltonian action of a Lie group G on a symplectic manifold (M, ω). If M admits a collective completely integrable system, then the action of G on M is multiplicity-free.*

For the groups $U(n)$ and $O(n)$ the action is multiplicity free if and only if M admits a collective completely integrable system (see Guillemin and Sternberg [1984b]).

The following theorem establishes a connection between multiplicity-free representations of Lie groups and the existence of collective completely integrable systems (see Guillemin and Sternberg [1984b]).

Theorem 1.27. *Suppose that a Lie group G acts on a manifold M. The induced action of G on T^*M is multiplicity free if and only if the representation T_g of G in $L^2(M)$ is multiplicity-free, where $T_g f(m) = f(g^{-1}m)$, $f \in L^2(M)$, $m \in M$.*

For compact Lie groups (G, K) the following three conditions are equivalent: 1) all G-invariant Hamiltonian systems on $T^*(G/K)$ are collective completely integrable; 2) the representation T_g in $L^2(G/K)$ has simple spectrum; 3) the algebra of G-invariant functions on $T^*(G/K)$ is commutative. There is a classification of simple compact Lie groups (G, K) which have these properties.

§ 5. The Topology of Surfaces of Constant Energy in Completely Integrable Hamiltonian Systems

5.1. The Multi-dimensional Case. Classification of Surgery of Liouville Tori. The experience accumulated in investigating various concrete Hamiltonian systems (see, for example, Pogosyan and Kharlamov [1979], Kharlamov [1983], Fomenko [1988a], Oshemkov [1986], [1987]) shows that an overwheming majority of the integrals of Hamiltonian systems discovered to date belong to a certain special class described in Definition 1.15. A point $x \in M_1$ is said to be a *regular* point of a mapping $f\colon M_1 \to M_2$ if $\dim df(x)(T_x M_1) = \dim M_2$ i.e. if $df(x)\colon T_x M_1 \to T_{f(x)} M_2$ is an epimorphism. Otherwise, x is said to be a *critical* point and its image $f(x)$ is called a critical value. Let N be the set of all critical points of f. The submanifold N is called *nondegenerate* if the second differential $d^2 f$ is nondegenerate on subspaces transversal to N, i.e. on such subspaces $V \subset T_x M$ that $T_x M = T_x N \oplus V$.

Definition 1.15. A smooth function f is called a *Bott function* if the critical points of f form nondegenerate critical submanifolds.

The general properties of such functions were studied by Bott [1954].

Throughout this section we shall only consider the integrability of a system on a single fixed nonsingular level surface of the Hamiltonian within the class of *Bott integrals*.

The topological theory of integrable systems outlined in this section has been recently developed by Fomenko [1985], [1986a, b] [1988a, b].

Let $v = \operatorname{sgrad} H$ be a Hamiltonian system on a symplectic manifold M^{2n}. Suppose that v is integrable, i.e. suppose there exist n independent (amost everywhere) smooth integrals $f_1, \ldots, f_n$ in involution, $f_1 = H$. Let $F\colon M^{2n} \to \mathbb{R}^n$ be the moment map corresponding to these integrals, i.e. $F(x) = (f_1(x), \ldots, f_n(x))$. Let $\Sigma = F(N)$ be the set of critical values (*the bifurcation diagram*) of the moment map. If $a \in \mathbb{R}^n$ is not a critical value (that is, $a \in \mathbb{R}^n \backslash \Sigma$), then its inverse image $B_a = F^{-1}(a) \subset M$ (called a nonsingular fiber) does not contain critical points of F, and therefore, by the Liouville theorem, each of its connected components is diffeomorphic to a torus T^n. Suppose, for simplicity, that the whole fiber B_a is compact. The corresponding assertions for noncompact fibers follow easily from the results obtained below. If $a \in \Sigma$, then the common level surface B_a of the integrals is singular (critical). As the point a is moved in $\mathbb{R}^n$, its inverse image (i.e., B_a) is deformed in some way. As long as a does not meet Σ, the fiber B_a is transformed by diffeomorphisms. In particular, any two fibers B_a and B_b for which the points a and b can be joined by a smooth curve $\gamma \subset \mathbb{R}^n \backslash \Sigma$ are diffeomorphic. But if the curve γ meets the set Σ at some point, then the fibers B_a and B_b can be different. If the point a pierces Σ then the fiber B_a undergoes topological surgery. The following general problem arises: Describe the topological surgery of Liouville tori arising at the moment when the point intersects the set Σ. We will show below that it is possible to classify the surgery in general position; all its types have a relatively simple form. It is clear that we must distinguish between two cases: a) $\dim \Sigma < n - 1$ and b) $\dim \Sigma = n - 1$. In case a), the set Σ does not separate $\mathbb{R}^n$; that is, any two points $a, b \in \mathbb{R}^n$ can be joined by a smooth curve $\gamma \subset \mathbb{R}^n \backslash \Sigma$. Consequently, all compact nonsingular fibers are diffeomorphic to one another and, in particular, consist of the same number of Liouville tori. Case b) is complicated. Here Σ will in general partition $\mathbb{R}^n$ into several open disjoint domains. Within each, the topology of the nonsingular fiber is its own. Thus, suppose $\dim \Sigma = n - 1$. We fix a point c on Σ and study the surgery of Liouville tori as a smooth curve γ pierces Σ at c. It suffices to consider a small neighbourhood $U = U(c)$ of the point c in $\mathbb{R}^n$. We shall study the "general position" case, i.e., the case when γ pierces Σ transversally at a point c on an $(n-1)$-dimensional smooth stratum of Σ; that is, we shall suppose that $U \cap \Sigma$ is a smooth $(n-1)$-dimensional submanifold of $\mathbb{R}^n$. In the general position case, we may assume that the set $N \cap F^{-1}(U)$ of critical points is a union of a finite number of smooth submanifolds of M stratified by the rank of dF. The concept of general position may be further sharpened as follows. Since we assume that Σ is an $(n-1)$-dimensional submanifold in a neighbourhood of c, we may assume that, in the neighborhood of some connected component B_c^0 of the singular fiber B_c, the last $n - 1$ integrals $f_2, \ldots, f_n$ are independent and the first integral $f_1 = H$ (the energy) becomes dependent on them on the submanifold

$T = N \cap B_c^0$ of critical points. Indeed, we restrict F to the submanifold $N \cap F^{-1}(U)$ which is, by the general position requirement, a union of a finite number of smooth submanifolds. Since the restriction of F to each stratum $N' \cap F^{-1}(U)$, including maximal ones, is a smooth map of a smooth submanifold, it follows that the map $dF(x)\colon T_xN' \to T_{F(x)}\Sigma$ is an epimorphism, and rank $dF(x) \geqslant n-1$ because dim $U \cap \Sigma =$ rank $dF(x) \leqslant n-1$. At the same time, rank $dF(x) \leqslant n-1$ because $x \in N$ is a critical point. Consequently, rank $dF(x) = n-1$. Therefore, we may assume that $f_2, \ldots, f_n$ are independent on B_c^0. Hence, f_1 becomes dependent on them on $T = N \cap B_c^0$.

We shall now consider five types of $(n+1)$-dimensional manifolds whose boundaries are tori.

1) We consider the "solid torus" $D^2 \times T^{n-1}$ with the torus T^{n-1} as "axis". Its boundary is the torus T^n. We call $D^2 \times T^{n-1}$ a *dissipative solid torus.*

2) We consider the product $T^n \times D^1$, which we shall call a *cylinder*. Its boundary is two tori T^n.

3) Let N^2 be a disk with two holes. We shall call the direct product $N^2 \times T^{n-1}$ an *oriented toroidal saddle* (or *trousers*). Its boundary is three tori T^n.

4) We consider all nonequivalent fibrations over T^{n-1} whose fiber is the interval $D^1 = [-1, +1]$. They are classified by elements α of the homology group $H_1(T^{n-1}, \mathbb{Z}_2) = \mathbb{Z}_2 \oplus \cdots \oplus \mathbb{Z}_2$ ($n-1$ times). Let Y_α^n denote the total space of the fibration corresponding to α, so that we have a fibration $D^1 \to Y_\alpha \to T^{n-1}$. Consider a new fibration $N^2 \to A_\alpha \to T^{n-1}$ with base T^{n-1} and fiber N^2. To construct it, we consider an interval D^1 on the disk N^2 which passes through the center of the disk and joins the centers of the deleted disks (holes). Then, to each fibration $Y_\alpha^n \to T^{n-1}$ we associate a new fibration $N^2 \to A_\alpha^{n+1} \to T^{n-1}$ by replacing the fiber D^1 by the fiber N^2. In addition, we require that the boundary of the manifold A_α^{n+1} be a union of tori. A particular case of such a fibration is the direct product $N^2 \times T^{n-1}$; i.e. a manifold of type 3) (see above). It is obtained only if $\alpha = 0$. But if $\alpha \neq 0$, the fibration A_α is nontrivial. For $\alpha \neq 0$, the boundary of the manifold Y_α^n is a torus T^{n-1}. We call A_α^{n+1} a nonoriented toroidal saddle if $\alpha \neq 0$. The boundary of A_α consists of two tori T^n. It is easy to see that, for $\alpha \neq 0$, all the manifolds A_α^{n+1} are diffeomorphic to one another. Therefore, we will write them as follows: $A_\alpha^{n+1} = N^2 \tilde{\times} T^{n-1}$ (fiber bundle). Thus, we obtain only two topologically distinct manifolds, $N^2 \times T^{n-1}$ and $N^2 \tilde{\times} T^{n-1}$. However, from the point of view of Hamiltonian mechanics, the manifolds A_α should be considered independently for different α.

5) Let $p\colon T^n \to K^n$ be a double cover of the nonorientable manifold K^n. All such coverings can be classified. We let K_p^{n+1} denote the mapping cylinder of p. It is clear that dim $K_p^{n+1} = n+1$ and $\partial K_p^{n+1} = T^n$.

The manifolds A_α^{n+1} and K_p^{n+1} can be formed by pasting together (along the boundary tori) manifolds of the first three types; that is, from a topological point of view, the "elementary blocks" consist only of the manifolds $D^2 \times T^{n-1}$, $T^n \times D^1$, and $N^2 \times T^{n-1}$.

We are now in a position to describe the five types of surgery of the torus T^n.

1) The torus is given as the boundary of a dissipative solid torus $D^2 \times T^{n-1}$ and then contracts to its "axis" T^{n-1}. We shall call this operation a limiting degeneration.

2) The two tori T_1^n and T_2^n that constitute the boundary of the cylinder $T^n \times D^1$ move along it towards one another and in the middle of the cylinder merge into one torus T^n.

3) The torus T^n that constitutes the lower boundary of the oriented saddle $N^2 \times T^{n-1}$ (trousers) rises upwards and in accordance with the topology of the manifold $N^2 \times T^{n-1}$ splits into two tori T_1^n and T_2^n.

4) The torus T^n which is one of the boundaries of the manifold A_α, where $\alpha \neq 0$, rises "upwards" through A_α and undergoes surgery in the middle, turning into a single torus that is the upper boundary of the manifold A_α.

5) The torus T^n is realized as the boundary of K_p^{n+1}. Upon deforming it inside K_p^{n+1} along the projection p, the torus T^n finally doubly covers the nonorientable manifold K^n after which it "disappears".

We now formulate the final definition of general position surgery of a Liouville torus. We fix the values of the last $n-1$ integrals $f_2, \ldots, f_n$ and consider the resulting $(n+1)$-dimensional surface X^{n+1}. The first integral $f_1 = H$ restricts to a smooth function f on X^{n+1}.

Definition 1.16. We say that surgery of the Liouville tori constituting a nonsingular fiber B_a is *general position surgery* if there is a neighbourhood of the torus T^n undergoing surgery in which X^{n+1} is compact and nonsingular and the restriction of the energy function $f = H$ to X^{n+1} is a Bott function in this neighbourhood.

Theorem 1.28 (A.T. Fomenko) (The classification theorem for surgery of Liouville tori). 1) *Fix an integrable system and let Σ be the bifurcation diagram of its moment map. If* $\dim \Sigma < h - 1$, *then all the nonsingular fibers B_a are diffeomorphic.*

2) *Suppose that* $\dim \Sigma = n - 1$. *Suppose that a nondegenerate Liouville torus T^n is moving along the joint nonsingular level surface X^{n+1} of the integrals $f_2, \ldots, f_n$, carried along by the changing value of the energy integral $f_1 = H$. This is equivalent to moving the point $a = F(T^n) \in \mathbb{R}^n$ along a smooth segment γ towards Σ. Suppose that the torus T^n undergoes topological surgery at some moment of time. This happens when (and only when) the torus T^n encounters critical points N of the moment map $F\colon M^{2n} \to \mathbb{R}^n$ along its path. In other words, the path γ transversally pierces an $(n-1)$-dimensional stratum of Σ with nonzero velocity at a point c. If the surgery is general position surgery, then it is a composition of the five canonical types* 1)–5) *of surgery listed above. In fact, from the topological point of view, only the first three types of surgery are independent; surgeries* 4) *and* 5) *are compositions of them.*

In case 1) the torus T^n first turns into (i.e., degenerates to) the torus T^{n-1} as the energy increases, and then disappears from the surface of constant energy $H = \text{const}$ (a limiting degeneration). In case 2) the two tori T_1^n and T_2^n first merge into one torus T^n as the energy increases, and then disappear from the surface $H = \text{const}$. In case 3) the torus T^n "breaks through" the critical energy level as the energy H increases and splits into two tori T_1^n and T_2^n on the surface $H = \text{const}$ (i.e. it "survives" the passage through the critical level). In case 4) the torus T^n "breaks through" the critical energy level as the energy H increases and turns again into a torus T^n (a nontrivial transformation by a double winding). In case 5) the torus T^n double covers the nonorientable manifold K^n, after which it disappears from the surface $H = \text{const}$. Changing the direction of motion of a Liouville torus gives the five inverse processes of bifurcation of the torus T^n.

Some of the surgery described in Theorem 1.28 was discovered earlier in concrete mechanical systems (see Kharlamov [1983] and Pogosyan and Kharlamov [1979]). For example, such is the surgery of tori in the Kowalewski and Goryachev-Chaplygin cases, see Kharlamov [1983]. The torus surgery found in this paper is in fact a composition of the surgery described in Theorem 1.28.

Theorem 1.28 was proved by Fomenko [1985], [1986a, b].

It is possible to describe explicitly the structure of the common level surface of the integrals.

Theorem 1.29 (A.T. Fomenko). *Let M^{2n} be a symplectic manifold and let $v = \operatorname{sgrad} H$ be a system which is integrable by means of smooth independent commuting integrals $H = f_1, f_2, \ldots, f_n$. Let X^{n+1} be a fixed, compact, nonsingular common level surface of the last $n-1$ integrals $f_2, \ldots, f_n$, and suppose that the restriction of H to X^{n+1} is a Bott function. Then X^{n+1} has the form*

$$X^{n+1} = m(D^2 \times T^{n-1}) + p(T^n \times D^1) + sA_\alpha^{n+1} + rK_p + q(N^2 \times T^{n-1}),$$

i.e. *it is obtained by pasting together the "elementary blocks" described above by some diffeomorphisms of their boundary tori. The number m is the number of limiting degenerations of the system v on the surface X^{n+1}.*

The nonorientable manifolds K^n which are minima or maxima of the Bott integral f on X^{n+1} can be explicitly described (A.V. Brailov, A.T. Fomenko). For $\alpha = 0, 1$ we let G_α denote the group of transformations of the torus $T^n = \mathbb{R}^n/\mathbb{Z}^n$ generated by the involution

$$R_\alpha(a) = \begin{cases} (-a_1, a_2 + \frac{1}{2}, a_3, \ldots, a_n), & \alpha = 0, \\ (a_2, a_1, a_3 + \frac{1}{2}, a_4, \ldots, a_n), & \alpha = 1, \end{cases}$$

where $a = (a_1, \ldots, a_n) \in \mathbb{R}^n/\mathbb{Z}^n$. Here we suppose that $n \geqslant 2$ for $\alpha = 0$ and $n \geqslant 3$ for $\alpha = 1$. The group G_α acts on T^n without fixed points. Consequently, the quotient set $K_\alpha^n = T^n/G_\alpha$ is a smooth manifold. The transformation R_α reverses orientation, so the manifold K_α^n is nonorientable. The manifolds K_0^n and K_1^n are not homeomorphic, because $H_1(K_0^n; \mathbb{Z}) = \mathbb{Z}^{n-1} \oplus \mathbb{Z}_2$ and $H_1(K_1^n; \mathbb{Z}) = \mathbb{Z}^n$. It follows from the definition of K_α^n that $K_0^n = K_0^2 \times T^{n-2}$ and $K_1^n = K_1^3 \times T^{n-3}$.

Theorem 1.30 (Brailov and Fomenko [1987]). *Let $f_1, \ldots, f_n$ be a complete, involutive set of smooth functions on M^{2n} and $F\colon M^{2n} \to \mathbb{R}^n$ the moment map, and* *let* $X^{n+1} = \{f_i(x) = c_i, i = 2, \ldots, n\}$ *be a common level surface. Suppose that the restriction* $f = f_1 | X^{n+1}$ *is a Bott function and* $K_p^n = \{f_i(x) = c_i, i = 1, \ldots, n\}$ *is a nonorientable minimum or maximum manifold of the function* f *(i.e. a critical fiber of the moment map). Then* K_p^n *is diffeomorphic to either* K_0^n $(n \geqslant 2)$ *or* K_1^n $(n \geqslant 3)$.

Let $v = \operatorname{sgrad} H$ be an integrable system on the manifold M^{2n} and let $f_1 = H$, $f_2, \ldots, f_n$ be a complete commuting set of integrals such that the restriction $f_1 = f | X^{n+1}$ to the common compact nonsingular level surface of the remaining $n - 1$ integrals $X^{n+1} = \{x \in M | f_2(x) = c_2, \ldots, f_n(x) = c_n\}$ is a Bott function. Let c_1 be a critical value of the function f on X^{n+1}. Let $B_c = F^{-1}(c)$ be the critical fiber of the moment map; i.e. let $B_c = \{f_1(x) = c_1\}$ be a critical level surface of f_1 on X^{n+1}. One may always assume that the set of critical points of f on B_c is connected.

Theorem 1.31 (Fomenko [1986b], [1988a]). *Each connected compact component* B_c^0 *of the critical fiber* B_c *is homeomorphic to a set which is of one of the following five types:* 1) *a torus* T^n; 2) *and* 3) *the nonorientable manifolds* K_0^n *and* K_1^n; 4) *a torus* T^{n-1}; *or* 5) *a cell complex obtained by fixing* $(n-1)$*-dimensional tori* T_1^{n-1} *in* T_1^n *and* T_2^{n-1} *in* T_2^n, *which realize nonzero generators of the homology groups* $H_{n-1}(T_1^n; \mathbb{Z})$ *and* $H_{n-1}(T_2^n; \mathbb{Z})$, *and pasting* T_1^n *and* T_2^n *together by identifying only the tori* T_1^{n-1} *and* T_2^{n-1} *(by means of a diffeomorphism). In this case the critical points of* f *in the critical fiber* B_c^0 *form a torus* T^{n-1} *(the result of pasting together the tori* T_1^{n-1} *and* T_2^{n-1}*) which is a "saddle" for the function* f.

Let us also mention the following realization theorem.

Theorem 1.32 (Brailov and Fomenko [1987]). *Let* X^{n+1} *be a smooth closed compact orientable manifold obtained by pasting together an arbitrary number of elementary manifolds (solid tori, cylinders, and trousers) by arbitrary diffeomorphisms of their boundry tori. Then there always exists a smooth compact symplectic manifold* M^{2n} *with boundary diffeomorphic to a disjoint union of several manifolds* $S^{n-1} \times T^n$ *and a complete involutive set of smooth functions* $f_1, \ldots, f_n$ *on* M^{2n} *such that* $X^{n+1} = \{x \in M^{2n} | f_2(x) = \cdots = f_n(x) = 0\}$.

5.2. The Four-dimensional Case. In this case we have classification theorems stated in the previous section. Moreover, new effects related to dimension four come into being. We shall describe them in more detail. Thus, let M^4 be a four-dimensional symplectic manifold and let $v = \operatorname{sgrad} H$ be a Hamiltonian system in M^4 with a smooth Hamiltonian H. Since H is an integral of the system, v can be restricted to an invariant three-dimensional surface Q of constant energy; i.e. $Q = \{x \in M | H(x) = \text{const}\}$. Since M is always orientable, Q is also orientable. We shall consider noncritical (nonsingular) surfaces Q, i.e., those for which $\operatorname{sgrad} H \neq 0$. Suppose that the system is Liouville integrable, i.e., suppose there exists an additional integral f which is independent of H (almost everywhere) and in involution with H. Restricting this integral to Q gives a smooth function f.

Definition 1.17. A Bott integral f on Q is said to be *orientable* if all its critical submanifolds are orientable. Otherwise, the integral f is said to be *nonorientable.*

A hypersurface Q^3 with Bott integral f will sometimes be called an integral surface. It turns out that, without any essential loss of generality, we can restrict oursevles to studying only orientable integrals. Namely, by considering the surfaces Q^3 to within a double cover we can always assume that the integral f is orientable.

Proposition 1.4 (Fomenko [1986a, b]). *Let Q^3 be a nonsingular compact surface of constant energy in M^4 and let f be a nonorientable Bott integral on Q. Then all the nonorientable critical submanifolds are homeomorphic to a Klein bottle and f attains either a (local) minimum or maximum on them. Let $U(Q)$ be a sufficiently small tubular neighbourhood of Q in M. Then there exists a double cover π: $(\tilde{U}(\tilde{Q}), \tilde{H}, \tilde{f}) \to (U(Q), H, f)$ (with fiber $\mathbb{Z}_2$), where $\tilde{U}(\tilde{Q})$ is a symplectic manifold with a Hamiltonian system $\tilde{v} = \operatorname{sgrad} \tilde{H}$ (where $\tilde{H} = \pi^*(H)$) which is integrable on $\tilde{Q} = \pi^{-1}(Q)$ by means of an orientable Bott integral $\tilde{f} = \pi^*(f)$. In addition, all critical Klein bottles on Q "develop" into critical tori T^2 on $\tilde{Q}$ (minima or maxima of $\tilde{f}$). The manifold $\tilde{U}(\tilde{Q})$ is a tubular neighbourhood of the surface $\tilde{Q}$.*

Thus, if f is a nonorientable integral on Q, then $\pi_1(Q) \neq 0$ and $\pi_1(Q)$ contains a subgroup of index 2. If, for example, $Q = S^3$ (a case which frequently occurs in mechanics), then any Bott integral f on S^3 is orientable.

Definition 1.18. Let γ be a closed integral curve (i.e., a periodic solution) of the system v on Q^3. We shall say that γ is *stable* if it has a tubular neighbourhood which is totally (without gaps) fibered into concentric two-dimensional tori which are invariant with respect to v and which envelop γ; that is, all integral curves close to γ are "packed" onto invariant tori with a common axis γ.

An integrable system need not have a stable periodic solution. An example is provided by the geodesic flow on a flat torus. It turns out that there exists a close connection between the following three objects: a) an additional Bott integral f on the surface Q b) the number of stable periodic solutions on Q, and c) the first integral homology group $H_1(Q; \mathbb{Z})$ (or the fundamental group $\pi_1(Q)$). Let $m = m(Q)$ be the number of stable periodic solutions of the system v on Q. Let $r = r(Q)$ be the number of critical submanifolds of the integral f on Q which are homeomorphic to a Klein bottle.

Theorem 1.33 (A.T. Fomenko). *Let M^4 be a smooth symplectic manifold (compact or not) and let $v = \operatorname{sgrad} H$ be a Hamiltonian field on M^4. Suppose that the system is integrable on some nonsingular compact three-dimensional surface $Q = \{H = \text{const}\}$ by means of a Bott integral f. Then the number m of stable periodic solutions of v on Q is bounded from below in terms of topological invariants of Q as follows.*

1) *If the integral f is orientable on Q, then* a) *$m \geqslant 2$ if the homology group $H_1(Q; \mathbb{Z})$ is finite, and* b) *$m \geqslant 2$ if the fundamental group $\pi_1(Q) = \mathbb{Z}$.*

2) *If the integral f is nonorientable on Q, then* a) $m + r \geqslant 2$ *if* $H_1(Q; \mathbb{Z})$ *is finite,* b) $m \geqslant 2$ *if* $H_1(Q, \mathbb{Z}) = 0$ *(here the group* $\pi_1(Q)$ *may be infinite),* c) $m \geqslant 1$ *if* $H_1(Q, \mathbb{Z})$ *is a finite cyclic group,* d) $m \geqslant 1$ *if* $\pi_1(Q) = \mathbb{Z}$ *or if* $\pi_1(Q)$ *is a finite group, and* e) $m \geqslant 2$ *if* $H_1(Q; \mathbb{Z})$ *is a finite cyclic group and Q does not belong to a series of manifolds of the form* $Q_0 = (S^1 \times D^2) + sA^3 + K^3$ *which are explicitly described below.*

In both cases 1) and 2), the integral f attains a local minimum or maximum on each of the stable periodic solutions of the system (or on the Klein bottles). If $H_1(Q; \mathbb{Z})$ is an infinite group (i.e. if the rank of H_1 is at least 1), then the system v may have no stable periodic solutions on Q at all.

This criterion is reasonably efficient, since verifying that an integral is a Bott function and computing the rank of $H_1(Q)$ do not usually present difficulties. For many integrable mechanical systems, the surfaces Q are diffeomorphic to either a sphere S^3, or projective space $\mathbb{R}P^3$, or $S^1 \times S^2$. For the equations of motion of a heavy rigid body, we may, after some factorization, assume that some Q are homeomorphic to $\mathbb{R}P^3$ (Kozlov [1980]). If the Hamiltonian H has an isolated minimum or maximum (an isolated equilibrium position of the system) on M^4, then all sufficiently close surfaces $Q = \{H = \text{const}\}$ are spheres S^3. Let $L_{p,q}$ be a lens space (the quotient of a sphere S^3 by the action of a cyclic group). We shall single out the cases which are of interest for Hamiltonian mechanics.

Proposition 1.5 (Fomenko [1986a, b]). *Suppose that* $v = \operatorname{sgrad} H$ *is integrable by means of a Bott integral f on a single constant energy surface Q which is homeomorphic to one of the following manifolds:* S^3, $\mathbb{R}P^3$, $S^1 \times S^3$, *or* $L_{p,q}$. *If f is orientable, then* $m \geqslant 2$, *i.e., the system v necessarily has at least two stable periodic solutions on Q. If f is nonorientable, then* $m \geqslant 2$ *if Q is homeomorphic to* S^3 *and* $m \geqslant 1$ *if Q is homeomorphic to* $\mathbb{R}P^3$, $S^1 \times S^2$, *or* $L_{p,q}$. *An integrable system on the sphere* S^3 *always has at least two stable periodic solutions.*

Thus, not only does an integrable system have two stable periodic solutions on spheres close to an isolated equilibrium position of the system (a minimum or maximum of H), but it has two such solutions on all expanding level surfaces of H as long as they remain homeomorphic to S^3. The criterion of Theorem 1.33 is sharp in the sense that there are examples of integrable systems for which $Q = \mathbb{R}P^3$ or $Q = S^3$ has exactly two (and no more) stable periodic solutions (Kozlov [1985]).

It follows from the results of Anosov [1982] and of Klingenberg and Takens (see Klingenberg [1978]) that there is an open, everywhere dense subset of the set of all geodesic flows on a smooth Riemannian manifold which consists of flows with no stable closed integral curves. Thus, the property that a geodesic flow does not have stable trajectories is the property of general position.

Corollary. *The geodesic flow of a Riemannian metric of general position on the sphere* S^3 *(i.e., a metric which does not have a stable closed geodesic) is not integrable in the class of smooth Bott integrals.*

We define here the rank R of the group $\pi_1(Q)$ as the smallest possible number of generators. If $R = 1$, then a system which is integrable by means of a Bott integral necessarily has at least one stable periodic solution on Q.

Proposition 1.6 (Fomenko [1986a, b]). *Let v be a Hamiltonian system which is integrable on a nonsingular compact three-dimensional constant energy surface Q by means of a Bott integral. If the system has no stable periodic solutions on Q, then $H_1(Q, \mathbb{Z})$ is not a finite cyclic group, and* rank $\pi_1(Q) \geqslant 2$, *and at least one of the generators of $\pi_1(Q)$ has infinite order.*

Consider the geodesic flow on a flat two-dimensional torus (with locally Euclidean metric). This flow is integrable in the class of Bott integrals and does not have closed stable trajectories. In this case, nonsingular surfaces Q are diffeomorphic to the torus T^3 and $H_1(Q; \mathbb{Z}) = \mathbb{Z}^3$.

Corollary (Fomenko [1986a, b]). *Let $v =$ sgrad H be a Hamiltonian system on M^4 and Q a nonsingular compact three-dimensional constant energy surface. Suppose that the system v has no stable periodic solutions on Q, and $H_1(Q; \mathbb{Z})$ is a finite cyclic group or* rank $\pi_1(Q) \leqslant 1$. *Then the system v is not integrable in the class of smooth Bott integrals on Q.*

The above theorems can be used to show that most smooth closed compact orientable three-dimensional manifolds cannot occur as constant energy surfaces in a Hamiltonian system which is integrable by means of Bott integrals (see Fomenko [1986b]). On the other hand, every compact orientable three-dimensional manifold is a constant energy surface of some smooth Hamiltonian system (see Fomenko [1988a]). This follows from a theorem of S.V. Matveev and A.T. Fomenko: Let M^3 be a smooth closed compact orientable three-dimensional manifold, and D^1 an interval. Then the direct product $M^3 \times D^1$ is a symplectic manifold, i.e. there always exists a smooth symplectic structure on $M^3 \times D^1$.

The results listed above are based on Theorem 1.29 which provides a topological classificaton of the surfaces Q of constant energy. In view of this theorem, in the four-dimensional case we have five types of simple manifolds out of which each such surface Q is built.

Type 1. The solid torus $S^1 \times D^2$; its boundary is the torus T^2.

Type 2. The cylinder $T^2 \times D^1$; its boundary is a union of two tori T^2.

Type 3. The oriented saddle (or "trousers") $N^2 \times S^1$, where N^2 is a two-dimensional disk with two holes. Its boundary is a union of three tori T^2.

Type 4. Consider a nontrivial fibration $N^2 \to A^3 \to S^1$ with base S^1 and fiber N^2. There are only two nonequivalent fibrations over S^1 whose boundaries are tori and whose fibers are N^2. These are $N^2 \times S^1$ (type 3) and A^3. We will call A^3 a nonorientable saddle; its boundary is a union of two tori T^2.

Type 5. The total space K^3 of the oriented twisted product of the Klein bottle K^2 and the interval. Its boundary is the torus T^2.

Remark 1.3. From a topological point of view, the manifolds A^3 of type 4 and K^3 of type 5 are not simple, since A^3 is obtained by pasting trousers to a solid torus by a diffeomorphism of the torus, and K^3 is obtained by pasting together two solid tori and trousers by a diffeomorphism.

Thus, only the first three of the five types of manifolds listed above are topologically independent. The last two are combinations of type 1–3. However, the manifolds A^3 and K^3 are of independent interest for the study of trajectories of the system v.

In contrast to the general case of Theorem 1.29, in the four-dimensional case the number of manifolds of type 1–5 has a precise geometrical description given in the following theorem.

Theorem 1.34 (Fomenko [1986b]) (The topological classificaton of three-dimensional surfaces of constant energy in integrable systems). *Let M^4 be a smooth symplectic manifold (compact or noncompact) and $v =$ sgrad H a Hamiltonian system which is Liouville integrable on a nonsingular compact three-dimensional surface Q of constant energy by means of a Bott integral f. Let m be the number of stable periodic solutions of the system v on Q (on which the integral f attains a strict local minimum or maximum), p the number of two-dimensional critical tori of the integral f (minima or maxima of the integral), q the number of critical circles of the integral f (unstable trajectories of the system) with orientable separatrix diagram, s the number of critical circles of the integral f (unstable trajectories of the system) with nonorientable separatrix diagram, and r the number of critical Klein bottles (minima or maxima). This is a complete list of all possible critical submanifolds of the integral f on Q. Then Q can be represented as the result of pasting together (by certain diffeomorphisms of the boundary tori) the elementary blocks described above*:

$$Q = m(S^1 \times D^2) + p(T^2 \times D^1) + q(N^2 \times S^1) + sA^3 + rK^3.$$

If the integral f is orientable, then the last summand does not appear, i.e. $r = 0$.

We recall the definition of a separatrix diagram. Let T be a critical submanifold of the integral f. *The separatrix diagram* $P(T)$ is the union of the integral curves of the field grad f which enter or leave T.

If we ignore the geometrical interpretation of the numbers m, p, q, s, r and pose the question of the simplest topological representation of Q, then the decomposition of Q can be simplified.

Theorem 1.35 (Fomenko [1986a, b]). *Let Q be a compact nonsingular surface of constant energy of the system $v =$ sgrad H and suppose that the system has a Bott integral on Q. Then Q admits the following topological representation*:

$$Q = m'(S^1 \times D^2) + q'(N^2 \times S^1),$$

where m' and q' are nonnegative integers. They are related to the numbers in Theorem 1.34 *as follows*: $m' = m + s + 2r + p$, *and* $q' = q + s + r + p$.

As in the multi-dimensional case, it is possible to describe the surgery of the level surfaces; we shall not go into details because no new phenomena occur here.

In this connection we shall consider four classes of three-dimensional manifolds:

Class (H) consists of surfaces Q^3 of constant energy in Hamiltonian systems which are integrable by means of Bott integrals. Class (Q) consists of manifolds of the form $m(S^1 \times D^2) + q(N^2 \times S^1)$. Based on intrinsic problems of three-dimensional topology Waldhausen [1967] investigated a class (W) of three-dimensional manifolds which he called Graphenmannigfaltigkeiten. S.V. Matveev considered the class (S) of three-dimensional manifolds on which there exists a smooth function g all of whose critical points are organized into nondegenerate circles and all of whose nonsingular level surfaces are unions of tori.

Theorem 1.36. *The four classes of three-dimensional manifolds described above coincide*, i.e., $(H) = (Q) = (W) = (S)$.

This theorem is the result of efforts of many authors: A.T. Fomenko, A.V. Brailov, H. Zieschang, S.V. Matveev, A.B. Burmistrova.

Thus, the class (H) of three-dimensional isoenergy surfaces of integrable systems (which coincides with the classes (Q), (W) and (S)) forms a "meager" subset, in some sense, of the class of all closed orientable three-dimensional manifolds. This follows from a theorem of Waldhausen for the class (W). Nevertheless, it is possible to construct three-dimensional oriented manifolds with any possible (for three-dimensional oriented manifolds) integral homology groups by pasting together "trousers" and solid tori (a remark of G. Mamedov). Thus, from the point of view of homology, (H) "coincides" with the class of all three-dimensional manifolds. In other words, it is impossible to distinguish the isoenergy surfaces among three-dimensional manifolds on the basis of homology groups only.

It turns out that the number of critical submanifolds of the integral f on Q can be bounded below by a universal constant which depends only on the group $H_1(Q, \mathbb{Z})$. Let $\beta = \operatorname{rank} H_1(Q, \mathbb{Z})$, and let ε be the number of elementary factors in the finite part $\operatorname{Tor}(H_1)$ of $H_1(Q, \mathbb{Z})$. If $\operatorname{Tor}(H_1)$ is written as an ordered sum of subgroups with the order of each subgroup dividing the order of the preceding one, then ε is the number of such summands.

Theorem 1.37 (Fomenko and Zieschang [1987], [1988]). *Let $Q \in (H)$, i.e. let Q be an isoenergy surface of a system which is integrable by means of a Bott integral. Let $Q = mI + pII + qIII + sIV + rV$ be the decomposition of Q as a sum of elementary manifolds (see above). Then $m + s + 2r \geqslant \varepsilon - 2\beta + 1$, $q \geqslant m + r - 2$ for $m + r + s + q > 0$. If $m = r = s = q = 0$ then $\varepsilon - 2\beta \leqslant 0$ and the equality $\varepsilon = 2\beta$ is actually attained for some pairs (Q, f). The case $m = r = s = q = 0$ is realized if and only if Q is a fibration over the circle with torus as fiber. If the integral is "completely orientable", i.e. if $s = r = 0$, we obtain a lower bound $m \geqslant \varepsilon - 2\beta + 1$ for the number m of stable periodic solutions of our integrable system.*

Typical properties of integrable Hamiltonian systems on "sufficiently complicated" isoenergy manifolds were studied by Fomenko and Zieschang [1987b], [1988].

Fig. 1

Surgery of the Liouville tori is conveniently depicted by means of diagrams. 1) A black circle with a single outgoing (incoming) edge of the graph will denote a minimal (maximal) critical circle of the integral. An edge of the graph depicts a one-dimensional family of tori which foliate the solid torus, a tubular neighbourhood of the critical circle (maximal or minimal). 2) An open circle with two outgoing (incoming) edges will denote a minimal (maximal) critical torus with two families of nonsingular Liouville tori emerging out of it. 3) A trefoil, i.e. a point at which three edges of the graph meet, will denote a connected neighbourhood of a critical saddle circle with orientable sepatrix diagram. Here one Liouville torus splits into two tori. 4) An asterisk (with one incoming and one outgoing edge) will denote a neighbourhood of a critical saddle circle with nonorientable separatrix diagram. Here the moving Liouville torus is "processed" into another torus by winding twice over it. 5) A circle with a dot inside it and with an outgoing (incoming) edge of the graph will denote a minimal (maximal) critical Klein bottle (see Fig. 1).

Each trefoil (type 3) describes either the splitting of a torus into two tori, or the merging of two tori into one depending on the orientation of the trefoil (two edges up or two edges down).

The indexation of the vertices of the graph by the numbers 1–5 chosen here is not fortuitous. In fact, there is a one-to-one correspondence between the five types of vertices and the five types of elementary manifolds listed earlier. It turns out that a small "neighbourhood" of a vertex of type i on the graph ($i = 1, 2, 3, 4, 5$) is homeomorphic (from the viewpoint of the surface Q) to the elementary manifold of type i.

We shall first consider the special case where the Bott integral f on Q is such that each connected component of each critical level of f contains precisely one critical submanifold. In this case all bifurcations of Liouville tori are elementary (canonical) and therefore are contained in the above list of surgery types 1)–5). One can construct a graph $\Gamma(Q, f)$ where each Liouville torus is depicted by a vertice and every surgery of Liouville tori in shown by the schemes 1–5 above (see Fig. 1). This graph allows one to visualize the evolution of Liouville tori inside the three-dimensional manifold Q under the variation of the value of the second integral.

The graph $\Gamma(Q, f)$ can also be constructed in the general case, i.e. if some of the critical levels of f contain more than one critical submanifolds. We shall describe this construction which is a constituent part of the theory of topological invariants of integrable systems developed by Fomenko [1988a, b].

Definition 1.19. We say that the Hamiltonian H is *nonresonance* on a given isoenergy surface Q^3 if in Q the Liouville tori on which the trajectories of the system v form a dense irrational winding are dense.

An attempt to study concrete known systems shows that on M^4 in the majority of cases Hamiltonians are nonresonance on almost all Q^3. In the case of M^{2n}, where $n > 2$, examples are known when H is resonance. This occurs, in particular, when the system is integrable in the noncommutative sense. (For an analysis of the basic cases of noncommutative integrability (starting with the works of E. Cartan) see Trofimov and Fomenko [1984].) We say that two integrable Hamiltonian systems v_1 and v_2 on the same isoenergy manifold Q are *topologically equivalent* if there exists a diffeomorphism $s: Q \to Q$, carrying the Liouville tori of the system v_1 into the Liouville tori of the system v_2. A general problem arises: classify the integrable Hamiltonian systems (on a given Q) up to topological equivalence. It turns out that this problem is also solved with the help of the topological invariant discovered by Fomenko [1988a, b]. The details can be found in Fomenko and Zieschang [1989].

Theorem 1.38 (Fomenko [1988b]). *Let v be a Hamiltonian system with nonresonance Hamiltonian H which is integrable by means of a Bott integral f on the compact nonsingular three-dimensional isoenergy surface Q. Then one can construct uniquely a graph $\Gamma(Q, f)$ with the following property: from the graph $\Gamma(Q, f)$ one can reconstruct uniquely (up to homeomorphism) the whole topological picture of the evolution and surgery (bifurcations) of Liouville tori inside the surface Q under variation of the value of the integral f.*

Let $f: Q \to \mathbb{R}$ be a Bott integral, $\alpha \in \mathbb{R}$, and let $f_\alpha = f^{-1}(\alpha)$ be a connected component of a level surface of the integral (singular or nonsingular). If $\alpha = a$ is a regular (noncritical) value for f, then f_a is the union of a finite number of tori. We denote the critical values for f by c, the connected component of the critical level surface of the integral by f_c, and the set of critical points of the integral f on f_c by N_c. As is proved in Fomenko [1986b], the connected components of the sets N_c can only be of the following types:

Type I, a minimax circle S^1 (a local minimum or maximum for f), then $N_c = f_c = S^1$.

Type II, a minimax torus T^2, then $N_c = f_c = T^2$.

Type III, a saddle critical circle S^1 with orientable separatrix diagram, then $N_c = S^1 \neq f_c$.

Type IV, a saddle critical circle S^1 with nonorientable separatrix diagram, then $N_c = S^1 \neq f_c$.

Type V, a minimax Klein bottle K^2, then $N_c = f_c = K^2$.

By $U(f_c)$ (where c is a critical value of the integral) we denote a regular connected closed ε-neighbourhood of the component f_c in the manifold Q^3. One can assume that $U(f_c)$ is a connected three-dimensional manifold whose boundary consists of a disconnected union of tori. As $U(f_c)$ one can take a connected component of the manifold $f^{-1}([c - \varepsilon, c + \varepsilon])$. One can say that $Q = \sum_c U(f_c)$, i.e., Q is obtained from all the manifolds $U(f_c)$ by pasting their boundaries by some diffeomorphisms of the bounding tori.

The sets $U(f_c)$ are divided into 5 types corresponding to the types 1–5 of the sets N_c. More precisely, we say that the set $U(f_c)$ has type I (respectively, types II and V) if it is a tubular neighbourhood of a connected set N_c of type I (respectively, of types II and V). Further, we say that $U(f_c)$ has type III if N_c consists of only critical saddle circles with orientable separatrix diagrams. Finally, the set $U(f_c)$ has type IV if N_c consists of critical saddle circles among which there is at least one with nonorientable diagram.

Theorem 1.39 (see Fomenko [1986b]). *Let Q be a compact nonsingular isoenergy surface of the system v with Hamiltonian H (not necessarily nonresonance) which is integrable by means of a Bott integral f. Then the manifolds $U(f_c)$ which occur in the decomposition $Q = \sum_c U(f_c)$ admit the following representation, depending on the type of the set $U(f_c)$:*

Type I: $U(f_c) = P_c^2 \times S^1$, where $P_c^2 = D^2$ (a disc).

Type II: $U(f_c) = P_c^2 \times S^1$, where $P_c^2 = S^1 \times D^1$ (a cylinder).

Type III: $U(f_c) = P_c^2 \times S^1$, where P_c^2 is a two-dimensional surface with boundary.

Type IV: $U(f_c) = P_c^2 \dot{\times} S^1$, where P_c^2 is a two-dimensional surface with boundary, and $P_c^2 \dot{\times} S^1$ is the total space of a Seifert bundle with base P_c^2 and fiber S^1 (cf. Fomenko [1988a] for a description).

Type V: $U(f_c) = P_c^2 \tilde{\times} S^1$, where $P_c^2 = \mu$ (a Möbius strip), and $\mu \tilde{\times} S^1$ denotes a fiber bundle (with boundary the torus T^2).

Corollary. *With each isoenergy surface Q (under the hypotheses of Theorem 1.39) one can associate uniquely (up to homeomorphism) a closed two-dimensional surface $P^2(Q, f) = \sum_c P_c^2$ obtained by pasting together the surfaces P_c^2 induced by the representation $Q = \sum_c U(f_c)$.*

Theorem 1.40 (A.T. Fomenko). *Let v be a Hamiltonian system which is integrable on Q by means of a Bott integral. Then there exists a unique (up to homeomorphism) canonical embedding $h(Q, f)$ of the graph $\Gamma(Q, f)$ in the surface $P^2(Q, f)$. If the Hamiltonian H is nonresonance on Q, then the triple (Γ, P, h) is independent of the choice of the second integral f. Namely, if f and f' are any Bott integrals of the system v, then the corresponding graphs $\Gamma(Q, f)$ and $\Gamma(Q, f')$ and the surfaces $P(Q, f)$ and $P(Q, f')$ are homeomorphic and the following diagram is commutative:*

$$\begin{array}{ccc} h\colon \Gamma & \longrightarrow & P \\ \wr\wr & & \wr\wr \\ h'\colon \Gamma' & \longrightarrow & P' \end{array}$$

Corollary. *In the nonresonance case the triple (Γ, P, h) is a topological invariant of the integrable case (Hamiltonian) itself and allows one to classify integrable Hamiltonians according to their topological type and complexity.*

We call this triple the isoenergy topological invariant of the integrable Hamiltonian. The division of the surface $P(Q)$ into domains defined by the graph $\Gamma(Q)$ is also a topological invariant. The surface $P(Q)$ is not necessarily embedded in Q.

We shall now describe an explicit construction of the invariants.

Let us first assume for simplicity that at each critical level f_c there is exactly one critical connected manifold N_c. In this case the constructon of the graph $\Gamma(Q, f)$ was described earlier (see also Fomenko [1986b]). Next we construct the graph Γ in general. Now at one critical level there may lie several critical manifolds. In contrast with ordinary Morse functions, the critical manifolds of a Bott integral lying at one level cannot generally be "spread out" at different levels by a small perturbation of the integral. A perturbation $\tilde{f}$ of the integral f may not be an integral. Let f_a be a level surface of the integral, i.e., $f_a \subset f^{-1}(a)$. If a is a regular value, then $f^{-1}(a)$ is the union of a finite number of tori. We represent them by points in $\mathbb{R}^3$ at level a, where the axis $\mathbb{R}$ is directed upwards. Varying a in the domain of regular values, we force these points to sweep out arcs, part of the edges of the future graph Γ. Let N_c be the set of critical points of f on f_c. We single out two cases: a) $N_c = f_c$, b) $N_c \subset f_c$ while $N_c \neq f_c$. All the possibilities for N_c are found in Fomenko [1986b]. We consider case a). Here only three types of critical sets are possible.

"*Minimax Circle*" *type.* Here $N_c = f_c$ is homeomorphic to a circle on which f achieves a local minimum or maximum. Its tubular neighbourhood $S^1 \times D^2$ is homeomorphic to a solid torus (in Q^3). As $a \to c$, nonsingular tori contract to the axis of the solid torus and for $a = c$ degenerate into a S^1. By convention we represent this situaton by a heavy black spot (vertex of the graph) into which (or out of which) one edge goes (comes) (Fig. 1).

"*Torus*" *type.* Here $N_c = f_c$ and is homeomorphic to a torus T^2 on which f achieves a local minimum or maximum. Its tubular neighbourhood is homeomorphic to a cylinder $T^2 \times D^1$. The boundary of the cylinder is two tori. As $a \to c$, they move toward one another and for $a = c$ they merge into one torus. We represent this situation by a white circle (vertex of the graph) into (or out of) which two edges of the graph go (or come).

"*Klein Bottle*" *type.* Here $N_c = f_c$ and is homeomorphic to a Klein bottle K^2 on which f achieves a local minimum or maximum. Its tubular neighbourhood $K^2 \tilde{\times} D^1$ is homeomorphic to the twisted product of K^2 by a segment. The boundary of $K^2 \tilde{\times} D^1$ is one torus. As $a \to c$, it tends to K^2 and double covers it for $a = c$. We represent this situaton by a white circle with a dot inside it (vertex of the graph) into (out of) which one edge of the graph goes (comes).

We consider case b). Here $N_c \subset f_c$ while $\dim N_c = 1$ and $\dim f_c = 2$. Then N_c is the disconnected union of disjoint circles. Each of them is a saddle for f. By convention, f_c can be represented by a planar horizontal square lying at level c in $\mathbb{R}^3$. Some edges of the graph run into it from below (for $a \to c$ and $a < c$), other edges of the graph leave upwards (when $a > c$). As a result, we have defined a graph A consisting of regular arcs (edges), some of which run into squares and some end at vertices of the three types described.

We fix a saddle value c and we construct the graph T_c showing precisely how the edges of the graph A interacting with f_c are joined. We fix a Riemannian metric on the manifold Q. On Q we consider the vector field $w = \operatorname{grad} f$. Its trajectories, going into critical points of the integral or issuing from them, are called separatrices. Their union is the separatrix diagram of the critical submanifold. From each saddle critical circle S^1 on f_c we emit its separatrix diagram. If it is orientable, then it can be obtained by pasting two planar rings (cylinders) along axial circles (Fomenko [1986b]). If it is nonorientable, then it can be obtained by pasting two Möbius strips along their axial circles. We consider noncritical values $c - \varepsilon$ and $c + \varepsilon$ close to c. The surfaces $f_{c-\varepsilon}$ and $f_{c+\varepsilon}$ consist of tori. The separatrix diagrams of the critical circles lying in f_c intersect the tori along circles (transversely) and divide the tori into union of domains which we call regular. At level $f_{c-\varepsilon}$ in each of them we choose a point and from these points we emit integral trajectories of the field w. They go past critical circles at level f_c and land in some other regular domains of the tori constituting $f_{c+\varepsilon}$. Obviously, in this way we get a homeomorphism between the open regular domains of $f_{c-\varepsilon}$ and the open regular domains of $f_{c+\varepsilon}$.

First we consider the orientable case, i.e., when all the separatrix diagrams are orientable (i.e., there are no Möbius strips). Since each nonsingular torus is a point on the graph A, one can join points at level $f_{c-\varepsilon}$ and at level $f_{c+\varepsilon}$ by arcs (segments) representing bundles of integral trajectories of the field w. We get a graph T_c. Its edges show us the motion of the open regular domains of the tori. The tori fall into pieces which afterwards rise (descend) and are regrouped into new tori. Each upper torus is composed of pieces of lower tori (and conversely). Now we consider the nonorientable case, i.e., when at least one of the critical circles on f_c has nonorientable diagram. On each torus approaching f_c we mark with asterisks the regular domains incident to the nonorientable separatrix diagrams (i.e., to the Möbius strips). We also mark with asterisks the corresponding edges of the graph. Thus, we construct a graph according to the scheme of the orientable case, after which we mark with asterisks those of its edges which represent motion of regular domains with asterisks. We denote the graph obtained by T_c. It is clear that the ends of edges of the graph T_c can be identified with ends of some edges of the graph A. Finally, we define the graph Γ to be the union (collage) $\Gamma = A + \sum_c T_c$ where $\{c\}$ are the critical saddle values of the integral.

Proposition 1.7 (Fomenko [1988b]). *Let f and f' be any two Bott integrals. Then under the homeomorphism $q(Q, f, f')\colon \Gamma(Q, f) \to \Gamma(Q, f')$ (cf. Theorem 1.40) the saddle subgraphs T_c for the integral f are carried homeomorphically into the saddle subgraphs T_c' for the integral f_c'. The asterisks of the graph Γ go into the asterisks of the graph Γ'. Vertices of the type "minimax circle" and "Klein bottle" of the graph Γ go into vertices of the same type (respectively) in the graph Γ'. Vertices of the "torus" type of the graph Γ can be mapped into ordinary interior points of edges of the graph Γ'. Conversely, ordinary interior points of edges of the graph Γ can be mapped into vertices of "torus" type in the graph Γ'.*

Definition 1.20. We call the triple $\Gamma(Q)$, $P(Q)$, $h(Q)$ the *isoenergy topological invariant* $I(H, Q)$ of the integrable Hamiltonian H on the given isoenergy surface Q. The collection of all triples $\{\Gamma(Q), P(Q), h(Q)\}$ for all Q will be called the *complete topological invariant* $I(H)$.

Of course, we consider homeomorphic triples to be equivalent. The complete invariant $I(H)$ now depends only on the Hamiltonian H.

Corollary. *If two integrable systems have nonhomeomorphic topological invariants, then the systems are not equivalent and it is impossible to establish, for example, a trajectory isomorphism between them.*

At the same time there exist analytically inequivalent integrable systems with identical topological invariants $I(H)$.

Now we construct a surface $P(Q, f)$. We define it to be the union (collage) of the form $P(A) + \sum_c P(T_c)$, where $P(A)$ and $P(T_c)$ are two-dimensional surfaces with boundary. We define $P(A) = (S^1 \times \operatorname{Int} A)^{\cdot} + \Sigma D^2 + \Sigma\mu + \Sigma S^1 \times D^1$. Here $\operatorname{Int} A$ is the union of all open edges of the graph A. Consequently, $S^1 \times \operatorname{Int} A$ is the union of open cylinders. The manifold $(S^1 \times \operatorname{Int} A)^{\cdot}$ is obtained from it by adding the boundary circles. Let an edge of the graph A end with a black vertex. Then we paste the disc D^2 to the corresponding bounding circle on the boundary of the manifold $(S^1 \times \operatorname{Int} A)^{\cdot}$. We denote the attachment of such discs by ΣD^2. Let two edges of the graph A meet a white vertex. It defines two bounding circles on $(S^1 \times \operatorname{Int} A)^{\cdot}$ to which we paste (attach) a cylinder $S^1 \times D^1$. We denote this operation by $\Sigma S^1 \times D^1$. Let an edge of the graph A end at a white vertex with a dot inside. To the bounding circle on $(S^1 \times \operatorname{Int} A)^{\cdot}$ corresponding to it we paste a Möbius strip μ. The operation is denoted by $\Sigma\mu$. Thus, ΣD^2, $\Sigma S^1 \times D^1$, and $\Sigma\mu$ correspond to minimax circles, tori, and Klein bottles. Now we construct $P(T_c) = P_c$. First we consider the orientable case when all critical circles on f_c have orientable separatrix diagrams. As proved by Fomenko [1986b], f_c is homeomorphic to the direct product $K_c \times S^1$, where K_c is a graph obtained from several circles by identifying certain pairs of points on them. Locally from each vertex of the graph K_c exactly 4 edges depart. In general we have the following proposition.

Proposition 1.8 (Fomenko [1986b]). *The complex f_c is obtained by pasting several two dimensional tori along circles realizing nonzero cycles γ on the tori. If several such circles are located on one torus, then they do not intersect. Circles along which the tori entering into f_c are tangent are critical for f. They are homologous and dissect f_c into the union of several flat annuli.*

Thus, the cycle γ is defined uniquely on f_c. On each torus in f_c we choose a circlar generator α which is supplementary to γ (meets γ in exactly one point). We call it an oval. One can assume that ovals are tangent to one another at points lying on critical circles of the integral. In the orientable case the union of the ovals gives the graph K_c. It will not necessarily be flat. Let x be a point of tangency of two ovals, i.e., a critical point for f. Then segments of the integral curves of the field w and the level line of the function f determine (on the two-dimensional disc with center at the point x, lying in Q and orthogonal to the critical circle on which the point x lies) near the point x a "coordinate cross" on each end of which there is an arrow indicating the direction of w. We construct such normal two-dimensional crosses at each vertex of the graph K_c. Different crosses are joined by segments which are parts of ovals. Now we join the ends of the crosses by thin strips which go along arcs of ovals. These strips consist of segments of integral curves of the field w which intersect the ovals orthogonally (outside of critical points). Arcs of ovals go along the axes of these strips. As a result we get a smooth two-dimensional surface with boundary. We mark the boundary circles corresponding to tori which approach f_c from below with the sign $-$. Those corresponding to tori which approach f_c from above we mark with the sign $+$. The number of negative (positive) tori is equal to the number of edges of the graph Γ approaching f_c from below (above). The surface obtained is denoted by $P(T_c) = P_c$. Its boundary circles are divided into two classes: lower (negative) ones and upper (positive) ones. The graph K_c is uniquely embedded in P_c^2 (up to a homeomorphism).

In the nonorientable case the surface P_c^2 is constructed in a similar way.

Now we can construct the whole surface $P(Q, f)$.

Clearly, there exists a one-to-one correspondence between the boundary circles of the surface $P(A)$ and the boundary circles of the union of the surfaces P_c. This correspondence is given by the edges of the graph A. We identify the corresponding circles with the help of homeomorphisms and we get a unique closed two-dimensional surface $P(Q, f)$ (orientable or nonorientable). On $P(Q, f)$ there is situated (uniquely up to a homeomorphism of the surface) a generally disconnected graph which we denote by $K(Q, f)$. We consider circles cutting in halves the cylinders which enter into the surface $S^1 \times \operatorname{Int} A$. One can assume that each of them has the form $S^1 \times p$, where p is the midpoint of the corresponding edge of the graph A. Now we define the graph K as the disjoint union of all graphs K_c and circles of the form $S^1 \times p$. The graph K has only vertices of multiplicity 4.

The construction of the surface $P(Q, f)$ is now complete.

Proposition 1.9 (Fomenko [1988b]). *If f and f' are any Bott integrals on Q, then $P(Q, f)$ is homeomorphic to $P(Q, f')$ (in the nonresonance case).*

We denote by K^* the graph dual to the graph K in the surface $P(Q, f)$. Its vertices are the centers of the domains into which the graph K divides P, and its edges are the arcs joining the vertices through the centers of edges of the graph K.

Proposition 1.10 (Fomenko [1988b]). *The graph $\Gamma(Q, f)$ coincides with the graph $K^*(Q, f)$. Consequently the graph $\Gamma(Q, f)$ admits an embedding $h(Q, f)$: $\Gamma(Q, f) \to P(Q, f)$ which is defined uniquely (up to homeomorphism of the surface) by the original integrable nonresonance system. The graph K divides the surface P into domains which are homeomorphic to a disk or an annulus or a Möbius strip.*

It follows that to every integrable nonresonance Hamiltonian system one can associate an integer called the genus of the system in Fomenko [1988b]. This is the genus of the surface $P(Q)$.

5.3. The Case of Four-dimensional Rigid Body. As an illustration of the general theory we shall give a complete description of the isoenergy surfaces for the integrable system describing the motion of a four-dimensional rigid body fixed at its center of mass (Euler's case). On the Lie algebra so(4) we consider a system of differential equations which is slightly more general then those describing the motion of n-dimensional rigid body with a fixed point (see Definition 1.1). The coadjoint orbits of the Lie group SO(4) in so(4) are distinguished by the invariants

$$h_1(X) = \sum_{i,j} x_{ij}^2,$$

$$h_2(X) = x_{12}x_{34} - x_{13}x_{24} + x_{14}x_{23},$$

where $X = \|x_{ij}\| \in \mathrm{so}(4)$. Generic orbits are homeomorphic to $S^2 \times S^2 = \{h_1 = p_1^2, h_2 = p_2\}$, $p_i = \mathrm{const}$, $|2p_2| < p_1^2$. As noted in Section 1.3, each coadjoint orbit has a canonical symplectic structure. With respect to this structure, we consider a Hamiltonian vector field on so(4) defined by the Hamiltonian

$$H = \sum_{i<j} \frac{b_i - b_j}{a_i - a_j} x_{ij}^2,$$

where $\Sigma a_i = \Sigma b_i = 0$. (The vector fields $\operatorname{sgrad} H$ are considered on each coadjoint orbit $\{\mathrm{Ad}_g f \mid g \in \mathrm{SO}(4)\}$, and combine into a vector field on so(4).)

Theorem 1.41. *On all generic coadjoint orbits of* SO(4) *the system* $\dot{X} = \operatorname{sgrad} H(X)$ *is a completely integrable Hamiltonian system. More precisely, the functions*

$$h_3(X) = \sum_{i<j} (a_i + a_j)x_{ij}^2,$$

$$h_4(X) = \sum_{i<j} (a_i^2 + a_i a_j + a_j^2)x_{ij}^2$$

form a complete involutive family on all generic coadjoint orbits of SO(4).

Remark 1. The construction of analogous completely integrable Hamiltonian systems on arbitrary semisimple Lie algebras is outlined in Section 3 of Chapter 2.

Remark 2. The system describing the motion of a 4-dimensional rigid body with a fixed point results from $\dot{X} = \operatorname{sgrad} H(X)$ in the particular case where $a_i = b_i^2$, $i = 1, 2, 3, 4$.

Thus, we can take $S^2 \times S^2$ for a four-dimensional symplectic manifold M^4, the above Hamiltonian for H, and any of the integrals h_3 or h_4 for the second additional (on Q^3) integral f. According to the general theory, the vertices of the

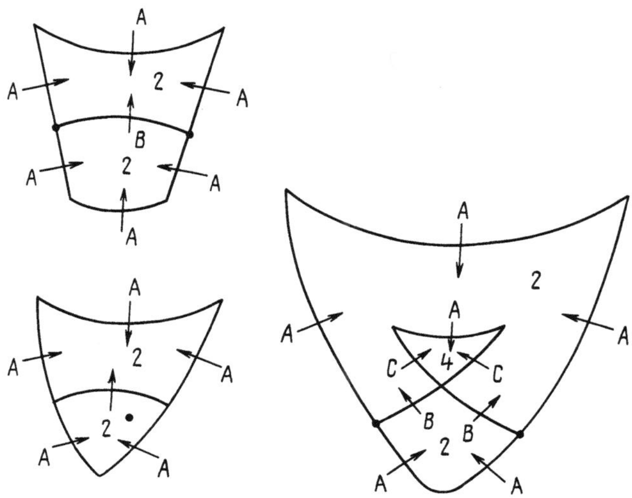

Fig. 2

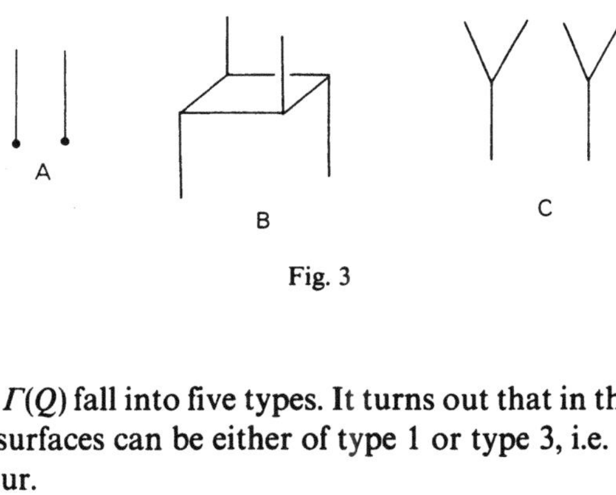

Fig. 3

graph $\Gamma(Q, f) = \Gamma(Q)$ fall into five types. It turns out that in this case the surgery of the isoenergy surfaces can be either of type 1 or type 3, i.e. surgery of types 2, 4, 5 does not occur.

The following two theorems are due to A.A. Oshemkov who applied the general theory outlined above in this special case.

Theorem 1.42. *Suppose that we are given a Hamiltonian system* $\dot{X} = \operatorname{sgrad} h_3(X)$ *on the Lie algebra* so(4). *Then the nonsingular orbits* $S^2 \times S^2$ *in* so(4) *are fibered into isoenergy surfaces* $Q^3 = \{h_3 = \text{const}\}$ *so that for all these surfaces* Q^3, *except possibly a finite number of them, the function* $f = h_4|S^2 \times S^2$ *is a Bott integral. Depending on the numerical values of the parameters* p_1 *and* p_2 *which determine a generic orbit* $S^2 \times S^2 = \{h_1 = p_1^2, h_2 = p_2\}$, *the bifurcation diagrams for the moment map* $F\colon h_3 \times h_4\colon S^2 \times S^2 \to \mathbb{R}^2$ *can only be of three types. Their explicit form is shown in Fig.* 2.

Numbers in Fig. 2 indicate the number of Liouville tori that constitute the inverse image $F^{-1}(y)$ for the points y in a given domain in $\mathbb{R}^2$. Surgery along the arrows in Fig. 2 has the form A, B, C shown in Fig. 3.

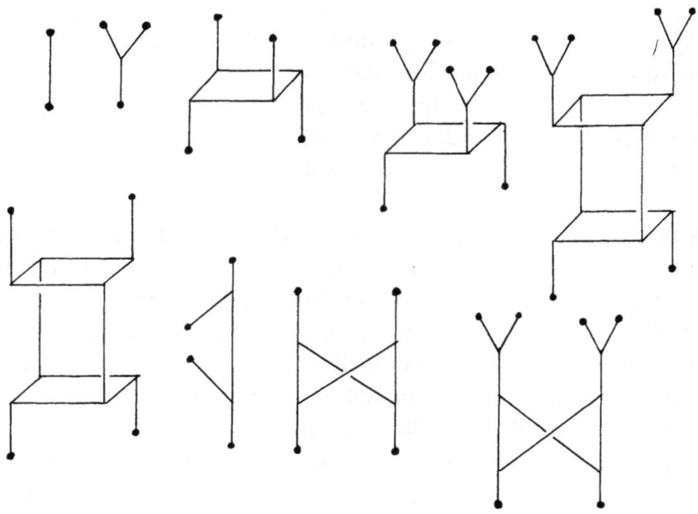

Fig. 4

Theorem 1.43 (Oshemkov). *Suppose that we are given a Hamiltonian system* $\dot{X} = \operatorname{sgrad} H(X)$ *with a general Hamiltonian H defined above on the Lie algebra* so(4). *We take* h_3 *or* h_4 *for a second integral f which is functionally independent of H almost everywhere. Then the nonsingular* (*i.e. generic*) *orbits* $S^2 \times S^2$ *in* so(4) *are fibered into the isoenergy surfaces* $Q^3 = \{H = \text{const}\}$ *such that f is a Bott integral for all these surfaces except, possibly, a finite number of them. A complete list of all nine connected graphs* $\Gamma(Q)$ *that describe the nonsingular surfaces Q occurring in this problem is given in Fig. 4. Any of these graphs can indeed occur for some Hamiltonian* (*for a suitable choice of the parameters in the Hamiltonian and the level surface of the second integral*).

We note that an arbitrary Hamiltonian H can be expressed as $H = k_1 h_1 + k_3 h_3 + k_4 h_4$ where the constants k_1, k_3, k_4 do not depend on x_{ij}. Therefore the bifurcation sets for the system with an arbitrary Hamiltonian can be obtained from the bifurcation set for the system with Hamiltonian h_3 by applying some nondegenerate linear transformations.

Some partial information on the structure of the bifurcation diagram for any compact Lie algebra g and for a complete involutive family $F_a = \{f_1, \ldots, f_N\}$ of the shifted coadjoint invariants of g was obtained by A.V. Bolsinov ($N = (\dim \mathfrak{g} + \operatorname{ind} \mathfrak{g})/2$). Let $\Sigma \subset \mathbb{R}^N$ denote the bifurcation diagram of the moment map $F = (f_1(x), \ldots, f_N(x))$, $F\colon \mathfrak{g} \to \mathbb{R}^N$.

Theorem 1.44 (A.V. Bolsinov). a) *Let* $Q = \{f_1 = c_1, \ldots, f_{N-1} = c_{N-1}\}$ *be a nonsingular compact level surface and let* Q_0 *be a connected component of Q. The* *function* f_N *has precisely two critical values* $c_{\min}$ *and* $c_{\max}$ *on* Q_0, *and the level surfaces* $\{f_N = c_{\min}\}$ *and* $\{f_N = c_{\max}\}$ *on* Q_0 *are* $(n-1)$*-dimensional tori, where* n *is the dimension of the generic Liouville torus.*

b) *The set* $F(\mathbb{R}^N)\backslash\Sigma$ *is arcwise connected. Any two regular fibers* $F^{-1}(b_1)$ *and* $F^{-1}(b_2)$ *where* $b_1, b_2 \in F(\mathbb{R}^N)\backslash\Sigma$ *can be continuously deformed one into another in the class of regular fibers of the moment map.*

Let us now formulate a program of further investigations in this direction. Since the topology of isoenergy surfaces can be depicted by means of a graph $\Gamma(Q)$ with 5 types of vertices, a list should be compiled, as complete as is possible, of all available integrable systems on four-dimensional manifolds M^4. Next, according to the scheme described earlier one can construct the graphs $\Gamma(Q)$. This will provide a qualitatively important and visualizable information on the topology of the integrable cases discovered to date.

Matveev, Fomenko and Sharko [1988] investigated connections between the class (H) of integrable isoenergy surfaces and the class (R) of three-dimensional manifolds which admit a round Morse function. An unexpected and profound relationship between the class (H) and the theory of three-dimensional closed hyperbolic manifolds (and the problem of calculating their volumes) was found by Matveev and Fomenko [1988a]. Also, as was shown by Matveev and Fomenko [1988b], the assumption that f is a Bott integral can be relaxed considerably and replaced by the assumption that f be only a "tame" integral. The class of "tame" integrals is much larger than that of Bott integrals.

It was pointed out by Fomenko [1988b] that by attaching some numerical marks to the invariant $I(H, Q)$, i.e. by considering the "marked" invariant $I(H, Q)^*$, one can obtain a criterion of topological equivalence of integrable Hamiltonian systems. This program, which was only sketched in Fomenko [1988b], was fully realized by Fomenko and Zieschang [1988]. We say that marked isoenergy topological invariants $I(H_1, Q_1)^*$ and $I(H_2, Q_2)^*$ are homeomorphic (or equal) if there is a homeomorphism taking the surface P_1^2 to the surface P_2^2, the graph Γ_1 to the graph Γ_2 and preserving all numerical marks (for their precise definition see Fomenko and Zieschang [1989]).

Theorem 1.45 (Fomenko and Zieschang [1989]) (Equivalence criterion for integrable systems). *Let* Q_1 *and* Q_2 *be two three-dimensional closed orientable isoenergy manifolds and let* v_1 *and* v_2 *be nonresonance Hamiltonian systems on them which are integrable by means of Bott integrals.*

a) *If the corresponding marked invariants* $I(H_1, Q_1)^*$ *and* $I(H_2, Q_2)^*$ *are not homeomorphic (are different), then the systems* v_1 *and* v_2 *are topologically inequivalent. The manifolds* Q_1 *and* Q_2 *in this case may be homeomorphic.*

b) *Conversely, suppose that the marked invariants of* v_1 *and* v_2 *are homeomorphic (equal). Then* 1) *the manifolds* Q_1 *and* Q_2 *are homeomorphic;* 2) *the systems* v_1 *and* v_2 *are topologically equivalent.*

Next, Fomenko and Zieschang [1989] introduce the concept of stably equivalent integrable systems. As a result, all integrable systems are divided into classes of stably equivalent systems. Fomenko and Zieschang [1989] have obtained a classification of integrable Hamiltonian systems up to stable equivalence (on fairly complicated irreducible isoenergy manifolds). In particular, it turns out that on any fairly complicated irreducible three-dimensional isoenergy manifold there is, up to stable equivalence, precisely one integrable Hamiltonian system (with a Bott integral). Further, the classes of stably equivalent integrable Hamiltonian systems (with Bott integrals) on fairly complicated irreducible isoenergy manifolds are classified by the fundamental groups of these manifolds. Presentations of these fundamental groups by generators and relations are given in Fomenko and Zieschang [1988].

Chapter 2
The Algebra of Hamiltonian Systems

§ 1. Representations of Lie Groups and Dynamical Systems

1.1. Symplectic Structures Associated with Representations. It is well known that the projective space $\mathbb{C}P^n$ carries a canonical symplectic structure (see, for instance, Arnol'd [1974] or Arnol'd and Givental' [1985]). We consider a representation $\rho\colon G \to GL(V)$ of a Lie group G on a finite-dimensional complex vector space V. There is an induced action of G on the projective space $P(V)$ associated with V. We denote by $G(z)$ the orbit of this action passing through a point $z \in P(V)$. It is now natural to ask when the symplectic structure of $P(V)$ induces a symplectic structure on the orbits of G in $P(V)$. Let us assume that G is compact, let T be a maximal torus in G, let $\mathfrak{g}$ and $\mathfrak{t}$ be the Lie algebras of G and T, respectively, and let $L^{\mathbb{C}}$ denote the complexification of a vector space L. We then have the root space decomposition $\mathfrak{g}^{\mathbb{C}} = \mathfrak{t}^{\mathbb{C}} + \sum_{\alpha \neq 0} \mathbb{C}E_\alpha$, $H_\alpha = [E_\alpha, E_{-\alpha}]$ (see, for instance, Helgason [1962], Jacobson [1962] or the Appendix). Let $\pi\colon V\backslash 0 \to P(V)$ be the natural projection which maps a vector $v \in V\backslash 0$ into the straight line passing through v and the origin.

Definition 2.1. Let $\rho\colon \mathfrak{g} \to \mathrm{End}(V)$ be a complex representation of a semisimple Lie algebra $\mathfrak{g}$. A vector $v \in V$ is called *symplectic* if a) v is a weight vector of weight λ for the representation ρ with respect to some Cartan subalgebra; b) if $\lambda(H_\alpha) = 0$ for some root α then $E_\alpha(v) = E_{-\alpha}(v) = 0$.

Theorem 2.1 (see Guillemin and Sternberg [1984a]). *Let G be a compact semisimple Lie group with an irreducible representation $\rho\colon G \to GL(V)$ on a com-*

plex vector space V. The orbit $G(z)$ through a point $z \in P(V)$ is symplectic relative to the restriction of the canonical symplectic form of $P(V)$ if and only if $z = \pi(v)$, where v is a symplectic vector in V with respect to ρ. There is only one orbit that is Kähler, and that is the orbit through $\pi(v)$ where v is a highest weight vector.

For more information on this problem see Guillemin and Sternberg [1984a].

1.2. Sectional Operators. For finite-dimensional Lie algebras there is a simple and natural method for constructing dynamical systems and symplectic structures on the orbits of arbitrary representations of the associated Lie groups. This method is based on the notion of sectional operator introduced by Fomenko [1980], [1981].

Let $\mathfrak{H}$ be a Lie algebra, H the corresponding Lie group, $\rho\colon \mathfrak{H} \to \mathrm{End}(V)$ a representation of $\mathfrak{H}$ on a linear space V, $\alpha\colon H \to GL(V)$ the associated representation of the group H, and $O(X)$ an orbit under the action of H on V, $X \in V$. A linear operator $Q\colon V \to \mathfrak{H}$, which we call *sectional*, gives rise to a vector field $\dot{X} = \rho(Q(X))(X)$. In some cases such an operator can also be used to define a symplectic structure on the orbits (see Fomenko [1980], [1981]).

Assume that H is a semisimple Lie group. For $a \in V$ let K be the annihilator of a with respect to the representation α, K' the orthogonal complement to K with respect to the Killing form of $\mathfrak{H}$, and $\xi, \eta \in T_X O(X)$ two tangent vectors to the orbit. There exist uniquely determined vectors $\xi', \eta' \in K'$ such that $\rho(\xi') = \xi$, $\rho(\eta') = \eta$. Given a sectional operator $C\colon V \to \mathfrak{H}$ we define a bilinear form $F_C = \langle C(X), [\xi', \eta'] \rangle$ where $[\xi', \eta'], C(X) \in \mathfrak{H}$. This form is defined on each orbit and is skew-symmetric. On the other hand, we have the vector field $\dot{X}$ constructed by using another sectional operator Q. We can now ask the natural queston: for which operator C is the 2-form F_C closed and nondegenerate, and for which operators Q and C is the flow $\dot{X}$ Hamiltonian with respect to F_C? In the case of symmetric spaces one can give sufficiently complete answers, see Fomenko [1980], [1981], Trofimov and Fomenko [1983b].

We shall describe a particular multi-parameter family of sectional operators which is of interest for applications. We construct a decomposition of $\mathfrak{H}$ into a sum of four vector subspaces $K + \tilde{B} + \tilde{R} + \tilde{P}$ and a similar decomposition of V into a sum of four subspaces $T + B + R + Z$. After this we define a sectional operator $Q\colon V \to \mathfrak{H}$ as a linear mapping which preserves these decompositions. In other words, the matrix of this operator is block-diagonal.

Choose an arbitrary point a in general position in V, i.e., such that the orbit of the group action passing through a has maximal dimension. Let K denote the annihilator of a in the Lie algebra $\mathfrak{H}$, i.e. $K = \operatorname{Ker} \Phi_a$ where the linear mapping $\Phi_a\colon \mathfrak{H} \to V$ is defined by $\Phi_a(h) = \rho(h)a$, $h \in \mathfrak{H}$. Since a is in general position, the dimension of K is minimal. For an arbitrary element $b \in K$ we consider the action of the operator $\rho(b)$ on V and denote $\operatorname{Ker} \rho(b) \subset V$ by M. Let K' be any algebraic complement to K in $\mathfrak{H}$, i.e. $\mathfrak{H} = K + K'$ and $K \cap K' = 0$. The choice of K' is not unique, and the freedom in varying this algebraic complement leads to the appearance of an additional family of parameters in the construction. Clearly, $a \in M$. By the definition of K', the mapping $\Phi_a: \mathfrak{H} \to V$ monomorphically takes K' to some plane $\Phi_a(K') \subset V$. Since $\Phi_a(K') = \Phi_a(\mathfrak{H})$, this plane does not depend on the choice of K' and is uniquely determined by the choice of a and the representation ρ. We assume that there is an element b in the annihilator such that V splits into the direct sum of the two subspaces M and $\operatorname{Im} \rho(b)$. For example, for such an element b we can take a semisimple element of K.

The plane $\Phi_a(K')$ intersects M and $\operatorname{Im} \rho(b)$ in planes which we denote by B and R', respectively. So we obtain a decomposition of $\Phi_a(K')$ into a direct sum of three planes $B + R' + P$, where B and R' are uniquely determined and the complementary plane P is non-unique and introduces another collection of parameters. We consider the acton of $\rho(b)$ on $\operatorname{Im} \rho(b)$. It is clear that $\rho(b)$ maps $\operatorname{Im} \rho(b)$ isomorphically onto itself. In particular, $\rho(b)$ is invertible on $\operatorname{Im} \rho(b)$. Let $\rho(b)^{-1}$ be its inverse. We set $R = \rho(b)^{-1}(R')$. Then $\rho(b): R \to R'$ and R is uniquely determined. Let Z be an algebraic complement to R in $\operatorname{Im} \rho(b)$. Then $\operatorname{Im} \rho(b) = Z + R'$, $R' \cong R$. Let T be a complement to B in M.

We have finally obtained a decomposition of V into the direct sum of four planes $V = T + B + R + Z$. Here R, B, M and $\operatorname{Im} \rho(b)$ are uniquely determined, but Z and T are non-unique, and this introduces still another family of parameters in the construction. If V is equipped with a scalar product, then Z and T can be uniquely determined as the orthogonal complements of the corresponding subspaces. Since K' is isomorphic to $\Phi_a(K')$ we have $K' = \tilde{B} + \tilde{R} + \tilde{P}$ where $\tilde{B} = \Phi_a^{-1}(B)$, $\tilde{R} = \Phi_a^{-1}(R)$, $\tilde{P} = \Phi_a^{-1}(P)$.

We now define a sectional operator $Q: V \to \mathfrak{H}$, $Q: T + B + R + Z \to K + \tilde{B} + \tilde{R} + \tilde{P}$ by setting

$$Q = \begin{bmatrix} D & 0 & 0 & 0 \\ 0 & \Phi_a^{-1} & 0 & 0 \\ 0 & 0 & \Phi_a^{-1}\rho(b) & 0 \\ 0 & 0 & 0 & D' \end{bmatrix}.$$

Here $D: T \to K$ and $D': Z \to \tilde{P}$ are arbitrary linear operators.

For coadjoint representations of Lie groups, the above construction of sectional operators can be made more precise. The construction that follows is due to V.V. Trofimov. Let $\mathfrak{g}$ be a Lie algebra,and $\mathfrak{H}$ its Abelian subalgebra. Consider the adjoint representation ad: $\mathfrak{H} \to \operatorname{End}(\mathfrak{g})$ and the restriction ad*: $\mathfrak{H} \to \operatorname{End}(\mathfrak{g}^*)$ of the coadjoint representation of $\mathfrak{g}$ to the subalgebra $\mathfrak{H}$. By the well-known theorem we have the decompositions

$$\mathfrak{g} = \mathfrak{g}_0 + \sum_{\lambda \neq 0} \mathfrak{g}_\lambda, \quad \mathfrak{g}^* = \mathfrak{g}_0^* + \sum_{\mu \neq 0} \mathfrak{g}_\mu^*$$

into root subspaces $\mathfrak{g}_\lambda = \{\eta \in \mathfrak{g} | (\operatorname{ad}_h - \lambda(h))^N \eta = 0 \text{ for any } h \in H \text{ for some } N\}$ and $\mathfrak{g}_\lambda^* = \{\eta \in \mathfrak{g}^* | (\operatorname{ad}_h^* - \lambda(h))^N \eta = 0 \text{ for any } h \in H \text{ for some } N\}$.

We fix a covector $a \in \mathfrak{g}^*$ and consider the mapping $\Phi_a(x) = \mathrm{ad}_x^*(a)$, $\Phi_a\colon \mathfrak{g} \to \mathfrak{g}^*$. If $a \in \mathfrak{g}_\lambda^*$, $b \in \mathfrak{g}_\mu^*$ then $\Phi_a(b) = \mathrm{ad}_b^*(a) \in \mathfrak{g}_{\lambda+\mu}$. The subspace $\mathfrak{g}_0$ is a subalgebra in $\mathfrak{g}$ and $\mathfrak{H} \subset \mathfrak{g}_0$.

Definition 2.2. A subalgebra $\mathfrak{H}$ in the Lie algebra $\mathfrak{g}$ is called *Hamiltonian* if

a) $\mathfrak{H}$ is commutative,

b) $\mathfrak{H} = \mathfrak{g}_0$, and

c) there is an element $a \in \mathfrak{g}_0^*$ such that the mapping $\Phi_a\colon \mathfrak{g}_\mu \to \mathfrak{g}_\mu^*$ is an isomorphism for any root $\mu \neq 0$.

Remark 2.1. If $\mathfrak{g}$ is semisimple, then $\mathfrak{H}$ is a Hamiltonian subalgebra if and only if H is a Cartan subalgebra (see the Appendix).

Construction. Let $\mathfrak{H}$ be a Hamiltonian subalgebra in $\mathfrak{g}$ and let $\mathfrak{g} = \mathfrak{g}_0 + \sum_{\alpha \neq 0} \mathfrak{g}_\alpha$ and $\mathfrak{g}^* = \mathfrak{g}_0^* + \sum_{\mu \neq 0} \mathfrak{g}_\mu$ be the corresponding root space decompositions. We fix an arbitrary self-adjoint linear operator $D\colon \mathfrak{g}_0^* \to \mathfrak{g}_0$ and choose an element $a \in \mathfrak{g}_0^*$ as in the definition of a Hamiltonian subalgebra. Let b be any element of $\mathfrak{g}_0$. Define a family of operators $C = C(a, b, D)\colon \mathfrak{g}^* \to \mathfrak{g}$ by

$$C = C(a, b, D) = \begin{pmatrix} D & 0 \\ 0 & \Phi_a^{-1}\,\mathrm{ad}_b^* \end{pmatrix}.$$

Here $\Phi_a^{-1}\,\mathrm{ad}_b^*\left(\sum_{\alpha \neq 0} x_\alpha\right)$ is, by definition, $\sum_{\alpha \neq 0} \Phi_a^{-1}\,\mathrm{ad}_b^*\, x_\alpha$ where $x_\alpha \in \mathfrak{g}_\alpha^*$.

The above construction gives all sectional operators known to date for which the corresponding Euler equations are completely integrable.

Definition 2.3. Let $\mathfrak{g}$ be a Lie algebra such that there exist sectional operators in the coadjoint representation given by the above construction. In this case the sectional operators $C(a, b, D)$ will be called the "*rigid body*" *operators* for $\mathfrak{g}$.

1.3. Integrals of Euler Equations. Shift of the Argument. Let $f(x)$ be a function given on a linear space V, and $a \in V$ a fixed vector. We construct a family of functions $f_\lambda(x) = f(x + \lambda a)$ on V, where λ is an arbitrary number ($\lambda \in \mathbb{R}$ if V is a linear space over $\mathbb{R}$). We say that the functions $f_\lambda(x)$ are obtained from $f(x)$ by the operation of shifting the argument. If f_λ can be expanded in a power series in λ (for instance, if $f(x)$ is a polynomial) then

$$f_\lambda(x) = \sum_{n=0}^{\infty} \lambda^n f_{\lambda,n}(x)$$

and the operation of shifting creates out of $f(x)$ a whole family of functions $\{f_{\lambda,n}(x)\}$.

In the theory of Hamiltonian systems, shift of the argument is used by virtue of the following theorem of Mishchenko and Fomenko [1978b]. This idea (for the Lie algebra $\mathrm{so}(n)$) first appeared in Manakov [1976].

Theorem 2.2 (Mishchenko and Fomenko [1976], [1978b]). *Let F and H be two functions on the dual space $\mathfrak{g}^*$ to a Lie algebra $\mathfrak{g}$ which are constant on* *coadjoint orbits (invariants of the coadjoint representation of the Lie group G associated to the Lie algebra $\mathfrak{g}$), let $a \in \mathfrak{g}^*$ be a fixed covector, and let $\lambda, \mu \in \mathbb{R}$ be any fixed numbers. Then the functions $F_\lambda(t) = F(t + \lambda a)$ and $H_\mu(t) = H(t + \mu a)$ are in involution on all coadjoint orbits with respect to the standard Kirillov symplectic structure.*

This theorem provides a complete involutive family of functions for a fairly broad class of Lie algebras, including semisimple Lie algebras (see Mishchenko and Fomenko [1976], [1978b]).

It turns out that the operation of shifting the argument also provides integrals of the Euler equations defined by the "rigid body" operators.

Theorem 2.3. *Let $C(a, b, D)$ be the "rigid body" operators for the Lie algebra $\mathfrak{g}$ and $\dot{X} = \mathrm{ad}^*_{C(X)}(X)$, $X \in \mathfrak{g}^*$, the Euler equations that correspond to the operators $C = C(a, b, D)$: $\mathfrak{g}^* \to \mathfrak{g}$. Then all functions of the form $F(X + \lambda a)$, $\lambda \in \mathbb{R}$, $X \in \mathfrak{g}^*$ are integrals of the Euler equations, where $F(X)$ is any invariant of the coadjoint action of the Lie group G associated to $\mathfrak{g}$, i.e. $F(X) = F(\mathrm{Ad}^*_g X)$ for every $g \in G$.*

1.4. Sectional Operators for Symmetric Spaces. For a symmetric space $M = G/H$, the above construction of sectional operators can be applied to the isotropy representation of the stationary subgroup H on the tangent space $T_H M$.

If one takes a semisimple Lie group H as the symmetric space, then it can be represented in the form $H \times H/H$ where the involution $\sigma: H \times H \to H \times H$ is given by $\sigma(x, y) = (y, x)$ (see Helgason [1962]). The corresponding decomposition of the Lie algebra $\mathfrak{g}$ has the form $\mathfrak{g} = \mathfrak{H} + V$ where $\mathfrak{H} = \{(X, X)\}$ is the eigenspace of the involution $d\sigma$ that corresponds to the eigenvalue 1 and $V = \{(X, -X)\}$ is the eigenspace of $d\sigma$ that corresponds to the eigenvalue -1. The 2-form F_C defined earlier in Section 1.2 becomes in this case (under a suitable choice of C) the canonical symplectic form on coadjoint orbits. Thus, by means of sectional operators the Kirillov form can be included in a multiparameter family of symplectic structures (see Fomenko [1980], [1981]).

If the symmetric space $M = G/H$ has maximal rank, then the 2-form F_C on V is induced by the curvature tensor of M. More precisely, $F_C(X; \xi, \eta) = 4\langle a', R(X, \xi)\eta\rangle$ where R is the curvature tensor and $a' \in \mathfrak{t}$ is a fixed vector.

A detailed description of symplectic structures F_C can be found in Fomenko [1980], [1981]. We note that there also exist other constructions of symplectic structures related to the curvature tensor (see, for instance, Kummer [1981], Trofimov [1984a]).

For a classification of Hamiltonian flows with respect to symplectic structure F_C see Fomenko [1980], [1981], Trofimov and Fomenko [1983b].

1.5. Complex Semisimple Series of Sectional Operators. For a complex semisimple Lie algebra $\mathfrak{g}$, the construction of "rigid body" operators outlined above takes the following form. Let $\mathfrak{t}$ be a Cartan subalgebra of $\mathfrak{g}$, and $\mathfrak{g} = \mathfrak{t} + V^+ + V^-$ the root space decomposition, where $V^+ = \sum_{\alpha>0} \mathbb{C}E_\alpha$ and $V^- =$

$\sum_{\alpha<0} \mathbb{C}E_\alpha$. Let $a, b \in \mathfrak{t}$, $a \neq b$, be any two regular elements of the Cartan subalgebra $\mathfrak{t}$. The operator ad_a preserves the root space decomposition. Since a is regular, ad_a is invertible on $V = V^+ + V^-$. The sectional operator $\varphi_{a,b,D}$: $\mathfrak{g} \to \mathfrak{g}$ is then defined by

$$\varphi_{a,b,D}(X) = \varphi_{a,b}(X') + D(t) = \mathrm{ad}_a^{-1} \mathrm{ad}_b X' + D(t),$$

where $X = X' + t$ is the unique decomposition of X into V- and $\mathfrak{t}$-components, and $D\colon \mathfrak{t} \to \mathfrak{t}$ is an arbitrary linear operator which is symmetric with respect to the Killing form. The operator $\varphi_{a,b,D}$ is parametrized by a, b, D. These operators were first considered by Mischenko and Fomenko [1976], [1978b], [1979a].

1.6. Compact and Normal Series of Sectional Operators. Every complex semisimple Lie algebra $\mathfrak{g}$ has a compact real form $\mathfrak{g}_u \subset \mathfrak{g}$. We recall that $\mathfrak{g}_u = \{E_\alpha + E_{-\alpha}, i(E_\alpha - E_{-\alpha}), iH_\alpha\}$ (see Helgason [1962]). Let $a, b \in i\mathfrak{t}_0$ (where $\mathfrak{t}_0$ is the real subspace in the Cartan subalgebra $\mathfrak{t}$ spanned by all the vectors $H_\alpha \in \mathfrak{t}$) be elements in general position. The preceding construction of the "rigid body" operators gives $\varphi = \varphi_{a,b,D}\colon \mathfrak{g}_u \to \mathfrak{g}_u$,

$$\varphi(X) = \varphi(X' + t) = \varphi_{a,b}(X') + D(t) = \mathrm{ad}_a^{-1} \mathrm{ad}_b(X') + D(t),$$

where $X = X' + t$ is the unique decomposition of X such that $t \in i\mathfrak{t}_0$, $X' \perp i\mathfrak{t}_0$.

In every compact form $\mathfrak{g}_u$ we consider the subalgebra $\mathfrak{g}_n$, the so-called normal compact subalgebra, which is spanned by the vectors $E_\alpha + E_{-\alpha}$ where α ranges over the set of all roots of $\mathfrak{g}$ with respect to the Cartan subalgebra $\mathfrak{t}$. All such subalgebras are described, for example, in Helgason [1962]. Since all the vectors $E_\alpha + E_{-\alpha}$ are eigenvectors for the operators φ of the compact series, by restricting them to $\mathfrak{g}_n$ we obtain the normal series of sectional operators. Thus, for $\varphi_{a,b}\colon \mathfrak{g}_n \to \mathfrak{g}_n$ we have

$$\varphi(X) = \mathrm{ad}_a^{-1} \mathrm{ad}_b(X), \quad X \in \mathfrak{g}_n, \quad a, b \in i\mathfrak{t}_0, \quad \alpha(a) \neq 0, \quad \alpha(b) \neq 0.$$

We note that the operators of the normal series require for their definition the elements $a, b \in \mathfrak{g}_u$ of some larger Lie algebra. A similar phenomenon also occurs in the construction of sectional operators on symmetric spaces (see Fomenko [1980], [1981]).

Hamiltonian systems of the normal series include the classical equations of motion of a multi-dimensional rigid body with a fixed point. To show this we take the Lie algebra $\mathrm{so}(n)$ and represent it as a normal subalgebra of $\mathrm{su}(n)$. We embed $\mathrm{su}(n)$ in the standard way in $\mathrm{u}(n)$ and consider two regular elements $a, b \in i\mathfrak{t}_0$ of the Cartan subalgebra $i\mathfrak{t}_0$ in $\mathrm{u}(n)$ (but not in $\mathrm{su}(n)$). If $a^2 = -ib^2$, we obtain a system of equations describing the motion of a rigid body with a fixed point. Moreover, the operators $\varphi_{a,b}$ of the normal series include the classical operator $\Psi(X) = IX + XI$ where I is a real diagonal matrix (see Section 1.6 of Chapter 1).

Theorem 2.4 (Mishchenko and Fomenko [1976], [1978b]). 1) *Let $\mathfrak{g}$ be a complex semisimple Lie algebra and $\dot{X} = [X, \varphi_{a,b}(X)]$ the Euler equations with an* *operator of the complex series. Then this system is Liouville completely integrable on generic orbits. If f is any invariant function on the Lie algebra $\mathfrak{g}$, then all the functions $f(X + \lambda a) = h_\lambda(X, a)$ are integrals of the flow $\dot{X}$ for any λ. Any two integrals $h_\lambda(X, a)$ and $g_\mu(X, a)$ constructed from functions $f, g \in I(\mathfrak{g})$ (where $I(\mathfrak{g})$ is the ring of invariant polynomials on $\mathfrak{g}$) are in involution on the orbits. The Hamiltonian $F = \langle X, \varphi(X) \rangle$ of the flow $\dot{X}$ also commutes with all integrals of the form $h_\lambda(X, a)$. From the set of such integrals we can choose a set of functionally independent integrals on generic orbits equal in number to half the dimension of the orbit. The integral F can be functionally expressed in terms of integrals of the form $h_\lambda(X, a)$.*

2) *Let $\mathfrak{g}_u$ be the compact real form of a semisimple Lie algebra $\mathfrak{g}$ and $\dot{X} = [X, \varphi X]$ a Hamiltonian system defined by an operator φ of the compact series. Then the set of functions of the form $f(X + \lambda a)$, where $f \in I(\mathfrak{g}_u)$, forms a complete commutative collection on generic orbits in $\mathfrak{g}_u$.*

3) *Let $\mathfrak{g}_n$ be the normal compact subalgebra of the compact Lie algebra $\mathfrak{g}_u$ and $\dot{X} = [X, \varphi X]$ a Hamiltonian system with an operator of the normal series. Then the set of functions $f(X + \lambda a)$, where $f \in I(\mathfrak{g}_u)$, forms a complete commutative collection of functions on generic orbits, and all these functions are first integrals for the Euler equations.*

Thus, the Euler equations associated with the complex semisimple, compact, and normal series of sectional operators are completely integrable Hamiltonian systems on all generic coadjoint orbits.

1.7. Sectional Operators for the Lie Algebra of the Group of Euclidean Motions. The "rigid body" operators exist for this Lie algebra, and the operation of shifting the invariants provides a complete involutive set of inegrals for the corresponding Euler equations.

Theorem 2.5 (see Trofimov and Fomenko [1983a, b]). 1) *The system of differential equations $\dot{X} = \mathrm{ad}^*_{Q(X)} X$, where $Q = C(a, b, D)$ is the "rigid body" operator on the Lie algebra $e(n)$ of the group of motions of the n-dimensional Euclidean space, is Liouville completely integrable on generic orbits.*

2) *Let f be an invariant function on $e(n)^*$. Then the functions $h_\lambda(X) = f(X + \lambda a)$ are integrals of motion for all $\lambda \in \mathbb{R}$. Any two integrals h_λ and g_μ are in involution on all coadjoint orbits of the group of Euclidean motions of $\mathbb{R}^n$, with the number of independent integrals of the relevant form being half the dimension of a generic orbit. If $\mathcal{O}$ is a coadjoint orbit of maximal dimension (generic orbit) then* codim $\mathcal{O} = [\frac{1}{2}(n + 1)]$.

We recall that the Euler equations on $e(n)^*$ are the same as the equations of motion by inertia of a multi-dimensional rigid body in ideal fluid (see Section 1.6 of Chapter 1).

1.8. Sectional Operators for the Lie Algebra $\Omega(\mathfrak{g})$. The definition of the Lie algebra $\Omega(\mathfrak{g})$ was given in Section 1.6 of Chapter 1.

If $\mathfrak{g}$ is one of the Lie algebras listed below, there exist "rigid body" operators for $\Omega(\mathfrak{g})$ (see Trofimov [1983a], [1985]).

Theorem 2.6 (Trofimov [1983a], [1985]). *Let $\mathfrak{g}$ be a complex semisimple Lie algebra, or a compact real form, or a compact normal subalgebra of a complex semisimple Lie algebra. Let $\dot{X} = \mathrm{ad}^*_{C(X)}(X)$ be the Euler equations on $\Omega(\mathfrak{g})^*$ associated with a "rigid body" operator. Then this system is Liouville completely integrable on all coadjoint orbits for the Lie group $\Omega(G)$ associated with $\Omega(\mathfrak{g})$. More precisely, let $F(x)$ be any smooth function on $\Omega(\mathfrak{g})^*$ which is invariant under the coadjoint action of $\Omega(G)$. Then all the functions $F(X + \lambda a)$, $\lambda \in \mathbb{C}(\mathbb{R})$, are first integrals of the Euler equations for every $\lambda \in \mathbb{C}(\mathbb{R})$. Any two such integrals $F(X + \lambda a), H(X + \mu a)$, $\lambda, \mu \in \mathbb{C}(\mathbb{R})$, are in involution on all orbits with respect to the Kirillov form. From this family of integrals one can choose functionally independent integrals equal in number to half the dimension of a generic coadjoint orbit.*

In the next section we shall discuss a more general construction which deals with tensor extensions.

1.9. Bi-Hamiltonian Properties of Euler Equations on Semisimple Lie Algebras. Let $\mathfrak{g}$ be a semisimple Lie algebra. In the ring $C^\infty(\mathfrak{g})$ of smooth functions on $\mathfrak{g}$ we define an additional Poisson bracket $\{f, g\}_a$. It is enough to define $\{f, g\}_a$ for the linear functions $f_u(x) = \langle x, u\rangle$, $f_v(x) = \langle x, v\rangle$ where $u, v \in \mathfrak{g}$ and $\langle a, b\rangle$ is the Killing form of $\mathfrak{g}$. We set $\{f_u, f_v\}_a(x) = \langle a, [u, v]\rangle$ and extend this Poisson bracket to the whole ring $C^\infty(\mathfrak{g})$ by requiring the Leibniz rule $\{f, gh\}_a = \{f, g\}_a h + \{f, h\}_a g$ for any $f, g, h \in C^\infty(\mathfrak{g})$. As is well known, the bracket is uniquely determined by these conditions. A bracket of this form was studied by Reyman [1980].

Theorem 2.7 (Meshcheryakov [1983]). *Let $\varphi\colon \mathfrak{g} \to \mathfrak{g}$ be a self-adjoint (with respect to the Killing form of $\mathfrak{g}$) linear operator. The Euler equation $\dot{X} = [X, \varphi(X)]$ is Hamiltonian with respect to the Poisson bracket $\{\ ,\ \}_a$ if and only if $\varphi = \varphi_{a,b,D}$ for some $b \in \mathfrak{H}$ and $D\colon \mathfrak{H} \to \mathfrak{H}$, where $\mathfrak{H}$ is a Cartan subalgebra of $\mathfrak{g}$ which contains a.*

§2. Methods of Constructing Functions in Involution

There are at present several methods of constructing involutive families of functions on the dual space $\mathfrak{g}^*$ of a Lie algebra $\mathfrak{g}$, i.e. commutative subalgebras in the Lie algebra $C^\infty(\mathfrak{g}^*)$ of smooth functions on $\mathfrak{g}^*$ with respect to the bracket $\{f, g\} = C^k_{ij} x_k \dfrac{\partial f}{\partial x_i}\dfrac{\partial g}{\partial x_j}$ (see Section 1.4 of Chapter 1; here C^k_{ij} is the structure tensor of $\mathfrak{g}$).

2.1. Inductive Construction of Integrable Dynamical Systems on Coadjoint Orbits (Chains of Subalgebras). The following construction was first described by Vergne [1972] in the context of geometric quantization (construction of polarizations) and by Trofimov [1979] in the context of integrable systems.

Construction. Suppose that a Lie algebra $\mathfrak{g}$ has a filtration of subalgebras $\mathfrak{g} \supset \mathfrak{g}_1 \supset \mathfrak{g}_2 \supset \cdots \supset \mathfrak{g}_q$. Then there arises a chain of mapping $\mathfrak{g}^* \to \mathfrak{g}_1^* \to \mathfrak{g}_2^* \to \cdots \to \mathfrak{g}_q^*$. Each mapping $\pi_i\colon \mathfrak{g}_i^* \to \mathfrak{g}_{i+1}^*$ is a restriction mapping of linear functionals on $\mathfrak{g}_i$ to $\mathfrak{g}_{i+1}$, i.e. $\pi_i(f) = f|_{\mathfrak{g}_{i+1}}$. Therefore, functions given on $\mathfrak{g}_i^*$ can be lifted to $\mathfrak{g}^*$, i.e. if $f \in C^\infty(\mathfrak{g}_i^*)$ then $\tilde{f} \in C^\infty(\mathfrak{g}^*)$, $\tilde{f} = f \circ \pi_1 \circ \pi_2 \circ \cdots \circ \pi_{i-1}$, where $\pi_i\colon \mathfrak{g}_i^* \to \mathfrak{g}_{i+1}^*$ is the mapping above.

Definition 2.4. Let $F \in C^\infty(\mathfrak{g}^*)$, where $\mathfrak{g}$ is a Lie algebra, and let $\mathrm{Ad}^*\colon G \to GL(\mathfrak{g}^*)$ be the coadjoint representation of the Lie group G associated to $\mathfrak{g}$. The function F is called an *invariant* if $F(\mathrm{Ad}_g^*\, x) = F(x)$ for all $x \in \mathfrak{g}^*$ and all $g \in G$. The function F is called a *semi-invariant* if $F(\mathrm{Ad}_g^*\, x) = \chi(g)F(x)$ for all $g \in G$ and all $x \in \mathfrak{g}^*$, where χ is a character of G.

Theorem 2.8 (Trofimov [1979]). *Consider a chain of subalgebras $\mathfrak{V} \supset \mathfrak{S}$. If the functions f and g on $\mathfrak{S}^*$ are in involution on all orbits of the representation Ad_S^*, where S is the Lie group corresponding to the Lie algebra $\mathfrak{S}$, then $\tilde{f}$ and $\tilde{g}$ are in involution on all orbits of the representation Ad_V^*, where V is the Lie group corresponding to $\mathfrak{V}$ and $\tilde{f} = f \circ \pi$, $\tilde{g} = g \circ \pi$, $\pi\colon \mathfrak{V}^* \to \mathfrak{S}^*$ is the restriction mapping.*

Remark 2.2. The method described in Theorem 2.8 is a generalization of a construction of Vergne [1972]. To construct global symplectic coordinates on orbits of maximal dimension of the coadjoint representation of nipotent Lie algebras, M. Vergne uses chains of ideals $\mathfrak{g}_1 \subset \mathfrak{g}_2 \subset \cdots \subset \mathfrak{g}$ such that $\dim \mathfrak{g}_i/\mathfrak{g}_{i+1} = 1$ and lifts to $\mathfrak{g}^*$ the invariants of the Ad^* representation of the ideals $\mathfrak{g}_i$ by means of the natural projection $\mathfrak{g}^* \to \mathfrak{g}_i^*$ $(i = 1, \ldots, \dim \mathfrak{g})$ which gives the required coordinates.

We note some particular cases which are frequently used for constructing involutive families of functions. The simplest version of the construction described in Theorem 2.8 is obtained in the case of the pair $\mathfrak{g} \supset \mathfrak{H}$, where $\mathfrak{H}$ is a maximal Abelian subalgebra of $\mathfrak{g}$ (see Trofimov [1979]). As invariants of the Abelian subalgebra we can take any of its elements, regarded as a function on the dual space. If $\mathfrak{g} \supset \mathfrak{g}' \supset \mathfrak{H}$, then to the invariants of $\mathfrak{H}$ we may add the semi-invariants of $\mathfrak{g}$.

Illustration. We consider a simple example where the above construction applies. Let $u(n)$ be the Lie algebra of the Lie group of unitary matrices. This Lie algebra has a standard filtration: $u(n) \supset u(n-1) \supset \cdots$. The method of chains of subalgebras then provides an involutive family of functions on $u(n)^*$. We take the invariants of $u(i)$ as functions on $u(i)^*$ that are lifted to $u(n)^*$. The functions on $u(n)^*$ obtained in this way have an important additional property: they make up a complete involutive family on $u(n)^*$.

2.2. Representations of Lie Groups and Involutive Families of Functions. The construction which follows was first described by Trofimov [1980].

Construction. Let $\mathfrak{g}$ be a Lie algebra, G the simply connected Lie group associated with $\mathfrak{g}$, $\mathfrak{g}^*$ the dual space to $\mathfrak{g}$, $H(\mathfrak{g}^*)$ the space of analytic functions on $\mathfrak{g}^*$, $\mathrm{Ad}^*: G \to GL(\mathfrak{g}^*)$ the coadjoint representation of G: $(\mathrm{Ad}_g^* f)(x) = f(\mathrm{Ad}_{g^{-1}} x)$. Let W be a finite-dimensional subspace in $H(\mathfrak{g}^*)$ invariant under Ad_G^*. We fix a basis $f_1, \ldots, f_s$ of W. Then every function $f \in W$ defines a set of covectors $c_*^i(f) \in \mathfrak{g}^*$ according to the following rule. Since W is invariant, we have $f(\mathrm{Ad}_g^* x) \in W$ for every $g \in G$ and hence we can write

$$f(\mathrm{Ad}_g^* x) = c_f^i(g) f_i(x).$$

This gives a set of linear functionals $c_*^i(f)$ on $\mathfrak{g}$:

$$c_*^i(f)(\xi) = \frac{d}{dt}\bigg|_{t=0} c_f^i(\exp t\xi) = dc_f^i(\xi),$$

i.e. $c_*^i(f)$ is the differential of c_f^i at the unit element of the group.

The skew gradients $\operatorname{sgrad} f$ of functions in W can easily be expressed in terms of the $c_*^i(f)$.

Proposition 2.1. *Under the assumptions above we have*

$$(\operatorname{sgrad} f)(x) = c_*^1(f) f_1(x) + \cdots + c_*^s(f) f_s(x).$$

We formulate the situation described above in the form of a separate definition.

Definition 2.5. A finite-dimensional subspace W of $H(\mathfrak{g}^*)$ with a distinguished set of functions $h_1, \ldots, h_p \in W$ is called an *S-representation* of G if a) W is invariant under the coadjoint action of G, b) $\langle c_*^k(h_j), dh_i(x)\rangle = 0$ on $\mathfrak{g}^*$.

Remark 2.3. Clearly, the condition $\langle c_*^k(h_j), dh_i(x)\rangle = 0$ does not depend on the choice of the basis $f_1, \ldots, f_s$ in W.

The following theorem reveals the connection between S-representations of Lie groups and functions in involution.

Theorem 2.9 (Trofimov [1980]). *Suppose that $(h_1, \ldots, h_s; W)$ is an S-representation of G. Then* a) *all the functions $h_1, \ldots, h_s$ are in involution on all coadjoint orbits of G;* b) *the shifts of the h_i are in involution on all coadjoint orbits of G, i.e. $\{h_i(x + \lambda a), h_j(x + \mu a)\} = 0$, $1 \leqslant i, j \leqslant s$, $\lambda, \mu \in \mathbb{R}$, where $a \in \mathfrak{g}^*$ is a fixed shift vector.*

Proposition 2.2. *If $h_1, \ldots, h_s$ are semi-invariants of the coadjoint action of G, then the set $\left(h_1, \ldots, h_s; W = \bigoplus_{i=1}^{s} \mathbb{R}h_i\right)$ is an S-representation of G.*

Remark 2.4. This proposition was proved by Arkhangelsky [1979] under the assumption that the semi-invariants are in involution. Actually, the assumption of the involutivity of the semi-invariants can be dropped, which is a fairly deep fact of the representation theory of Lie groups.

Theorem 2.2 of Mishchenko and Fomenko can be derived from Theorem 2.9. We have the following proposition.

Proposition 2.3. *If $h_1, \ldots, h_s$ are invariants of the coadjoint action of the Lie group G, then $\left(h_1, \ldots, h_s; W = \bigoplus_{i=1}^{s} \mathbb{R}h_i\right)$ is an S-representation of G.*

The operation of shifting the argument can be combined with the operation of lifting functions used in Section 2.1 (see Trofimov [1980]).

Proposition 2.4. *Let $\mathfrak{k}$ be a subalgebra of $\mathfrak{g}$ and $V \subset H(\mathfrak{g}^*)$ a finite-dimensional subspace which is invariant under Ad_G^*. A function $f \in V$ determines the covectors $c_*^i \in \mathfrak{g}^*$ for a choice of basis in V. If the lift $\tilde{g} \in H(\mathfrak{g}^*)$ of a function $g \in H(\mathfrak{k}^*)$ is constant along these covectors, then $\tilde{g}$ is in involution with f.*

For some classes of Lie groups there is an algorithm which allows us to construct new S-representations out of known ones (see Trofimov [1984b]).

Remark 2.5. The operation of shifting the argument can be extended to arbitrary finite-dimensional algebras (not necessarily commutative or skew commutative) so that the resulting functions are integrals for some analogues of Euler equations. For an arbitrary algebra, however, there is no longer a symplectic or Poisson structure and therefore one cannot speak about complete integrability of the corresponding equations (see Trofimov [1982b]).

To construct complete involutive families of functions one can also use nonlinear shifts of the argument. We shall state one useful proposition of this type which deals with nonlinear shifts.

Theorem 2.10 (Reyman [1980], Ratiu [1982]). *Let $\mathfrak{g}$ be a Lie algebra endowed with a nondegenerate $\mathfrak{g}$-invariant bilinear form (ξ, η). Suppose that the functions f and h satisfy $[\operatorname{grad} f(\xi), \xi] = 0$, $[\operatorname{grad} h(\xi), \xi] = 0$ for all $\xi \in \mathfrak{g}$. We denote $f_\lambda(\xi, \eta) = f(\xi + \lambda\eta + \lambda^2\varepsilon)$, $h_\mu(\xi, \eta) = h(\xi + \mu\eta + \mu^2\varepsilon)$ where ε is a fixed element of $\mathfrak{g}$ and λ, μ are arbitrary parameters. Then for all λ and μ the functions $f_\lambda(\xi, \eta)$ and $h_\mu(\xi, \eta)$ are in involution with respect to the canonical Poisson bracket on the coadjoint orbits of the Lie algebra $\Omega(\mathfrak{g}) = \mathfrak{g} \otimes (\mathbb{R}[x]/(x^2))$.*

Remark 2.6. There have been many investigations on Lie algebras with a nondegenerate invariant bilinear form. For a classification of such Lie algebras see Medina and Revoy [1983–1984].

2.3. Involutive Families of Functions on Semidirect Sums. Let $\mathfrak{k} = \mathfrak{g} \,{}_\rho\!+ V$ be a semidirect sum of a complex Lie algebra $\mathfrak{g}$ and a linear space V with respect to an irreducible representation $\rho\colon \mathfrak{g} \to \mathrm{End}(V)$. We identify the dual space to $\mathfrak{k} = \mathfrak{g} \,{}_\rho\!+ V$ with $\mathfrak{g}^* + V^*$ by setting $\mathfrak{g}^* = V^\perp$, $V^* = \mathfrak{g}^\perp$. Consider the shifts of invariants of $\mathfrak{k}$ by an element $a \in \mathfrak{g}^* \subset \mathfrak{k}^*$. It turns out that this collection can be completed with functions $v \in V$ where $v(f) = f(v)$, $f \in V^*$; namely, we have the following assertion which is due to A.V. Brailov.

Theorem 2.11. *The functions $\{f_i(x + \lambda a)\} \cup V$, $i = 1, \ldots, k$, $\lambda \in \mathbb{R}$, where $f_1, \ldots, f_k$ are invariants of $\mathfrak{k}$, form an involutive set of functions on $\mathfrak{k}^*$.*

Let us discuss the completeness problem for the involutive family constructed above. Let $\mathrm{St}(v)$ denote the stationary subalgebra of $v \in V$ with respect to the representation $\rho^*\colon \mathfrak{g} \to \mathrm{End}(V^*)$ which is dual to ρ. The following assertion is due to A.V. Brailov.

Theorem 1.12. *Linear functions on V^* combined with the shifts of invariants of $\mathfrak{k}$ by a covector from $\mathfrak{g}^*$ provide a complete involutive set of functions on $\mathfrak{k}^*$ if and only if the shifted invariants of $\mathrm{St}(v)$ provide a complete involutive set of functions on $\mathrm{St}(v)^*$, where v is a generic element of V.*

As a special case we have the following proposition.

Proposition 2.5. *If $\mathrm{St}(v) = 0$, then the elements of V regarded as linear functions on V^* form a complete involutive set of functions on $\mathfrak{k}^*$.*

Remark 2.7. It is known that $\mathrm{St}(v) = 0$ if and only if $\dim \mathfrak{g} < \dim V$ (see Andreev et al. [1967]). This case was discussed by Pevtsova [1982], i.e. a complete involutive set of functions was produced. The general case will be discussed in the next section.

2.4. The Method of Tensor Extensions of Lie Algebras. This method was first suggested by Trofimov [1982a], [1983a], [1985], and later developed by Brailov [1983] and Le Ngok T'euen [1985]. Let A be a commutative associative (finite-dimensional) algebra with unit. The tensor product $\mathfrak{g} \otimes A$ is a Lie algebra with commutator $[g \otimes a, h \otimes b] = [g, h] \otimes ab$ for all $g, h \in \mathfrak{g}$, $a, b \in A$. The problem that arises here is as follows: given a complete involutive family of functions on $\mathfrak{g}^*$ construct a complete involutive family of functions on $(\mathfrak{g} \otimes A)^*$. This problem was first studied by Trofimov [1982a] for the algebras $A = \mathbb{R}[x_1, \dots, x_n]/(x_1^{m_1+1}, \dots, x_n^{m_n+1})$ where $\mathbb{R}[x_1, \dots, x_n]$ is the polynomial ring in the variables $x_1, \dots, x_n$ and $(x_1^{m_1+1}, \dots, x_n^{m_n+1})$ is the ideal generated by $x_1^{m_1+1}, \dots, x_n^{m_n+1}$. Subsequently this problem was discussed by A.V. Brailov for algebras with Poincaré duality.

1. The Ring $A = \mathbb{R}[x_1, \dots, x_n]/(x_1^{m_1+1}, \dots, x_n^{m_n+1})$. Let ε_i denote the image of x_i in A under the natural mapping $\mathbb{R}[x_1, \dots, x_n] \to A$, and let $e_1, \dots, e_r$ be a basis of $\mathfrak{g}$. Then the vectors $\varepsilon_1^{\alpha_1} \dots \varepsilon_n^{\alpha_n} e_j$, $0 \leqslant j \leqslant r$, $0 \leqslant \alpha_i \leqslant m_i$, $1 \leqslant i \leqslant n$, form a basis of $\mathfrak{g} \otimes A$. We denote the dual basis of $\mathfrak{g}^*$ by e^i; it is defined by $\langle e^i, e_j \rangle = \delta^i_j$ where $\langle f, x \rangle$ is the value of $f \in \mathfrak{g}^*$ at the vector $x \in \mathfrak{g}$. Coordinates in $\mathfrak{g} \otimes A$ in the dual basis to $\varepsilon_1^{\alpha_1} \dots \varepsilon_n^{\alpha_n} e_j$, $1 \leqslant j \leqslant r$, $0 \leqslant \alpha_i \leqslant m_i$, $1 \leqslant i \leqslant n$, are denoted by $x(\alpha_1, \dots, \alpha_n)_j$, and coordinates in $\mathfrak{g}^*$ in the dual basis to e_i, $1 \leqslant i \leqslant r$, by x_i, $1 \leqslant i \leqslant r$.

We introduce new variables in $\mathfrak{g} \otimes A$:

$$z_i = \sum_{\substack{\alpha_j + \beta_j = m_j \\ 1 \leqslant j \leqslant n}} \varepsilon_1^{\alpha_1} \dots \varepsilon_n^{\alpha_n} x(\beta_1, \dots, \beta_n)_i.$$

Description of the algorithm (a) (V.V. Trofimov). Let $F(x_1, \dots, x_r)$ be an analytic function on $\mathfrak{g}^*$ (or on an open set in $\mathfrak{g}^*$) with values in $\mathbb{R}$. Expanding

$F(x_1, \ldots, x_r)$ in a Taylor series and substituting z_i for x_i, we obtain a finite sum, since sufficiently high powers of the ε_i vanish. Thus, $F(z_1, \ldots, z_r)$ is a well-defined function on $(\mathfrak{g} \otimes A)^*$ with values in A; it can be represented in the following form:

$$F(z_1, \ldots, z_r) = \sum_{\substack{0 \leqslant \alpha_j \leqslant m_j \\ 1 \leqslant j \leqslant n}} \varepsilon_1^{\alpha_1} \ldots \varepsilon_n^{\alpha_n} F_{\alpha_1, \ldots, \alpha_n}(x(\beta_1, \ldots, \beta_n)_i),$$

because $\varepsilon_1^{\alpha_1} \ldots \varepsilon_n^{\alpha_n}$ is a basis of the algebra $A = \mathbb{R}[x_1, \ldots, x_n]/(x_1^{m_1+1}, \ldots, x_n^{m_n+1})$; here $F_{\alpha_1, \ldots, \alpha_n}(x(\beta_1, \ldots, \beta_n)_i) \in \mathbb{R}$. By definition, the algorithm $(\mathfrak{a})$ takes the function F on $\mathfrak{g}^*$ to the set of functions $F_{\alpha_1, \ldots, \alpha_n}$, i.e. $\mathfrak{a}(F) = \{F_{\alpha_1, \ldots, \alpha_n}\}$.

Theorem 2.13 (Trofimov [1982a], [1983b]). *Suppose that the functions $F_1(x)$, $\ldots, F_N(x)$ defined on $\mathfrak{g}^*$ are in involution with respect to the Kirillov form on all coadjoint orbits of the Lie group G associated with the Lie algebra $\mathfrak{g}$. Then all the functions $\mathfrak{a}(F_1) \cup \cdots \cup \mathfrak{a}(F_N)$ that result from the algorithm $(\mathfrak{a})$ are in involution with respect to the Kirillov form on all coadjoint orbits of the Lie group $G \otimes A$ associated with the Lie algebra $\mathfrak{g} \otimes A$. If $F_1, \ldots, F_N$ are functionally independent on $\mathfrak{g}^*$, then all functions in the family $\mathfrak{a}(F_1) \cup \cdots \cup \mathfrak{a}(F_N)$ are functionally independent on $(\mathfrak{g} \otimes A)^*$. If the family $F_1, \ldots, F_N$ is complete, then so is the family $\mathfrak{a}(F_1) \cup \cdots \cup \mathfrak{a}(F_N)$.*

2. Algebras with Poincaré Duality. A partial generalization of the algorithm $(\mathfrak{a})$ to the case of algebras with Poincaré duality was obtained by Brailov [1983]; the generalized algorithm acts on polynomials defined on $\mathfrak{g}^*$.

We consider a graded commutative algebra $A = A_0 \oplus \cdots \oplus \cdots \oplus A_n$, $A_i A_j \subset A_{i+j}$, $i + j \leqslant n$.

Definition 2.6. Suppose that $\dim A_n = 1$ and that ω is a linear functional on A that vanishes identically on A_i for $i \leqslant n - 1$ but does not vanish on A_n. Let $\beta(a, b) = \omega(ab)$ be a symmetric bilinear form on A. If β is nondegenerate, then A is called an *algebra with Poincaré duality*.

It is easy to see that $A_0 = \mathbb{R}$ for such algebras. Let $\varepsilon_1 = 1$. We choose an arbitrary basis $\varepsilon_2, \ldots, \varepsilon_{j_1}$ of A_1, then an arbitrary basis $\varepsilon_{j_1+1}, \ldots, \varepsilon_{j_2}$ of A_2, and so on, up to the grading $\frac{n}{2}$. In $A_{n/2}$ we choose a basis in which the matrix of β is diagonal. In the spaces A_i, $i > \frac{n}{2}$, we choose bases that are dual with respect to β to those already chosen in $A_{(n/2)-i}$. Altogether, we obtain a homogeneous basis $\varepsilon_1, \ldots, \varepsilon_N$ of A which is self-dual with respect to β: $\beta(\varepsilon_{\omega i}, \varepsilon_j) = \delta_{ij}$, where ω is the transposition.

Let $e_1, \ldots, e_m$ be a basis of $\mathfrak{g}^*$, and $x_1, \ldots, x_m$ the corresponding coordinates. The linear functions $x_1, \ldots, x_m$ on $\mathfrak{g}^*$ can naturally be regarded as elements of the Lie algebra $\mathfrak{g}$. The elements $x_i^j = x_i \otimes \varepsilon_j$ of the Lie algebra $\mathfrak{g} \otimes A$ can be regarded as linear coordinate functions on $(\mathfrak{g} \otimes A)^*$. For a polynomial function $P(x)$ on $\mathfrak{g}^*$

$$P(x_1, \ldots, x_m) = \sum_{k=0}^{\infty} \sum_{i_1, \ldots, i_k} P_{i_1, \ldots, i_k} x_{i_1} \ldots x_{i_k} = P_I x_I$$

and for $1 \leqslant j \leqslant N$ we define a polynomial function $P^{(j)}(x)$ on $(\mathfrak{g} \otimes A)^*$:

$$P^{(j)}(x_1^1, \ldots, x_m^N) = \sum_{k=0}^{N} \sum_{\substack{i_1, \ldots, i_k \\ j_1 \ldots j_k}} P_{i_1, \ldots, i_k}(\varepsilon_j, \varepsilon_{j_1} \ldots \varepsilon_{j_k}) x_{i_1}^{\omega j_1} \ldots x_{i_k}^{\omega j_k} = P_I(\varepsilon_j, \varepsilon_J) x_I^{\omega J},$$

here $(\varepsilon_j, \varepsilon_{j_1} \ldots \varepsilon_{j_k})$ is the projection of $\varepsilon_{j_1} \ldots \varepsilon_{j_k}$ onto the basis vector ε_j. The family of polynomial functions on $(\mathfrak{g} \otimes A)^*$ of the form $P^{(j)}$, $P \in \mathfrak{F}$, $1 \leqslant j \leqslant N$, is denoted by $\mathfrak{F}^A$.

Theorem 2.14 (see Brailov [1983]). *If $\mathfrak{F}$ is an involutive family of functions on $\mathfrak{g}^*$, then $\mathfrak{F}^A$ is an involutive family of functions on $(\mathfrak{g} \otimes A)^*$. If $\mathfrak{F}$ is a complete involutive family and the number of independent polynomial invariants of $\mathfrak{g}$ is equal to the index, then $\mathfrak{F}^A$ is a complete involutive family on $(\mathfrak{g} \otimes A)^*$. If $\mathfrak{F}$ contains a non-degenerate quadratic form, then so does $\mathfrak{F}^A$.*

3. Frobenius Algebras. Le Ngok T'euen has extended the above results to the case of Frobenius algebras.

Definition 2.7. An associative commutative algebra A with unit is called a *Frobenius algebra* if there is on A a nondegenerate scalar product (x, y) which satisfies the invariance condition $(ab, c) = (a, bc)$ for all $a, b, c \in A$.

To construct a generalization of the above algorithms we need an alternative description of Frobenius algebras.

Consider a finite-dimensional associative commutative algebra A with unit, $\dim A = N$. We fix a basis $\varepsilon_1, \ldots, \varepsilon_N$ in A and a linear involutive operator $\Omega\colon A \to A$ which preserves the basis ε_i, $1 \leqslant i \leqslant N$, i.e. $\Omega\varepsilon_i = \varepsilon_{\omega i}$, $1 \leqslant i \leqslant N$, where ω is a permutation of indices: $\omega\colon \{1, 2, \ldots, N\} \to \{1, 2, \ldots, N\}$. Define a scalar product $\langle x, y\rangle$ on A for which $\varepsilon_1, \ldots, \varepsilon_N$ is an orthonormal basis, i.e. $\langle \varepsilon_i, \varepsilon_j\rangle = \delta_{ij}$.

Definition 2.8. An associative commutative algebra with unit is called *self-adjoint* with respect to the triple $(A, \Omega, \varepsilon_i)$ if the scalar product $\langle x, y\rangle$ satisfies $\langle \varepsilon_{\omega i}, \varepsilon_j \varepsilon_k\rangle = \langle \varepsilon_{\omega j}, \varepsilon_i \varepsilon_k\rangle$, $1 \leqslant i, j \leqslant N$ (here $\varepsilon_1, \ldots, \varepsilon_N$ is an orthonormal basis with respect to $\langle x, y\rangle$).

Theorem 2.15. *An associative commutative algebra A is self-adjoint if and only if A is a Frobenius algebra.*

We now define an algorithm which transforms a function P on $\mathfrak{g}^*$ into a function on $(\mathfrak{g} \otimes A)^*$. Any polynomial on $\mathfrak{g}^*$ has the form

$$P = \sum_{k=0}^{\infty} \sum_{i_1, \ldots, i_k = 1}^{n} P_{i_1 \ldots i_k} x_{i_1} x_{i_2} \ldots x_{i_k} = P_I x_I,$$

where $I = (i_1, i_2, \ldots, i_k)$, $P_I = P_{i_1 \ldots i_k}$, $x_I = x_{i_1} \ldots x_{i_k}$. If the ε_i are a basis of A and the x_j are coordinates on $\mathfrak{g}^*$, then let x_i^α denote the function $x_i \otimes \varepsilon_\alpha$. We define a function $\tilde{P}$ on $(\mathfrak{g} \otimes A)^*$:

$$\begin{aligned}\tilde{P} &= \sum_{k=0}^{\infty} \sum_{i_1,\dots,i_k=1}^{n} \sum_{j_1,\dots,j_k=1}^{N} P_{i_1 i_2 \dots i_k} \varepsilon_{j_1} \varepsilon_{j_2} \dots \varepsilon_{j_k} x_{i_1}^{\omega j_1} x_{i_2}^{\omega j_2} \dots x_{i_k}^{\omega j_k} \\ &= P_I \varepsilon_J x_I^{\Omega J}\end{aligned}$$

For any element $a \in A$ we define $P^a = (\Omega a, \tilde{P})$, i.e., $P^a = (\Omega a, P_I \varepsilon_J x_I^{\Omega J}) = (\Omega a, \varepsilon_J) P_I x_I^{\Omega J}$.

We shall only be interested in such extensions for which the Poisson bracket has the following properties: for any polynomials P, Q on $\mathfrak{g}^*$ we have the relation $\{P^a, Q^b\}_{(\mathfrak{g} \otimes A)^*} = \{P, Q\}_{\mathfrak{g}^*}^{\pi(a,b)}$ where $a, b \in A$ and $\pi: A \times A \to A$ is some mapping.

Theorem 2.16 (Le Ngok T'euen). *Let A be an associative commutative algebra with unit and $(A, \Omega, \varepsilon_i)$ its associated triple, and let $\pi: A \times A \to A$ be some mapping. Given a polynomial function P on the dual space $\mathfrak{g}^*$ to $\mathfrak{g}$ and an element $a \in A$ we define the extension P^a of P with respect to the triple $(A, \Omega, \varepsilon_i)$. If the identity $\{P^a, Q^b\}_{(\mathfrak{g} \otimes A)^*} = \{P, Q\}_{\mathfrak{g}^*}^{\pi(a,b)}$ holds for all polynomials P, Q on $\mathfrak{g}^*$, then $\pi(a, b) = ab$.*

We have the following basic property of the algorithm of extending functions from $\mathfrak{g}^*$ to $(\mathfrak{g} \otimes A)^*$ (see Le Ngok T'euen [1985]).

Theorem 2.17. *Let A be an associative commutative algebra with unit. A triple $(A, \Omega, \varepsilon_i)$ such that for any Lie algebra $\mathfrak{g}$ and polynomials P, Q the extended functions satisfy $\{P^a, Q^b\} = \{P, Q\}^{ab}$ exists if and only if A is a Frobenius algebra.*

§ 3. Completely Integrable Euler Equations on Lie Algebras

3.1. Euler Equations on Semisimple Lie Algebras. On every semisimple Lie algebra there exists a complete involutive family of functions. Such a family was constructed in Theorem 2.4 in connection with the integration of the equations of motion of a rigid body (see Mishchenko and Fomenko [1976], [1978b]).

This theorem can be reinforced: a complete involutive family exists on every semisimple orbit in a semisimple Lie algebra, in particular the orbit need not be generic. This result was first announced by Dao Chong Tkhi [1978] but his proof contained some gaps. These gaps were finally filled by A.V. Brailov.

Theorem 2.18. *Let $\mathcal{O}$ be a semisimple orbit in a semisimple Lie algebra $\mathfrak{g}$, and let $J_1, \dots, J_r$ be a complete set of invariants of the adjoint representation. Let $J_{i,a}^j$ denote the coefficient of λ^j in the polynomial $J_i(X + \lambda a)$. If a is an element in general position, one can always choose among the $J_{i,a}^j$ a set of functions whose number is half the dimension of the orbit and which are functionally independent on the orbit. (The orbit here is not required to be generic.)*

Euler equations on the Lie algebra so(4) were studied by Adler and van Moerbeke [1982], Bogoyavlensky [1983], Novikov and Schmeltser [1981],

Veselov [1983], Reyman and Semenov-Tian-Shansky [1986]. In particular, a criterion was found for the existence of an additional quadratic integral for the Euler equations with a quadratic Hamiltonian.

A general construction of sectional operators which uses a filtration of the algebra by subalgebras was proposed by Bogoyavlensky [1984].

In the semisimple case one can construct complete involutive families not only by means of shifted invariants but also by using chains of subalgebras (see the example of $u(n)$ in Section 2.1).

3.2. Euler Equations on Solvable Lie Algebras. Solvable Lie algebras, as well as semisimple ones, give rise to various physically meaningful systems. Some of these systems which are generalizations of the classical Toda lattice (see Toda [1967]) were found by Bogoyavlensky [1976].

An important class of solvable Lie algebras consists of nilpotent Lie algebras. Complete involutive families on nilpotent Lie algebras were constructed by Vergne [1972].

The basic example of non-nilpotent solvable Lie algebras in provided by Borel subalgebras of semisimple Lie algebras:

$$\mathfrak{bg} = \bigoplus_i \mathbb{R}h_i \oplus \sum_{\alpha>0} \mathbb{R}e_\alpha,$$

where $\{h_i, e_\alpha\}$ is a Chevalley basis of a semisimple Lie algebra $\mathfrak{g}$ (see Helgason [1962], Jacobson [1962]). A complete involutive family on $(\mathfrak{bg})^*$ was constructed by Trofimov [1979] [1980]; he exploited both shifted invariants and chains of subalgebras.

To obtain a complete involutive family on $\mathfrak{bg}^*$ one must know ind $\mathfrak{bg}$. The answer is given in the following theorem.

Theorem 2.19 (Trofimov [1979], [1980], [1983b]). *Let $\mathfrak{g}$ be a simple Lie algebra, $\mathfrak{bg}$ a Borel subalgebra of $\mathfrak{g}$, and w_0 the element of maximal length in the Weyl group. If $\mathcal{O}$ is an orbit of maximal dimension for the representation Ad^*_{BG}, then* codim $\mathcal{O} = \frac{1}{2}$ card A *where $A = \{\alpha_i \in \Delta | (-w_0)\alpha_i \neq \alpha_i\}$, Δ is the set of simple roots for $\mathfrak{g}$ and card S is the cardinality of the set S.*

The papers mentioned above contain an explicit description of a complete involutive family of functions on $\mathfrak{bg}^*$.

Complete involutive families on certain subalgebras of $\mathfrak{bg}$ were constructed by Le Ngok T'euen [1983].

3.3. Euler Equations on Non-solvable Lie Algebras with a Non-trivial Radical

Theorem 2.20 (A.V. Bolsinov). *Let $\mathfrak{k} = \mathfrak{g} \underset{\rho}{+} V$ be the semidirect sum of a simple complex Lie algebra $\mathfrak{g}$ and a linear space V with respect to an irreducible representation $\rho\colon \mathfrak{g} \to \mathrm{End}(V)$. Then on the dual space $\mathfrak{k}^*$ there exists a complete involutive family of functions.*

The proof of this theorem uses the method described in Section 2.3.

In connection with the integration of Kirchhoff's equations in Section 1.7 we constructed a complete involutive family on the Lie algebra $e(n)$ of the group of Euclidean motions in $\mathbb{R}^n$ (see Trofimov and Fomenko [1983a, b]).

For the purpose of integration of finite-dimensional analogues of the equations of magnetohydrodynamics (see Section 1.8), a complete involutive family was constructed for the Lie algebra $\Omega(\mathfrak{g})$. In a similar way, one can produce complete involutive families on the tensor extensions $\mathfrak{g} \otimes A$ where A is a Frobenius algebra (see Section 2.4).

Complete involutive families on some semidirect sums were constructed by Pevtsova [1982].

A complete involutive family for the Lie algebra $e(n)$ was constructed by Belyaev [1981] in connection with the multi-dimensional Lagrange case of motion of a heavy rigid body in a gravitational field. The Lagrange case was also studied by Ratiu [1981], [1982].

3.4. Integrable Systems and Symmetric Spaces. Earlier we have already discussed one method of constructing dynamical systems related to symmetric spaces. This method is based on the notion of sectional operator. More details concerning the construction and integration of these systems can be found in Fomenko [1981], Trofimov and Fomenko [1983b]. In this section we outline another construction of Hamiltonian systems related to symmetric spaces which is due to Fordy, Wojciechowski and Marshall [1986]. Their scheme gives Hamiltonian systems with Hamiltonian $H = 2^{-1} \sum_{k=1}^{n} p_k^2 + V(q_1, \ldots, q_n)$ where the potential V is a polynomial of degree 4. The importance of such potentials lies in their use as the simplest nonlinear approximation to any even potential in the neighbourhood of an equilibrium point. Only a few integrable quartic potentials are known to date, for example, $\left(\sum_k q_k^2\right)^2$ and $\left(\sum_k q_k^2\right)^2 - \sum_k w_k q_k^2$ are such potentials.

We consider the symmetric space related to a symmetric Lie algebra $(\mathfrak{g}, \sigma)$, and let $\mathfrak{g} = \mathfrak{k} \oplus \mathfrak{m}$ be the corresponding decomposition of the symmetric semisimple Lie algebra $\mathfrak{g}$ (see the Appendix). We have the inclusions $[\mathfrak{k}, \mathfrak{k}] \subset \mathfrak{k}$, $[\mathfrak{k}, \mathfrak{m}] \subset \mathfrak{m}$, $[\mathfrak{m}, \mathfrak{m}] \subset \mathfrak{k}$. In the case of a Hermitian symmetric space there exists an element $A \in \mathfrak{k}$ such that $\mathfrak{k} = C_{\mathfrak{g}}(A) = \{b \in \mathfrak{g} \mid [A, b] = 0\}$. Let $\mathfrak{H}$ be a Cartan subalgebra of $\mathfrak{g}$. The element A can be chosen to lie in $\mathfrak{H}$. We have $\mathfrak{m} = \mathfrak{m}^+ \oplus \mathfrak{m}^-$, $[A, \mathfrak{k}] = 0$, $[A, X^\pm] = \pm a X^\pm$ with a being the same value for all $X^\pm \in \mathfrak{m}^\pm$.

For $Q(x, t) \in \mathfrak{m}$, consider the linear spectral problem $\psi_x = (\mu A + Q)\psi$ where μ is the spectral parameter, t the time parameter and the time dependence of $Q(x, t)$ is defined by $\psi_t = P(x, t, \mu)\psi$.

The compatibility conditions for the two equations above (together with $\mu_t = 0$) have the form

$$Q_t = P_x - [\mu A + Q, P].$$

Let $e_{\pm\alpha}$ be a basis for $\mathfrak{m}^\pm$ and Λ a constant diagonal matrix. We set

$$P = \mu^2 A + \mu Q + \frac{1}{a}\sum_{\alpha}(q_x^{\alpha} e_{\alpha} - r_x^{-\alpha} e_{-\alpha}) - \frac{1}{a}\sum_{\alpha,\beta} q^{\alpha} r^{-\beta}[e_{\alpha}, e_{-\beta}] + \Lambda,$$

$$Q = \sum_{\alpha}(q^{\alpha} e_{\alpha} + r^{-\alpha} e_{-\alpha}).$$

The compatibility conditions then become

$$aq_t^{\alpha} = q_{xx}^{\alpha} + \sum_{\beta,\nu,\delta} R^{\alpha}_{\beta,\gamma,-\delta} q^{\beta} q^{\gamma} r^{-\delta} + \omega_{\alpha} q^{\alpha},$$

$$-ar_t^{-\alpha} = r_{xx}^{-\alpha} + \sum_{\beta,\gamma,\delta} R^{-\alpha}_{-\beta,-\gamma,\delta} r^{-\beta} r^{-\gamma} q^{\delta} + \omega_{\alpha} r^{-\alpha},$$

where $R^{\alpha}_{\beta,\gamma,-\delta}$ is the curvature tensor of the symmetric space and the ω_{α} are linear combinations of the eigenvalues of Λ.

The integrable quartic potentials correspond to the stationary flows of the above compatibility equations. The stationary flows are Hamiltonian systems with energy given by

$$H = \sum_{\alpha,\beta} g^{\alpha,-\beta} p_{\alpha} s_{-\beta} + \frac{1}{2}\sum_{\alpha,\beta,\gamma,\delta} R_{-\alpha,\beta,\gamma,-\delta} r^{-\alpha} q^{\beta} q^{\gamma} r^{-\delta} + \sum_{\alpha,\beta} \omega_{\alpha} g_{\alpha,-\beta} q^{\alpha} r^{-\beta},$$

where $p_{\alpha} = \sum_{\beta} g_{\alpha,-\beta} r_x^{-\beta}$, $s_{-\beta} = \sum_{\alpha} g_{\alpha,-\beta} q_x^{\alpha}$, and $g_{\alpha,-\beta} = \operatorname{tr} e_{\alpha} e_{-\beta}$ is the metric on the symmetric space. These equations admit a Lax representation $\frac{\partial L}{\partial x} = [\mu A + Q, L]$. More interesting integrable systems are obtained by reduction $r^{-\alpha} = -q^{\alpha}$ ($s_{-\alpha} = -p_{\alpha}$). The canonical equations then become

$$q_{xx}^{\alpha} = \sum_{\beta,\gamma,\delta} R^{\alpha}_{\beta,\gamma,-\delta} q^{\beta} q^{\gamma} q^{\delta} - \omega_{\alpha} q^{\alpha}.$$

There are four infinite series of Hermitian symmetric spaces. Some of the simpler potentials associated to them are listed below:

$$V(q_1, \dots, q_n) = \frac{1}{2}\sum_{i=1}^{n} \omega_i q_i^2 + \frac{1}{2}\left(\sum_{i=1}^{n} q_i^2\right)^2,$$

$$V(q_1, \dots, q_4) = \frac{1}{2}\sum_{i=1}^{4} \omega_i q_i^2 + \frac{1}{2}\left(\sum_{i=1}^{4} q_i^2\right)^2 - (q_1 q_3 - q_2 q_4)^2,$$

$$V(q_1, q_2, q_3) = \frac{1}{2}\sum_{i=1}^{3} \omega_i(\delta_{i2} + 1) q_i^2 + \frac{1}{2}\left(\sum_{i=1}^{3} (\delta_{i2} + 1) q_i^2\right)^2 - (q_1 q_3 - q_2^2)^2,$$

$$V(q_1, \dots, q_6) = \frac{1}{2}\sum_{i=1}^{6} \omega_i q_i^2 + \frac{1}{2}\left(\sum_{i=1}^{6} q_i^2\right)^2 - (q_3 q_5 - q_2 q_6 - q_1 q_4)^2,$$

$$V(q_1, \dots, q_4) = \frac{1}{2}\sum_{i=1}^{4} \omega_i q_i^2 + \frac{1}{2}\left(\sum_{i=1}^{4} q_i^2\right)^2 - (q_1 q_3 + q_2 q_4)^2.$$

The first two potentials correspond to the class AIII symmetric spaces: $SU(m+n)/S(U(m) \times U(n))$. The third potential corresponds to the class CI:

Sp(n)/U(n) for the case $n = 2$. The fourth potential corresponds to the class $DIII$: SO($2n$)/U(n) for $n = 4$. The fifth potential corresponds to the class BDI: SO($n + 2$)/SO(2) × SO(n) for $n = 4$.

Theorem 2.21. *Hamiltonian systems that correspond to the potentials listed above are completely integrable.*

3.5. Theorem on the Completeness of Shifted Invariants. The study of Hamiltonian systems on Lie algebras leads to a natural question: How can one describe those Lie algebras for which the shifted invariants of the coadjoint representation provide a complete involutive family of functions on generic orbits? In some sense, this problem is solved by the following theorem of A.V. Bolsinov. Let $I(\mathfrak{g})$ denote the space of invariants of the coadjoint representation of the Lie group G associated with the Lie algebra $\mathfrak{g}$.

Theorem 2.22. *Consider a complex Lie algebra* $\mathfrak{g}$ *and a regular covector* $x \in \mathfrak{g}^*$. *For* $a \in \mathfrak{g}^*$, *we define the space* $M = \{df(x + \lambda a) | \lambda \in \mathbb{C}, f \in I(\mathfrak{g})\}$. *We have* $\dim M = \frac{1}{2}(\dim \mathfrak{g} + \operatorname{ind} \mathfrak{g})$, *i.e. the shifts of the invariants by the element* a *are complete at the point* x, *if and only if* $a + \lambda x$ *is a regular element for all* $\lambda \in \mathbb{C}$.

This theorem implies the following basic completeness criterion for the shifted coadjoint invariants.

Theorem 2.23 (Bolsinov [1988]). *Let* $\mathfrak{g}$ *be a finite-dimensional Lie algebra,* $\mathfrak{g}^*$ *the dual space of* $\mathfrak{g}$, *and* $S = \{y \in \mathfrak{g}^* | \dim \operatorname{Ann}(y) > \operatorname{ind} \mathfrak{g}\}$ *the set of singular elements in* $\mathfrak{g}^*$. *The shifts of the invariants by an arbitrary regular covector* $a \in \mathfrak{g}^*$ *form a complete involutive family on* $\mathfrak{g}^*$ *if and only if* $\operatorname{codim} S \geqslant 2$.

The completeness of the shifted invariants on singular orbits can be derived from the following theorem.

Theorem 2.24. *If* $x \in \mathfrak{g}^*$ *is a singular element,* $\operatorname{codim} S \geqslant 2$ *and* $\operatorname{ind} \operatorname{Ann}(x) = \operatorname{ind} \mathfrak{g}$, *then there exists a regular covector* $a \in \mathfrak{g}^*$ *such that the invariants shifted by* a *form a complete involutive family on the singular orbit* $\mathcal{O}(x)$.

Corollary. *The method of shifting the argument provides complete involutive families on every coadjoint orbit for the Lie group* SL(n, $\mathbb{C}$).

Corollary. *Let* $\mathfrak{g} = \mathrm{sl}(n, \mathbb{C}) \,{}_{\varphi}{+}\, \mathbb{C}^{nk}$ *be the semidirect sum where* φ *is the sum of* k *copies of the minimal representation of* sl(n, $\mathbb{C}$) *on* $\mathbb{C}^n$. *The shifts of the coadjoint invariants by a regular covector form a complete involutive family on* $\mathfrak{g}^*$ *if and only if either* n *is not a multiple of* k *or* $k = 1$.

Theorem 2.23 is in fact a consequence of the following simple construction. Let A and B be two skew-symmetric (degenerate) forms on $\mathbb{C}^n$, let $R = \max\limits_{\lambda, \mu \in \mathbb{C}} \operatorname{rank}(\lambda A + \mu B)$ and let $M \subset \mathbb{C}^n$ be the subspace spanned by the kernels of those forms $\lambda A + \mu B$ whose rank equals R. Then M is isotropic with respect to all forms $\lambda A + \mu B$. Moreover, M is a maximal isotropic subspace with respect to a form $\lambda A + \lambda B$ of rank R if and only if all the forms $\lambda A + \mu B$ have rank R

(except for $\lambda = \mu = 0$). Note that the forms A and B can be regarded as Poisson structures on $(\mathbb{C}^n)^*$ with constant coefficients. It turns out that the previous observations can easily be generalized to the case of an arbitrary compatible Poisson pair; Theorem 2.23 is then a special case for the pair of Poisson brackets $\{f, g\}(x) = x([df, dg])$ and $\{f, g\}_a(x) = a([df, dg])$.

General Construction. Let $\{\ ,\ \}_0$ and $\{\ ,\ \}_1$ be a pair of analytic compatible (degenerate) Poisson brackets in a domain $U \subset \mathbb{R}^n$, i.e. any linear combination of these brackets is again a Poisson bracket. Let S be the two-dimensional space spanned by $\{\ ,\ \}_0$ and $\{\ ,\ \}_1$ in the whole space of skew-symmetric tensor fields of type (2, 0). Consider the family F of functions which are central functions for those brackets in S which have rank R, where $R = \max\limits_{x \in U, C \in S} \operatorname{rank} C(x)$. It can be shown that f is an involutive family. It may happen that central functions are not globally defined, so the family F is considered locally. The following result, which generalizes Theorem 2.23, was obtained under some restrictions independently by A.V. Brailov and A.V. Bolsinov.

Theorem 2.25. *The involutive family F is complete at the point $x \in U$ if and only if* rank $A(x) = R$ *for every nonvanishing tensor field $A \in S^{\mathbb{C}}$, where $S^{\mathbb{C}}$ is the complexification of S.*

Appendix

We give a summary of the necessary facts from the theory of Lie groups and Lie algebras.

Definition. Let G be a smooth manifold endowed with a group structure. G is called a *Lie group* if the mapping $G \times G \to G$ defined by $(a, b) \to ab^{-1}$ is smooth.

Given a vector $v \in T_eG$ we can transfer it to other points of the group G by means of left translations and obtain a vector field on G. More precisely, for $a \in G$ we set $L_a(g) = ag$; since $(L_a)^{-1} = L_{a^{-1}}$, this is a diffeomorphism of G. Let $\xi_a = (dL_a)_e(v)$, $L_a \colon G \to G$. The field obtained in this way is left-invariant: $(dL_a)_b\xi_b = \xi_{ab}$ for all $a, b \in G$.

The space of left-invariant vector fields is a Lie algebra under the commutator of vector fields. This Lie algebra is finite-dimensional: its dimension is equal to the dimension of G. It is called the *Lie algebra* of G and is denoted by $\mathfrak{g}$.

A left invariant vector field ξ on G generates a globally defined group of diffeomorphisms. A *one-parameter subgroup* of a Lie group G is by definition a smooth homomorphism $\xi \colon \mathbb{R}^1 \to G$. It is easily verified that a left-invariant field ξ generates a one-parameter subgroup of G.

Thus the Lie algebra of G can be defined in any of the four equivalent ways using a) one-parameter subgroups, b) tangent vectors at the identity of the group, c) left-invariant vector fields, d) left-invariant $\mathbb{R}^1$-actions. All one-parameter subgroups can be combined into a single universal mapping Exp: $\mathfrak{g} \to G$ where $\mathrm{Exp}(tX)$: $\mathbb{R}^1 \to \mathfrak{g} \to G$ determines a one-parameter subgroup with velocity vector X at the identity $e \in G$.

Definition. The group G acts on itself by conjugation $(g_1, g_2) \to g_1 g_2 g_1^{-1}$. This action induces a linear action on the Lie algebra: $\mathrm{Ad}\colon G \to GL(\mathfrak{g})$, called the *adjoint action* of G: $\mathrm{Ad}_g\, \xi = d(I_g)_e \xi$, $I_g(x) = gxg^{-1}$.

The adjoint action of the group induces the *adjoint representation of the Lie algebra*: $\mathrm{ad} = d(\mathrm{Ad})_e\colon \mathfrak{g} \to \mathrm{End}(\mathfrak{g})$. We have $\mathrm{ad}_X(Y) = [X, Y]$ where $[X, Y]$ is the commutator in $\mathfrak{g}$. In Hamiltonian mechanics we need a different representation. Let $\mathfrak{g}^*$ be the dual space of $\mathfrak{g}$, i.e. the space of linear mappings $f\colon \mathfrak{g} \to \mathbb{R}$. We define the coadjoint action $\mathrm{Ad}^*\colon G \to GL(\mathfrak{g}^*)$ by $(\mathrm{Ad}_g^* f)(x) = f(\mathrm{Ad}_{g^{-1}} x)$. The differential of this representation is called the *coadjoint representation* of the Lie algebra $\mathfrak{g}$: $\mathrm{ad}^*\colon \mathfrak{g} \to \mathrm{End}(\mathfrak{g}^*)$. We have $(\mathrm{ad}_X^* f)(Y) = f([X, Y])$, $X, Y \in \mathfrak{g}$, $f \in \mathfrak{g}^*$. In the theory of Hamiltonian systems a distinguished role is played by the orbits of the coadjoint representation Ad^*. These orbits possess a natural symplectic structure and every homogeneous symplectic manifold is a coadjoint orbit for some Lie group (up to a covering).

We note that the orbits of the representation Ad^* in general are different from the orbits of Ad. If there is a nondegenerate inner product (X, Y) on $\mathfrak{g}$ such that $(\mathrm{Ad}_g X, \mathrm{Ad}_g Y) = (X, Y)$, then the adjoint and the coadjoint representations are equivalent.

This latter condition is fulfilled, in particular, for the so-called semisimple Lie algebras. To describe this class of Lie algebras we give the following definitions.

Definition. Let $\mathfrak{g}$ be a Lie algebra, and let $[\mathfrak{g}, \mathfrak{g}]$ denote the subspace of $\mathfrak{g}$ spanned by all commutators $[x, y]$. The Lie algebra $\mathfrak{g}$ is called *solvable* if $\mathfrak{g}^{(n)} = 0$ for some n, where $\mathfrak{g}^{(n)} = [\mathfrak{g}^{(n-1)}, \mathfrak{g}^{(n-1)}]$ and $\mathfrak{g}^{(1)} = [\mathfrak{g}, \mathfrak{g}]$.

Definition. A Lie algebra $\mathfrak{g}$ is called *semisimple* if it has no solvable ideals except $\{0\}$.

Theorem. *A Lie algebra $\mathfrak{g}$ is semisimple if and only if its Killing form $(X, Y) = \mathrm{tr}\, \mathrm{ad}_X\, \mathrm{ad}_Y$ is nondegenerate.*

Definition. A Lie algebra $\mathfrak{g}$ is *simple* if it is not commutative and has no ideals except $\{0\}$ and $\mathfrak{g}$.

Every semisimple Lie algebra is the direct sum of simple ideals.

In every Lie algebra $\mathfrak{g}$ there is a maximal solvable ideal $\mathfrak{r}$ which contains all solvable ideals of $\mathfrak{g}$. The ideal $\mathfrak{r}$ is called the radical of $\mathfrak{g}$. The quotient algebra $\mathfrak{g}/\mathfrak{r}$ is semisimple.

Theorem. *Any Lie algebra $\mathfrak{g}$ contains a semisimple subalgebra $\mathfrak{S}$ such that $\mathfrak{g} = \mathfrak{S} \oplus \mathfrak{r}$ (the direct sum of vector spaces).*

Definition. A Lie algebra $\mathfrak{g}$ is *nilpotent* if $[\ldots[x_1, x_2], \ldots, x_n] = 0$ for all $x_1, \ldots, x_n \in \mathfrak{g}$.

Let $\mathfrak{g}$ be a Lie algebra, $\mathfrak{g}^*$ the dual space to $\mathfrak{g}$, $\pi: \mathfrak{g} \to \mathrm{End}(V)$ a representation of $\mathfrak{g}$ on a vector space V. For a linear function $\lambda \in \mathfrak{g}^*$, we set $V_n^\lambda = \{\xi \in V | (\pi(x) - \lambda(x))^n \xi = 0$ for all $x \in \mathfrak{g}\}$ and let V^λ be the union of all V_n^λ, $n = 1, \ldots$. A linear form $\lambda \in \mathfrak{g}^*$ is called a *weight* of the representation π if $V^\lambda \neq \{0\}$. The latter condition is equivalent to $V_1^\lambda \neq \{0\}$. Nonzero elements $\xi \in V_1^\lambda$ are called the *weight vectors* of weight λ for the representation π.

Let $V_0 = \sum_{\lambda \in \mathfrak{g}^*} V^\lambda$; this is a direct sum. If $\mathfrak{g}$ is nilpotent and the ground field is algebraically closed, then $V_0 = V$. In that case V^λ is an invariant subspace of V for every λ. There is a basis for V^λ in which all the operators $\pi(x) - \lambda(x)$ $(x \in \mathfrak{g})$ are strictly triangular.

If $\mathfrak{H}$ is a subalgebra of $\mathfrak{g}$, the weights of the adjoint representation $\pi(x) = \mathrm{ad}_x (x \in \mathfrak{H})$ are called the *roots* of the Lie algebra $\mathfrak{g}$ with respect to $\mathfrak{H}$. The corresponding weight vectors $\xi \in \mathfrak{g}$ are called *root vectors* (with respect to $\mathfrak{H}$).

Definition. A nilpotent subalgebra of a Lie algebra $\mathfrak{g}$ is called a *Cartan subalgebra* if it coincides with its normalizer.

Suppose that $\mathfrak{H}$ is a nilpotent subalgebra of $\mathfrak{g}$. Consider the representation $\pi(x) = \mathrm{ad}_x$ of $\mathfrak{H}$ on $\mathfrak{g}$ and the subspace $\mathfrak{g}^\lambda$ of $\mathfrak{g}$ for any $\lambda \in \mathfrak{H}^*$. Assume that for any $x \in \mathfrak{H}$ the endomorphism ad_x can be reduced to triangular form. Then we have the following assertions. The space $\mathfrak{g}$ decomposes as $\mathfrak{g} = \bigoplus_{\lambda \in \mathfrak{H}^*} \mathfrak{g}^\lambda$, so that $[\mathfrak{g}^\lambda, \mathfrak{g}^\mu] \subset \mathfrak{g}^{\lambda+\mu}$ and, in particular, $[\mathfrak{g}^0, \mathfrak{g}^\mu] \subset \mathfrak{g}^\mu$. The subspace $\mathfrak{g}^0$ is a subalgebra of $\mathfrak{g}$ which contains $\mathfrak{H}$. The subalgebra $\mathfrak{H}$ is a Cartan subalgebra if and only if $\mathfrak{g}^0 = \mathfrak{H}$. If $\mathfrak{g}^0$ is nilpotent, then $\mathfrak{g}^0$ is a Cartan subalgebra of $\mathfrak{g}$. If $\mathfrak{H}$ is a Cartan subalgebra of $\mathfrak{g}$, then $\mathfrak{H}$ is a maximal nilpotent subalgebra of $\mathfrak{g}$.

The root space decomposition $\mathfrak{g} = \bigoplus_{\lambda \in \mathfrak{H}^*} \mathfrak{g}^\lambda$ has the following properties with respect to the Killing form: a) $(\mathfrak{g}^\lambda, \mathfrak{g}^\mu) = 0$ if $\lambda \neq -\mu$; b) $(\mathfrak{H}, \mathfrak{g}^\lambda) = 0$ if $\lambda \neq 0$; c) $(X, Y) = \sum_{\alpha \in \mathfrak{H}^*} (\dim \mathfrak{g}^\alpha)\alpha(X)\alpha(Y)$ for $X, Y \in \mathfrak{H}$.

The coefficients a_i of the polynomial $\det(t - \mathrm{ad}_X) = t^n + a_{n-1}(X)t^{n-1} + \cdots + a_1(X)t$ are polynomial functions on $\mathfrak{g}$. If $X \in \mathfrak{g}$ then $\dim \mathfrak{g}^0(X)$ is the least integer p such that $a_p(X) \neq 0$, where $\mathfrak{g}^0(X) = \bigcup_{n \geqslant 0} \mathrm{Ker}(\mathrm{ad}_X)^n$. Let s denote the least integer such that a_s does not vanish identically. The number s is called the rank of $\mathfrak{g}$. The element X is *regular* if $a_s(X) \neq 0$ and *singular* otherwise.

For a regular element X of a Lie algebra $\mathfrak{g}$ over an algebraically closed field, the subalgebra $\mathfrak{H} = \mathfrak{g}^0(X)$ is the unique Cartan subalgebra that contains X.

If $\mathfrak{H}$ and $\mathfrak{k}$ are two Cartan subalgebras of a Lie algebra $\mathfrak{g}$ over an algebraically closed field, then there is an automorphism α of $\mathfrak{g}$ such that $\alpha(\mathfrak{H}) = \mathfrak{k}$. The dimension of a Cartan subalgebra is equal to the rank of $\mathfrak{g}$ and coincides with the index of $\mathfrak{g}$.

Let $R(\mathfrak{g}, \mathfrak{H})$ denote the set of all roots of the Lie algebra $\mathfrak{g}$ with respect to a Cartan subalgebra $\mathfrak{H}$. This set has the following properties.

Theorem. *Let* $\mathfrak{g}$ *be a semisimple Lie algebra over an algebraically closed field,* $\mathfrak{H}$ *its Cartan subalgebra,* $R = R(\mathfrak{g}, \mathfrak{H})$ *its root system,* (X, Y) *the Killing form. The root space decomposition has the following properties:*

a) *The root space decomposition has the form* $\mathfrak{g} = \mathfrak{H} \oplus \left(\bigoplus_{\alpha \in R} \mathfrak{g}^\alpha \right)$ *and* $\dim \mathfrak{g}^\alpha = 1$ *for every root* $\alpha \in R$.

b) *The Lie algebra* $\mathfrak{H}$ *is commutative. If* $h \in \mathfrak{H}$ *and* $x \in \mathfrak{g}^\alpha$, *then* $[h, x] = \alpha(h)x$. *For any* $\alpha, \beta \in R$ *we have* $[\mathfrak{g}^\alpha, \mathfrak{g}^\beta] \subset \mathfrak{g}^{\alpha+\beta}$. *If* $\alpha \in R$, *then* $-\alpha \in R$ *and* $\mathfrak{H}_\alpha = [\mathfrak{g}^\alpha, \mathfrak{g}^{-\alpha}]$ *is a one-dimensional subspace in* $\mathfrak{H}$ *which contains a uniquely determined element* h_α *such that* $\alpha(h_\alpha) = 2$.

c) *If* $\alpha, \beta \in \mathfrak{H}^*$ *and* $\alpha + \beta \neq 0$, *then the subspaces* $\mathfrak{g}^\alpha$ *and* $\mathfrak{g}^\beta$ *are orthogonal with respect to the Killing form. The restriction of the Killing form to the subspace* $\mathfrak{g}^\alpha \times \mathfrak{g}^{-\alpha}$ *(in particular, to* $\mathfrak{H} \times \mathfrak{H}$*) is nondegenerate. If* $x, y \in \mathfrak{H}$, *then* $(x, y) = \sum_{\alpha \in R} \alpha(x)\alpha(y)$.

d) *The elements of* R *span the whole of* $\mathfrak{H}^*$.

e) *If* $\alpha \in R$, *then the vector space* $\mathfrak{G}_\alpha = \mathfrak{H}_\alpha + \mathfrak{g}^\alpha + \mathfrak{g}^{-\alpha}$ *is a subalgebra of* $\mathfrak{g}$. *For any* $E_\alpha \in \mathfrak{g}^\alpha \backslash \{0\}$, *there is a unique element* $E_{-\alpha} \in \mathfrak{g}^{-\alpha}$ *such that* $[E_\alpha, E_{-\alpha}] = h_\alpha$.

The root system of a semisimple Lie algebra has the following properties. Let $\mathfrak{g}$ be a semisimple Lie algebra over an algebraically closed field, $\mathfrak{H}$ its Cartan subalgebra, $R = R(\mathfrak{g}, \mathfrak{H})$ its root system and $\alpha, \beta \in R$. The scalar $a_{\alpha\beta} = \beta(h_\alpha)$ is an integer. The set of integers $t \in \mathbb{Z}$ such that $\beta + t\alpha \in R \cup \{0\}$ fill an interval $[-t', t'']$ where $t', t'' \geqslant 0$. We have $a_{\beta\alpha} = t' - t''$ and $\beta - a_{\beta\alpha}\alpha \in R$. If $\beta - \alpha \notin R \cup \{0\}$, then $a_{\beta\alpha} \leqslant 0$, $t' = 0$ and $t'' = -a_{\beta\alpha}$. If $\beta + \alpha \in R$, then $[\mathfrak{g}^\alpha, \mathfrak{g}^\beta] = \mathfrak{g}^{\alpha+\beta}$. The only roots proportional to α are α and $-\alpha$.

There is an abstract definition of a root system which in particular is satisfied by the root system R.

Definition. A finite system R of nonzero vectors in a real Euclidean space V is called a *root system* if a) the numbers $c(\alpha, \beta) = 2(\alpha, \beta)/(\alpha, \alpha)$ are integers, b) $\alpha - c(\alpha, \beta)\beta \in R$ for all $\alpha, \beta \in R$.

A subsystem $S \subset R$ is called a *base* for R if every root $\alpha \in R$ can be uniquely expressed in the form $\sum_i n_i\alpha_i (\alpha_i \in S)$ where the n_i are integers all positive or all negative. A root α is called *positive* if all $n_i \geqslant 0$ and *negative* if all $n_i \leqslant 0$. Accordingly, $R = R_+ \cup R_-$ where $R_+(R_-)$ is the set of all positive (negative) roots in R. If $\alpha, \beta \in S$ then $(\alpha, \beta) \leqslant 0$.

A base for R can be constructed by introducing a lexicographical ordering of the linear span L of R. We fix a basis of L and for two vectors $\xi, \eta \in L$ with components ξ_i, η_i we set $\xi > \eta$ if $\xi_1 > \eta_1$, or $\xi_1 = \eta_1$ but $\xi_2 > \eta_2$, etc. A root $\alpha > 0$ is called *simple* if it cannot be written as a sum of two roots $\beta > 0, \gamma > 0$. The subsystem S of all simple roots is a base for R.

The role of simple roots in the theory of semisimple Lie algebras is explained by the following fact: the system of simple roots allows to reconstruct the whole root system of a semisimple Lie algebra, and the root system uniquely determines the Lie algebra itself. There is a complete classification of complex simple Lie algebras. These are: $\mathrm{sl}(n, \mathbb{C})$, the Lie algebra of traceless $n \times n$ matrices x with commutator $[x, y] = xy - yx$; $\mathrm{so}(2n + 1, \mathbb{C})$, the Lie algebra of skew-symmetric $(2n + 1) \times (2n + 1)$ matrices with commutator $[x, y] = xy - yx$; $\mathrm{sp}(n, \mathbb{C})$, the Lie algebra of matrices $x \in \mathrm{gl}(2n, \mathbb{C})$ such that $fx^t + xf = 0$ where

$$f = \begin{pmatrix} 0 & I \\ -I & 0 \end{pmatrix}, \quad I_{ij} = \delta_{i,n-j+1}, \quad i, j = 1, \ldots, n,$$

with commutator $[x, y] = xy - yx$; $\mathrm{so}(2n, \mathbb{C})$, the Lie algebra of skew-symmetric $2n \times 2n$ matrices with commutator $[x, y] = xy - yx$. These Lie algebras are also denoted by A_{n-1}, B_n, C_n, D_n respectively. Besides these series there are 5 exceptional Lie algebras which will not be described here.

Let $\mathfrak{g}$ be a Lie algebra over the complex numbers. A real subalgebra $\mathfrak{g}_0$ of $\mathfrak{g}$ is called a *real form* of $\mathfrak{g}$ if the canonical mapping of the complexification $\mathfrak{g}_0^{\mathbb{C}}$ of $\mathfrak{g}_0$ into $\mathfrak{g}$ is an isomorphism.

All real forms $\mathfrak{g}_0$ are in one-to-one correspondence with those involutions $\sigma\colon \mathfrak{g} \to \mathfrak{g}$ of $\mathfrak{g}$ for which $\sigma^2 = \mathrm{id}$, $\sigma(X) = X$ if $X \in \mathfrak{g}_0$, $\sigma(A + B) = \sigma(A) + \sigma(B)$, A, $B \in \mathfrak{g}$, $\sigma(\lambda A) = \bar{\lambda}\sigma(A)$, $\sigma([A, B]) = [\sigma(A), \sigma(B)]$.

Definition. A Lie algebra $\mathfrak{g}$ over $\mathbb{R}$ is said to be *compact* if it possesses an invariant scalar product. A subalgebra $\mathfrak{k}$ of $\mathfrak{g}$ is said to be *compactly embedded* if there is a $\mathfrak{k}$-invariant scalar product on $\mathfrak{g}$.

A real form $\mathfrak{g}_0$ of a complex Lie algebra is called a *compact real form* if $\mathfrak{g}_0$ is a compact real Lie algebra. We shall describe a construction of compact real forms for semisimple complex Lie algebras. Let $\mathfrak{H}$ be a Cartan subalgebra of $\mathfrak{g}$ and R the root system of $\mathfrak{g}$ relative to $\mathfrak{H}$. Then $\mathfrak{g} = \mathfrak{H} + \mathbb{C}E_\alpha + \mathbb{C}E_{-\alpha} + \cdots$, $[h, E_\alpha] = \alpha(h)E_\alpha$ for $h \in \mathfrak{H}$, $[E_\alpha, E_{-\alpha}] = h_\alpha$, $(E_\alpha, E_{-\alpha}) = -1$, $[E_\alpha, E_\beta] = N_{\alpha\beta}E_{\alpha+\beta}$ if $\alpha, \beta, \alpha + \beta \in R$, where $N_{\alpha,\beta} = N_{-\alpha,-\beta}$ and $N_{\alpha,\beta}N_{-\alpha,-\beta}$ is positive, so that $N_{\alpha,\beta} \in \mathbb{R}$. We set $\mathfrak{H}_r = \sum_{\alpha \in R} \mathbb{R}h_\alpha$; then for $\alpha \in \mathfrak{H}_r$ and $\alpha \in R$ we have $\alpha(h) \in \mathbb{R}$. We fix some lexicographical ordering and set

$$C_\alpha = E_\alpha + E_{-\alpha}, \quad S_\alpha = i(E_\alpha - E_{-\alpha}),$$

$$\mathfrak{g}_u = i\mathfrak{H}_r + \sum_{\alpha > 0} (\mathbb{R}C_\alpha + \mathbb{R}S_\alpha).$$

The subalgebra $\mathfrak{g}_u$ is a compact real form for $\mathfrak{g}$. A basis $\{E_\alpha, \alpha \in R\}$ of $\mathfrak{g}$ modulo $\mathfrak{H}$ such that $[h, E_\alpha] = \alpha(h)E_\alpha$, $[E_\alpha, E_{-\alpha}] = -h_\alpha \in \mathfrak{H}_r$, $[E_\alpha, E_\beta] = N_{\alpha,\beta}E_{\alpha+\beta}$ for α, β, $\alpha + \beta \in R$ is called a *Weyl basis* for $\mathfrak{g}$.

Definition Let $\{E_\alpha\}$ be a Weyl basis for a semisimple Lie algebra $\mathfrak{g}$. The real subalgebra $\{E_\alpha, E_{-\alpha}, h_\alpha\}$ in $\mathfrak{g}$ is called a *normal subalgebra* and is denoted by $\mathfrak{g}_n$.

The theory of semisimple Lie algebras is closely related to the theory of symmetric spaces. We shall recall some simple constructions of this theory.

We consider a connected Lie group G and an involutive automorphism $\sigma\colon G \to G$, i.e. $\sigma^2 = \mathrm{id}$, $\sigma \neq \mathrm{id}$. Let G_σ denote the closed subgroup of G consisting of all fixed points of σ and G_σ^0 its connected component containing the identity.

Proposition. *Suppose that G is a connected compact Lie group and H a subgroup of G such that $G_\sigma \supset H \supset G_\sigma^0$ for some involutive automorphism $\sigma: G \to G$. Then the homogeneous space G/H with the Riemannian metric induced by the bi-invariant metric of G is a Riemannian symmetric space.*

Every symmetric space can be identified as G/H where $G_\sigma \supset H \supset G_\sigma^0$ for a suitable involutive automorphism σ of a Lie group G. To every symmetric space one can associate a pair $(\mathfrak{g}, \sigma)$ where $\mathfrak{g}$ is the Lie algebra of G and σ an involutive automorphism of $\mathfrak{g}$. The set $\mathfrak{k} = \{X \in \mathfrak{g} | \sigma(X) = X\}$ is a compact subalgebra of $\mathfrak{g}$; together with $\mathfrak{p} = \{X \in \mathfrak{g} | \sigma(X) = -X\}$ it satisfies the commutation relations $[\mathfrak{k}, \mathfrak{k}] \subset \mathfrak{k}, [\mathfrak{k}, \mathfrak{p}] \subset \mathfrak{p}, [\mathfrak{p}, \mathfrak{p}] \subset \mathfrak{k}$. The pair $(\mathfrak{g}, \sigma)$ is called an *orthogonal symmetric Lie algebra.* Associating an orthogonal symmetric Lie algebra to the symmetric space reduces the classification problem for symmetric spaces to that for orthogonal symmetric Lie algebras.

A Riemannian symmetric space is said to be *irreducible* if the associated orthogonal symmetric Lie algebra is such that $\mathfrak{g}$ is semisimple, $\mathfrak{k}$ contains no nontrivial ideals of $\mathfrak{g}$, and $\mathfrak{k}$ is a maximal proper subalgebra of $\mathfrak{g}$.

Definition. An operator $A: V \to V$ on a linear space V is said to be *semisimple* if for every invariant subspace $L \subset V$ there is an invariant subspace $M \subset V$ such that $V = L \oplus M$. An operator A is said to be nilpotent if $A^N = 0$ for some N.

Let $\mathfrak{g}$ be a semisimple Lie algebra. For an element $x \in \mathfrak{g}$ the following conditions are equivalent:

a) The endomorphism $\mathrm{ad}_x: \mathfrak{g} \to \mathfrak{g}$ is semisimple (nilpotent).

b) There is an exact finite-dimensional representation ρ of $\mathfrak{g}$ for which the endomorphism $\rho(x)$ is semisimple (nilpotent).

c) For any finite-dimensional representation α of $\mathfrak{g}$ the endomorphism $\alpha(x)$ is semisimple (nilpotent).

Definition. Let $\mathfrak{g}$ be a semisimple Lie algebra. An element $x \in \mathfrak{g}$ which satisfies the above conditions is called *semisimple* (respectively, *nilpotent*).

Every element $z \in \mathfrak{g}$ of a semisimple Lie algebra $\mathfrak{g}$ can be uniquely written as $z = s + n$ where s is semisimple, n is nilpotent and $[s, n] = 0$.

The orbit of the adjoint representation passing through a semisimple element is called a *semisimple orbit.*

References*

Abraham, R., Marsden, J. [1978]: Foundations of Mechanics (2nd ed.). Benjamin Cummings: New York, 641 pp., Zbl.393.70001

Adler, M., van Moerbeke, P. [1982]: The algebraic integrability of the geodesic flow on $SO(4)$. Invent. Math. *67*, 297–331, Zbl.539.58012

Andreev, E.M., Vinberg, E.B., Elashvili, A.G. [1967]: Orbits of greatest dimension in semisimple linear Lie groups. Funkts. Anal. Prilozh. *1*, No. 4, 3–7. English transl.: Funct. Anal. Appl. *1*, 257–261 (1967), Zbl.176,303

Anosov, D.V. [1982]: On typical properties of closed geodesics. Izv. Akad. Nauk SSSR, Ser. Mat. *46*, No. 4, 675–709. English transl.: Math. USSR. Izv. *21*, 1–29 (1983), Zbl.512.58014

Arkhangel'sky, A.A. (= Arkhangel'skij, A.A.) [1979]: Completely integrable hamiltonian systems on the group of triangular matrices. Mat. Sb., Nov. Ser. *108*, No. 1, 134–142. English transl.: Math. USSR, Sb. *36*, 127–134 (1980), Zbl.425.58013

Arnol'd, V.I. [1966]: Sur la géometrie differentielle des groupes de Lie de dimension infinie et ses applications a l'hydrodynamique des fluides parfaites. Ann. Inst. Fourier *16*, 319–361, Zbl.148,453

Arnol'd, V.I. [1969]: The Hamiltonian character of the Euler equations of the dynamics of a rigid body in an ideal fluid. Usp. Mat. Nauk *24*, No. 3, 225–226 (Russian), Zbl.181.542

Arnol'd, V.I. [1974]: Mathematical Methods of Classical Mechanics. Nauka: Moscow, 431 pp. English transl.: Graduate Texts in Mathematics *60*. Springer-Verlag: New-York – Berlin – Heidelberg 1978, 462 pp., Zbl.386.70001

Arnol'd, V.I., Givental', A.B. [1985]: Symplectic geometry. In: Dynamical Systems IV. Itogi Nauki Tekh., Ser. Sovrem. Probl. Mat., Fundam. Napravleniya *4*, 5–139. English transl. in: Encyclopaedia of Mathematical Sciences *4*, 1–136. Springer-Verlag: Berlin – Heidelberg – New York 1990, Zbl.592.58030

Atiyah, M.F. [1982]: Convexity and commuting Hamiltonians. Bull. Lond. Math. Soc. *14*, 1–15, Zbl.482.58013

Belyaev, A.V. [1981]: On the motion of a multidimensional body with a fixed point in a gravitational field. Mat. Sb., Nov. Ser. *114*, No. 3, 465–470. English transl.: Math. USSR, Sb. *42*, 413–418 (1982), Zbl.486.70011

Berezin, F.A. [1967]: Some remarks on the universal enveloping algebra of a Lie algebra. Funkts. Anal. Prilozh. *1*, No. 2, 1–14. English transl.: Funct. Anal. Appl. *1*, 91–102 (1968), Zbl.227.22020

Berezin, F.A., Perelomov, A.M. [1980]: Group-theoretical interpretation of equations of Korteweg-de Vries type. Funkts. Anal. Prilozh. *14*, No. 2, 50–51 English transl.: Funct. Anal. Appl. *14*, 119–121 (1980), Zbl.477.35078

Bogoyavlensky, O.I. (= Bogoyavlenskij, O.I.) [1976]: On perturbation of the periodic Toda lattice. Commun. Math. Phys. *51*, 201–209

Bogoyavlensky, O.I. [1983]: New algebraic constructions of integrable Euler's equations. Dokl. Akad. Nauk SSSR *268*, No. 2, 277–280. English transl.: Sov. Math., Dokl. *27*, 60–63 (1983), Zbl.545.58019

Bogoyavlensky, O.I. [1984]: Integrable equations on Lie algebras arising in problems of mathematical physics. Izv. Akad. Nauk SSSR, Ser. Mat. *48*, No. 5, 883–938. English transl.: Math. USSR, Izv. *25*, 207–257 (1985), Zbl.583.58012

Bolsinov, A.V. [1988]: A completeness criterion for a family of functions in involution constructed by the argument shift method. Dokl. Akad. Nauk SSSR *301, No. 5*, 1037—1040. English transl.: Sov. Math., Dokl. *38*, 161–165 (1989)

Bott, R. [1954]: Non-degenerate critical manifolds. Ann. Math., II. Ser. *60*, 248–261, Zbl.58,91

* For the convenience of the reader, references to reviews in Zentralblatt für Mathematik (Zbl.), compiled using the MATH database, have, as far as possible, been included in this bibliography.

Brailov, A.V. [1983]: Involutive sets on Lie algebras and extension of the ring of scalars. Vestn. Mos. Univ., Ser. I, No. 1, 47–51. English transl.: Mosc. Univ. Math. Bull. *38*, No. 1, 57–61 (1983), Zbl.516.17003

Brailov, A.V. [1985]: Complete integrability with non-commuting integrals of some Euler equations. Primenenie Topologii v Sovrem. Anal., Nov. Global'nom Anal. 1985, 22–41, English transl.: Global analysis-studies and applications II, Lect. Notes Math. *1214*, 21–39 (1986), Zbl.569.58022

Brailov, A.V., Fomenko, A.T. [1987]: The topology of integral submanifolds of completely integrable Hamiltonian systems. Mat. Sb., Nov. Ser. *134*, No. 3, 375–385. English transl.: Math. USSR, Sb. *62*, No. 2, 373–383 (1989), Zbl.647.58025

Cartan, E. [1927]: Leçons sur les Invariants Intégraux. Hermann: Paris, 210 pp.

Dao Chong Tkhi [1978]: The integrability of the Euler equations on homogeneous symplectic manifolds. Mat. Sb., Nov. Ser. *106*, No. 2, 154–161. English transl.: Math. USSR, Sb. *34*, 707–713 (1978), Zbl.394,53024

Dixmier, J. [1974]: Algèbres Enveloppantes. Gautier-Villars: Paris – Bruxelles – Montreal, 387 pp., Zbl. 308.17007

Dubrovin, B.A., Fomenko, A.T., Novikov, S.P. [1979]: Modern Geometry – Methods and Applications. Nauka: Moscow, 759 pp. English transl. in two parts: Part I: Graduate Texts in Mathematics *93*, Springer-Verlag: New York – Berlin – Heidelberg – Tokyo 1984, 464 pp. Part II: Graduate Texts in Mathematics *104*, Springer-Verlag: New York – Berlin – Heidelberg – Tokyo 1985, 430 pp., Zbl.433.53001

Dubrovin, B.A., Krichever, I.M., Novikov, S.P. [1985]. Integrable systems. I. In: Dynamical Systems IV. Itogi Nauki Tekh., Ser. Sovrem. Probl. Mat., Fundam. Napravleniya *4*, 179–248. English transl.: Encyclopaedia of Mathematical Sciences *4*, 173–280. Springer-Verlag: Berlin – Heidelberg – New York 1990, Zbl.591.580013

Fomenko, A.T. [1980]: Symplectic group structures on homogeneous spaces. Dokl. Akad. Nauk SSSR *253*, No. 5, 1062–1067. English transl.: Sov. Math., Dokl. *22*, 226–231 (1980), Zbl.508.58019

Fomenko, A.T. [1981]: On symplectic structures and integrable systems on symmetric spaces. Mat. Sb., Nov. Ser. *115*, No. 2, 263–280. English transl.: Math. USSR, Sb. *43*, 235–250 (1982), Zbl.467.58014

Fomenko, A.T. [1983a]: Differential Geometry and Topology. Supplementary Chapters. Moscow State University: Moscow. English transl.: Contemporary Soviet Mathematics, Consultants Bureau. [Plenum]: New York – London 1987, Zbl.517.53001

Fomenko, A.T. [1983b]: The integrability of some Hamiltonian systems. Ann. Global Anal. Geom. *1*, No. 2, 1–10, Zbl.528.58018

Fomenko, A.T. [1984a]: Algebraic properties of some integrable Hamiltonian systems. Topology, general and algebraic topology, and applications, Proc. Int. Conf., Leningrad 1982, Lect. Notes Math. *1060* (Faddeev, L.D., Mal'cev, A.A. (Eds.)), 246–257, Zbl.534.58029

Fomenko, A.T. [1984b]: Algebraic structure of certain integrable Hamiltonian systems. Global analysis – studies and applications I, Lect. Notes Math. *1108* (Borisovich, Yu.G., Gliklikh, Yu.E. (Eds.)), 103–127, Zbl.554.58027

Fomenko, A.T. [1985]: Topology of three-dimensional manifolds and integrable mechanical systems. In: Fifth Tiraspol Sympos. General Topology Appl., Kishinev, 235–237 (Russian)

Fomenko, A.T., [1986a]: Morse theory of integrable Hamiltonian systems. Dokl. Acad. Nauk SSSR *287*, 1071–1075. English transl.: Sov. Math., Dokl. *33*, 502–506 (1986), Zbl.623.58009

Fomenko, A.T., [1986b]: The topology of surfaces of constant energy of integrable Hamiltonian systems, and obstructions to integrability. Izv. Akad. Nauk SSSR Ser. Mat. *50*, No. 5, 1276–1307. English transl.: Math. USSR, Izv. *29*, 629–658 (1987), Zbl.619.58023

Fomenko, A.T., [1986c]: Complete integrability of some classical Hamiltonian systems. In: Monogenic functions and mappings, Collect. Sci. Works, Kiev, 1982, 3–19, Transl., II. Ser., Am. Math. Soc. *133* (1986), 79–90, Zbl.532.58009

Fomenko, A.T. [1987]: Qualitative geometric theory of integrable systems, the classification of isoenergetic surfaces and bifurcation of Liouville tori for critical energy values. In: Geometriya:

Teoriya Osobennostej v Nelinejnykh Uravneniyakh, Nov. Global'nom Anal. 1987, 118–139 English transl.: Global analysis – studies and applications III, Lect. Notes Math. *1334*, 221–245 (1988), Zbl.662.58022

Fomenko, A.T. [1988a]: Symplectic Geometry. Moscow Univ. Press: Moscow, 413 pp. English transl.: Gordon and Breach: New York etc. (1988), Zbl.716.53005

Fomenko, A.T. [1988b]: Topological invariants of Liouville integrable Hamiltonian systems. Funkts. Anal. Prilozh. *22*, No. 4, 38–51. English transl.: Funct. Anal. Appl. *22*, 286–296 (1988), Zbl.697.58021

Fomenko, A.T. (ed.) [1992]: Topological Classification of Integrable Systems. Adv. Sov. Math. *6*, American Math. Society

Fomenko, A.T., Trofimov, V.V. [1988]: Integrable Systems on Lie Algebras and Symmetric Spaces. Gordon and Breach: New York etc.

Fomenko, A.T., Zieschang, H. [1987]: On the topology of the three-dimensional manifolds arising in Hamiltonian mechanics. Dokl. Akad. Nauk SSSR *294*, No. 2, 283–287. English transl.: Sov. Math., Dokl. *35*, 529–534 (1987)

Fomenko, A.T., Zieschang, H. [1988]: On typical topological properties of integrable Hamiltonian systems. Izv. Akad. Nauk. SSSR, Ser. Mat. *52*, No. 2, 378–407. English transl.: Math. USSR, Izv. *32*, 385–412 (1989), Zbl.647.58018

Fomenko, A.T., Zieschang, H. [1989]: Classification of stably equivalent integrable Hamiltonian systems and a criterion for their topological equivalence. (In press)

Fordy, A.P., Wojciechowski, S., Marshall, I., [1986]: A family of integrable quartic potentials related to symmetric spaces. Phys. Lett. A *113*, 395–400

Fuks, D.B. [1984]: Cohomology of Infinite-Dimensional Lie Algebras. Nauka: Moscow, 272 pp. English transl.: Contemp. Sov. Math, New York (1986), Zbl.592.17011

Gelfand, I.M., Cherednik, I.V. [1983]: The abstract Hamiltonian formalism for classical Yang-Baxter bundles. Usp. Mat. Nauk *38*, No. 3, 3–21. English transl.: Russ. Math. Surv. *38*, No. 3, 1–22 (1983), Zbl.536.58006

Gromov, M. [1971]: A topological technique for the construction of solutions of differential equations and inequalities. In: Actes Congr. Int. Math. 1970. 2, 221–225, Zbl.237.57019

Guillemin, V., Sternberg, S. [1980]: The moment map and collective motion. Ann. Phys. *127*, 220–253, Zbl.453.58015

Guillemin, V., Sternberg, S. [1982]: Convexity properties of the moment mapping. Invent. Math. *67*, 491–513, Zbl.503.58017

Guillemin, V., Sternberg, S. [1984a]: Symplectic Techniques in Physics. Cambridge Univ. Press: Cambridge, 468 pp., Zbl.576.58012

Guillemin, V., Sternberg, S. [1984b]: Multiplicity-free spaces. J. Differ. Geom. *19*, 31–56, Zbl.548.58017

Haefliger, A. [1971]: Lecture on a theorem of Gromov. Proc. Liverpool Singularities Symp. 1969–1970, 128–141, Zbl.222.57020

Helgason, S. [1962]: Differential Geometry and Symmetric Spaces. Academic Press: New York – London 486 pp., Zbl.111,181

Holm, D.D., Kupershmidt, B.A. [1983]: Nonlocal formulation of ideal magnetohydrodynamics. Physica *D7*, 330–333

Jacobson, N. [1962]: Lie Algebras. Interscience Publishers: New York – London, 331 pp., Zbl.121,275

Kazhdan, D., Kostant, B., Sternberg, S. [1978]: Hamiltonian group actions and dynamical systems of Calogero type. Commun. Pure Appl. Math. *31*, 491–508, Zbl.368.58008

Kharlamov, M.P. [1983]: Topological analysis of classical integrable systems in the dynamics of a rigid body. Dokl. Akad. Nauk SSSR *273*, No. 6, 1322–1325. English transl.: Sov. Math., Dokl. *28*, 802–805 (1983), Zbl.561.58021

Kirillov, A.A. [1978]: Elements of the Theory of Representations (2nd ed.). Nauka: Moscow. 343 pp. English transl. (of the first edition): Grundlehren der Mathematischen Wissenschaften *220*, Springer-Verlag: New York – Berlin – Heidelberg 315 pp., Zbl.264.22011

Klingenberg, W. [1978]: Lectures on Closed Geodesics. Springer-Verlag: Berlin – Heidelberg – New York, 227 pp., Zbl. 397.58018

Kozlov, V.V. [1980]: Methods of Qualitative Analysis in the Dynamics of a Rigid Body. Moscow State University: Moscow, 230 pp. (Russian), Zbl.557.70009

Kozlov, V.V. [1985]: Calculus of variations in the large and classical mechanics. Usp. Mat. Nauk *40*, No. 2, 33–60. English transl.: Russ. Math. Surv. *40*, No. 2, 37–71 (1985), Zbl.557.70025

Kummer, M. [1981]: On the construction of the reduced phase space of a Hamiltonian system with symmetry. Indiana Univ. Math. J. *30*, 281–291, Zbl.425.70019

Kupershmidt, B.A. [1985]: Discrete Lax equations and differential-difference calculus. Asterisque *123*, 212 pp., Zbl.565.58024

Le Ngok T'euen [1983]: Commutative collections of functions on orbits of finite-dimensional Lie algebras in general position. Usp. Mat. Nauk *38*, No. 1, 179–180. English transl.: Russ. Math. Surv. *38*, No. 1, 204–205 (1983), Zbl.538.58018

Le Ngok T'euen [1985]: Complete involutive collections of functions on extensions of Lie algebras connected with Frobenius algebras. Tr. Semin. Vectorn. Tenzorn. Anal. Prilozh. Geom. Mekh. Fiz. *22*, 69–106, Zbl.599.17013

Magri, F. [1978]: A simple model of the integrable Hamiltonian equation. J. Math. Phys. *19*, 1156–1162, Zbl.383.35065

Manakov, S.V. [1976]: Note on the integration of Euler's equations of the dynamics of an *n*-dimensional rigid body. Funkts. Anal. Prilozh. *10*, No. 4, 93–94. English transl.: Funct. Anal. Appl. *10*, 328–329 (1977), Zbl.343.70003

Marsden, J., Weinstein, A., [1974]: Reduction of symplectic manifolds with symmetry. Rep. Math. Phys. *5*, 121–130, Zbl.327.58005

Maslov, V.P. [1965]: Theory of Pertubations and Asymptotic Methods. Moscow State University: Moscow, 554 pp. French transl.: Dunod: Paris 1972, Zbl.247.47010. 2nd Russian ed. Nauka: Moscow 1988, Zbl.653.35002

Matveev, S.V., Fomenko, A.T. [1988a]: Constant energy surfaces of Hamiltonian systems, enumeration of three-dimensional manifolds in increasing order of complexity, and computation of volumes of closed hyperbolic manifolds. Usp. Mat. Nauk *43*, No. 1, 5–22. English transl.: Russ. Math. Surv. *43*, No. 1, 3–24 (1988), Zbl.671.58008

Matveev, S.V., Fomenko, A.T. [1988b]: Morse-type theory for integrable Hamiltonian system with tame integrals. Mat. Zametki *43*, No. 5, 663–671. English transl.: Math. Notes *43*, No. 5, 382–386, Zbl.671.58017

Matveev, S.V., Fomenko, A.T., Sharko, V.V. [1988]: Round Morse functions and isoenergy surfaces of integrable Hamiltonian systems. Mat. Sb., Nov. Ser. *135*, No. 3, 325–345. English transl.: Math. USSR, Sb. *63*, No. 2, 319–336 (1989), Zbl.673.58023

McDuff, D. [1984]: Examples of simply connected symplectic non-Kählerian manifolds. J. Differ. Geom. *20*, 267–277, Zbl.567.53031

Medina, A., Revoy, P. [1983–1984]: Algèbres de Lie et produit scalaire invariant. Séminaire de Géometrie différentielle. Univ. des Sciences et Techniques du Languedoc, 63–93, also appeared in: Ann. Sci. Ec. Norm. Super., IV Ser. *18*, 553–561 (1985), Zbl.592.17006

Meshcheryakov, M.V. [1983]: A characteristic property of the inertia tensor of a multi-dimensional rigid body. Usp. Mat. Nauk *38*, No. 5, 201–202. English transl.: Russ. Math. Surv. *38*, No. 5, 156–157 (1983), Zbl.534.58026

Mishchenko, A.S., Fomenko, A.T. [1976]: The integration of the Euler equations on semisimple Lie algebras. Dokl. Akad. Nauk SSSR *231*, No. 3, 536–538. English transl.: Sov. Math., Dokl. *17*, 1591–1593 (1976), Zbl.392,58001

Mishchenko, A.S., Fomenko, A.T. [1978a]: A generalized Liouville method for the integration of Hamiltonian systems. Funkts. Anal. Prilozh. *12*, No. 2, 46–56. English transl.: Funct. Anal. Appl. *12*, 113–121 (1978), Zbl.396.58003

Mishchenko, A.S., Fomenko, A.T. [1978b]: The Euler equations on finite-dimensional Lie groups. Izv. Akad. Nauk SSSR, Ser. Mat. *42*, No. 2, 396–415. English transl.: Math. USSR, Izv. *12*, 371–389 (1978), Zbl.383.58006

Mishchenko, A.S., Fomenko, A.T. [1979a]: The integrability of the Euler equations on semisimple Lie algebras. Tr. Semin. Vectorn. Tenzorn. Anal. Prilozh. Geom. Mekh Fiz. *19*, 3–94. English transl.: Sel. Math. Sov. *2*, 207–291 (1982), Zbl.452.58015

Mishchenko, A.S., Fomenko, A.T. [1979b]: Symplectic Lie group actions. In: Algebraic topology, Proc. Symp., Aarhus 1978, Lect. Notes Math. *763* (Dupont, J.L., Madsen, I.H. (Eds.)) 504–538, Zbl.439.58015

Mishchenko, A.S., Fomenko, A.T. [1981]: Integration of Hamiltonian systems with non-commutative symmetries. Tr. Semin. Vectorn. Tenzorn. Anal. Prilozh. Geom. Mekh. Fiz. *20*, 5–54 (Russian), Zbl.473.58015

Moser, J. [1975]: Three integrable Hamiltonian systems connected with isospectral deformations. Adv. Math. *16*, 197–220, Zbl.303.34019

Narasimhan, M., Ramanan, S. [1961]: Existence of universal connections. Am. J. Math. *83*, 563–572, Zbl.114,382

Ness, L. [1984]: A stratification of the null cone via the moment map. Am. J. Math. *106*, 1281–1329, Zbl.604.14006

Novikov, S.P. [1982]: The Hamiltonian formalism and a many-valued analogue of Morse Theory. Usp. Mat. Nauk *37*, No. 5, 3–49. English transl.: Russ. Math. Surv. *37*, No. 5, 1–56 (1982), Zbl.571.58011

Novikov, S.P., Schmeltser, I. [1981]: Periodic solutions of the Kirchhoff equations for the free motion of a rigid body in an ideal fluid and the extended Lyusternik – Shnirel'man – Morse theory. Funkts. Anal. Prilozh. *15*, No. 3, 54–66. English transl.: Funct. Anal. Appl. *15*, No. 3, 197–207 (1982), Zbl.571.58009

Oshemkov, A.A. [1986]: Bott integrals of some integrable Hamiltonian systems. In: Geometry, Differential Equations and Mechanics. Moscow State University: Moscow, 115–117 (Russian)

Oshemkov, A.A. [1987]: The topology of isoenergy surfaces and bifurcation diagrams for the integrable cases of the dynamics of a rigid body on $SO(4)$. Usp. Mat. Nauk *42*, No. 6, 199–200. English transl.: Russ. Math. Surv. *42*, No. 6. 241–242 (1987), Zbl.648.58016

Pevtsova, T.A. [1982]: Symplectic structure of the orbits of the coadjoint representation of Lie algebras of type ExG. Usp. Mat. Nauk *37*, No. 2, 225–226. English transl.: Russ. Math. Surv. *37*, No. 2, 245–246 (1982), Zbl.497.22010

Pogosyan, T.I., Kharlamov, M.P. [1979]: Bifurcation sets and integral manifolds in the problem on the motion of a rigid body in a linear force field. Prikl. Mat. Mekh. *43*, No. 3, 419–428. English transl.: J. Appl. Math. Mech. *43*, 452–456 (1980), Zbl.467.70009

Ratiu, T. [1981]: The Neumann problem as a completely integrable system on an adjoint orbit. Trans. Am. Math. Soc. *264*, 321–329, Zbl.475.58006

Ratiu, T. [1982]: Euler-Poisson equations on Lie algebras and the N-dimensional heavy rigid body. Am. J. Math. *104*, 409–448, Zbl.509.58026

Reyman, A.G. [1980]: Integrable Hamiltonian systems connected with graded Lie algebras. Leningr. Otd. Mat. Just. Steklova 95, 3–54. English transl.: J. Sov. Math. *19*, 1507–1545 (1982), Zbl.488.70013

Reyman, A.G., Semenov-Tian-Shansky, M.A., [1986]: A new integrable case of the motion of the four-dimensional rigid body. Commun. Math. Phys. *105*, 461–472, Zbl.606.58029

Stepin, A.M. [1981]: Integrable Hamiltonian systems. I. Methods of integrating Hamiltonian systems. II. Series of integrable systems. In: Qualitative methods of investigating nonlinear differential equations and nonlinear oscillations, Collect. Sci. Works, Kiev, 116–142, 143–170 (Russian), Zbl.496.70030, Zbl.496.70031

Sternberg, S. [1964]: Lectures on Differential Geometry. Prentice-Hall: Englewood Cliffs, 390 pp., Zbl.129,131

Sulanke, R., Wintgen, P. [1972]: Differentialgeometrie und Faserbündel. Deutscher Verlag der Wissenschaften: Berlin, 325 pp. VEB Birkhäuser: Basel – Stuttgart, Zbl.271.53035

Symes, W.W., [1980]: Hamiltonian group actions and integrable systems. Physica *D1*, 339–374

Thimm, A. [1981]: Integrable geodesic flows on homogeneous spaces. Ergodic Theory Dyn. Syst. *1*, 495–517, Zbl.491.58014

Thurston, W. [1976]: Some simple examples of symplectic manifolds. Proc. Am. Math. Soc. *55*, 467–468, Zbl.324.53031

Tischler, D. [1977]: Closed 2-forms and embedding theorem for symplectic manifolds. J. Differ. Geom. *12*, 229–235, Zbl.386.58001

Toda, M. [1967]: Wave propagation in anharmonic lattices. J. Phys. Soc. Jap. *23*, 501–506

Trofimov, V.V. [1979]: The Euler equations on Borel subalgebras of semisimple Lie algebras. Izv. Akad. Nauk SSSR, Ser. Mat. *43*, No. 3, 714–732. English transl.: Math. USSR, Izv. *14*, 653–670 (1979), Zbl.409.58014

Trofimov, V.V. [1980]: Finite-dimensional representation of Lie algebras and completely integrable systems. Mat. Sb., Nov. Ser. *111*, No. 4, 610–621. English transl.: Math. USSR, Sb. *39*, 547–558 (1981), Zbl.435.58013

Trofimov, V.V. [1982a]: Completely integrable geodesic flows of left-invariant metrics on Lie groups connected with commutative graded algebras with Poincaré duality. Dokl. Akad. Nauk SSSR *263*, 812–816. English transl.: Sov Math. Dokl. *25*, 449–453 (1982), Zbl.509.58025

Trofimov, V.V. [1982b]: First integrals of the Galerkin approximations for the equations of magnetohydrodynamics. In: Some problems of applied mathematics and computer software. Moscow State University: Moscow, 7–8 (Russian)

Trofimov, V.V. [1983a]: Extensions of Lie algebras and Hamiltonian systems. Izv. Akad. Nauk SSSR, Ser. Mat. *47*, No. 6, 1303–1321. English transl.: Math. USSR, Izv. *23*, No. 3, 561–578 (1984), Zbl.547.58024

Trofimov, V.V. [1983b]: Semi-invariants of the coadjoint representation of Borel subalgebras of simple Lie algebras. Tr. Semin. Vectorn. Tenzorn. Anal. Prilozh. Geom. Mekh. Fiz. *21*, 84–105. English transl.: Sel. Math. Sov. *8*, No. 1, 31–56 (1989), Zbl.521.17004

Trofimov, V.V. [1984a]: Symplectic structures on automorphism groups of symmetric spaces. Vestn. Mosk. Univ., Ser. I 1984, No. 6, 31–33. English transl.: Mosc. Univ. Math. Bull. *39*, No. 6, 44–47 (1984), Zbl.594.58004

Trofimov, V.V. [1984b]: Methods for constructing *S*-representations. Vestn. Mosk. Univ., Ser. I 1984, No. 1, 3–9. English transl.: Mosc. Univ. Math. Bull. *39*, No. 1, 1–9 (1984), Zbl.561.17004

Trofimov, V.V. [1985]: Flat symmetric spaces with noncompact groups of motions and Hamiltonian systems. Tr. Semin. Vectorn. Tenzorn. Anal. Prilozh. Geom. Mekh. Fiz. *22*, 162–174 (Russian), Zbl.604.58030

Trofimov, V.V., Fomenko, A.T. [1983a]: Dynamical systems on orbits of linear representations of Lie groups and the complete integrability of some hydrodynamic systems. Funkts. Anal. Prilozh. *17*, No. 1, 31–39. English transl.: Funct. Anal. Appl. *17*, No. 1, 23–29 (1983), Zbl.521.58031

Trofimov, V.V., Fomenko, A.T. [1983b]: Non-invariant symplectic group structures and Hamiltonian flows on symmetric spaces. Tr. Semin. Vektorn. Tenzorn. Anal. Prilozh. Geom. Mekh. Fiz. *21*, 23–83. English transl.: Sel. Math. Sov. 7., No. 4. 356–414 (1988), Zbl.521.58030

Trofimov, V.V., Fomenko, A.T. [1984]: Liouville integrability of Hamiltonian systems on Lie algebras. Usp. Mat. Nauk *39*, No. 2, 3–56. English transl.: Russ. Math. Surv. *39*, No. 2, 1–67 (1984), Zbl.549.58024

Vergne, M. [1972]: La structure de Poisson sur l'algèbre symmétrique d'un algèbre de Lie nilpotente. Bull. Soc. Math. Fr. *100*, 301–335, Zbl.256.17002

Veselov, A.P. [1983]: On integrability conditions for the Euler equations on $SO(4)$. Dokl. Akad. Nauk SSSR *270*, No. 6, 1298–1300. English transl.: Sov. Math., Dokl. *27*, 740–742 (1983), Zbl.539.58013

Vinogradov, A.M., Kupershmidt, B.A. [1977]: The structure of Hamiltonian mechanics. Usp. Mat. Nauk *32*, No. 4, 175–236. English transl.: Russ. Math. Surv. *32*, No. 4, 177–243 (1977), Zbl.365.70016

Vishik, S.V., Dolzhansky, F.V. (= Dolzhanskij, F.V.) [1978]: Analogues of the Euler–Lagrange equations and equations of magnetohydrodynamics connected with Lie groups. Dokl. Akad. Nauk SSSR *238*, No. 5, 1032–1035. English transl.: Sov. Math., Dokl. *19*, 149–153 (1978), Zbl.393.22019

Waldhausen F. [1967]: Eine Klasse von 3-dimensionalen Mannigfaltigkeiten. I. Invent. Math. *3*, 308–333, Zbl.168,445

Author Index

Subject Index

MIX
Papier aus verantwortungsvollen Quellen
Paper from responsible sources
FSC® C105338

If you have any concerns about our products,
you can contact us on
ProductSafety@springernature.com

In case Publisher is established outside the EU,
the EU authorized representative is:
Springer Nature Customer Service Center GmbH
Europaplatz 3, 69115 Heidelberg, Germany

Printed by Libri Plureos GmbH
in Hamburg, Germany